T0392467

ELECTRIC MOTOR DRIVES AND THEIR APPLICATIONS WITH SIMULATION PRACTICES

ELECTRIC MOTOR DRIVES AND THEIR APPLICATIONS WITH SIMULATION PRACTICES

V. INDRAGANDHI
Associate Professor, School of Electrical Engineering, VIT, Vellore, Tamilnadu, India

R. SELVAMATHI
Associate Professor, Department of Electrical and Electronics Engineering, AMC Engineering College, Bangalore, India

V. SUBRAMANIYASWAMY
Professor, School of Computing, SASTRA Deemed University, India

Academic Press is an imprint of Elsevier
125 London Wall, London EC2Y 5AS, United Kingdom
525 B Street, Suite 1650, San Diego, CA 92101, United States
50 Hampshire Street, 5th Floor, Cambridge, MA 02139, United States
The Boulevard, Langford Lane, Kidlington, Oxford OX5 1GB, United Kingdom

ISBN: 978-0-323-91162-7

For Information on all Academic Press publications visit our website at https://www.elsevier.com/books-and-journals

Publisher: Charlotte Cockle
Acquisitions Editor: Lisa Reading
Editorial Project Manager: Michelle Fisher
Production Project Manager: Manju Thirumalaivasan
Cover Designer: Miles Hitchen

Typeset by Aptara, New Delhi, India

Contents

About the authors

Dr. V. Indragandhi completed BE in Electrical and Electronics Engineering from Bharadhidasan University in the year 2004. She received the ME degree in Power Electronics and Drives from Anna University and was awarded Gold Medal for the achievement of the University's first rank. Subsequently, she was awarded a doctorate of philosophy at Anna University, Chennai. At present, she is working as an Associate Professor in the School of Electrical Engineering, VIT, Vellore, Tamil Nadu. She has been engaged in teaching and research work for the past 15 years in the area of Power Electronics and Renewable Energy Systems.

She has authored 110 research articles in leading peer-reviewed international journals and published articles in referred high-impact factor journals. She filed three patents with her research ideas. Recently, her book was published, and details are given as *Software Tools for the Simulation of Electrical Systems*, ISBN 9780128194164, Academic Press, 2020. She organized many international conferences/workshops in collaboration with the top leading universities in the world successfully. Also, she received a travel grant from DST for attending a conference at NTU, Singapore. Moreover, she received funds for two research projects under the VIT SEED grant scheme. She visited many countries for her research discussions and collaborations as well. Indragandhi received the best researcher award from NFED, Coimbatore, and from VIT for her quality publications.

R. Selvamathi completed her graduate program in Electrical and Electronics Engineering from Madurai Kamaraj University and her post-graduate program in Applied Electronics from Anna University. She was awarded a doctorate of philosophy in the field of Transformerless inverter for Solar Applications at Vellore Institute of Technology, Vellore, in the year 2020. At present, she is working as an Associate Professor at AMC Engineering College, Bangalore. She has more than 16 years of teaching experience. Her research interests include Power Electronics and Renewable energy systems.

V. Subramaniyaswamy is currently working as Professor in the School of Computing, SASTRA Deemed University, India. In total, he has 18 years of experience in academia. He has received the BE degree in Computer Science and Engineering and MTech degree in Information Technology from Bharathidasan University, India, and Sathyabama University, India, respectively. He received a PhD degree from Anna University, India, and continued the extension work with the Department of Science and Technology support as a Young Scientist award holder. He has contributed papers and chapters for many high-quality technology journals and books being edited by internationally acclaimed professors and professionals. He has published more than 150 papers in reputed international journals and conferences. He is on the reviewer board of several international journals and has been a program committee member for several international/national conferences and workshops. He also serves as a Guest Editor for various special issues of reputed international journals. He is working as a Research Supervisor and a Visiting Expert at various universities in India. He has filed five patents with his research ideas and produced four PhD candidates as well. His technical competencies lie in Recommender Systems, Smart Grid, Internet of Things, Machine Learning, Big Data Analytics, and Renewable Energy Systems.

Preface

Simulation is the first step toward implementation. For the investigation of novel system designs, retrofits to existing systems, and suggested modifications to operating regulations, simulation is a powerful tool. A valid simulation is both an art and a science to carry out. Learning electrical science software tools provide an interactive platform for numerical computation, visualization, and programming. Students must also work with a variety of circuit topologies and design circuits that are appropriate for the applications. They will need a step-by-step tutorial to help them simulate circuits.

The dynamic performance of open- and closed-loop AC and DC drives is assessed using computer models of electric machines presented in this book. Because of their inherent integration of vectorized system representations in block diagram form, numerical analysis methods, a graphical depiction of time evolutions of signals, and simple implementation of controller and power electronic excitation functionality, the Simulink/MATLAB and PSIM implementations are used. Simulink models of drive assemblies may be created very easily by merging input–output block representations of the many components that make up the system. The simplicity with which the effects of parameter changes and changes in system configurations and control techniques may be observed makes this approach a strong design tool.

The basics of electric motor drives and their applications, as well as their simulation using MATLAB and PSIM, are covered in-depth in this book. It teaches engineers and students how to simulate various electric drives and their applications, which helps them to enhance their software skills. It is aimed particularly at aspiring engineers, researchers, and industrial engineers working in the subject of power electronics and drives. This book should also appeal to practicing engineers looking for a quick introduction to simulation software in a variety of settings.

This book, on a popular topic, is directed at engineering students who use or plan to use simulation software tools for DC and AC drives and their control. Furthermore, this book is unique in that it covers the creation of simulation models for power electronic converter fed drives using all available software, including MATLAB and PSIM. Also covered are operation, simulation, and applications, with numerous examples and step-by-step instructions. Furthermore, the last chapter of this book covers drives utilized in water pumping, which will be of greater use to readers conducting energy system study. Furthermore, each chapter includes complete simulated circuits of actual problems with procedures as well as unresolved problems for practice.

There are nine chapters in the book. The first chapter covers the fundamentals of electric drives and their major components, as well as the MATLAB/Simulink library. Chapter 2 introduces the concept of converter-fed DC drives and includes simulations in MATLAB/Simulink. Chapter 3 walks you through the steps of simulating chopper-fed dc drives in MATLAB/Simulink. In Chapter 4, you will learn how to regulate induction motors with different approaches utilizing PSIM and MATLAB/Simulink. Chapter 5 covers voltage source inverter and current source inverter fed synchronous motors, as well as load and line commutated cycloconverter fed synchronous motor drive in MATLAB/Simulink. Brushless DC motor drives using various control approaches, including MATLAB/Simulink and PSIM circuits, are examined in Chapter 6. Permanent magnet synchronous motor control is covered in Chapter 7 using both MATLAB/Simulink and PSIM circuits. The role of electric motor drives in electric vehicles is discussed in Chapter 8. MATLAB simulation is often used to discuss different motor drives in electric vehicles. Future motor driving trends in electric vehicle applications are also discussed. The use of drives in the production of drinking water, distribution, irrigation, and wastewater canalization systems is detailed in Chapter 9. MATLAB simulations are available for solar PV-based water pumping systems.

Acknowledgments

Dr. V. Indragandhi wishes to thank her husband, Mr. Arunachalam M., and daughter Subiksha for their motivation and lovable support to finish her work. She dedicates the book to her father and mother, who is the backbone for all her successes, and special thanks to her brother for his guidance and special thanks to Sachith, Girish, and Moulish for cherishing her always.

Dr. R. Selvamathi wishes to thank her husband Mr. Srinivasan and son Navatej, for their constant support in being patient and giving her all the love, time, and space to finish her work. She dedicates the book to her father Mr. Ramachandran and mother Ms. Kalyani who laid the foundation for all her successes and special thanks to her brother Mr. Sundararajan.

Dr. V. Subramaniyaswamy would like to take this opportunity to acknowledge those people who helped in completing this book. He is thankful to his wife, Raja Brindha N., for her constant support during writing. He would like to express his special gratitude to his sons S. Jai Girish and S. Jai Moulish for being patient and giving him all the love, time, and space to finish the work. He wishes to thank his family members A. Subiksha and V. Indragandhi for their constant support. He dedicates the book to his father Mr. Vairavasundaram, and his mother Mrs. Chellammal, who laid the foundation for his successes.

CHAPTER 1

Introduction to electric drives and MATLAB drive blocks

Contents

1.1 Introduction to electric drives

A operation of an electrical machine is known as an electric drive. As a key source of electricity, this drive uses a prime mover such as a gasoline engine, otherwise diesel, steam turbines, otherwise gas, electrical, and hydraulic motors. These prime movers can provide mechanical energy to the drive system for motion control. An electric drive can be constructed using an electric drive motor and a complex control mechanism to control the rotation shaft of the motor. At the moment, software is all that is needed to manage this. As a consequence, regulation is more accurate, and the drive principle is often easy to implement.

When we hear the words "electric motor" or "electric generator," we always assume that the speed at which these devices rotate is solely determined by the applied voltage and frequency of the source current. However, by using the principle of drive, an electrical machine's rotational speed can be precisely regulated.

Electric Motor Drives and Their Applications with Simulation Practices.
DOI: https://doi.org/10.1016/B978-0-323-91162-7.00001-1

The biggest benefit of this approach is that motion control could be conveniently programmed with the aid of a drive. Electrical drives, to put it simply, are the mechanisms that govern the movement of electrical devices. A traditional drive system consists of an electric motor (or several) and a complex control system that regulates the motor shaft's rotation. This monitoring can now be accomplished effectively with the help of tech. As a result, the control becomes much more precise, and this drive principle often offers ease of use.

Industrial motor drive: Electric motors can be used in a variety of products, including washing machines and refrigerators, as well as means of transportation such as automobiles and aircraft. Electrical motors are responsible for all of the everyday conveniences we take for granted. Historical personalities and business leaders such as Werner von Siemens, Thomas Alva Edison, Nikola Tesla, and George Westinghouse invented the first motors in the early nineteenth century. Everyday life will be impossible to conceive without electric motors.

However, the motors in the industry that enabled the assembly line conveyor belts used to assemble consumer goods and the motors built into the automatic welders used in the automobile industry are perhaps even more significant. Motors are also used in a variety of other industries, including medicine, aerospace, and renewable energy. Electrical motors are used in two different applications. The first scenario involves motors that are wired to the grid and run at a constant rotor rpm. The second is the vast range of applications in which motors must be controlled at various speeds and torques. This application, also known as a variable frequency drive (VFD), necessitates the installation of a power conversion unit between the grid and the engine.

1.2 Importance of electric drives

Electrical drives have numerous advantages over other drives. The advantages of electrical drives are as below:

- They last longer than most types of drive systems.
- They are pollution-free and they do not use flue gases.
- Which is more cost-effective.
- There is no requirement for gasoline storage or transportation.
- It is very effective.
- There are a variety of speed control options available.
- They take up less space.
- It is a dependable and cost-effective power source.
- It can be managed from a distance.

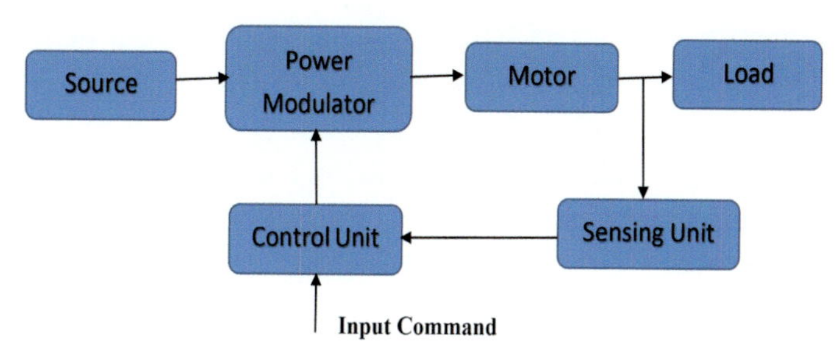

Figure 1.1 Electric drives block diagram.

- Speed, torque, and power are only a few of the parameters available.
- Transmission lines can carry electrical energy over long distances.

Industrial motor drives: Modern global manufacturing is built on the basis of motor drives. They are well covered, but they play an important role in sectors that contribute to a more prosperous environment and way of life. Drives are most often used today to power fans, pumps, and compressors. Around 75% of all drives in use around the world are for these purposes.

AC drives, adjustable speed drives, adjustable frequency drives, VFD, variable speed drives, frequency converters, inverters, and power converters are all names for motor drives. Drives can regulate the speed of an electrical motor by changing the frequency and voltage of the power supply, allowing them to improve process control, minimize energy consumption, produce energy effectively, or maximize the operation of different applications that rely on electric motors. Drives will also monitor the ramp-up and ramp-down of a motor at start-up and stop, respectively. This reduces mechanical tension in motor control systems while also improving ride efficiency in elevators, for example.

Motor drives may also be used to convert energy from natural and renewable sources such as the sun, wind, and tides and send it to the power grid or use it locally. Motor drives are used in hybrid systems to incorporate traditional energy sources with energy storage to create overall energy efficiency strategies.

1.3 Basic block diagram of electric drives

The illustration depicts the electrical drive's block diagram (Fig. 1.1). The electrical motor is the component of the electrical load, such as fans, generators, trains, and so on. The speed and torque requirements of an electrical load are calculated. For the load drive, the motor that best matches the load's capability is selected.

Parts of electrical drive—The main parts of the electrical drives are the power modulator, motor, controlling unit, and sensing units. Their parts are explained below in detail.

Power modulator—The power modulator controls the source's output power. It directs power from the source to the motor in such a way that the motor transmits the load's necessary speed-torque characteristic. The excess current pulled from the source is reversed during temporary operations such as starting, braking, and speed reversal. The excessive current drawn from the source can cause it to overload or drop in voltage. As a result, the source and motor currents are limited by the power modulator.

The power modulator transforms energy according to the motor's needs, for example, if the source is DC and the motor is an induction motor, the power modulator converts DC to AC. It also determines whether the engine is in motoring or braking mode.

Control unit—The power modulator, which works at low voltage and power levels, is regulated by the control unit. The power modulator is also regulated by the control unit. It also produces commands for the motor's safety and the power modulator. An input order signal from an input to the control device that adjusts the operating point of the drive.

Sensing unit—It detects motor current and speed as well as other drive parameters. It is primarily used for either safety or closed-loop service.

Source—Any AC or DC source in the device may be used as a source. In most locations, 1- or 3-phase, 50 Hz AC is used in the drive mechanism. Low-power drives are typically powered by a single-phase supply, whereas high-power drives are typically powered by a three-phase supply. Any of the drives are also driven by a battery.

Electrical motor—In most systems, a motor is used to transform electrical energy into electrical energy. Induction motors are used in electric drives. Synchronous motors, DC motors, stepper motors, and reluctance motors are all examples of motors. Induction and synchronous motors were previously only used for constant-speed applications. However, owing to advancements in new computing devices, AC motors are now used in variable speed drives. In battery drives, DC motors are used.

Load—Something that absorbs energy is referred to as a load. Fans, blowers, generators, robots, and computers are examples of equipment that perform a specific purpose. Motors and sources may be selected based on load specifications. Industrial, residential, rural, and other types of loads are all categorized in this way.

1.4 Applications of electric drives

1.4.1 Machine tool drives

A plain, more or less constant speed drive is required by many machine tools. Squirrel-cage induction motors with basic manual controls are sufficient for this. Gearboxes and stepped pulleys are used to transmit power to cutting instruments or work parts at a finite range of speeds. The use of a gear system causes vibrations and noise, which reduces machining precision. If step-less speed control is used, it can improve system timing and finish.

To achieve the optimal smooth speed modulation, electro-hydraulic and electromagnetic controls are used. These methods have been partly replaced by thyristor-controlled drives. Because of their high cost, thyristorized drives are currently only used in specialized machine tools in our region.

In certain situations, changing poles is preferable to moving gears. If more than two speeds are needed, two-pole changing windings, such as 4/8 and 6/12 poles, can be used in the same stator slot to provide speeds in the ratio of 1: 1.5: 2: 3. Small vertical drilling machines benefit from such an engine.

In certain situations, such as in woodworking machinery, speeds exceeding 3000 rpm are expected. A direct drive is favored over a geared drive, but with a 50 Hz supply, only commutator machines can do this. If there are many of these drives, an induction form frequency-changer should be installed to provide high frequencies (100 Hz and/or 150 Hz). This enables the use of comparatively inexpensive and durable squirrel cage motors at speeds reaching 6000 and 9000 rpm, respectively.

Variable speed is used by some drives. The Ward-Leonard design is superior for a larger speed range, but it has now been surpassed by a less expensive solution, such as a rectifier-DC motor hybrid. A field-controlled DC shunt motor is sufficient for a smaller speed range, and if a DC supply is not usable, a free-firing rectifier may be used. Variable-speed drives may also use AC commutator motors with induction regulators or brush-shifting speed modulation.

Fast frequency, three-phase induction motors with speeds up to 180,000 rpm and relatively high output forces in limited sizes are used for internal grinding spindles for horological purposes, drilling of printed circuit boards, and other applications. For high-speed grinder drives, typical ratings are 1.2 kW at 120,000 rpm and 0.5 kW at 150,000 rpm for high-precision horological applications.

Braking torques are needed in some applications. Many motors used in machine-tool drives have small enough ratings that capacitors can be used for braking without being too large or expensive. The Ward-Leonard set's regenerative braking, in addition to its other advantages, is particularly useful when there is a lot of resistance, such as in planning-machine drives.

1.4.2 Cranes and hoist motor

The prime requirement of a crane or hoist or lifting motor is that it should produce high starting torque and be able to carry out several transitions. Cranes favor DC series or compound wound motors because of their fast-starting torque and smooth speed control. DC compound wound, Dc series wound, and Ac slip-ring induction motors (with rotor resistance control) are used for raising, traveling, and reversing, as well as conveying and hoisting.

Hoist motors are equipped with special electromechanical brakes that, by the use of springs, are designed to support the load in the event of a power outage. When control is restored, the brakes are released by energizing a solenoid attached across the motor's terminals. Such motors are half- or one-hour rated, producing starting torque twice that of full-load torque, and are durable enough to withstand the extreme stresses to which they are subjected.

1.4.3 Lifts

High smooth accelerated torque (twice the full-load torque at start), high overload power and pull-out torque, a high degree of quiet, and a moderate speed are all required. It is possible to typically expect between 150 and 200 h of operation before failure for 1 h-rated motors. The vast majority of DC wound and AC slip-ring induction motors are made using lead-acid batteries. Forced induction and variable speed AC commutator motors, induction repulsion and variable induction commutation are both. In single-phase assembly, a shunt motor, a commutator may be mounted directly to the motor. The shunt field will either vary or be regulated to provide constant speed depending on the amount of current draw or the current draw with respect to voltage changes (Ward Leonard). copper slip-ring polyphase slip-ring induction motors have a limiting speed that can be regulated by resistance and varies according to copper failure (I^2R loss).

1.4.4 Lathes, milling, and grinding machines

Typically, lathes are powered by squirrel cage induction motors with constant rpm. Occasionally, variable-speed AC motors or variable-speed DC motors are used. Milling machines are usually powered by squirrel cage induction motors with constant speed control. Larger machines, such as planer-style milling machines, have individual motors for each milling head and each feed motion. The common rule is to drive the headstock with adjustable speed motors and the wheels with either constant or adjustable speed motors. Frequently, separate motors are used to feed the wheel in relation to the work.

Grinding machinery differs significantly in terms of construction. The general rule is to drive the headstock with adjustable speed DC motors, the wheels with constant speed squirrel cage or adjustable speed DC motors, and the traverse with constant speed squirrel cage motors. Bench, pedestal, and centreless grinders are driven by constant speed squirrel cage induction motors with the grinding wheels mounted directly on the motor shaft extension.

1.4.5 Planers

A planer is made up of a bed and a platen that rolls forward and backward on the bed. A clamped tool that is stationary on the platen prepares the job; at the end of the stroke, the tool swings slightly to create a new cut. Due to the slow cutting stroke and the rapid return stroke, variable speed, reversing motors are needed. Where a direct current supply is used, a shunt or compound wound motor is used; when an alternating current supply is used, a slip-ring induction motor is used.

1.4.6 Punches, presses, and shears

The high peak loads and high starting torque are unique features. These are often fitted with flywheels to meet transient power demands. Motors used in these devices should have a drooping speed-torque characteristic that enables the motor to slow down when confronted with heavy loads. These applications include high-slip squirrel cage induction motors, slip-ring induction motors, or DC cumulative compound motors.

1.4.7 Frequency converters

Converters are powered by squirrel cage induction motors and synchronous motors. Nowadays, static frequency changers are being used more often, owing to recent advancements in the field of silicon-driven rectifiers.

1.4.8 Air compressors

Induction motors with slip rings and synchronous motors are typically used in large-scale units. If engineered properly, the latter type can be used for power factor correction in addition to driving air compressors. Squirrel cage induction motors are only suitable for use with small compressors.

1.4.9 Electric traction

DC series motors are designed to be simple and compact, with fast-starting torque and smooth speed control. They are ideal for all forms of services but are especially well suited for residential services that need a high degree of acceleration. Single-phase AC compensated series motors are widely used for mainline work because they have similar speed-current and speed-torque characteristics to DC series motors. These motors are not suited for suburban services with regular stops due to their low starting torque.

1.4.10 Pumps

Typically, drip-proof or fully sealed surface-cooled polyphase induction motors are used, which are often placed on a standard bedplate and are directly coupled to the pump. Occasionally, the engine is placed on a flange. As the speed exceeds the limit of the set motor speed, a V-belt drive is used. Squirrel cage motors equipped with reduced voltage starters are used in centrifugal pumps where the needed starting torque is approximately 40%–50% of the full-load torque. Slip-ring induction motors are used to start reciprocating pumps that need between 100% and 200% of full-load torque.

1.4.11 Refrigeration and air conditioning

In refrigeration's vapor compression scheme, the motor driving the compressor is thermostatically regulated. When the engine is restarted, it must operate the compressor against a high head load. As a result, the motor used should be capable of producing starting torque equal to or greater than two to two and a half times the full-load torque.

For small units, single-phase, 230V, capacitor-type induction motors with D-O-L starters are usually used. For big systems, squirrel cage induction motors or slip-ring induction motors with a high torque rating are used. A synchronous motor driving a turbo-compressor may be suitable for very large plants, particularly if power factor correction is also needed.

1.4.12 Belt conveyors

Sand and gravel are moved using belt conveyors. Acceleration of large loads is expected. We use double cage induction motors with direct-on-line starters or wound rotor induction motors with a normal starting current and a high starting torque. These motors must be fully sealed and surface cooled due to the presence of grit and dust in the atmosphere.

1.4.13 Woodworking machinery

Unless there is a possibility of the motor being covered in sawdust, screen-protected motors are typically used. In locations where a motor can become buried in sawdust, a fully sealed surface cooled motor is used. If a higher speed than 3000 revolutions per minute is desired, induction motors in combination with an induction style frequency converter or commutator motors may be used.

1.4.14 Printing machinery

Squirrel cage induction motors are used in printing equipment for continuous speed work such as guillotines, moving platens, and other small devices. Single-phase capacitor motors can be used where line power and starting torque are appropriate. Two squirrel cage motors—one for each stroke—are used in rolling presses that involve a slow forward stroke and a short return stroke.

Other applications, such as rotary presses, may include DC compound or three-phase slip-ring induction motors with rotor resistance control or AC commutator motors. To inching the press, a pony or barring motor is used to maintain a steady crawling pace.

1.4.15 Petrochemical industries

Fluid processing machinery is widely used in the refining and chemical industries. Pumps have been powered by induction motors with flow control provided by throttling valves as necessary. According to figures, a typical refinery wastes 30% or more of its electrical energy due to the throttling effect of control valves. Adjustable-speed pumping, on the other hand, is very economical due to the lower energy usage and operating costs associated with multispeed induction motors or inverter-fed induction motors.

Other critical activities within the petrochemical industry's manufacturing processes include gas liquefaction, compression, refrigeration, and heat recovery. Compressor drive systems are critical components of such plants. The majority of drives run at a steady speed and are equipped with a 4- or 6-pole motor and step-up gear.

Synchronous motors with cylindrical rotors equipped with both excitation and starting windings (damper bars) are chosen for high ratings. These motors are engineered specifically to accommodate the oscillating torques produced during asynchronous acceleration as a result of the rotor's magnetic and electrical anisotropy and special cooling conditions.

1.4.16 Sugar mills

A centrifuge is used in sugar mills to separate crystallized sugar from syrup collected from a steam evaporator and to dry it out using centrifugal power. Charging, intermediate rotating, spinning, regenerative and reverse current braking, and plough plugging are all functions involved. All of these tasks must be executed at varying speeds, ranging up to 1: 30.

The motors used for this purpose are usually four-speed, pole-changing motors with two sets of stator windings. This configuration enables them to achieve synchronous speeds of 1500/750/214/107 rpm or 1000/500/214/107 rpm. They are not only capable of supplying the desired set speeds of operation but also of restoring some energy to the supply mains during regen-erative braking, which is achieved by transitioning from a lower pole to a higher pole operation.

The operation of reverse current braking results in a ploughing speed of about 50 rpm. Automatic duty cycle management is accomplished by feeding the control equipment directly from the output of a tach generator coupled to the centrifuge motor.

Centrifuge motors are constructed differently from other motors. They are installed vertically to ensure proper coupling with the centrifuge shaft. Larger air spaces can be used in the motors to compensate for potential rotor oscillations about the vertical axis. Motor insulation must be moisture resistant in order to work effectively in a humid climate. Thermal elements found in the windings shield the engine from overheating.

These elements, which are often referred to as sensotherms in the trade, work a few degrees below the winding's maximum permitted temperature. Thermal components operate in one of two ways: they either directly trip the motor or provide visible and audible warning signals that the service cycle has been completed. Following this, the next cycle cannot begin until the motor's temperature has returned to normal.

1.4.17 Cement mills

The starting torque of big cement plant mill motors is limited to 125% of full-load torque, while the pull-out torque is limited to approximately 240%

of full-load torque. Typically, 6.6 kV slip-ring induction motors with liquid resistance starters are used. Gearboxes are used to achieve the desired mill speed of approximately 15 rpm, and power factor correction is accomplished by the use of high voltage capacitors of sufficient reliability, mechanical capacitor control switchgear, and circuit breakers.

Owing to the high ratings (over 3000 kW) needed for raw and cement mill drives and the scarcity of large size gearboxes and motors, twin drives are used—the two motors used in twin drives must be nearly identical, as must their liquid resistance starters.

The rotary kiln is an integral part of the cement manufacturing process. The engines used to power the kilns range in power from 100 to 1000 kW. The kiln's optimum speed is approximately 1 rpm, and the required speed range is approximately 1: 10. The needed starting torque could be between 200% and 250% of the rated torque. Initially, variable speed AC commutator motors were used for kiln drives; however, AC commutator motors have been phased out in favor of Ward-Leonard drives due to their high cost and maintenance requirements. Nowadays, DC motors with static power supply are widely used. To keep up with the growing capacity of kilns, the current trend is to use twin motor DC drives in kiln applications.

Usually, crusher drives have a starting torque of 160% of the full-load torque and a pull-out torque of 200%–250% of the full-load torque. Typically, slip-ring induction motors are used in crushers. Typically, the motors are engineered to endure locked rotor current for 1 min when operating without any external resistance introduced into the rotor circuit—a key function given the frequency at which crushers get stuck.

For fan drives, the starting torque requirement is approximately 120% of the full-load torque, while the pull-out torque requirement is between 200% and 250% of the full-load torque. Slip-ring induction motors with variable speed are used in these drives. Typically, cast iron grid resistance controls are used to start and monitor the rpm. For ease of cleaning, the slip rings and brush gears are fully enclosed and held external to the motor enclosures.

Typically, air compressor motors are rated between 300 and 450 kW. Induction motors with squirrel cages or wound rotors may be used depending on the power requirements. Enclosures are usually TEFC, and speeds range between 1000 and 750 rpm.

1.4.18 Mining work

Electric motors used in coal mines are grouped into two types: auxiliary motors, which are used to power auxiliaries such as compressors, fans, conveyors, pumps, and hoists, and mine face motors, which are used on continuous miners, drills, shuttle buses, cutting machines, and loaders, among others. The engines in the two categories are very different. Auxiliary motors are usually modified general-purpose automotive motors, while face motors are custom-built to meet unique requirements. Auxiliary motors' duties are often well specified and consistent, while face motors' duties include random loading and a variety of high-shock loads.

These motors can normally be used in small, induction, low- or medium-power equipment for coal drills and gas drills, at 125 V, at 2860 rpm, and 50 amps, with a flameproof enclosure. Occasionally, high-frequency motors (150 or 200 Hz) are used in conjunction with a frequency changer located in the drill column.

Slip-ring induction motors with water rheostats are typically used to power the colliery winder, which is used to raise and lower coal, as well as people and other loads. Initially, DC rheostatic braking is used; mechanical braking is used only at the end of the winding cycle. Although motors are supplying the lower capacity from a Ward Leonard generator are required for precise headroom control, synchronous driving of the winders, the drum speed and location is of the winders is important. The current trends call for the use of the thyristorsisterate electrolytic components, in which the thyristorsitic current is controlled by the circuit

The motors employed for haulage (mine-winches) operate under an extraordinary set of mechanical stresses, including repeated restarting and shutdowns, as well as reverse action. When the ship picks up the anchor, variously varied stresses and motions are brought to bear on it. Therefore, slip-ring induction motors with disc resistance controllers are usually preferred. All the other drum machine components and rotor resistances are contained in their own protective fire-proof casings.

Radial flow fans are also found in coal mines and are mainly in the following two varieties of ventilation fans: axial and pulley. Until relatively recently, induction motors have about five poles, which generally made a significant amount of torque. Current practice prefers four-low-high speed four-pole squirrel cage motors because of their compact design.

There are two kinds of pumps: a triplex pump powered by a high-speed motor and a centrifugal pump that is directly connected to the motor, both are driven by the motor and the high gear. For the purpose of a stand-still dynamic testing, which involves a system with a rotor whose self-tapping feature is engaged, a significant starting torque is required, resulting in a ratio of 200% torque in Nm being needed. Squirrel cage motors need only about

40% of the motor's rated torque to operate, which makes them very popular for driving centrifugal pumps, because of their simple control system. In industry, people are more likely to have high levels of electromagnetic exposure in their environments, which presents a greater health hazard for electric motors. Motors used in mining must be flameproof, and all commutators and slip rings must be contained inside flameproof chambers with an inch metal to the metal flange.

1.4.19 Textile mills

Historically, textile mills were usually powered by diesel generators and steam turbines. Due to the design of the boiler, engine, and so on, consideration of an electric drive was required. At the time, AC capacity was favored. This was due to the mill's electricity prices. Following that, group drive machines were created, but with recent advances, the need for individual driving became apparent.

Standard open or fully sealed motors cannot be used in textile mills due to environmental (fluff and dust fly appearance, high humidity, limited airflow, etc.), operational (wide voltage variations, constant starting and stopping, erratic activity resulting in rapid shift in torque and power requirement), and drivability considerations. To obtain a rapid pick-up time, the loom motor's starting torque should be high (2–2.5 times of the rated torque). The loom motor is driven by a reciprocating mechanism that generates torque and current pulses. Furthermore, the loom motor is subjected to repeated starts and stops. This results in a significant temperature increase, which is mitigated by the motor's large thermal dissipation ability.

This is fully sealed, surface-cooled engines. Due to the high level of fluff in the atmosphere, the housing and shields must have a smooth surface finish to prevent fluff from accumulating on the motor surface. The motor's insulation must be capable of withstanding high moisture content. These motors have a higher slip rate or a slower speed than normal motors in order to provide a flywheel effect that smooths out current pulsations. Motors used to drive looms for light fabrics such as silk, rayon, cotton, and nylon have ratings of 0.37, 0.55, 0.75, 1.1, and 1.5 kW; whereas motors used to make hard fabrics (wool and canvas) have ratings of 2.2 and 3.7 kW. Typically, engines have six to eight poles.

1.4.20 Woollen mills

Loom motors are fully enclosed and capable of producing three times the torque of a full-load motor when started directly or when started light and operated by clutch. Motors and switchgear in the dyeing segment should be fully sealed and painted with acid-proof paint due to the presence of fumes.

1.4.21 Paper mills

A paper mill is divided into two distinct processes: pulp production and paper production. The drives used in both of these procedures are very distinct. There are two ways to make pulp: mechanically or chemically. The former process entails grinding approximately 1-m-long logs of wood on massive grind-stones. Grinders run at nearly constant speeds between 200 and 300 revolutions per minute and can be started under light load conditions. As a result, synchronous motors with constant speed and geared drives are the most appropriate. Typically, pulp production by solely mechanical means consumes more than half of a paper mill's total power requirement. As a result, large-capacity grinders powered by 3000–4000 kW motors are typically known to be economical.

Pulp may also be made by chipping wood logs into several centimeter-long pieces and treating them with alkalies in conjunction with other raw materials such as hay, rags, and so on. The substance is constantly beaten during the chemical treatment. Wood choppers have erratic load characteristics and a high moment of inertia, which varies according to the size of the disc on which the chopper's knives are mounted. Typically, beaters are expected to begin with a broad load.

The motor used for chipping, pounding, refining, and storage may have a rating of several hundred to thousand kilowatts, depending on the scale of the mill. But for beaters, these drives use synchronous engines. Due to the fact that beaters always need speeds less than 200 rpm and a high starting torque, wound rotor induction motors are the best choice for such drives.

The paper manufacturing machine must mould sheets, remove moisture from sheets, dry sheets, press sheets, and reel up sheets. There are two kinds of drives used in the manufacture of paper from pulp: line shaft and sectional. The different parts of the paper machine are powered by a line shaft that runs the length of the machine. Cone pulleys and belt configurations transmit power from the line shaft to the different parts of the system via right-angled gear reductions. Typically, electric motors are used to drive the transmission shaft. Both AC and DC drives can be used to achieve almost zero-loss speed modulation. Only the AC commutator motor with shunt characteristics is ideal for providing a cost-effective speed control mechanism in AC drive.

However, since the speed of an AC commutator motor is load-dependent, its use as a drive for a paper manufacturing machine with strict specifications for constant speed is not recommended. Additionally, the motor's speed range (typically 1: 3) and the amount of power needed have a significant effect on its size. In contrast to a DC drive, the AC commutator motor's open-loop speed control is slow, as speed is varied by changing an induction regulator and moving the brush rocker. The speed of a paper manufacturing machine is operated using DC drives by varying the armature voltage of a separately excited DC motor. Rotary converters or static converters are used to transform AC to variable DC voltage.

Each part of the papermaking machine has its own electrical motor in sectional drive. Variation of the supply voltage may be used to regulate the output of the paper making process. By changing the field excitation of any motor, the speed of that motor can be varied in relation to other motors. Line shaft drives have less strict requirements than sectional drives. It has a host of significant disadvantages to the sectional drive system. The lower cost of electrical equipment needed for a line shaft drive system is more than compensated by the increased cost of mechanical equipment and its upkeep.

1.4.22 Ship-propulsion

But for small boats, both ships are propelled by electric motors. In electric ship propulsion, the prime mover (steam turbine or diesel engine) powers an alternating current or direct current generator, which supplies electricity to an alternating current or direct current motor mounted on the propeller shafts. Steam turbines with AC equipment are typically used on large ships needing more than 2500 kW of electricity, whereas diesel engines with DC equipment are typically used on smaller vessels.

Three-phase induction or synchronous motors are used in an AC configuration. Speed is regulated in part by applied voltage variation and in part by frequency variation. Additionally, induction motors' speeds can be adjusted using a pole-changing system, which results in two economical speeds. Reversal is accomplished by altering the supply's step chain. The ship's power is provided by turboalternators operating at 2200 or 6600 V.

Speed modulation and reversal were accomplished in the DC system by adjusting the ground excitation of the generator field in the Ward-Leonard system. The ship's power is provided by diesel-electric generators operating at a low voltage of 650 volts.

Instead of a constant voltage grid, propeller motor armatures can be fed by a constant current system when their fields are supplied by a constant voltage supply. This type of supply device is intrinsically resistant to overloads and delivers the best-protected torque possible without possibility of current disruption is governed by field power. This is ideal for small vessels, such as tugs and ferries, that require maximum manoeuvrability in congested waters.

Propeller shafts are directly coupled to diesel engines in very small vessels through slip couplings. The propeller shaft's speed can be varied by adjusting the DC excitation, and hence the coupling's slip.

1.4.23 Rolling mills

Steel mills' primary feature is rolling steel, a procedure that reduces the cross section of the metal while increasing its length proportionately. Generally, rolling mills are graded according to the content they produce. Direct drive is chosen if the mill's speed allows. In general, direct drive (direct-connected motor) is not desirable for speeds less than 150 rpm and outputs less than 750 kW. Where the mills' speeds are too slow for direct driving, gear drive is used. It is customary to use machine-cut, double helical gears for this type of work, while single helical gears are sometimes used, and provisions are made to hold the thrust provided by such gears (Fig. 1.2). Spur gearing is only used on low-speed drives with pitch line speeds of less than 450 m per minute. Although the use of gears causes losses and cancels out the reliability gains of high--speed motors, in many situations the actual cost of construction is smaller, and there are additional benefits, such as a higher power factor for induction motors and often a decrease in available space.

If required, the flywheel should be mounted between the gear and the turbine, if the mill's speed allows, so that the gear does not have to transmit the flywheel's peak loads. Where the mill shaft operates at a speed that precludes economical flywheel construction, the flywheel can be mounted on the gear unit's pinion shaft. This allows the wheel's peripheral speed to be comparatively high and its weight to be reduced, but the gears must be wide enough to transmit the mill's peak loads. Due to their intrinsic features, DC motors are ideally suited for rolling mills. Motors for reversing mills must have a high starting torque, a large speed range, accurate speed control, the ability to tolerate overload and pull-out torques up to three times the rated value, and excellent commutation within the speed range. Acceleration

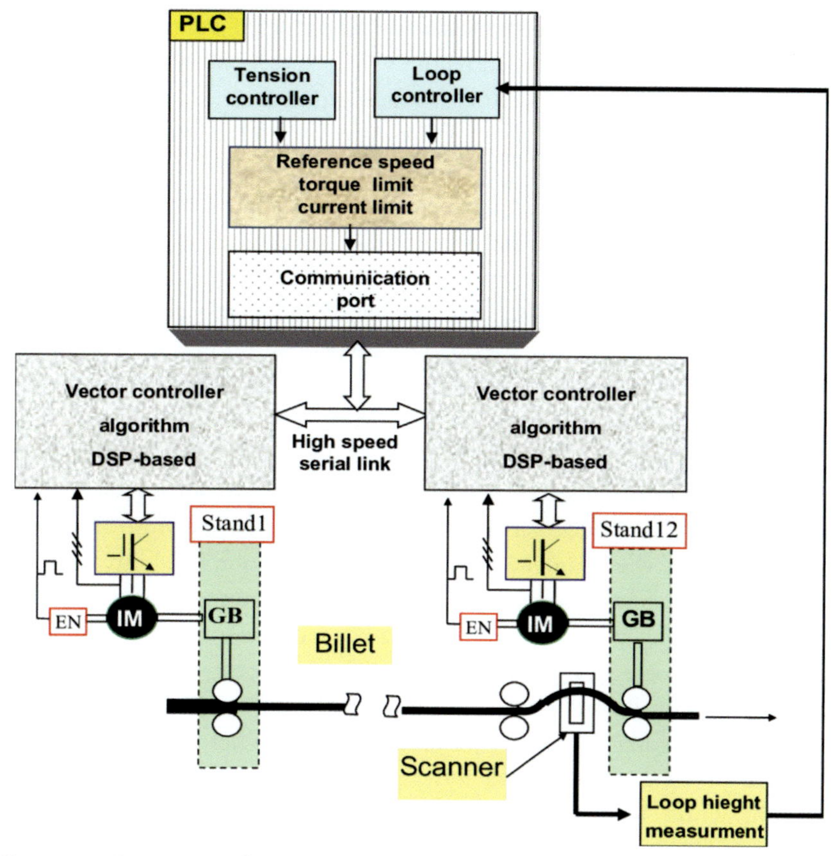

Figure 1.2 Drive system for sugar mills.

from zero to base speed, then to peak speed, followed by reversal from top speed forward to top speed backward, must be accomplished in a matter of seconds. The armature's moment of inertia must be as minimal as practicable. The need for a higher constant torque at low speed during the initial few passes can be met by varying the applied voltage to the armature while maintaining a constant field. On the other hand, by strengthening the field, it is possible to achieve the low torque and higher speed necessary for rolling action in final passes. As a result, steel mill motors are fully insulated, forced ventilated, have a higher class of insulation, a smaller diameter, and a longer range.

In general, the use of DC motors raises the initial cost and adds losses as a result of the need for converting machines, but the benefits obtained by such motors far outweigh these extra losses and costs. A grid-controlled mercury-arc rectifier, a silicon-controlled rectifier (SCR), or a Ward-Leonard device can be used to supply DC motors for rolling mill drive.

The primary benefits of a grid-controlled mercury-arc rectifier are increased reliability, increased overload potential, stepless voltage regulation from 0% to 100%, and rapid response to very rapid changes in operating variables. However, it has the disadvantage of producing higher current peaks in the power supply, which are followed by large voltage variations. The silicon-powered rectifier has the benefit of being compact but has the disadvantage of lacking overload capability. As a result, grid-controlled mercury-arc rectifiers or SCRs are used to provide rolling mills with an adequate AC supply network. The primary benefits of using the Ward-Leonard range are the ability to vary the speed within 100% of rated speed, the ability to perform electric braking by using the mill motor as a generator, and the ability to reduce variations in the power demand from the supply grid. However, the Ward-Leonard system has a poor reliability, a longer reaction time, and needs more maintenance. Synchronous motors are used to drive constant speed mills. They are advantageous in applications where power factor correction is desired or slow-speed motors are needed. These are especially well-suited for continuous mills where the load is held for comparatively long periods of time or for drives where the length of the material prevents the use of a flywheel and, as a result, the motor must bear the whole load. Synchronous motors can be designed more cheaply than induction motors at low speeds. Slip-ring induction motors are ideal for roughing and re-rolling mills that do not require very complex speed control. Their performance is poor due to energy loss due to rotor resistance. Additionally, there is a sharp increase in motor speed as the material leaves the rolling stands. By using a cascaded induction motor, the disadvantages of the slip-ring induction motor with variable rotor resistance can be overcome.

The output of the low-frequency rotor is rectified using a silicon rectifier, and the rectified power is forced to act against an adjustable back emf produced by a separately excited DC motor mounted on the same shaft, which converts the rectified power to mechanical power. Slip-ring induction motors now have a shunt speed-torque characteristic, allowing for speed modulation without sacrificing power in the secondary resistances. When the motor is started with maximum DC excitation on, the starting torque is approximately four times the rated torque. Where the target speed difference is less than 25%, the cascading arrangement becomes more cost-effective than DC motors.

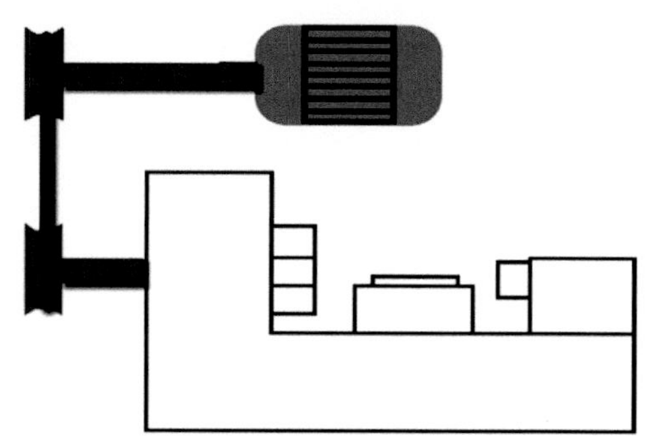

Figure 1.3 Individual drive.

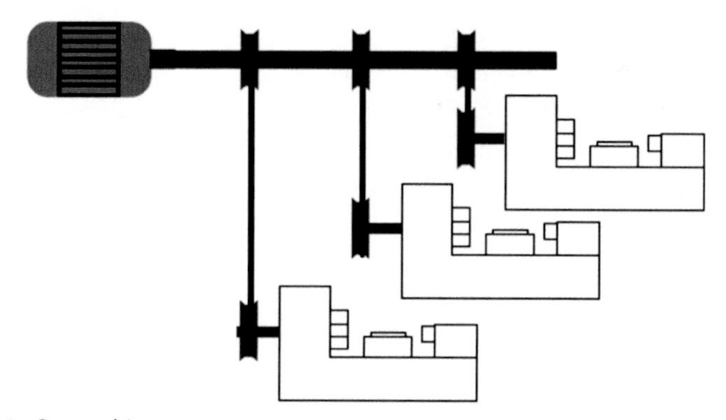

Figure 1.4 Group drive.

1.5 Classification of electric drives

Electric drives are typically divided into three categories according to their growth, namely group, individual, and multimotor electric drives.

When a single motor is used to drive or actuate a single function and the motor performs all of the tasks associated with the load, the drive is referred to as an independent drive (Fig. 1.3). For instance, a single machine can perform all operations associated with running a lathe. Transmission devices can be needed if these operations must be done at varying speeds. Due to power loss, performance can degrade over a period of many operations. In certain cases, the drive motor and driving load may be combined into a single assembly.

When several systems or devices are arranged on a single shaft and are driven or actuated by a single motor, the structure is referred to as a group drive or shaft drive (Fig. 1.4). The various mechanisms that are related can operate at varying speeds. As a result, the shaft is fitted with several stepped pulleys and belts for attachment to specific loads. Since all connected loads do not occur simultaneously, a single computer with a rating less than the amount of all connected loads may be used in this form of drive. This makes the drive economical, despite the apparent high cost of the shaft with stepped pulleys.

Due to the following drawbacks, this approach is seldom used in modern drive systems and has gained historical interest:

- The drive's performance is poor as a result of losses in many transmission mechanisms.
- If the engine needs maintenance or replacement, the entire drive mechanism must be shut down.
- The position of the mechanical equipment being powered is determined by the shaft, and its structure is restricted.
- The operation of the machine is not really secure.
- The noise level is very high at the job location.

Each movement of the mechanism is handled by a separate drive motor in a multimotor vehicle. The machine is composed of multiple independent drives, each of which is used to power a separate device. This kind of drive is used in complex machine tools, traveling cranes, and rolling mills, among other applications. Automatic monitoring methods can be used, and each operation can be carried out optimally (Fig. 1.5).

Additionally, we will discuss mechanical, hydraulic, and electrical/electronic drives (eddy-current coupling, rotating DC, DC converters, and variable-frequency AC).

As shown in Fig. 1.6, the mechanical drive utilizes variable-pitch pulleys. Typically, the pulleys are spring-loaded and can be expanded or contracted in diameter with a hand crank. The mechanical drive is always driven by an alternating current supply—typically three-phase alternating current. The three-phase alternating current is then supplied to the fixed-speed AC motor. This drive unit's output speed can be adjusted by adjusting the diameter of one or both pulleys. Variable speed operates on the same basis as a 15-speed bicycle's gears. Shifting gears allow the chain to fall through a sprocket of a larger or smaller diameter. When this occurs, the same input power is used to reach a faster or slower speed.

Hydraulic drives have long been the workhorse in many metal processing and industrial applications. The compact size of the hydraulic motor makes it suitable for applications requiring high strength in confined spaces. Indeed, the hydraulic motor is 1/4–1/3 the size of an electric motor of comparable capacity. A hydraulic drive is depicted in Fig. 1.7.

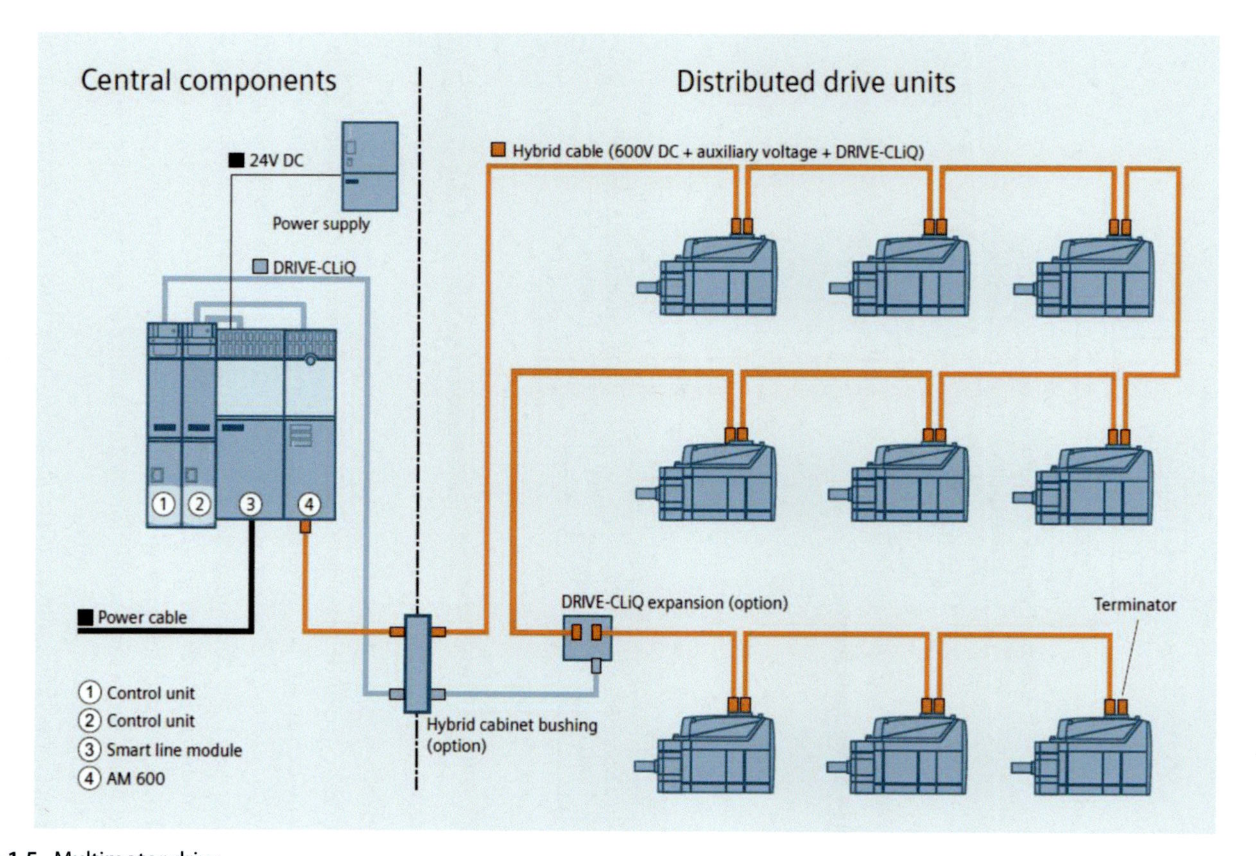

Figure 1.5 Multimotor drive.

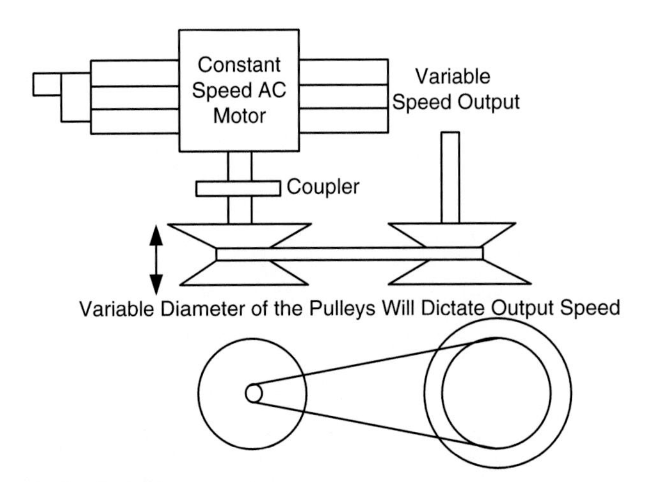

Figure 1.6 Mechanical drive.

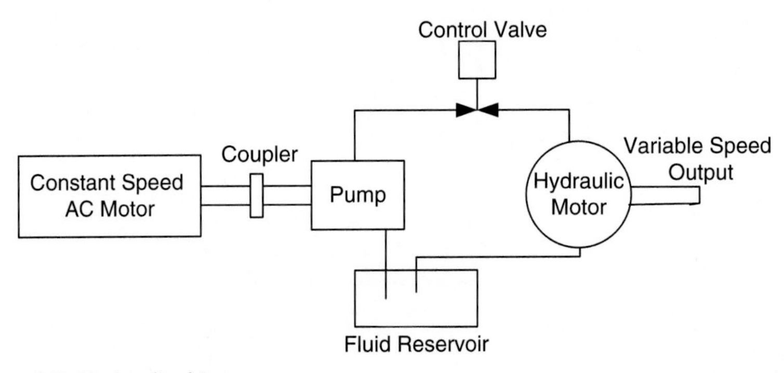

Figure 1.7 Hydraulic drives.

Eddy-current drives originated in the heavy machinery industry. Eddy-current drives are an excellent option for grinding wheels. This device converts alternating power to direct current, allowing for variable shaft speeds based on the amount of power transferred. Fig. 1.8 illustrates a straightforward eddy current drive mechanism.

This system dates back to the mid-1940s. The system also gained the name M-G set, which stands for motor-generator set. As seen in Fig. 1.9, that description is quite accurate.

Since the 1940s, direct current drives have been the cornerstone of manufacturing. At the time, vacuum tubes provided the technology for power transfer. In the 1960s, vacuum tubes gave way to solid-state systems. The SCR or thyristor power conversion unit is now used in modern

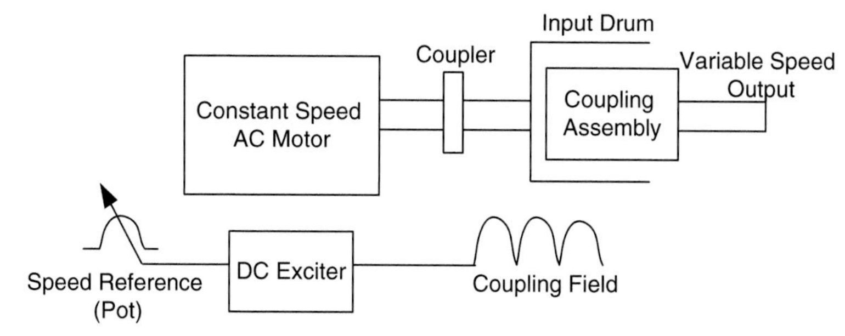

Figure 1.8 Eddy-current drives.

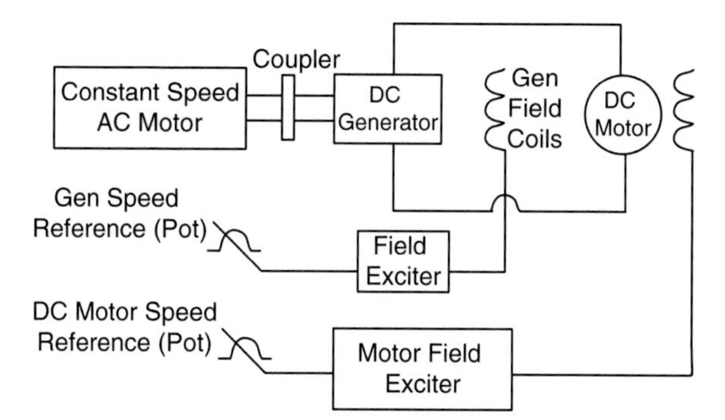

Figure 1.9 Motor-generator set.

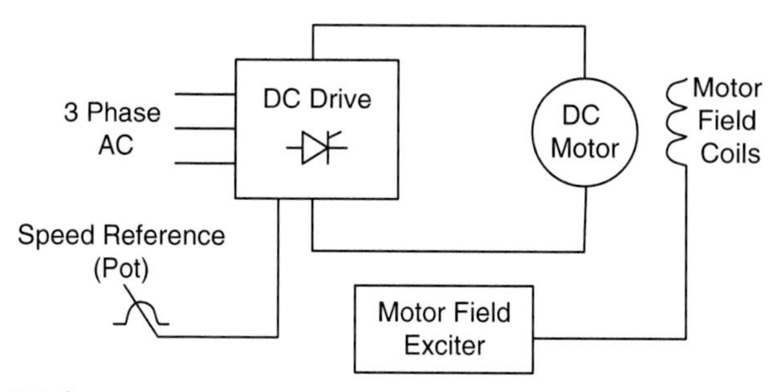

Figure 1.10 DC drive.

electronic direct current drives. The components of a basic DC drive mechanism are depicted in Fig. 1.10.

Presently, three distinct types of AC drive technologies are available. While each method of power conversion is somewhat different, the end result is the use of a variable-speed AC induction motor. Both AC drives take an alternating current input, convert it to direct current, and then convert direct current to a variable alternating current output using a system called

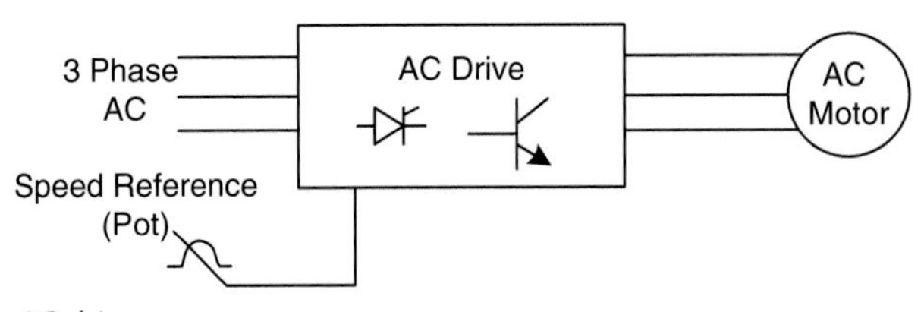

Figure 1.11 AC drive.

an inverter (i.e., inverts DC back to AC voltage). For the purposes of this segment, we will address a generic air conditioning trip (Fig. 1.11).

1.6 Introduction to MATLAB/Simulink

MATLAB is a high-performance programming language designed for use in technical computing. It combines computation, simulation, and programming in a simple-to-use environment in which problems and solutions are represented using standard mathematical notation. Math and calculation, data processing, discovery, and simulation are all popular applications.

Simulink is a block diagram environment that enables the simulation of several domains and model-based design. It enables the design of embedded systems at the device level, emulation, automated code creation, and continuous test and verification. Simulink is a graphical modeling and simulation environment that includes a graphical editor, flexible block libraries, and solvers for modeling and simulating dynamic systems. It is compatible with MATLAB, allowing you to embed MATLAB algorithms into models and export simulation results for further study in MATLAB (Fig. 1.12).

Simscape electrical (formerly Sim Power Systems and Sim Electronics) is a library of mechanical, mechatronic, and electrical power system components for modeling and simulating electronic, mechatronic, and electrical power systems. It includes semiconductor models, motor models, and part models for electromechanical actuation, smart grids, and green energy systems. These modules can be used to test analog circuit architectures, design mechatronic devices with electric drives, and conduct grid-level analysis of electrical power generation, conversion, transmission, and usage.

Simscape electrical enables you to design control systems and evaluate their performance at the device level. Simulink enables you to parameterize your models through MATLAB variables and expressions and to construct control systems for electrical systems. Utilize modules from the Simscape family of products to incorporate mechanical, hydraulic, thermal, and other physical structures into the model (Fig. 1.13). Simscape electrical facilitates

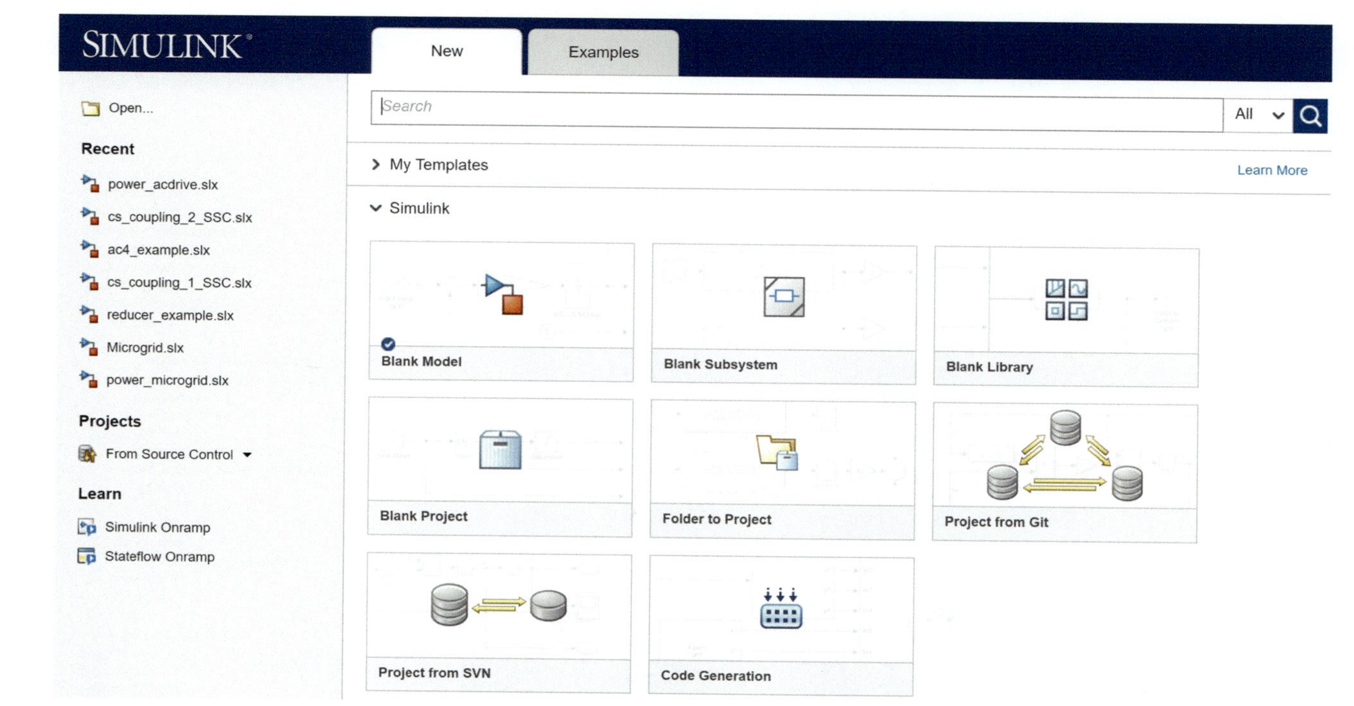

Figure 1.12 Simulink.

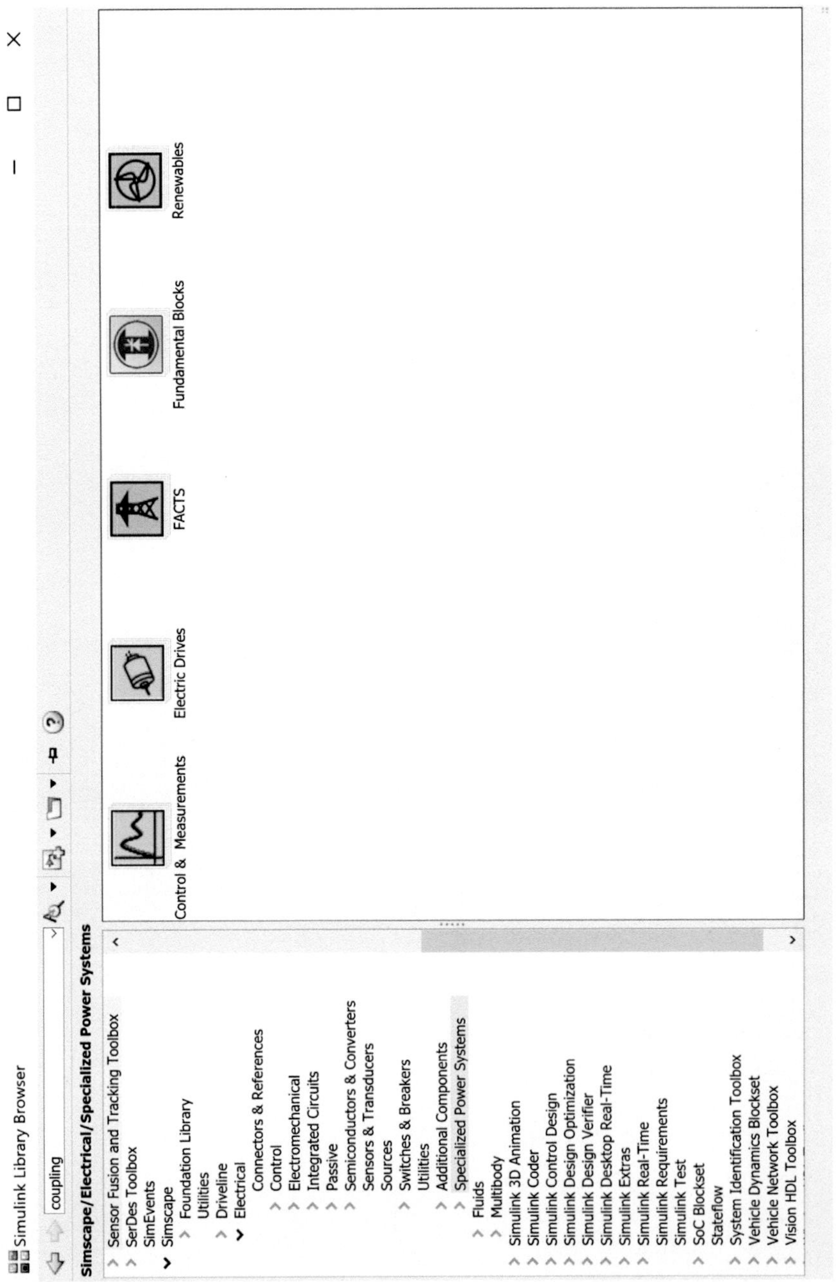

Simulink Library Browser
coupling
Simscape/Electrical/Specialized Power Systems
Sensor Fusion and Tracking Toolbox
SerDes Toolbox
SimEvents
Simscape
Foundation Library
Utilities
Driveline
Electrical
Connectors & References
Control
Electromechanical
Integrated Circuits
Passive
Semiconductors & Converters
Sensors & Transducers
Sources
Switches & Breakers
Utilities
Additional Components
Specialized Power Systems
Fluids
Multibody
Simulink 3D Animation
Simulink Coder
Simulink Control Design
Simulink Design Optimization
Simulink Design Verifier
Simulink Desktop Real-Time
Simulink Extras
Simulink Real-Time
Simulink Requirements
Simulink Test
SoC Blockset
Stateflow
System Identification Toolbox
Vehicle Dynamics Blockset
Vehicle Network Toolbox
Vision HDL Toolbox
Control & Measurements
Electric Drives
FACTS
Fundamental Blocks
Renewables

Figure 1.13 Simscape electrical.

Specialized Power Systems

Model electrical power systems using specialized components and algorithms

Electrical Sources and Elements
AC and DC sources, breakers, transformers, RLC branches and loads, transmission lines

Motors and Generators
Asynchronous and synchronous machines, motors, excitation systems

Power Electronics
Thyristors, diodes, bridges

Sensors and Measurements
Current, voltage, and impedance sensors, specialized measurement blocks

Control and Signal Generation
Pulse generators, filters, signal transformation blocks

Electric Drives
AC drives, DC drives, shafts, speed reducers, batteries, fuel cells

Power Electronics FACTS
Phasor type compensators

Renewable Energy Systems
Wind turbine models

Interface to Simscape
Blocks that connect Simscape™ Electrical™ Specialized Power Systems and Simscape electrical circuits

Simulation and Analysis
Simulation performance and analysis tools and techniques

Figure 1.14 Specialized power systems library.

C-code creation for the deployment of models to other simulation environments, including hardware-in-the-loop systems. As shown in Fig. 1.14, a dedicated power library—electric drive—includes all of the drives available for Simscape simulation.

1.7 Other software's used for electric drives simulation

Ansys: It provides a comprehensive workflow, from concept design to extensive electromagnetic, thermal, and mechanical studies of the engine. The motor's electromagnetic, thermal, stress, and vibro-acoustics simulations using Ansys tools result in a high-fidelity, precise, and durable design that is designed for performance, cost, and power.

Maxwell: An electromagnetic field simulation solver for electric machines and electromechanical devices. Solve static, frequency-domain, and time-varying electric fields.

Motor-CAD: A template-based modeling tool for performing multi-physics studies on electric motors across their entire torque-speed operating

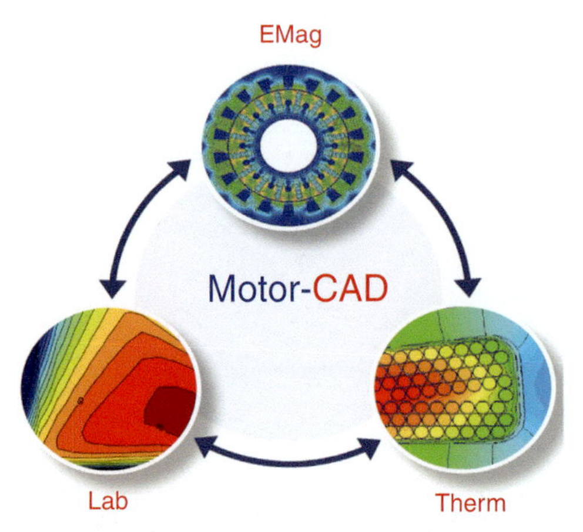

Figure 1.15 Motor CAD.

range with the goal of optimizing EV output, power, and scale. Motor-CAD is a four-module program suite. It enables rapid design and verification of electric motors in terms of output, scale, and power. The multiphysics computations have been optimized for speed and usability to greatly reduce the development time (Fig. 1.15).

OPAL-RT: It includes electrical motor models based on FPGA and Simulink, such as PMSM, IM, SRM, BLDC, DC, and AC. OPAL-RT is also well-known for customizing real-time versions with unusual engines. When measuring electrical motor controls, high motor fidelity provides engineers with richer harmonics performance, practical current saturation behaviors, and reliable torque results, allowing for greater test coverage and accuracy. Mechanical dynamics are usually 10 times slower than electrical dynamics conducted on a CPU, allowing for greater interface versatility.

Simplorer/twin builder is a multiphysics circuit simulator that can insert a single schematic electric, mechanical, hydraulic, and thermal part and combine it with mathematical operations defined in terms of state space, block diagram, state machines, and scripting algorithm. Simplorer/twin builder provides a perfect platform for combining control drives (power electronics) and electromagnetic devices (FEM Maxwell models), as well as the ability to use ROMs from most Ansys brands. Simplorer/twin builder is a very versatile and efficient framework for multiphysics device analysis due to the ability to import Modelica, SPICE, and FMU/FMI modules, the ability to expand component libraries using VHDL-AMS and C++ scripting, and the coupling with the third party software like Simulink.

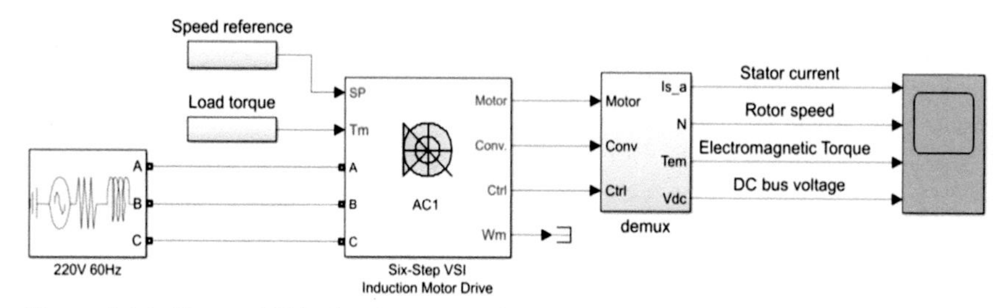

Figure 1.16 Six step VSI induction motor drive.

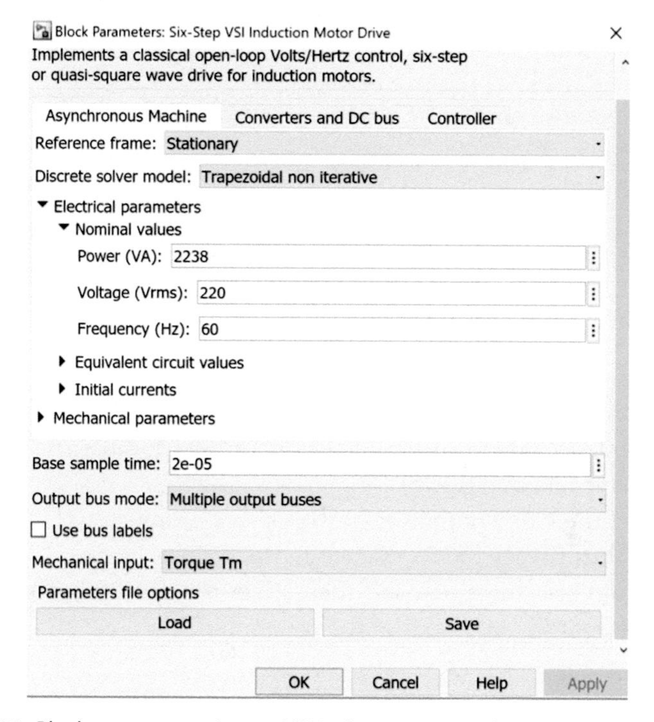

Figure 1.17 Block parameters six step VSI induction motor drive.

1.8 Retune the drive parameters

1. Open any one example: ac1_example/ ac2_example/ ac3_example (Fig. 1.16). Type the same name in the command window. The respective file will get open.
2. Open the drive, the parameters are set for a 200 3hp motor.
3. Double-click the six-step VSI Induction Motor Drive block and select the **Asynchronous Machine** tab. Copy the parameters of the motor rating which you need for your simulation into the drive's mask (Fig. 1.17).

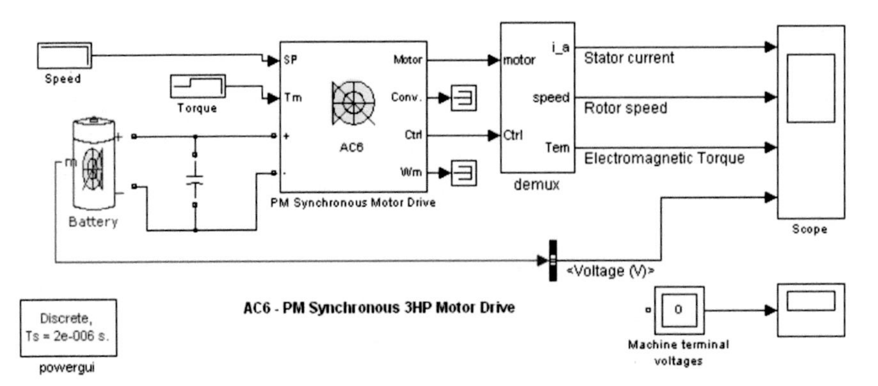

Figure 1.18 Battery connected PM synchronous motor drive.

4. In the same way the flux regulator, speed regulator, and DC bus voltage parameters can also be retuned in the MATLAB.

1.9 Modify a drive block

1. Save any readily available model like ac6_example2.
2. In the diagram, delete the three-phase source. Replace it with a 300Vdc/1Ah/NiMH Battery block and a 100 μF capacitor block connected in parallel (Fig. 1.18).
3. Connect the Conv output of the block to a terminator block. Remove the DC bus voltage blocks that are in the Demux block.
4. Add a bus selector block and then select the **voltage (V)** bus signal coming from the m output of the battery block.

1.10 Electric drives library

Under electric drives library in MATLAB, we have fundamental drives blocks with some categories like control (Table 1.1), electrical (Table 1.2), hybrid (Table 1.3). AC drives (Table 1.4), DC drives (Table 1.5), mechanical shaft and speed reducers (Table 1.6) and extra sources (Table 1.7) are also available under the electric drives library.

Inputs

The block has two inputs: Nm and Nl (Fig. 1.19).

The first input, Nm, is the speed (rpm) of the driving end of the shaft.

The second input, Nl, is the speed (rpm) of the load connected to the second end of the shaft.

Table 1.1 Fundamental drives block-control.

Bridge firing unit (AC)	Implement six pulse firing unit with notch filters for three phase thyristor bridge
Bridge firing unit (DC)	Generate gate signals for single-or three-phase thyristor bridge
Current controller (brushless DC)	Implement current/torque controller model for brushless DC machine
Current controller (DC)	Implement PI current controller model for DC machine
Direct torque controller	Implement direct torque and flux controller (DTC) model
Field-oriented controller	Implement a field-oriented controller model based on indirect or feedforward vector control strategy
Regulation switch	Implement torque controller, and switch that selects between torque or speed regulation for DC motor drives
Six-step generator	Implement pulse generator for six-step VSI AC motor drive
Space vector modulator	Implement space vector modulator for PWM VSI AC motor drive
Speed controller (AC)	Implement speed controller model for vector-controlled AC motor drives
Speed controller (scalar control)	Implement speed-controller model for scalar controlled AC drives
Vector controller (PMSM)	Implement vector controller model for permanent magnet synchronous machines (PMSM)
Vector controller (SPIM)	Implement vector controller model for single-phase induction motor (SPIM)
Voltage controller (WFSM)	Implement vector controller model for a wound-field synchronous machine (WFSM)
Voltage controller (DC bus)	Implement DC bus voltage controller for thyristor bridge rectifier

Table 1.2 Fundamental drives block-electrical.

Chopper	Implement DC Chopper model for DC motor drives
Circulating current inductors	Implement circulating current inductors for four-quadrant thyristor bridge converter
Inverter (five-phase)	Implements five-phase inverter model for five-phase PM synchronous motor drive
Inverter (three-phase)	Implements a three-phase inverter model for AC Motor Drives
Thyristor converter	Implements Single or three-phase thyristor converter for DC motor drives

Table 1.3 Fundamental drives block-hybrid.

Active rectifier	Implement three-phase active (PWM) rectifier model for AC motor drives
DC bus	Implement DC bus model that includes resistive braking chopper

Table 1.4 AC drives.

Brushless DC motor drive	Implement brushless DC motor drive using permanent magnet synchronous motor (PMSM) with trapezoidal back electromotive force (BEMF)
DTC induction motor drive	Implement direct torque and flux control (DTC) induction motor drive model
Field-oriented control induction motor drive	Implement field-oriented control (FOC) induction motor drive model
Five-phase PM synchronous motor drive	Implement five-phase permanent magnet synchronous motor vector control drive
PM synchronous motor drive	Implement permanent magnet synchronous motor (PMSM) vector control drive
Self-controlled synchronous motor drive	Implement self-controlled synchronous motor drive
Single-phase induction motor drive	Implement single-phase induction motor drive
Six-Step VSI induction motor drive	Implement six-step inverter fed induction motor drive
Space vector PWM VSI induction motor drive	Implement space vector PWM VSI induction motor drive

Table 1.5 DC drives.

Four-quadrant chopper DC drive	Implement four-quadrant chopper DC drive
Four-quadrant single-phase rectifier DC drive	Implement four-quadrant single-phase rectifier DC drive
Four-quadrant three-phase rectifier DC drive	Implement four-quadrant three-phase rectifier DC drive
One-quadrant chopper DC drive	Implement one-quadrant chopper DC drive (buck converter topology)
Two-quadrant chopper DC drive	Implement two-quadrant chopper DC drive (buck-boost converter topology)
Two-quadrant single-phase rectifier DC drive	Implement two-quadrant single-phase rectifier DC drive
Two-quadrant three-phase rectifier DC drive	Implement two-quadrant three-phase rectifier DC drive

Table 1.6 Shaft and speed reducers.

Mechanical shaft	Implement mechanical shaft
Speed reducer	Implement speed reducer

Table 1.7 Extra sources.

Battery	Generic battery model
CCCV battery charger	Constant-Current constant-voltage battery charger
Fuel cell stack	Implement generic hydrogen fuel cell stack model
Supercapacitor	Implement generic supercapacitor model

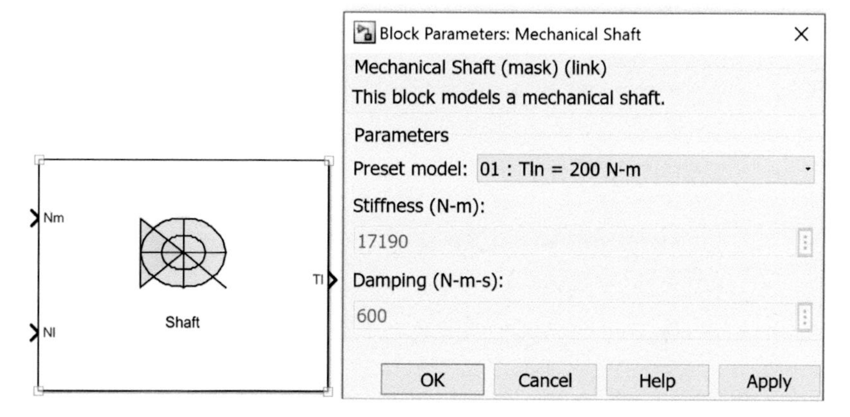

Figure 1.19 Mechanical shaft parameters.

Outputs

The block has one output: Tl.

The Tl output is the torque transmitted from the driving end of the shaft to the load.

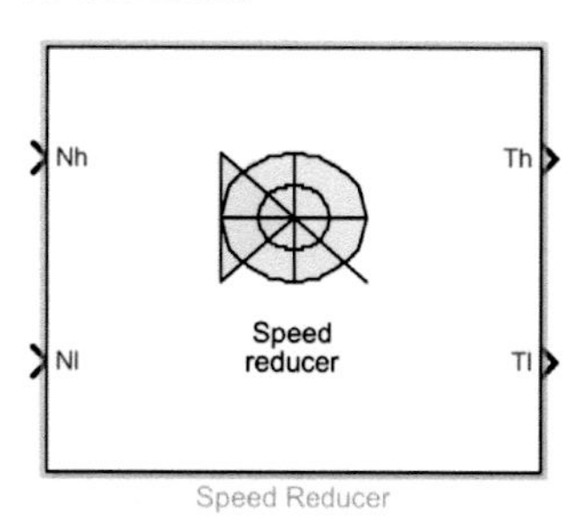

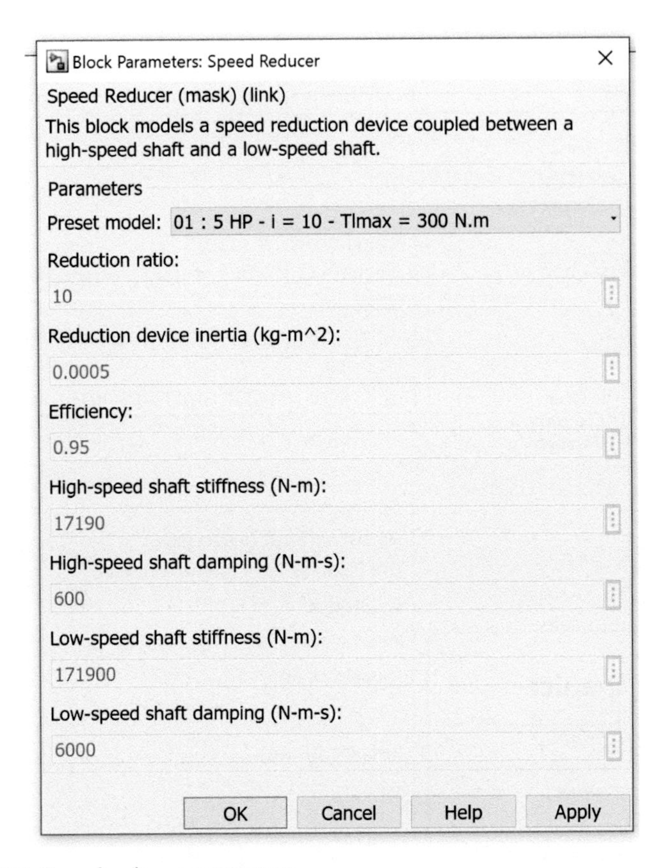

Figure 1.20 Speed reducer parameters.

Inputs

The block has two inputs: Nh and Nl (Fig. 1.20).

The first input, Nh, is the speed (rpm) of the driving end of the high-speed shaft.

The second input, Nl, is the speed (rpm) of the loaded end of the low-speed shaft.

Outputs

The block has two outputs: Th and Tl.

The Th output is the torque transmitted by the high-speed shaft to the reduction device.

The Tl output is the torque transmitted by the low-speed shaft to the load.

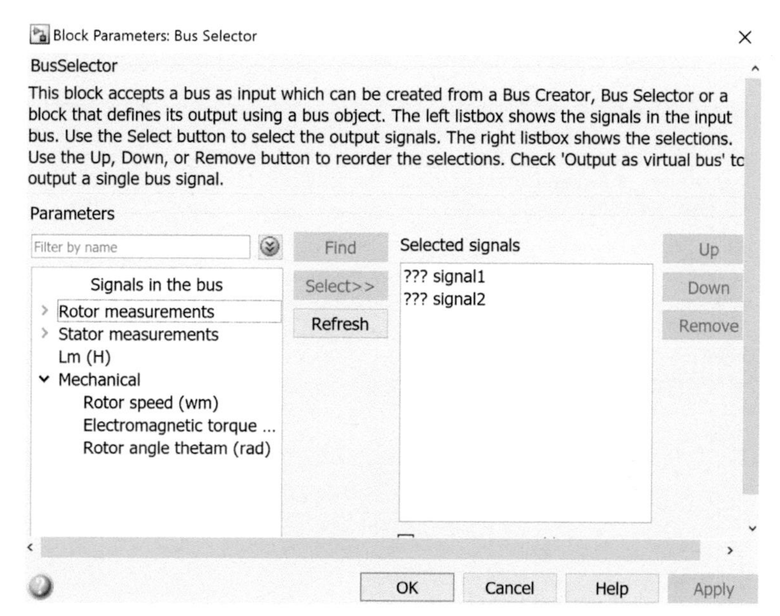

Figure 1.21 Bus selector parameters.

1.11 Mechanical coupling of two motor drives

1. Decide the motor need to be coupled, here induction motor drive is coupled with Rectifier fed DC drive.
2. Take the AC drive from Simulink library: Simscape/Electrical/Specialized Power Systems/Electric Drives/AC Drives.
3. The ABC terminals are connected to the three-phase source. It is taken from Simulink library: Powerlib/Electrical Sources/Three-phase Source.
4. Take the DC drive from Simulink library: Electric Drive lib/DC Drives/Four Quadrant Single Phase Rectifier DC Drive.
5. SP is the speed or torque set point. The speed setpoint can be a step function.
6. Tm or Wm: The mechanical input: load torque (Tm) or motor speed (Wm). For the mechanical rotational port (S), this input is deleted.
7. Motor: The motor measurement vector. This vector allows you to observe the motor's variables using the Bus Selector block (Figs. 1.21 and 1.22).
8. Conv: The three-phase converters measurement vector. This vector contains: The DC bus voltage, the rectifier output current and the inverter input current.

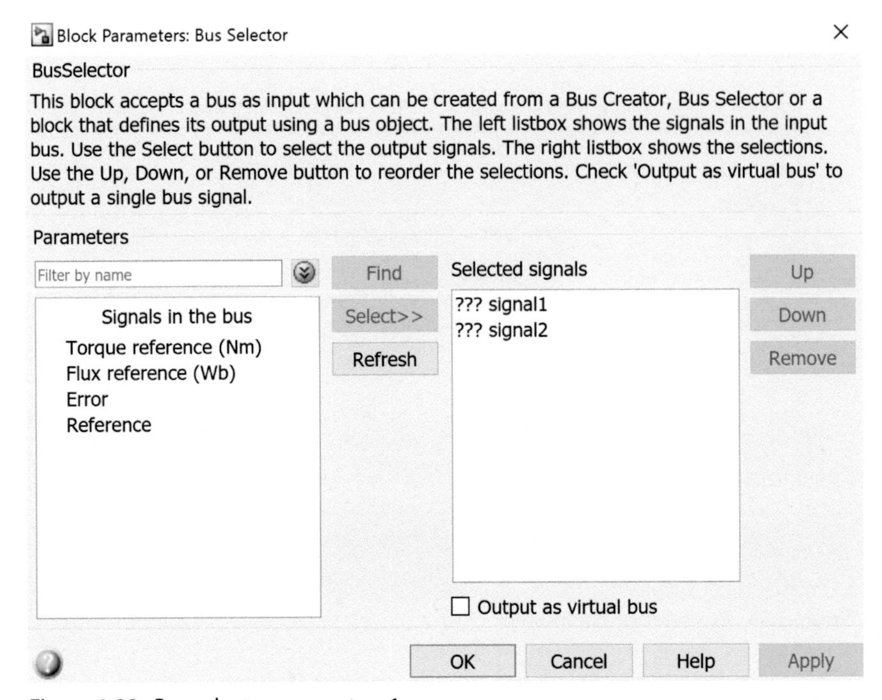

Figure 1.22 Bus selector parameters 1.

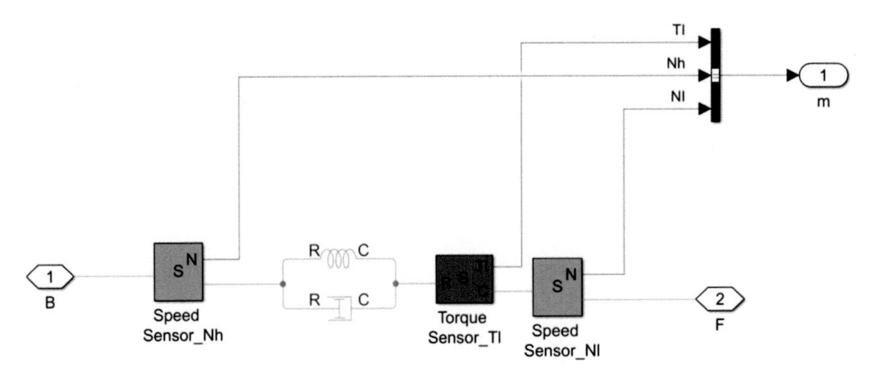

Figure 1.23 Mechanical coupling.

9. Ctrl: The controller measurement vector. This vector contains: The torque reference, The speed error (difference between the speed reference ramp and actual speed) and The speed reference ramp or torque reference.
10. The mechanical coupling (Figs. 1.23 and 1.24) represents the torsional dynamics of a mechanical shaft when the torque is transferred from the motor to the load.

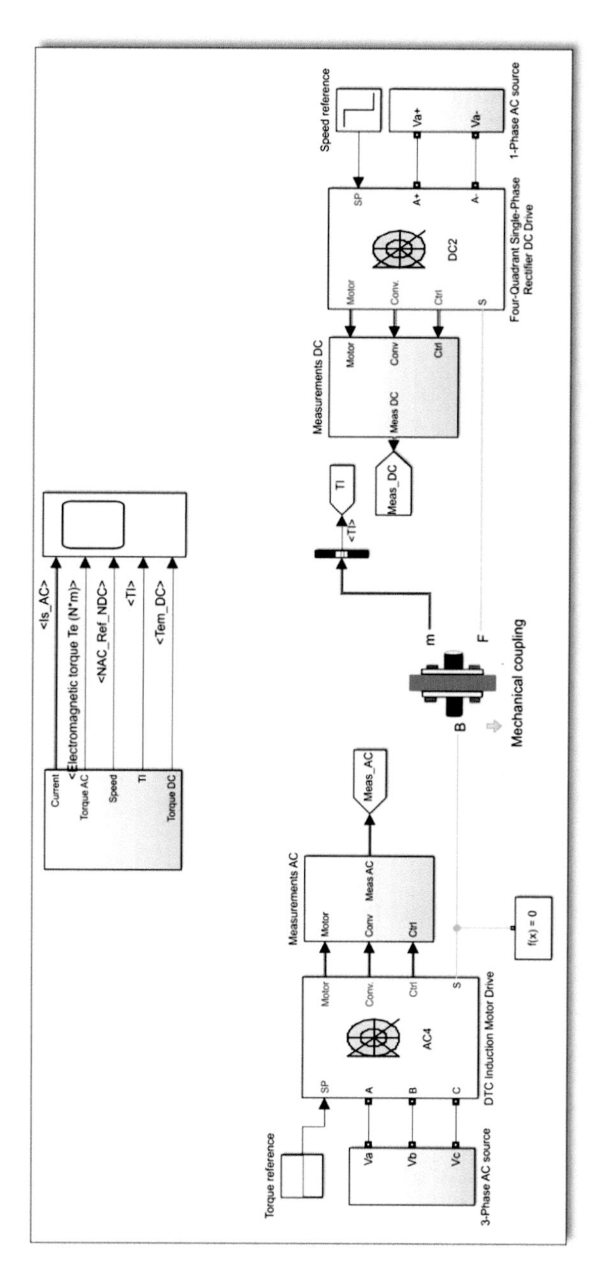

Figure 1.24 Mechanical coupling of two motors.

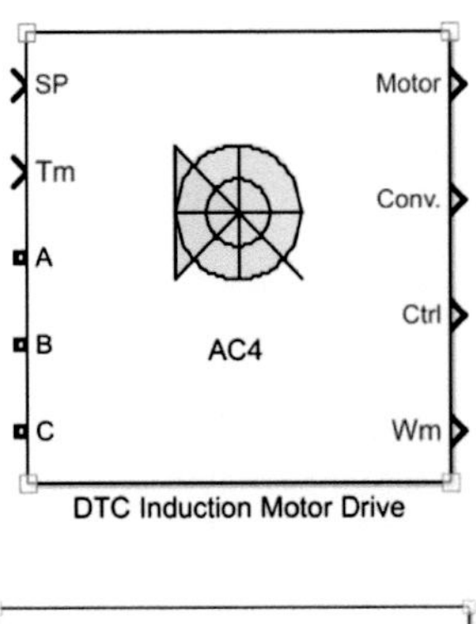

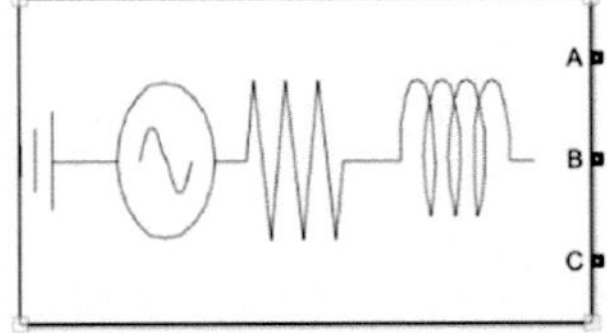

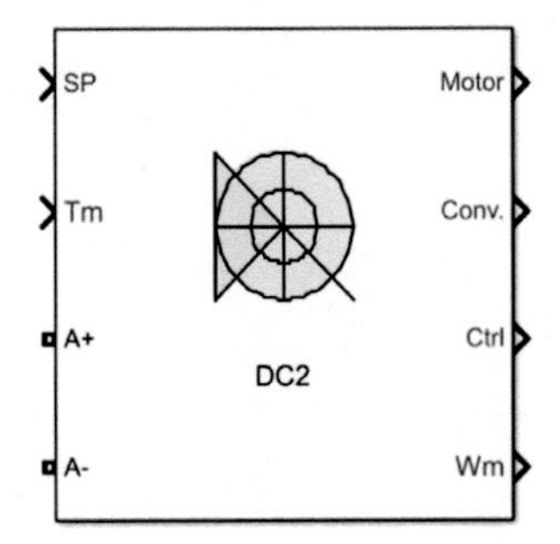

1.12 Various control methods of electric drive

There are two kinds of control systems in a control system: open-loop control systems and closed-loop control systems. In an open-loop control system, the output has little effect on the input, that is, the controlling phenomenon is independent of the output. In a closed-loop control system, the output is fed back to the input terminal, which determines the amount of input to the system, for example, if the output is greater than a predetermined value, the input is reduced and vice versa. Feedback loops or closed-loop power in electrical drives must meet the following conditions.

1. Security.
2. Improvements in reaction time.
3. To increase steady-state precision.

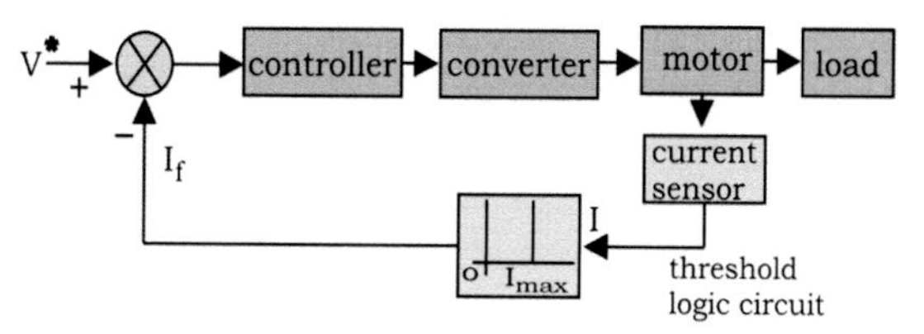

Figure 1.25 Current control.

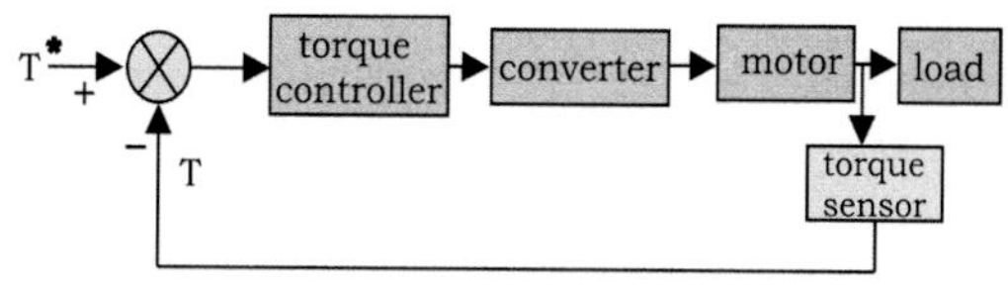

Figure 1.26 Torque control.

In the following discussions, we will look at various closed-loop configurations used in electrical drives, regardless of the form of supply they are supplied, that is, DC or AC.

Control of current limit: We know that if precautionary steps are not taken at startup, there is a risk of a large current flowing through the motor circuit. A current limit controller is mounted to limit the current and feel the current fed to the engine. The feedback loop has little effect on the drive's normal operation, so if the current reaches the fixed safe limit, the feedback loop unlocks and reduces the current to below the safe limit. When the current falls below the safe limit, the feedback loop deactivates again, and current control is achieved (Fig. 1.25).

Closed-loop torque control: This type of torque controller is seen mainly in battery-operated vehicles like cars, trains, etc., the accelerator present in the vehicles is pressed by the driver to set the reference torque T. The actual torque T follows the T* which is controlled by the driver via accelerator (Fig. 1.26).

Closed-loop speed control: Speed control loops are perhaps the most widely used feedback loops for drives. If we first see the block diagram of this loop then it will be a lot easier for us to understand.

The diagram shows two control loops, which can be referred to as an inner loop and an outer loop. The inner current control loop keeps the converter and motor current or torque within a stable range. By using functional examples, we can now appreciate the role of the control loop and push. Assume the reference speed Wm* increases and there is a positive error Wm, indicating that the speed must be increased.

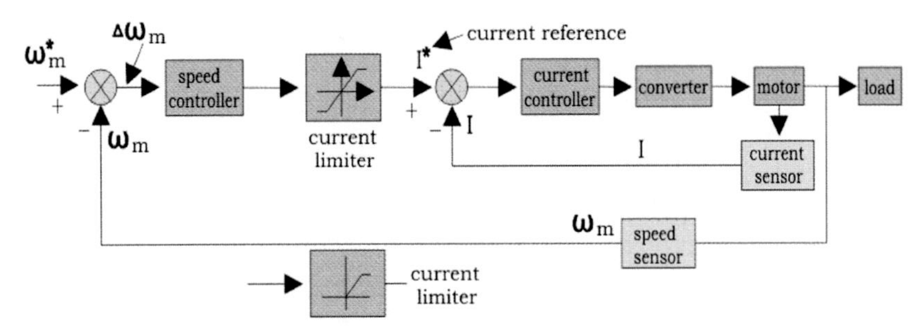

Figure 1.27 Speed control.

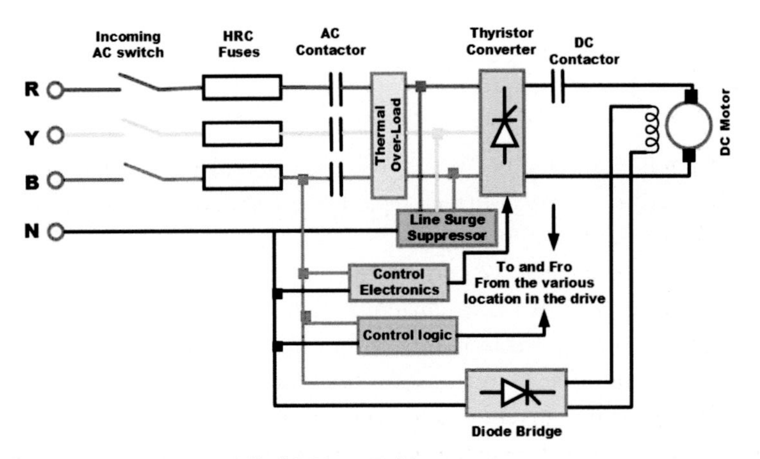

Figure 1.28 Converter-controlled DC motor drive.

The inner loop now raises the current while remaining under the full permissible current. The driver then accelerates, and as the speed approaches the target speed, the motor torque equals the load torque, and the reference speed Wm decreases, indicating that there is no need for more acceleration but that deceleration is needed, and braking is performed by the speed controller at the highest permissible current (Fig. 1.27). So, during speed regulation, the mechanism constantly switches from motoring to braking and back to motoring for the motor's smooth operation and working.

Consider the following two drive mechanisms. One uses a converter-controlled DC motor, while the other uses an inverter-fed AC motor. Fig. 1.28 depicts the converter-controlled DC motor drive mechanism. Fig. 1.29 depicts the GTO inverter-controlled induction motor drive. The major components of the electrical power system are listed below.

1. New incoming AC Switchgear consists of the following components: It consists of a switch fuse unit and an AC control contractor with ranges of

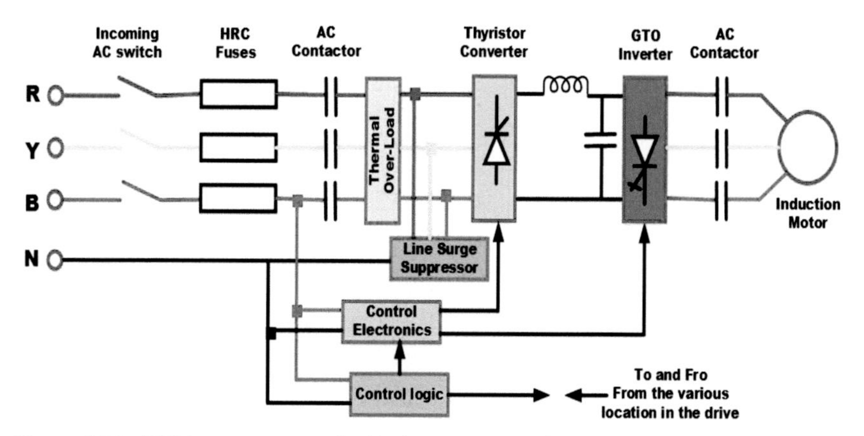

Figure 1.29 GTO inverter-controlled induction motor drive.

up to 660V and 800A. The switchgear replaces the standard contractor with a bar-mounted contractor, and an air circuit breaker is used as an incoming switch. The contractor with a bar installed increases the range to 1000V and 1200A.

It employs an HRC fuse with a 660V, 800A rating. Thermal overload is used in the AC switchgear to prevent the device from overloading. The switchgear contractor is also replaced by a moulded case circuit breaker.

2. Power converter/inverter assembly: This assembly is made up of two main components: power and control circuitry. Semiconductors, heat sinks, semiconductor fuses, surge suppressors, and cooling fans are also part of the power electronics blocks. Control electronics are made up of a triggering circuit, its own controlled power supply and drive circuit, and an isolation circuit. The power flow to the motor is controlled and regulated by the driving and isolation circuit. When the drive is in a closed loop, it has a controller as well as current and speed feedback loops. The control device has three port separation, including the power source, inputs, and outputs, all of which are adequately insulated.
3. Line surge suppressors: It prevents the semiconductor converter from voltage spikes generated in the line as a result of the load switching on and off on the same line. The voltage spikes are suppressed by the line surge suppressor and the inductance. As the incoming circuit breaker works and breaks the current supplied to the pit, the line surge suppressor consumes a certain amount of trapped energy. When the power modulator is not a semiconductor, the line surge suppressor is not necessary.

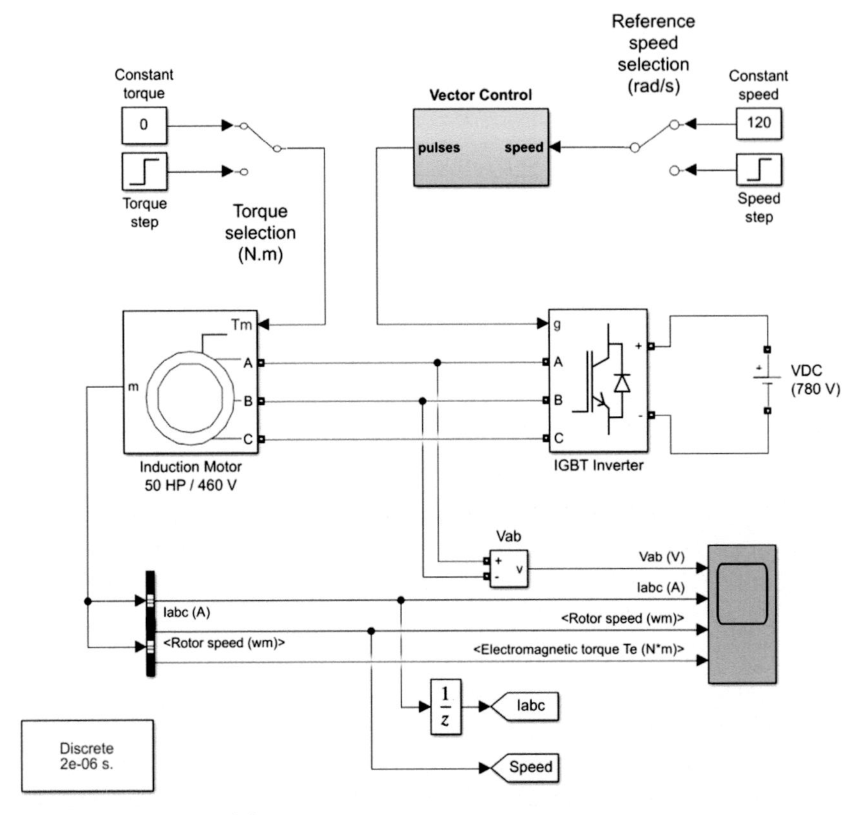

Figure 1.30 Drive model.

4. Control logic: It is used for interlocking and sequencing various drive system operations in regular, fault, and emergency conditions. The interlocking safeguards the system against abnormal and dangerous procedures, while the sequencing safeguards the multiple drive operations, such as starting, stopping, reversing, jogging, and so on, which are performed in a preplanned series. The programmable logic controller is used for dynamic interlocking and sequence operations.

1.13 Building your own drive

Although the electric drives library contains models of motor drives widely used in the industry, you might have some specific requirements leading you to build your own motor drive model. Here builds the field-oriented-control motor drive, very similar to the AC3 model.

Need to write the basic equations related to FOC drive system (Fig. 1.30).

Open the power_acdrive model and save it as case 3 in your working directory so that you can make further modifications without altering the original file.

The next figure shows the power_acdrive model in which blocks from Simscape Electrical Specialized Power Systems and Simulink libraries are used to model the induction motor drive.

The current regulator consists of three hysteresis controllers and is built with Simulink blocks.

The conversions between abc and dq reference frames are executed by the abc_to_dq0 Transformation and dq0_to_abc Transformation blocks.

The rotor flux is calculated by the Flux_Calculation block.

The rotor flux position (Θe) is calculated by the Teta Calculation in Vector Control Block.

The stator quadrature-axis current reference (iqs*) is calculated by the iqs*_Calculation block.

The stator direct-axis current reference (ids*) is calculated by the id*_Calculation block.

The speed controller is of proportional-integral type and is implemented using Simulink blocks.

Simulating the drive: In order to increase simulation speed, this model is discretized using a sample time of 2 μs. The variable Ts = 2e-6 automatically loads into your workspace when you open this model. This sample time Ts is used both for the power circuit (Ts specified in the Powergui) and the control system.

Run the simulation.

The motor voltage and current waveforms as well as the motor speed and torque are displayed on four axes of the scope connected to the variables Vab, Iabc, ωm, and Te.

Starting the drive: You can start the drive by specifying [1,0,0,0,0,0,0,0] as the initial conditions for the Asynchronous Machine block (initial slip = 1 and no currents flowing in the three phases).

Steady-state voltage and current waveforms: When the steady state is attained, you can stop the simulation and zoom on the scope signals.

Speed regulation dynamic performance: You can study the drive dynamic performance (speed regulation performance versus reference and load torque changes) by applying two changing operating conditions to the drive: a step change in speed reference and a step change in load torque.

1.14 Summary

This chapter provides the basic concepts of electric drives and important components of drive. The MATLAB/Simulink library details are provided with appropriate screenshots taken from MATLAB software. The applications of electric drive and control methods are presented in detail. The retune drive parameters and building user own drive procedures are given with all the steps.

1.15 Review Questions

1. Draw the block diagram of electric drive with neat sketch and explain the working of each block.
2. List the industrial applications of electric drive and explain the working of drives in each application.
3. Write the step by step procedure to build own drive in Simulink environment.
4. What are the software's used for simulation of electric drives.
5. How to control the electric drives in closed loop, explain it with proper diagram.
6. Give the step-by-step procedure to mechanically couple two electric motor drives.
7. How to modify the readily available drive block in Simulink.
8. Name all the electric drive libraries available in Simulink.
9. Explain the importance in electric drive in today's world.
10. What do you meant by electric drives.
11. Classify electric drives.
12. How to retune electric drives in Simulink environment.

CHAPTER 2

Converter fed DC drives with simulation

Contents

2.1 Introduction

To power a direct current (DC) motor, a DC drive transforms an alternating current (AC) into DC. Where good dynamic response and steady-state performance are needed, DC motors are widely used in variable-speed drives and position control systems. Robotic drives, printers, machine tools, process rolling mills, the paper and textile industries, and many other industries are examples. Since the commutators are integrated into the motor, controlling a DC motor, especially one that is separately excited, is quite simple. If the field, current is kept constant, the motor-developed torque is proportional to the armature current thanks to the commutators brush. Classical control theories can then be easily extended to the design of a drive system's torque and other control loops. A DC chopper, which is a single-stage DC to DC conversion unit, can provide variable voltage to the armature of a DC motor for speed control. Present limit control or time ratio control may be used to achieve voltage variance at the load terminals. The chopper in the former is regulated such that the load current varies between two

Electric Motor Drives and Their Applications with Simulation Practices.
DOI: https://doi.org/10.1016/B978-0-323-91162-7.00004-7

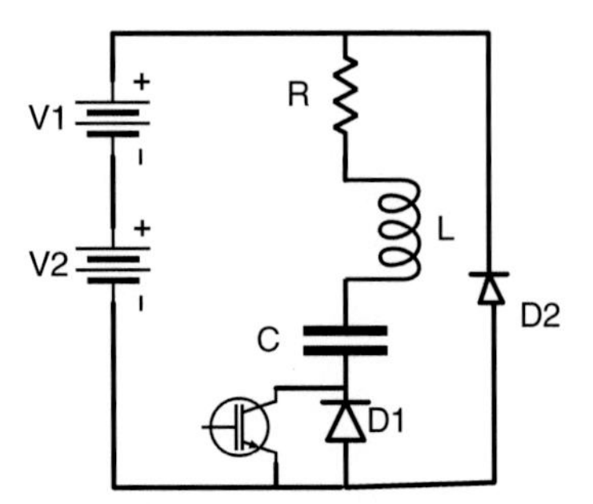

Figure 2.1 Chopper fed DC motor. *DC*, direct current.

limits, as previously discussed. The chopper is switched off when the current exceeds the upper limit, disconnecting the motor from the supply. The load current decays as it freewheels through the freewheeling diode. The chopper is turned on when it reaches the lower limit, connecting the motor to the supply. A constant average current is preserved. When the chopper is powered by Time Ratio Control (TRC), the chopper's TON/TOFF ratio changes. If (TON+TOFF) is kept constant, the operation would be at a fixed frequency. To regulate voltage, only TON is changed. With TON kept constant and (TON+TOFF) varied, the operation will be performed at a variable frequency. However, a fixed frequency TRC is commonly used due to many advantages of simplicity. Both separately excited and series motors are regulated by chopper circuits.

Chopper circuits advantages over phase-controlled converters:

1. The production has a small amount of ripple material. The current ratios of peak/average and rms/average are both small. This boosts commutation and reduces the motor's harmonic heating. There are also less pulsating torques.
2. The chopper is powered by batteries at a steady voltage. The power factor issue does not arise at all. As the angle is delayed, the traditional phase control approach has a lower power factor. This means that the chopper draws less current than an AC/DC phase-driven converter.
3. The circuit is straightforward and can be tweaked to provide regeneration.
4. The control circuit is straightforward.

The chopper, however, could be more expensive than a phase-controlled converter due to the forced commutation used. Fig. 2.1 shows a chopper-fed DC motor, whereas Fig. 2.2 shows the output.

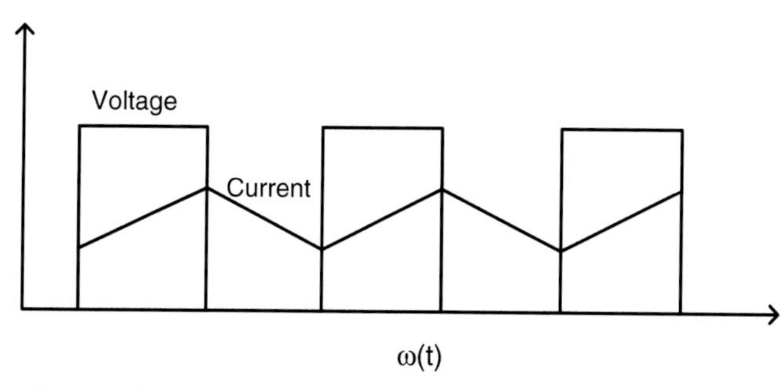

Figure 2.2 Chopper fed DC motor. *DC*, direct current.

Variable-speed drives depend heavily on DC motors, which have a variety of characteristics.

- DC motors have a high starting torque and can be controlled over a large speed range; speed control methods are usually easier and less costly than those used in AC drives.
- In today's industrial drives, DC motors play a major role.
- In variable speed drives, both series and separately excited DC motors are commonly used, but series motors are typically used for traction applications.
- DC motors are not ideal for very high-speed applications due to commutators, and they need more maintenance than AC motors.
- AC motor drives are becoming more competitive with DC motor drives due to recent developments in power conversions, control techniques, and microcomputers.
- While AC drives are expected to become more common in the future, DC drives are still widely used in many industries. It may take many decades for DC drives to be fully replaced by AC drives.

Controlled rectifiers produce a variable dc output voltage from a fixed AC input voltage, while a DC–DC converter produces a variable DC voltage from a fixed DC input voltage.

- Powered rectifiers and DC–DC converters revolutionized modern industrial control equipment and variable-speed drives with power levels varying from fractional horsepower to several megawatts, thanks to their ability to supply a continuously variable DC voltage.
- Regulated rectifiers are commonly used to regulate the speed of DC motors.
- A diode rectifier followed by a DC–DC converter is another choice.

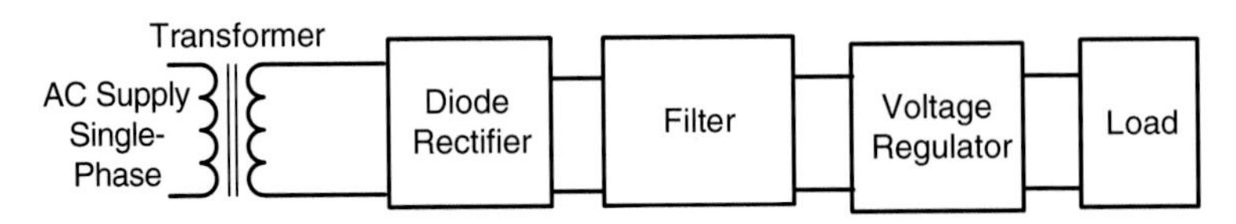

Figure 2.3 Genetic scheme of rectifier.

2.2 Uncontrolled converter fed DC drives

The effects on the rectifier behavior of the inductive components of the load and of the AC line will be investigated. The use of passive filters to reduce the harmonic content (ripple) of the voltage and current at the output of the rectifier will be discussed.

Magnetic fields cause electrons or other charged particles to travel into orbits or trajectories in particle accelerators. The magnetic field strength needed to achieve the desired effects is proportional to the energy of the particles. Electromagnets, either traditional hot or superconducting, are commonly used. The magnets' excitation current can range from a few amperes for small orbit correction coils to hundreds or thousands of amperes for large orbit correction coils. The power converters required to cover such a wide current range have a wide range of structures and characteristics, and many solutions for the same power requirement are frequently available.

Since the currents in the magnets must be varied according to the energy of the particles, or at the very least ramped from the turn-on values to their final values (which is critical if the load—a magnet string—has a long time constant), the rectifiers use thyristor-based structures or mixed structures. The inductive components of the load and the AC line will be investigated for their effects on rectifier behavior. It will be addressed how to use passive filters to reduce the harmonic content (ripple) of the voltage and current at the rectifier's output.

Performance parameters: Before starting to examine different topologies for single-phase or multiphase rectifiers, we should define some parameters. These parameters are needed to compare the performances among the different structures. The genetic scheme of the rectifier is shown in Fig. 2.3.

The DC voltage on the load is the average over the period T of the output voltage of the rectifier:

$$V_{dc} = \frac{1}{T} \int V_L(t)dt \tag{2.1}$$

Similarly, it is possible to define the RMS voltage on the load:

$$V_L = \sqrt{\frac{1}{T} \int V_L^2(t) dt} \tag{2.2}$$

The ratio of the two voltages is the form factor (FF):

$$FF = \frac{V_L}{V_{DC}}$$

This parameter is quite important since it is an index of the efficiency of the rectification process. Having assumed the load to be purely resistive, it is possible to define the currents as

$$i_L(t) = \frac{V_L(t)}{R_L} \tag{2.3}$$

$$I_{DC} = \frac{V_{DC}}{R_L} \tag{2.4}$$

$$I_L = \frac{V_L}{R_L} \tag{2.5}$$

The rectification ratio (η), also known as rectification efficiency, is expressed by

$$\eta = \frac{P_{DC}}{P_L + P_D} \tag{2.6}$$

were,

$$P_{DC} = V_{DC}.I_{DC} \tag{2.7}$$

$$P_L = V_L.I_L \tag{2.8}$$

$$P_D = R_D.I_L^2 \tag{2.9}$$

$$\eta = \frac{V_{DC}.I_{DC}}{V_L.I_L + R_D.I_L^2} \tag{2.10}$$

Another important parameter used to describe the quality of the rectification is the ripple factor (RF). It indicates the smoothness of the voltage waveform at the rectifier's output (we have to keep in mind that our goal is to obtain a voltage and a current in the load as steady as possible). The RF is defined as the ratio of the load voltage's effective AC component to the

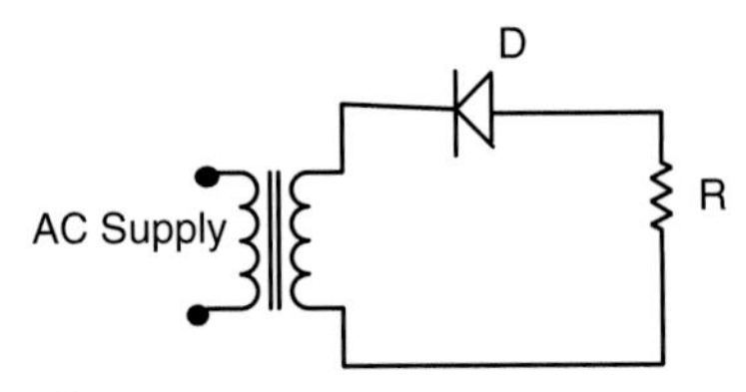

Figure 2.4 Half-wave rectifier.

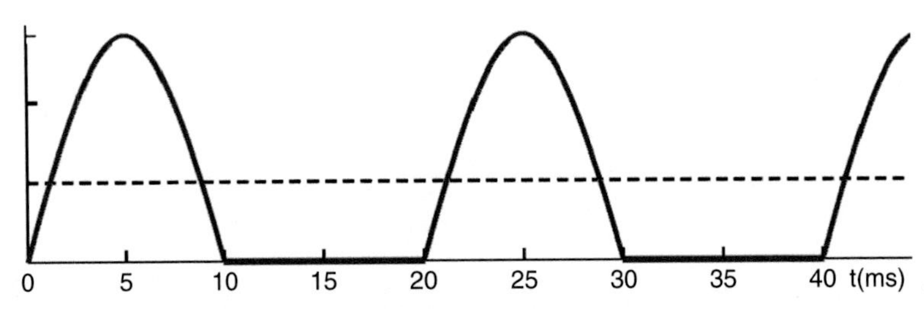

Figure 2.5 Voltage and current waveform.

DC voltage:

$$RF = \frac{\sqrt{V_L^2 - V_{DC}^2}}{V_{DC}} = \sqrt{FF^2 - 1} \tag{2.11}$$

The main topologies of the rectifiers are of two types, they are as follows:

a. Single-phase system.

b. Multiphase system.

2.2.1 Single-phase system

a) *Half-wave rectifier*

This is the simplest structure shown in Fig. 2.4. Only one diode is placed at the secondary of the transformer.

Fig. 2.5 depicts the waveform. The voltage on the load is proportional to the current since the load is a resistance. The term half-wave rectifier derives from the fact that the rectification process occurs only during half-periods. Because the load current iL(t) always circulates in the same direction in the secondary winding, it is also known as single-way. We get the following results using the definitions stated in the preceding section:

$$V_{DC} = \frac{1}{T} \int V_L(t)dt = \frac{1}{2\pi} \int_0^{\pi} V_s \sin(\varpi t)dt = \frac{V_s}{\pi} \tag{2.12}$$

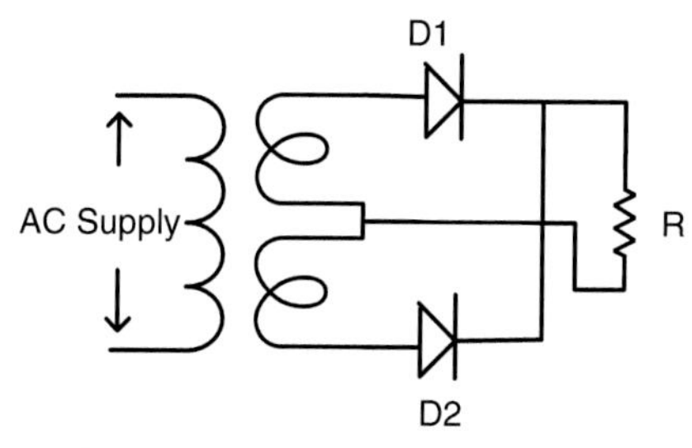

Figure 2.6 Full-wave rectifier.

And, similarly, we can calculate the other parameters:

$$I_{DC} = \frac{V_{DC}}{R_L} = \frac{V_S}{2.R_L} = I_S \tag{2.13}$$

The current in the transformer's secondary can only flow when the diode is conducting, hence it is equal to the current in the load:

$$FF = \frac{V_L}{V_{DC}} = \frac{\pi}{2} \tag{2.14}$$

$$\eta = \left(\frac{1}{FF}\right)^2 = \frac{4}{\pi^2} = 0.405 \tag{2.15}$$

$$RF = \sqrt{FF^2 - 1} = 1.21 \tag{2.16}$$

The poor performance of this rectifier is also confirmed by the utilization of the transformer. From Eq. (2.14), we get

$$TUF = 0.323 \text{ (or } TUF = 0.286\text{)}.$$

The use of the transformer confirms the rectifier's poor performance. We get from Eq. (2.14)

b) *Full-wave rectifier—center-tapped*

Two diodes could be used to exploit both half of the secondary AC voltage waveform, and a return path for the current can be established by adding a tap at the secondary winding's center. The center-tapped rectifier is what it is called. Fig. 2.6 depicts a full-wave rectifier.

During the positive half-wave of the voltage, diode D1 conducts. The negative half of the diode D2 conducts. The current always passes from the diodes' common point, through the load, and back to the transformer's central tap. The rectification occurs throughout the voltage period, as depicted in Fig. 2.6. A full-wave rectifier is what this is. Full-wave rectifier waveforms are given in Fig. 2.7.

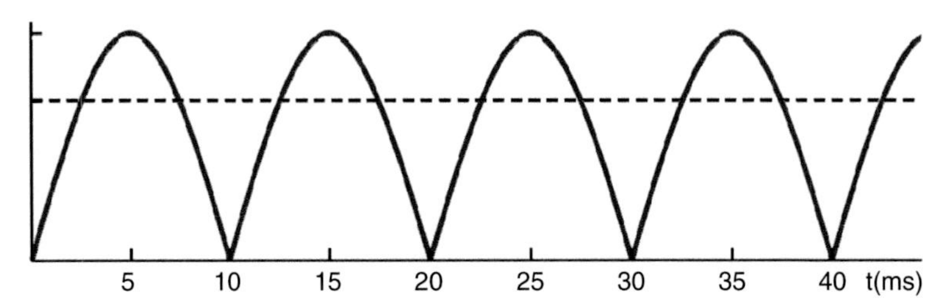

Figure 2.7 Wave form of full wave rectifier.

It must be observed that the current flows in the same direction across the two parts of the secondary winding in this scenario as well. As a result, this is also a one-way construction.

We derive the following findings using the definitions from the previous section and the symmetries:

$$V_{DC} = \frac{1}{T} \int V_L(t)dt = \frac{2}{2\pi} \int_0^{\pi} V_s \sin(\varpi t)dt = \frac{2.V_s}{\pi} \tag{2.17}$$

$$V_L = \sqrt{\frac{1}{T} \int_0^T V_L^2(t)dt} = \sqrt{\frac{1}{\pi} \int_0^{\pi} V_S^2 \sin^2(\varpi t)dt} = \frac{V_S}{\sqrt{2}} \tag{2.18}$$

$$I_{DC} = \frac{V_{DC}}{R_L} = \frac{2.V_S}{\pi.R_L} \tag{2.19}$$

$$FF = \frac{V_L}{V_{DC}} = \frac{\pi}{2.\sqrt{2}} = 1.11 \tag{2.20}$$

$$\eta = \left(\frac{1}{FF}\right)^2 = 0.81 \tag{2.21}$$

$$RF = \sqrt{FF^2 - 1} = 0.483 \tag{2.22}$$

Because it is a single-way topology, both secondary windings have a DC, resulting in a low TUF.

TUF = 0.671 (or 0.572 TUF).

2.2.2 Multiphase system

a. *Three-phase current regulator and diode rectifier*

Thyristors with VRRM and VDRM greater than 6500 V (or more) and on state average currents surpassing 1200 A or even 3000 A are available (e.g., Powerex TBK0 or FT1500AU-240 or EUPEC T2871N or T2563N). However, other applications, such as RF klystrons, necessitate even greater

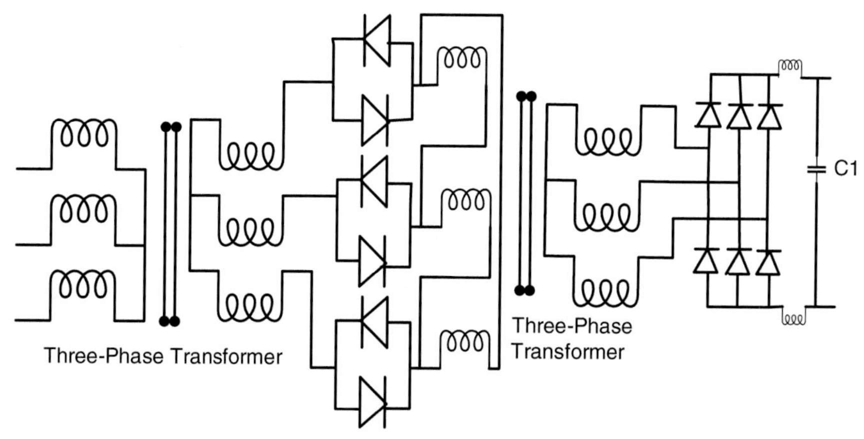

Figure 2.8 Three-phase current regulator and diode rectifier.

voltages of up to 100 kV . When thyristors are connected in series to handle the high voltage, the difficulty of their simultaneous operation arises, firing: a very good It is necessary to ensure equalization of the trigger pulses as well as the voltage drop on the stack. The voltage on the components may reach damaging levels if certain thyristors in the stack also are conducting while others have been turning on.

As a result, it is preferable to employ diode stacks, which are inherently commuting devices that do not require a trigger. A preregulation using thyristor on the AC side is utilized to maintain the ability to control the output voltage, as shown in Fig. 2.8.

b. *Three-phase uncontrolled or controlled bridge and linear regulator*

Quite often, there is a requirement to supply a large number of low-power loads (up to a few kilowatts) or loads that require higher dynamics than a simple line commutated rectifier can provide, or there is a requirement to supply loads with diverse characteristics with the same converter (this is the typical case of multipurpose spare power supplies).

The DC power for the channels connected to the distribution rails is provided by a diode rectifier followed by a passive filter for harmonic reduction. Each channel is a transistor linear regulator (or PWM DC/DC module) that supplies power to a different load. Positive and negative DC voltage is provided by two bridges connected in series and using the same point as a return line to power four quadrants (bipolar in voltage and current) transistor regulators.

Because the delay angle is large at low currents and the ripple on the DC side is considerable, using a thyristor bridge alone is not the best solution for supplying loads that demand variable currents or loads with diverse characteristics (as in the case of reserve power supplies). A diode rectifier is not really suited because it produces a fixed DC voltage, and at low currents, the majority of the output voltage falls on the regulator's transistors, increasing losses.

High dynamics, high efficiency, and low ripple are provided by a fully regulated rectifier, followed by a filter with a large reservoir capacitor and a series linear regulator. Such power supplies are currently accessible as conventional products on the market.

2.3 Controlled converter fed DC drives

The speed of a DC motor can be easily adjusted by adjusting its supply voltage using phase-regulated rectifiers. The field or the armature circuit can both be controlled with this control. Because the time constant of the field is substantially bigger than that of the armature, the motor reaction with armature control is faster than with field control. Field control is typically used for speeds over the rated value, while armature control is utilized for speeds below the rated value. When an AC supply is available, controlled rectifiers, either single-phase or multiphase, are commonly utilized in DC driving applications. By altering the triggering angle of the thyristor or any other power semiconductor device that could be provided to a DC motor, the single-phase or multiphase AC is converted to DC to produce a variable DC source, and therefore the speed of the motor may be regulated.

Where strong dynamic response and steady-state performance are required, direct-current motors are widely utilized in variable-speed drives and position control systems. Robotic drives, printers, machine tools, process rolling mills, the paper, and textile industries, and many more industries are examples. Because the commutators are integrated into the motor, controlling a DC motor, particularly one that is individually stimulated, is quite simple. If the field current is kept constant, the motor-developed torque is proportional to the armature current thanks to the commutators brush. Classical control theories can thus be simply applied to the design of a drive system's torque and other control loops.

The main difference between a thyristor-controlled rectifier and a chopper is that the former always flows through the supply, whereas the latter only flows from the supply terminals for a portion of each cycle.

Because a single-switch chopper based on a transistor, MOSFET, or IGBT can only give positive voltage and current to a DC motor, it is limited

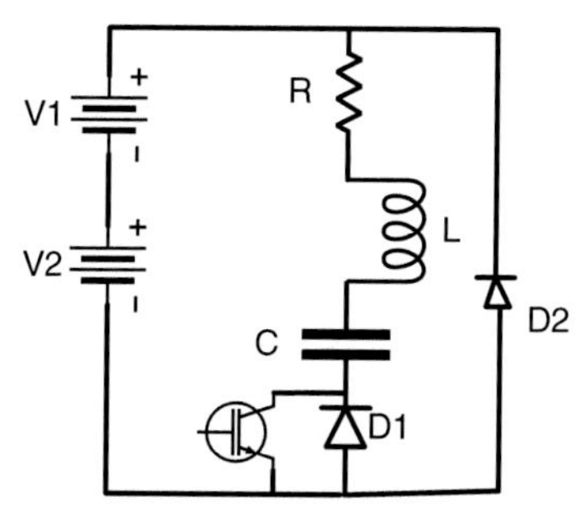

Figure 2.9 Chopper fed DC motor. *DC*, direct current.

to quadrant 1 driving. More sophisticated circuitry, comprising two or more power switches, is necessary when regenerative and/or quick speed reversal is required, resulting in the greater cost.

This subject only has an output voltage range of 0 E, where E is the battery voltage, hence this sort of chopper is only acceptable if the motor voltage is lower than the battery voltage. A "step-up" chopper with an extra inductance as an intermediary energy store is employed when the motor voltage is higher than the battery voltage.

2.3.1 Performance of chopper-fed DC motor drives

The DC motor operated almost as well when fed from a phase-controlled rectifier as it did when fed with pure DC, as we saw before. The chopper-fed motor is, if anything, superior to the phase-controlled motor since a high chopping frequency can reduce armature current ripple. A chopper is a high-speed on/off semiconductor switch that connects and disconnects the load from the source quickly. o Choppers are used to generate variable DC voltage from a fixed DC source. Self-commutated devices like MOSFETs, Power transistors, IGBTs, GTOs, and IGCTs are used in choppers because they can be commutated by a low-power control signal and do not require a communication circuit. They can also be operated at a higher frequency with the same rating. Both individually excited and Series circuits are controlled by chopper circuits. Fig. 2.9 shows a DC motor that is fed by a chopper.

The armature voltage waveform's shape tells us that when the transistor is turned on, the battery voltage V is applied directly to the armature, and the course of the armature current is traced at this time. The transistor is switched "off" for the rest of the cycle, and the current freewheels via the diode. The armature voltage is capped at (nearly) zero when the current is freewheeling across the diode.

The average armature voltage (V dc), which is determined by the percent of the whole cycle time (T) for which the transistor is "on," determines the motor's speed. If T on = kT and T off = (1 − k)T are used to define the on and off timings, the average voltage is simply given by

$$V_{dc} = kV \tag{2.23}$$

This shows that the on-time ratio, k, is used to control speed. The higher waveform in the current waveforms represents full load, that is, the average current (I dc) delivers the motor's full rated torque. The steady-state speed will remain the same, but the new mean steady-state current will be halved, as shown by the lower dotted curve, if the load torque on the motor shaft is lowered to half the rated torque and the resistance is insignificant. We should note, however, that while the load determines the mean current, the ripple current remains unaffected, as explained below.

The equation regulating the current during the "on" time if resistance is ignored is

$$V = E + L\left(\frac{d_i}{d_t}\right) \quad (23)$$

$$\frac{d_i}{d_t} = \left(\frac{1}{L}\right)(V - E) \tag{2.24}$$

Because V is larger than E, the current gradient (di /dt) is positive, as shown in Fig. 2.13C. The battery is giving electricity to the motor during this "on" phase. Some of the energy is transferred to mechanical output power, but some of it is also stored in the inductance's magnetic field. As the current (I) rises, more energy is stored, which is provided by 1=2Li 2.

The equation regulating the current during the "off" time is 0 = E +

$$L\left(\frac{d_i}{d_t}\right) \tag{2.25}$$

$$\frac{d_i}{d_t} = \left(\frac{-E}{L}\right) \tag{2.26}$$

2.3.2 Hard-switching converters for DC drive

Over the last few decades, traditional pulse width modulated (PWM) converters have been employed in switched mode. When power switches function in hard-switching mode, they must, however, cut off the load current during the turn-on and turn-off time intervals. Due to the stressful switching scenario of power electrical devices, the phrase "hard-switching" was coined. Fig. 2.10 depicts the switching behavior of a hard-switching

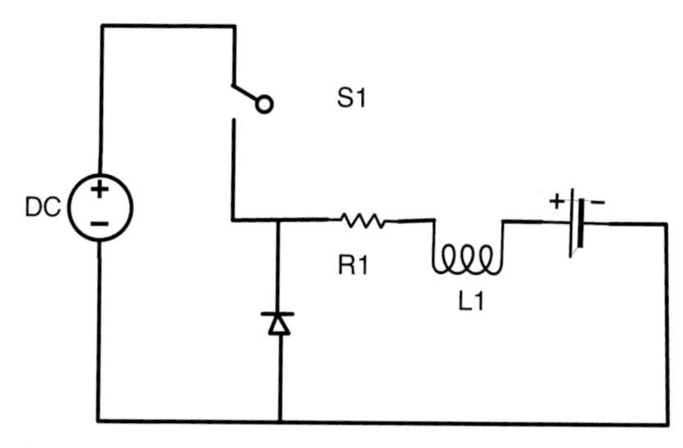

Figure 2.10 Hard switching converter.

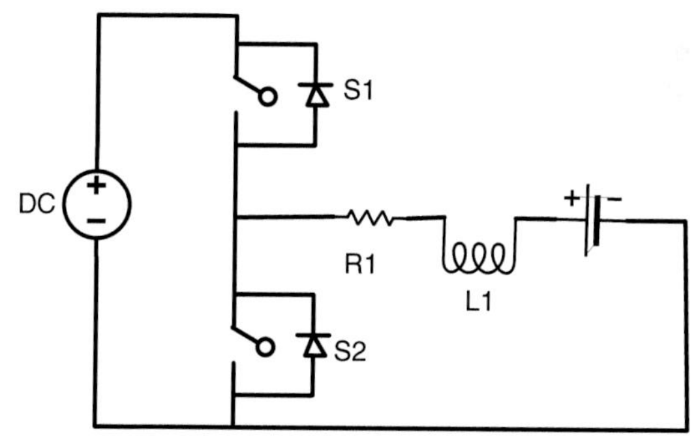

Figure 2.11 Two-quadrant hard-switching DC motor. *DC*, direct current.

device. During the turn-on or turn-off procedure, the power device is exposed to high voltage and current at the same time, resulting in severe switching stress and losses.

Under hard-switching settings, the reduction of stress and losses is a phenomenon that must be addressed. As a result, including dissipative passive snubber into power circuits is a common approach for redirecting stress and losses to these passive snubber circuits. The dv/dt and di/dt can be reduced as a result of this addition. Another component targeted at reducing switching losses is lowering the maximum switching frequency of the converters, as switching losses are related to switching frequency. The power converters' normal switching frequency is limited, often ranging between 20 and 50 kHz. Furthermore, due to the presence of stray capacitances and inductances in the power circuits, there are significant transient effects in the converters, resulting in an increase in electromagnetic interference (EMI) concerns. Fig. 2.11 shows a hard-switching converter.

A BJT, thyristor, MOSFET, or IGBT can be used as the switching device. Because it can only feed positive voltage and current into the DC motor's armature, it can only operate in the first quadrant, which means driving. The equations that control continuous conduction operation are

$$V_0 = D.V_{in} \tag{2.27}$$

$$V_r = V_{in}\sqrt{D(1-D)} \tag{2.28}$$

$$I_{SW} = D\frac{V_{in} - e_b}{R_a} - \frac{t}{T} \tag{2.29}$$

$$I = \frac{\frac{V_{in}}{R_a}\left(1 - e^{-D.T/t}\right)\left(1 - e^{-(1-D).\frac{T}{t}}\right)}{1 - e^{-T/t}} \tag{2.30}$$

where Vo, eb, Vr, I¯sw, I¯D, Ip-p, T, D are the average output voltage, back EMF voltage, output AC ripple voltage, average switch current, average diode current, peak to peak output ripple current, switching period, and duty cycle, respectively.

A two-quadrant hard-switching DC motor drive is shown in Fig. 2.11. The SW1 and D1 components constitute the first-quadrant chopper, via which the energy from the Vin source is transmitted to the DC motor, resulting in the motoring mode of operation. The SW2 and D2 components, in turn, comprise a fourth-quadrant operation in which energy is supplied from the Vin source to the DC motor, resulting in the regeneration mode of operation, when the circuit is in a state of continuous conduction The polarity of the maximum and minimum peak output currents, as well as the polarity of the mean value of the output current, affect the average diodes and switches currents. As a result, estimating these average current values is more difficult. The average currents flowing through the SW1 and D1 components can be computed when the converter is in the first quadrant. The currents running through the SW2 and D2 components, on the other hand, are zero.

An H-bridge or four-quadrant DC chopper is shown in Fig. 2.12. Because there are four switches, multiple control methods are used to produce four-quadrant output voltage and current using bidirectional current and voltage. To reduce distortion, complementary switching components are utilized in each leg in all techniques (T1 or T4 can be in the "on" state but not both, and T2 and T3 but not both). Around the zero-current output, the described operation philosophy provides current continuity.

When two separate two-quadrant choppers need to be operated, this H-bridge can be handled in the same way. The key restriction is that switches in the same leg cannot be activated at the same time. T1 and T4 switches

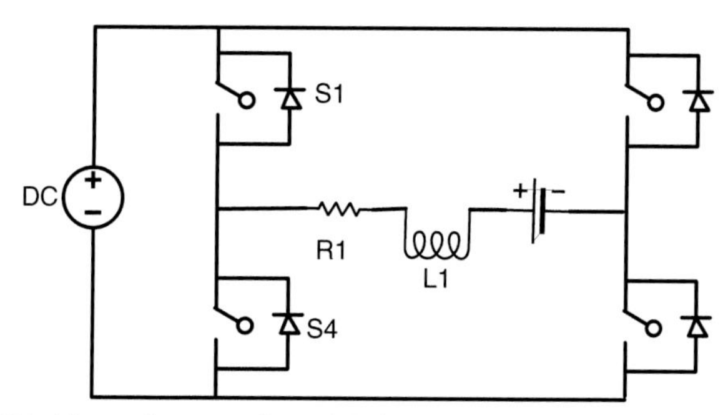

Figure 2.12 H-bridge or four-quadrant DC chopper. *DC*, direct current.

are connected to D1 and D4 diodes to form the first chopper. The first chopper operates in the first and second quadrants, with positive output current and bidirectional output voltage, that is, a second chopper is created by connecting T2 and T3 switches to D2 and D3 diodes, resulting in a third and fourth quadrant operation. A negative output current is accessible in this scenario, with a bidirectional output voltage of Vo. The operation of the DC motor into the four quadrants is ensured by the two two-quadrant choppers indicated above, which are incorporated in the same topology. Unifying the functioning of all four switches is another control mechanism for the H-bridge converter.

Using the preceding control mechanisms, the converter's output voltage can be bipolar or multilayer, depending on whether zero output voltage loops are used or not. In the event of a bipolar output, the ripple current value is increased. However, with minimal crossover distortion, a faster current reversal is facilitated. The operation is unaffected by the direction of the output current. The three-level output voltage and the bipolar output voltage are two H-bridge control strategies that can be explored based on the above. The following formulae can be used to define the average output voltage and the AC ripple voltage in the case of a bipolar output voltage:

$$V_r = \sqrt{2}V_{in}\sqrt{(2D - 1)(1 - D)} \tag{2.31}$$

2.3.3 Soft-switching converters for DC drives

When DC to DC converters are hard-switched, the electronic switches run at full load when the switches are turned on or off. This type of action puts a strain on the semiconductors, resulting in higher losses. The switching frequency of the converters causes a linear rise in these losses. Furthermore, the big dv/dt and di/dt cause greater EMI issues. The above

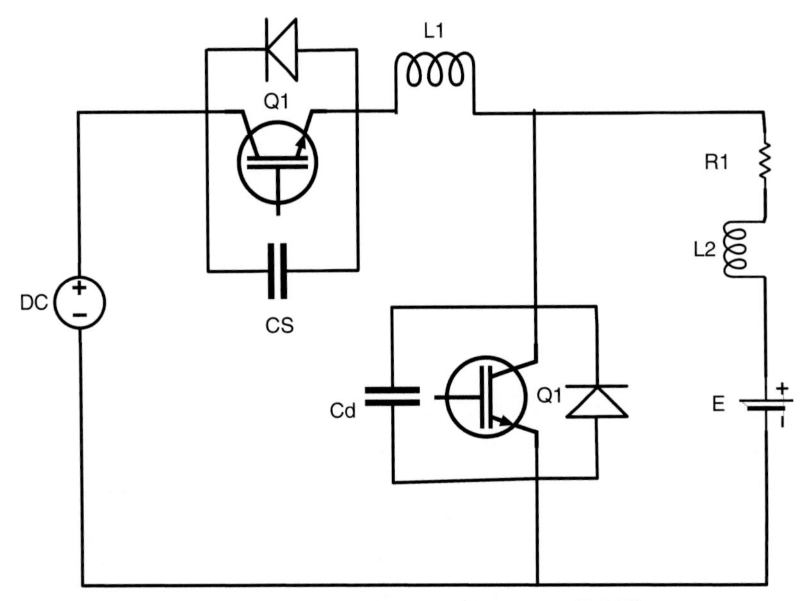

Figure 2.13 Two-quadrant (2Q) zero-voltage multiresonant (ZVMR) converter.

atypical situations result from the demand for smaller and lighter converters, resulting in improved power density.

Changing the converter's switch state when the current flowing through it or the voltage across it is zero at switching instant is one technique for minimizing the above-mentioned flaws. This can be accomplished by using LC resonant circuits to shape the voltage or current waveforms and force the power device to flip to a zero-voltage (ZV) or zero-current (ZC) state. The well-known "resonant soft-switching" converters are made possible by the use of LC resonant circuits. Soft-switching DC–DC topologies for DC motor drives have been discovered and allowing for driving and regenerative braking, which is enabled by the bidirectional power flow. The functioning of soft-switching topologies is based on resonant elements, which have the following characteristic impedance and angular frequency:

$$Z = \sqrt{{L_r}/{C_r}}, \tag{2.32}$$

$$\omega = \sqrt{{1}/{L_r C_r}} \tag{2.33}$$

A two-quadrant (2Q) zero-voltage multiresonant (ZVMR) converter is shown in Fig. 2.13. A typical 2Q-PWM DC drive, two resonant capacitors, and a resonant inductor make up this architecture. It's used in systems where

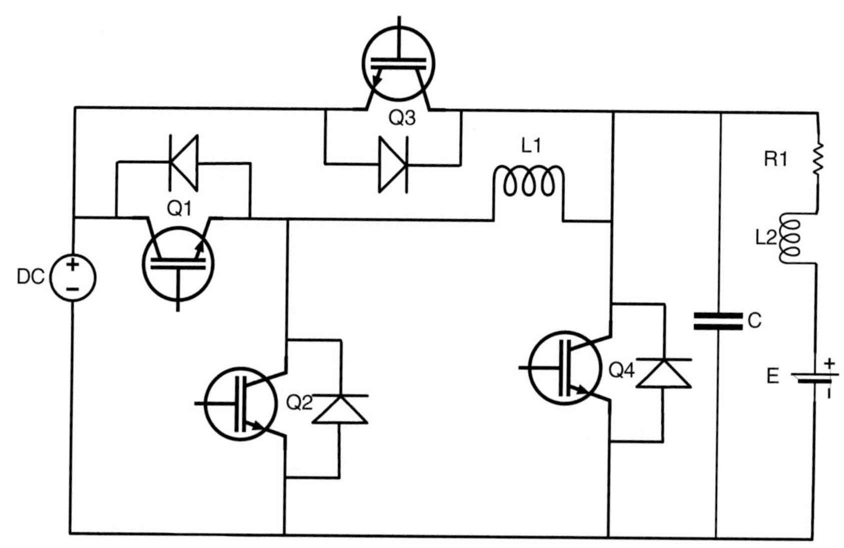

Figure 2.14 2Q zero-voltage-transition (*ZVT*) converter. *2Q*, two-quadrant.

DC motors' driving and regenerative braking modes need to be realized, allowing for bidirectional electric power transfer. One advantage of this soft-switching converter is that it can operate at high switching frequencies, often above 100 kHz, while maintaining low ripple current values in the DC motor and minimal switching losses due to zero-voltage-switching for all switches. This converter also supports load fluctuation and full voltage conversion ranges.

This ZVMR converter makes advantage of all of the built-in diodes and can absorb all main parasitic signals. It is also worth noting that ZVR technology is particularly desirable for MOSFET-based converters. This is due to the fact that MOSFETs have extraordinarily large capacitive voltage turn-on losses. The 2Q-ZVMR manages short-circuit down to no-load conditions without any further steps because it behaves as a constant source after achieving the maximum output current value.

Due to circulating energy and conduction losses, the power rating of the semiconductors (MOSFET) associated with the MR cell is higher than the traditional two-quadrant PWM DC drive. A 2Q zero-voltage-transition (ZVT) converter is shown in Fig. 2.14, and it can function in both driving and regenerative braking modes of DC motor drives, allowing for bidirectional electric power flow. The 2Q-ZVT converter, unlike traditional 2Q-PWM DC drives, requires additional components such as a resonant capacitor, two auxiliary switches, and a resonant inductor.

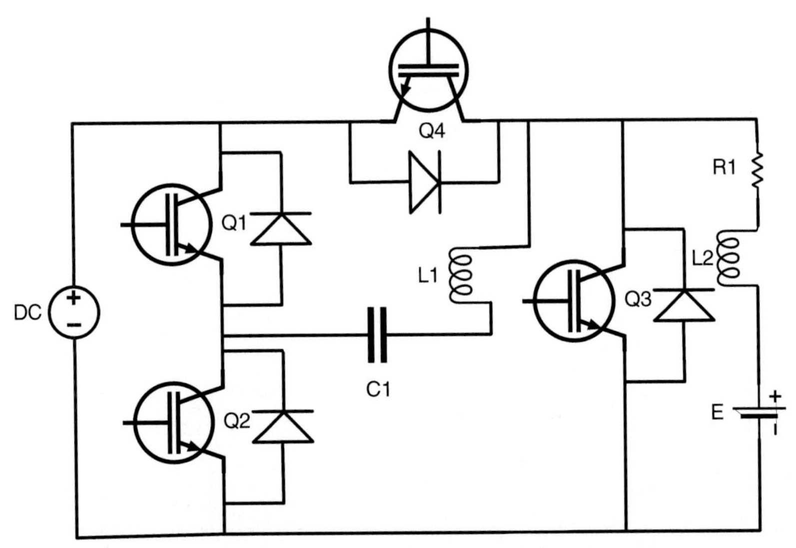

Figure 2.15 2Q zero-current-transition (*ZCT*) converter. *2Q*, two-quadrant.

The 2Q-ZVT converter has a number of advantages, including unity device current and voltage stress both driving and regenerative modes of operation, zero voltage switching for all diodes and primary switches, the same resonant tank for both backward and forward power flows, complete utilization of all power switch built-in diodes, and a simple circuit design. The 2Q-ZVT converters can operate at high switching frequencies, high efficiency, and high power density due to the aforesaid features.

In addition, for this converter type, a proper control system to control the semiconductor switches is critical. A digital signal processor can be used to implement this control system (DSP). A 2Q zero-current-transition (ZCT) converter is shown in Fig. 2.15. This converter, like the two previous topologies, may function in driving and regenerative braking modes. This design requires a resonant capacitor, a resonant inductor, and two auxiliary switches in comparison to its PWM counterpart.

This converter type has several advantages, including zero-current switching for all main and auxiliary diodes and switches, the same resonant tan for both backward and forward power flows, full utilization of all built-in diodes of the power switches, minimal current and voltage strains, low cost, and a simple circuit topology. Due to these qualities, medium frequency switching characteristics in the range of 50 kHz, high efficiency, and high power density are deployed.

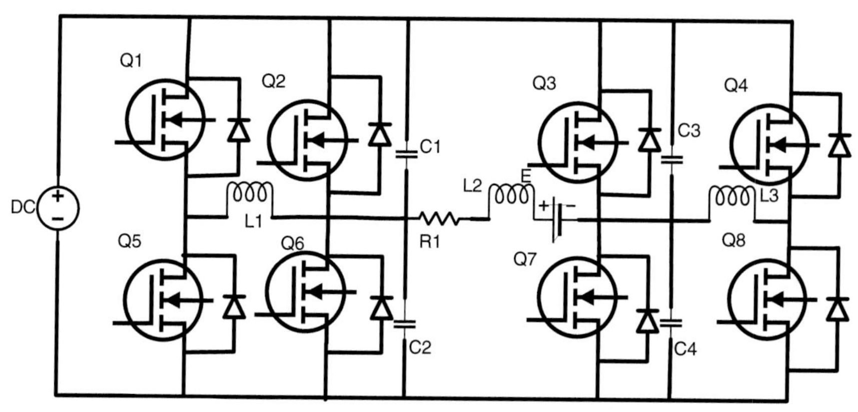

Figure 2.16 Four-quadrant zero-voltage-transition (*4Q-ZVT*) converter.

The 2Q-ZCT converter is typically used in medium-power DC motor applications ranging from a few kW to a few megawatts, with IGBTs serving as the main switching power devices (insulated-gate bipolar transistors). Inductive turn-off and diode reverse recovery turn-on losses are common in IGBTs.

A four-quadrant zero-voltage-transition (4Q-ZVT) converter is shown in Fig. 2.16. In this type of converter, MOSFET power semiconductors are commonly employed. It also allows for forward and reversing operation of DC motor drives, as well as driving and regenerative braking. Simple circuit topology, unity current and voltage stress, and zero voltage switching for both main and auxiliary diodes and switches are all advantages of the 4Q-ZVT. Furthermore, complete usage of all power switch built-in diodes may be obtained, lowering overall hardware costs, and it employs the same resonant tank for both backward and forward power flows. Because of the aforementioned qualities, the converter has a high efficiency and power density. Using two sets of resonant tanks, a zero-voltage switching action can be accomplished. The auxiliary switches Sa and Sa' for soft switching S1 and S4 are part of the first set, which also contains the resonant capacitors Ca/2, the inductor La, and the auxiliary switches Sa and Sa'. The auxiliary switches Sb and Sb' for soft switching S2 and S3 are included in the second set, which also comprises the resonant capacitors Cb/2, inductor Lb, and auxiliary switches Sb and Sb'. There are two 2Q-ZVT converters that fed the DC motor at the same time, as shown in Fig. 2.17.

A 4Q-ZCT (four-quadrant zero-current-transition) converter is shown in Fig. 2.17. IGBTs serve as the basic power elements for this converter,

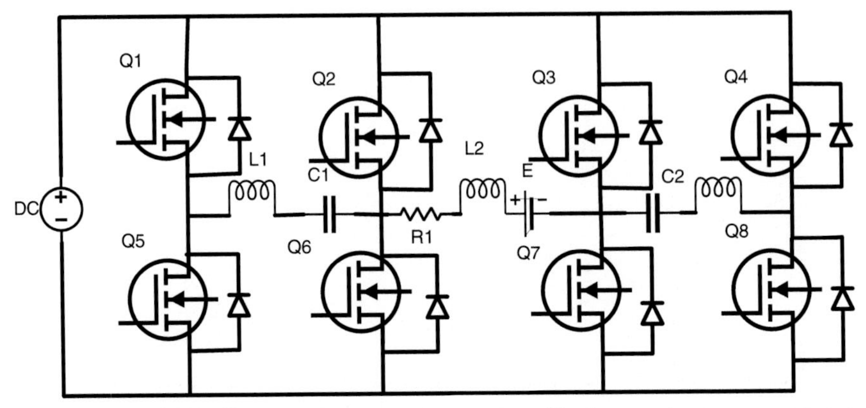

Figure 2.17 4Q-ZCT (four-quadrant zero-current-transition) converter.

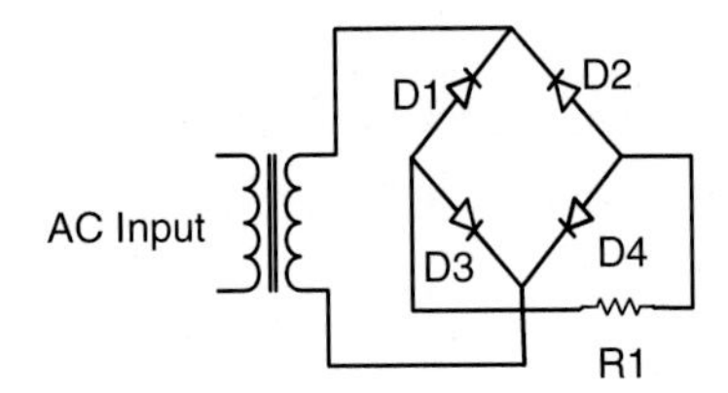

Figure 2.18 Bridge rectifier.

which can be employed for driving and regenerative braking for DC motors, as well as forward and reversal operations.

ZCS operation necessitates the use of two sets of resonant tanks. A resonant capacitor Ca, an inductor La, and the auxiliary switches Sa and Sa' for soft switching of S1 and S4 make up the first tank. In order to perform soft switching of S2 and S3, the second one incorporates a resonant capacitor Cb, an inductor Lb, and the auxiliary switches Sb and Sb'. There are two 2Q-ZCT converters feeding the DC motor at the same time, as can be seen in Fig. 2.18. The 4Q-ZCT shares the same qualities as its forerunner, the 2Q-ZCT. The 4Q-ZCT architecture is suitable for DC motors up to 5kW in power, and its semiconductors are capable of operating at high frequencies.

2.4 Modeling of full-bridge rectifier fed DC motor in Simulink bridge rectifier

A bridge rectifier is an AC to DC converter that converts the mains AC input to DC output. Bridge rectifiers are commonly used in power supply to supply the required DC voltage for electronic components and devices.

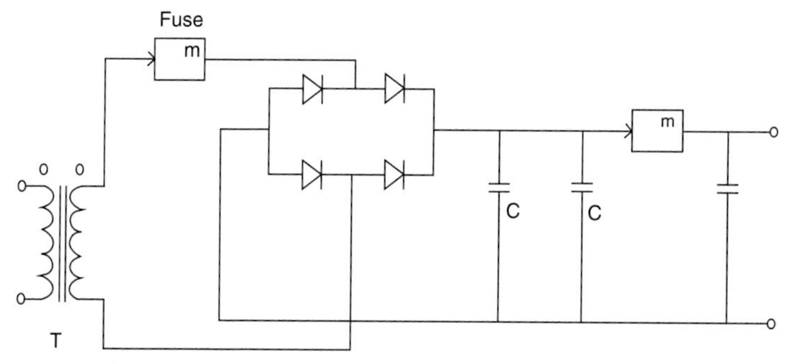

Figure 2.19 Bridge rectifier.

They can be made out of four or more diodes or any other type of controlled solid-state switch. Fig. 2.18 depicts the model.

A suitable bridge rectifier is chosen based on the load current requirements. When choosing a rectifier power supply for an acceptable electronic circuit's application, consider component ratings and specifications, breakdown voltage, temperature ranges, transient current rating, forward current rating, mounting requirements, and other factors.

2.4.1 Construction of a bridge rectifier

Fig. 2.19 depicts the bridge rectifier construction. This circuit can be built with four diodes, D1, D2, D3, and D4, as well as a load resistor (RL). To efficiently convert AC to DC, these diodes can be connected in a closed-loop arrangement. The lack of an exclusive center-tapped transformer is the key advantage of this design. As a result, both the size and the cost will be lowered.

The o/p DC signal can be obtained across the RL once the input signal is applied across the two terminals A & B. A load resistor is linked between two terminals, C and D, in this case. Two diodes can be arranged in such a way that the electricity is transmitted by two diodes throughout every half cycle. During the positive half cycle, pairs of diodes such as D1 and D3 will conduct electric current. D2 and D4 diodes, on the other hand, will carry electric current throughout a negative half cycle.

The fundamental advantage of a bridge rectifier over a full-wave rectifier with a center-tapped transformer is that it provides nearly twice the output voltage. However, because a center-tapped transformer is not required, this circuit resembles a low-cost rectifier. The bridge rectifier circuit schematic includes a transformer, a diode bridge, filters, and regulators, among other devices. In general, the combination of all of these blocks is referred to as a regulated DC power supply, and it is used to power various electronic gadgets.

The first stage of the circuit is a step-down transformer, which reduces the amplitude of the input voltage. To step-down the AC mains 230V to 12V AC power, most electronic projects require a 230/12V transformer. A diode-bridge rectifier, which uses four or more diodes depending on the type of bridge rectifier, is the following stage. When selecting a diode or other switching device for a matching rectifier, various device characteristics must be considered, such as peak inverse voltage, forward current If, voltage ratings, and so on. By conducting a series of diodes for every half cycle of the input signal, it is responsible for producing unidirectional or DC to the load.

Filtering is required since the output from the diode bridge rectifiers is pulsing and must be produced as a pure DC. Filtering is typically accomplished by connecting one or more capacitors across the load, as shown in the diagram below, where the wave is smoothed. The output voltage has an impact on the capacitor rating. The voltage regulator in the last step of this controlled DC supply keeps the output voltage constant. Assume the microcontroller operates at 5V DC, but the output after the bridge rectifier is roughly 16V; a voltage regulator is required to reduce this value and maintain a constant level—regardless of voltage variations on the input side.

2.4.2 Working principle

As previously stated, a single-phase bridge rectifier is made up of four diodes that are connected across the load. We must analyze the following circuit for demonstration purposes in order to comprehend the bridge rectifier's operating principle, which is represented in Fig. 2.20.

D1 and D2 are forward biased during the positive half cycle of the input AC waveform diodes, while D3 and D4 are reverse biased. When the voltage rises beyond the threshold level of the diodes D1 and D2, the load current flows through them, as illustrated in the path of the red line in the diagram below.

The diodes D3 and D4 are forward biased during the negative half cycle of the input AC waveform, whereas D1 and D2 are reverse biased. When the D3 and D4 diodes begin to conduct as illustrated in the diagram, load current begins to flow through them.

The load current direction is the same in both circumstances, that is, up to down as illustrated in the diagram - therefore unidirectional, which signifies

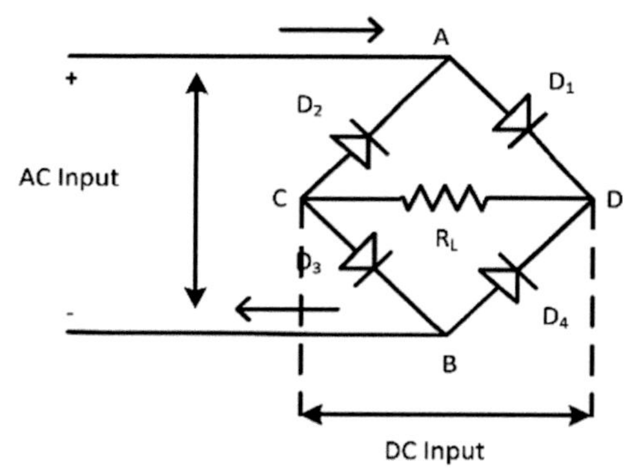

Figure 2.20 Working principle.

DC. The input AC is thus converted to a DC using a bridge rectifier. This bridge wave rectifier's output at the load is pulsating, but producing a pure DC requires an extra filter, such as a capacitor. Different bridge rectifiers can be operated in the same way, but controlled rectifiers require thyristor triggering to direct the current to the load.

2.4.3 Creating a Simulink model in MATLAB

To get the elements in Simulink, we need to browse it from Simulink library, as depicted in Fig. 2.21. Follow the below steps to open library.

To open the Library Browser, use slLibrary Browser.

To only load the Simulink block library, use this command: load_system Simulink.

To start Simulink without opening the Library Browser or Start Page, use start_simulink, which is faster.

Once the library is opened, it appears as below.

Once the Simulink library is opened, search for the sources and the block required. Click on the block and right-click and then add the block to the file. The entire power electronics element is under simscape or you can search for the element in search box. Here an example for Bridge rectifier in Simulink MATLAB is shown. The circuit and waveform is depicted in Fig. 2.22 and Fig. 2.23

2.5 Single-phase fully controlled converter fed separately excited DC motor drive

Fig. 2.24 shows the basic circuit for a single-phase independently excited DC motor drive. A semi-converter or full-converter controls the armature

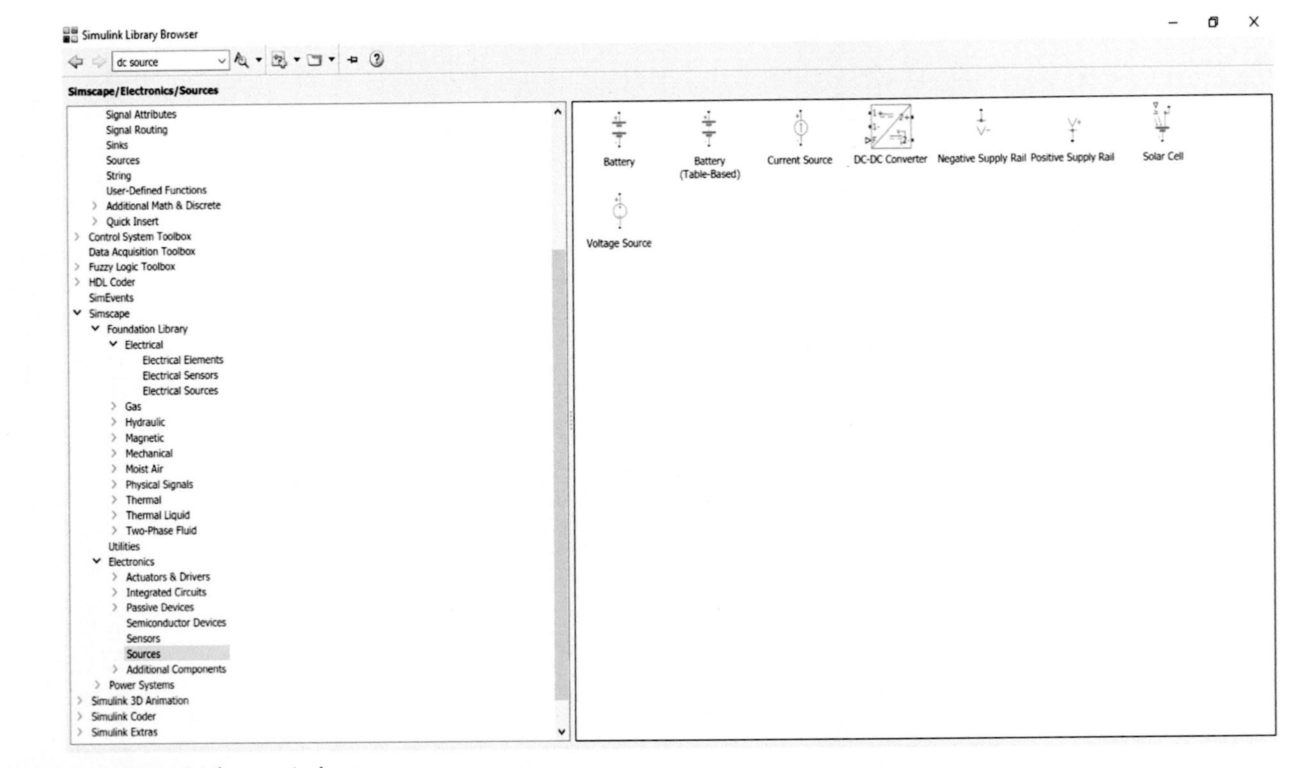

Figure 2.21 MATLAB Library window.

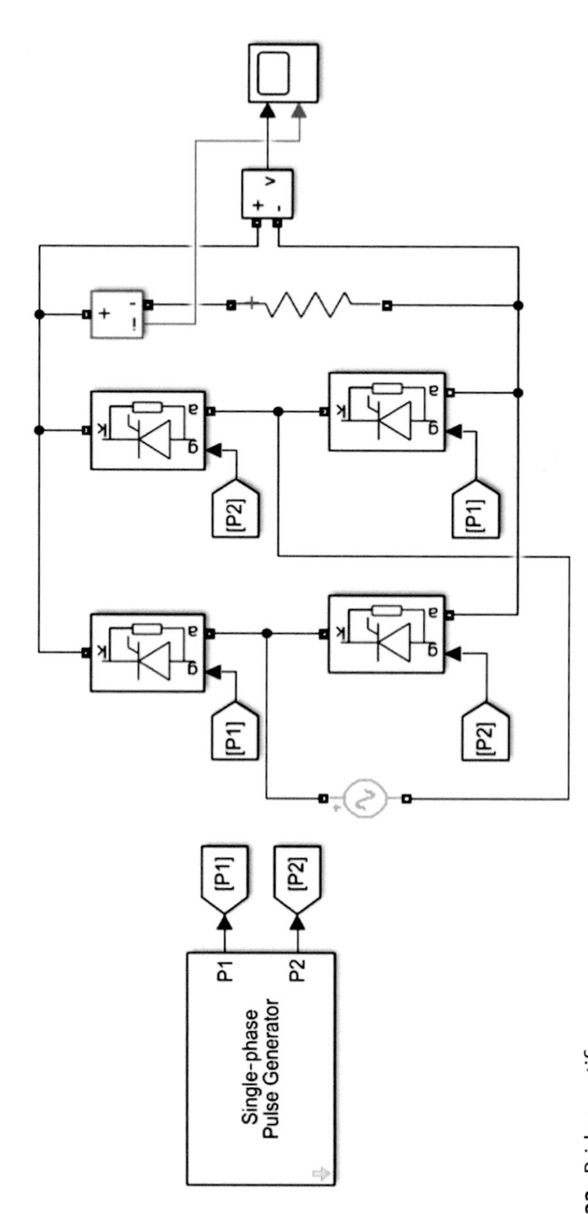

Figure 2.22 Bridge rectifier.

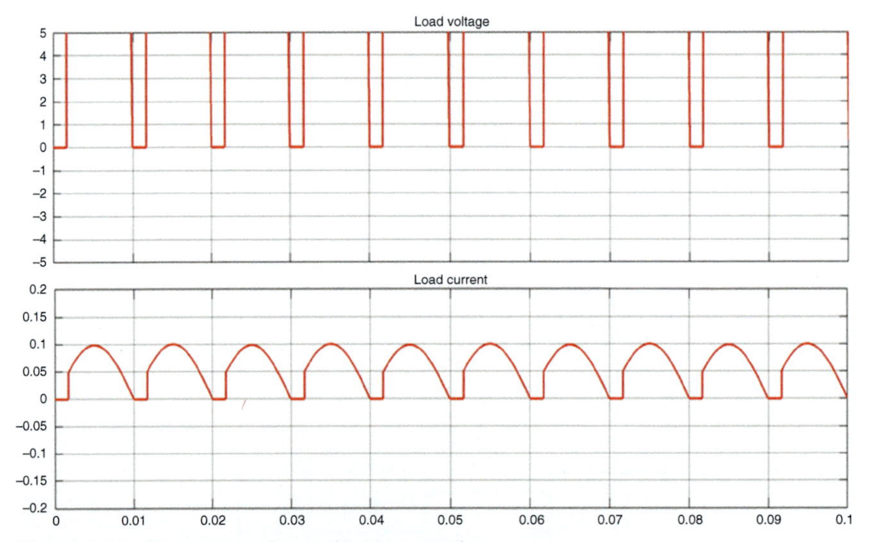

Figure 2.23 Output waveform of bridge rectifier.

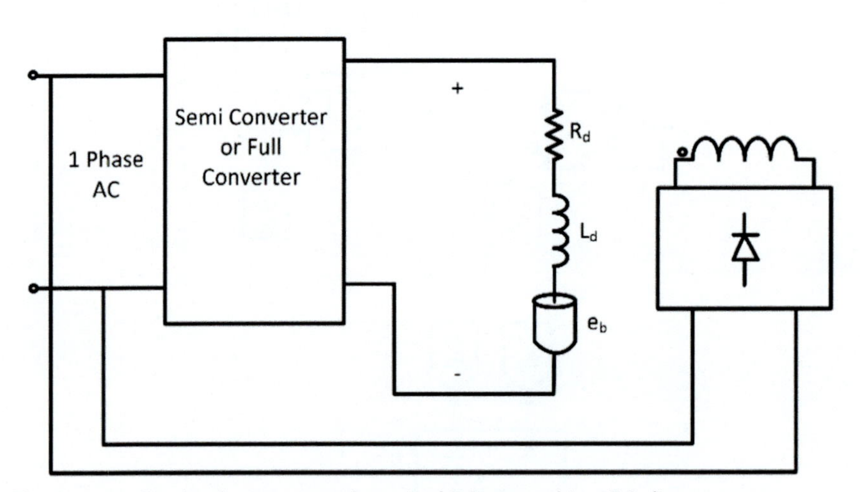

Figure 2.24 Single-phase separately excited DC motor drive. *DC*, direct current.

voltage, while a diode bridge feeds the field circuit from the AC supply. The thyristors in the converters prevent the motor current from reversing. The average output voltage (Ea) is always positive when semi-converters are utilized. As a result, power flow (Eala) from the AC supply to the DC load is always positive. Regeneration or reverse power transfer from the motor to the AC supply is not possible in drive system semi-converters.

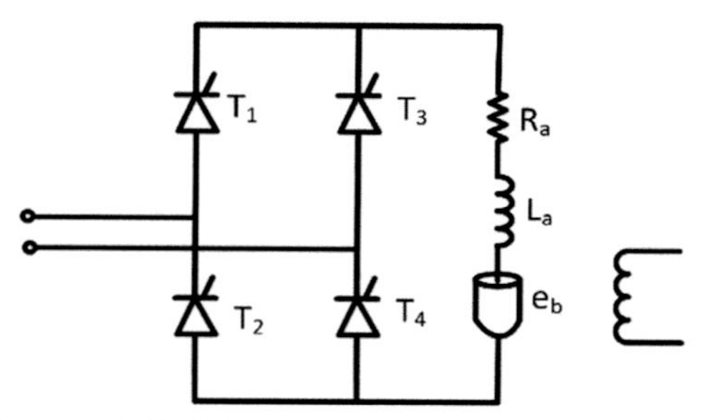

Figure 2.25 Single-phase fully fed DC drive. *DC*, direct current.

When the thyristor blocks in semi-converters, free-wheeling (i.e., dissipation of armature inductance energy through the free-wheeling channel) occurs. Low and medium-horsepower applications are served by single-phase full-wave drives, as shown in Fig. 2.24. When it comes to open-loop firing angle control, such drives have poor speed regulation. However, effective regulation can be performed via armature voltage or tachometer feedback.

2.5.1 Circuit of single phase fully fed DC drive

A full-wave converter powers the Single Phase Separately Excited DC Motor Drives. Fig. 2.25 depicts a full-wave drive with a fully regulated converter. The firing of the thyristors in all positions provides negative average voltages, allowing power to flow from the load to the supply. Regenerative braking can be used to efficiently stop the motor. This is feasible with a steady current. Because the load current flows throughout both positive and negative half-cycles, the average current value in the half-wave drive is higher. For a certain armature heating, the torque capability increases. Here, the peak-to-average and rms-to-average ratios are improved. There is less speed oscillation. Because of the enhanced current conduction in the load, the speed regulation improves. The load has two pulses with a pulse frequency of 2f. In this scenario, the ripple amplitude is less. The fluctuations in speed have been reduced. This has an impact on the speed regulation. The performance is improved by adding inductance to the armature. It lowers the ripple content, lowers the chance of discontinuous conduction, and enhances speed regulation, among other things. The performance of the drive on the line side is likewise affected by this inductance. Because of the possibility of a discontinuous load current, the harmonic content of the line current is higher at lower values of inductance. The harmonic factor reduces as the inductance increases. With more inductance, the peak value of current drops. This boosts the ability to commute. The Simulink and output waveform is depicted in Figs. 2.26 and 2.27.

2.6 1-phase half-controlled converter fed separately excited DC motor

A single-phase half-wave circuit is used to power the single-phase separately excited DC motor drives, as shown in Fig. 2.28. The advantages of this drive are its low cost and simplicity. Only one quadrant operation is possible. It is not feasible to regenerate. The thyristor's conduction angle is extremely tiny, resulting in a very low average current. At the rated rms current, the torque created is very tiny, resulting in torque loss. The ratio of rms to average current is likewise higher. The motor current is never constant. The frequency of the ripple is the same as the frequency of the supply.

The performance is improved by using a freewheeling diode across the load. The speed limit is really low. The motor gets power in pulses at low speeds, and when the load is high, the motor may chug. The oscillation in speed is fairly high. Because of the DC component of the load current, the supply transformer is premagnetized. This drive can only be used at low power levels. The Simulink and waveform of Single Phase Separately Excited DC Motor is depicted in Figs. 2.29 and 2.30.

2.7 Three-phase fully controlled converter fed separately excited DC motor

In large-power DC motor drives, three-phase controlled rectifiers are used. A three-phase controlled rectifier produces more voltage pulses per supply frequency cycle. This keeps the motor current constant and reduces the need for filters. For three-phase regulated rectifiers, the number of voltage pulses each cycle is determined by the number of thyristors and their connections. The armature circuit is coupled to the output of a three-phase controlled rectifier in three-phase drives. For high-power applications up to megawatts, three-phase drives are used. The armature voltage has a larger ripple frequency than single-phase drives, requiring less inductance in the armature circuit to reduce armature current ripple. Up to 1500KW drives, three-phase complete converters are employed in industrial applications. It's a quadrant converter with two quadrants, that is, the average output voltage can be positive or negative, but the average output current can never be negative.

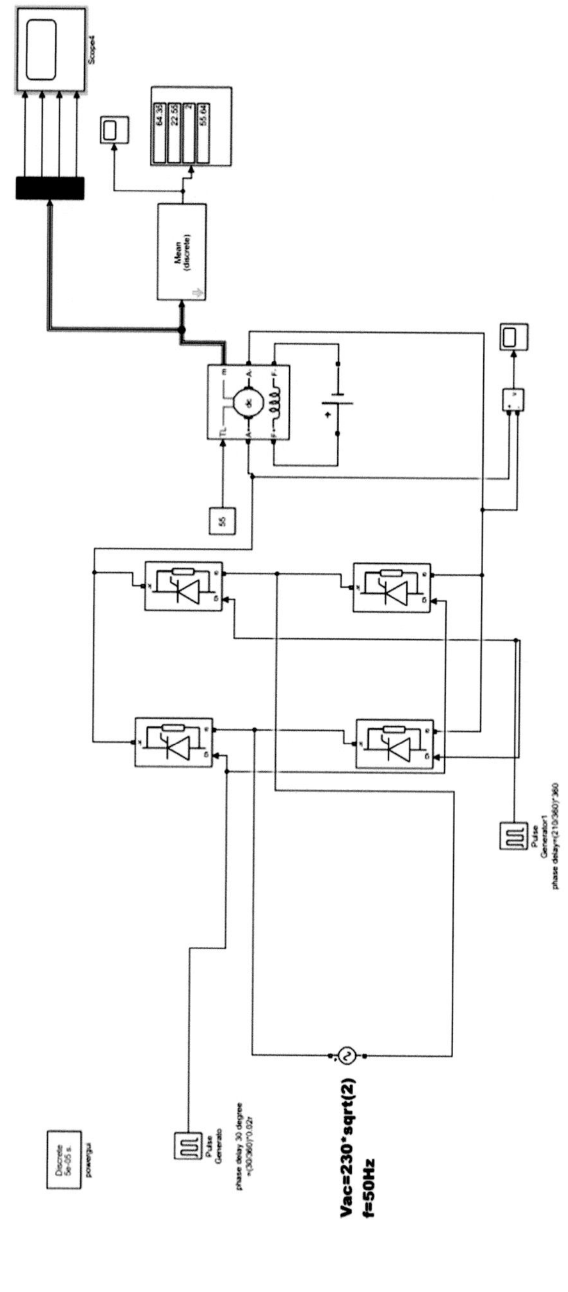

Figure 2.26 Simulink of single phase fully fed DC drive. *DC*, direct current.

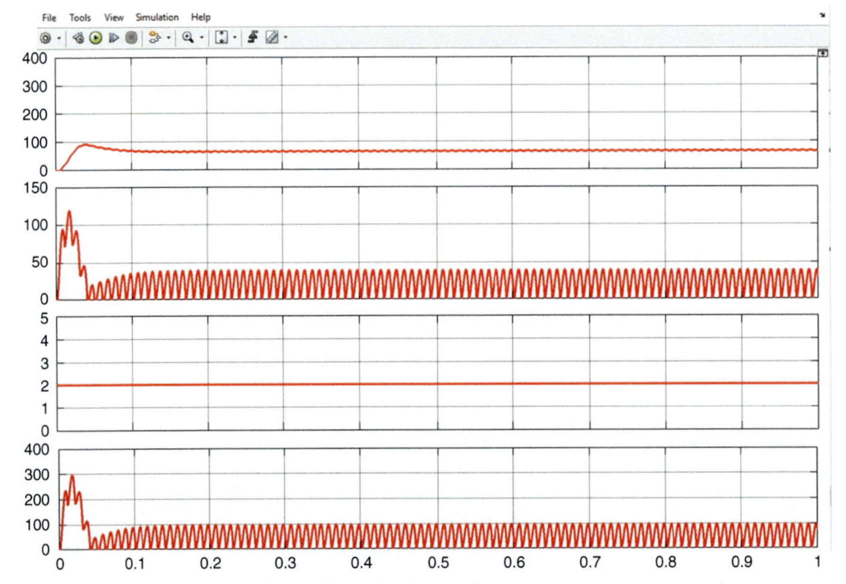

Figure 2.27 Output waveform of single-phase drive.

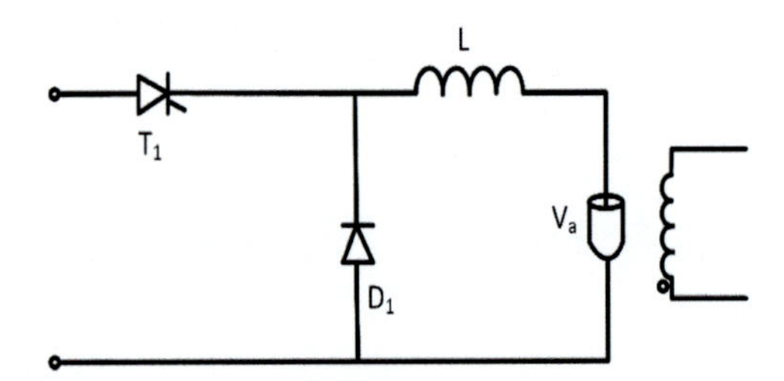

Figure 2.28 Single-phase separately excited DC motor. *DC*, direct current.

Fig. 2.31 shows a three-phase complete converter bridge circuit linked across the armature terminals. For firing angle delay 00 900, the circuit functions as a three-phase AC to DC converter, and for 900 1800, it functions as a line commutated inverter. When power regeneration is necessary, a three-phase full converter supplied DC motor is used, which operates in two quadrants.

The controlled rectifier is made up of six thyristors that are placed in three legs, each with two series thyristors. A three-phase power supply is linked to the center points of three legs. Although the transformer is not required, it does give benefits such as voltage level change, electrical isolation,

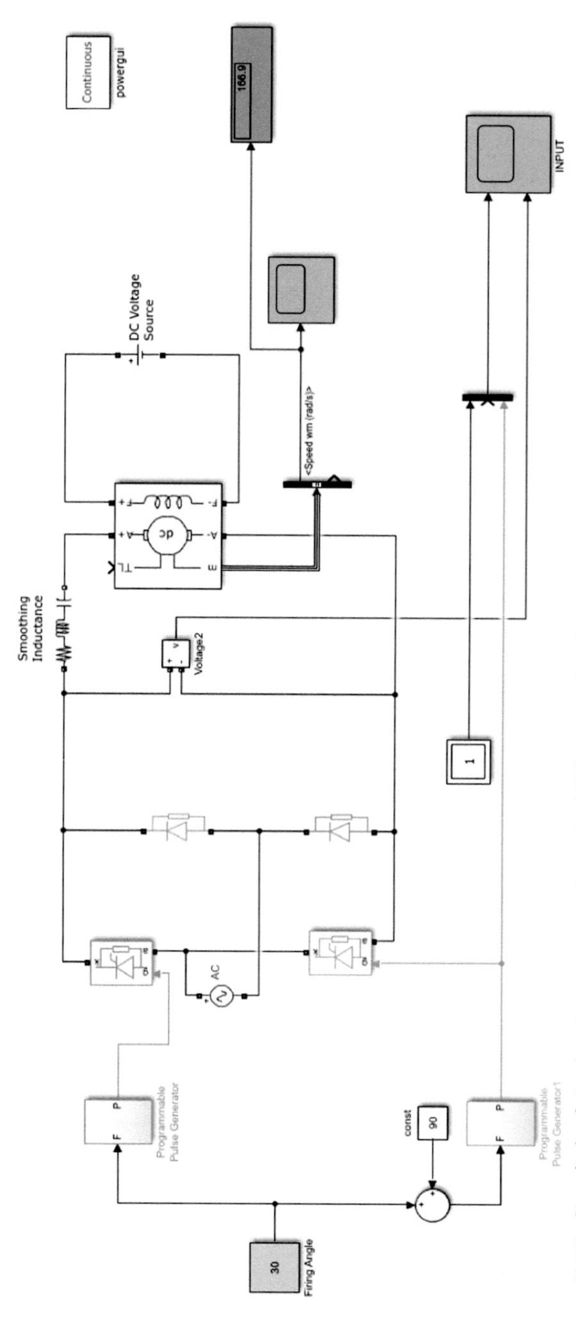

Figure 2.29 Simulink of single-phase separately excited DC motor. *DC*, direct current.

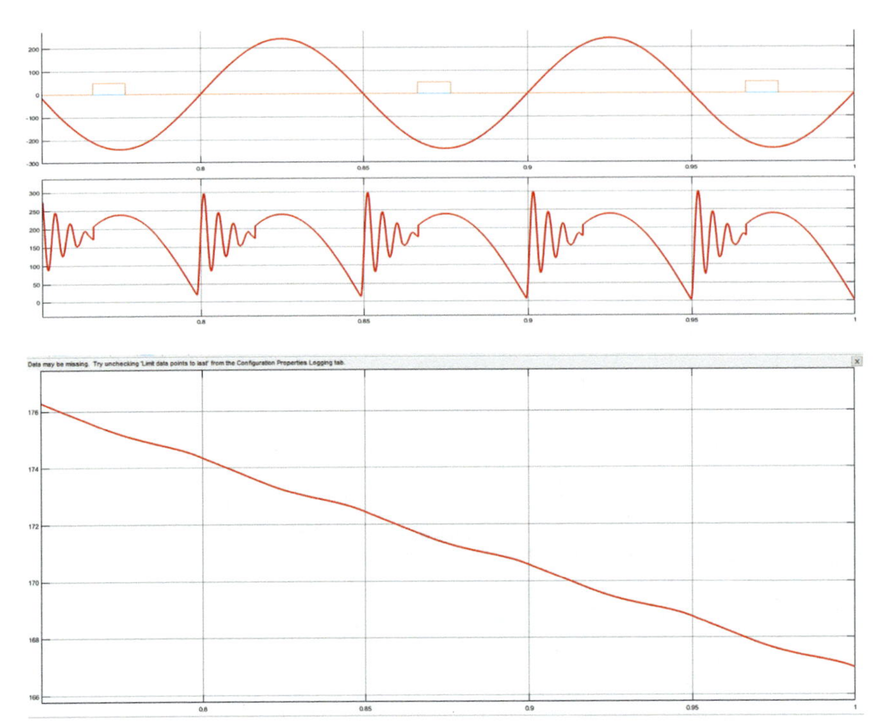

Figure 2.30 Output wave form of single-phase separately excited DC motor. *DC*, direct current.

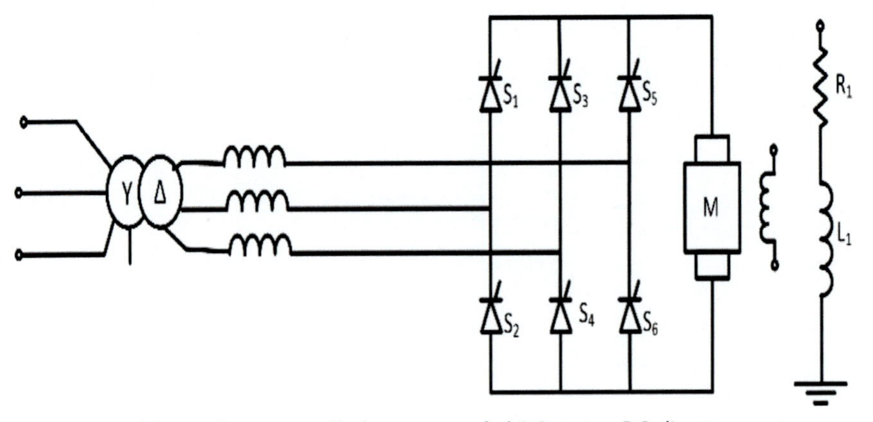

Figure 2.31 Three-phase controlled converters fed DC motor. *DC*, direct current.

and phase shift from the main. To contribute load current id in a three-phase bridge, one device from the positive group (Q1 Q3 Q5) and another from the negative group (Q4 Q6 Q2) must conduct simultaneously. For the specified conduction interval, each thyristor is generally equipped with pulse train firing. The firing angle control of the thyristors can control the motor's speed. Figs. 2.32 and 2.33 depict the Simulink and waveform of three phase-controlled converters fed DC motor.

2.8 Three-phase half-controlled converter fed separately excited DC motor

The motor load is linked between the converter positive terminal (cathodes of all thyristors) and the supply neutral in a three-phase half-wave converter, as shown in Fig. 2.34. From the zero crossings of the input voltages, the firing angle is also defined to be zero. Because of the DC component in the line current, this converter is rarely used in reality for motor ratings of 10 to 50 hp.

For all AC-to-DC converter circuits, a firing angle of zero degrees generates the maximum output DC voltage. Each thyristor carries current for 120 s before switching to nonconduction for the next 240 s. The firing angle can be adjusted between 0° and 180°. The output DC voltage becomes negative at > 90°, although the motor current remains positive and continuous. This suggests that the converter is operating in the fourth quadrant of the V–I plane, where it is inversion mode. The motor continuously supplies power to the AC source through the converter in this mode of operation. Regenerative conversion is the name for this style of operation. An overhauling motor, for example, can feed its energy to the AC mains in this manner. Controlled braking is, however, feasible in the case of the overhauling motor. Fig. 2.35 depicts the Simulink model of three-phase half-wave converter.

2.9 Pulse width modulation converter fed DC drives

A power semiconductor switch (typically a MOSFET) drives a magnetic element (transformer or inductor) whose rectified output creates a DC voltage in a switch-mode converter. Efficiencies of more over 90% are frequent, roughly double that of a linear regulator.

The DC output current of a switch-mode converter varies in response to load variations. Pulse width modulation (PWM) is a frequently used method for controlling the output power of a power switch by altering the ON and OFF timings. The duty cycle is the ratio of ON time to switching period time. The three different PWM duty cycle changes are 10%, 50%, and 90%. The duty cycle and power handling have very little to do with one another. Instead, the output voltage is regulated by adjusting the duty cycle.

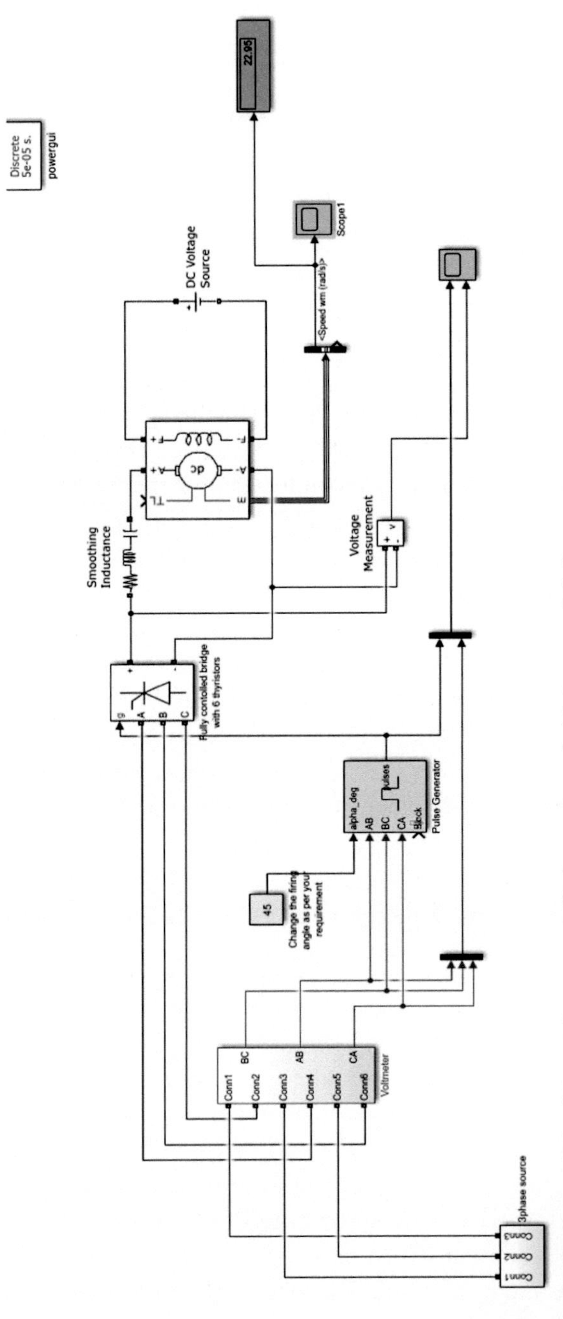

Figure 2.32 Simulink of three-phase controlled converters fed DC motor. *DC*, direct current.

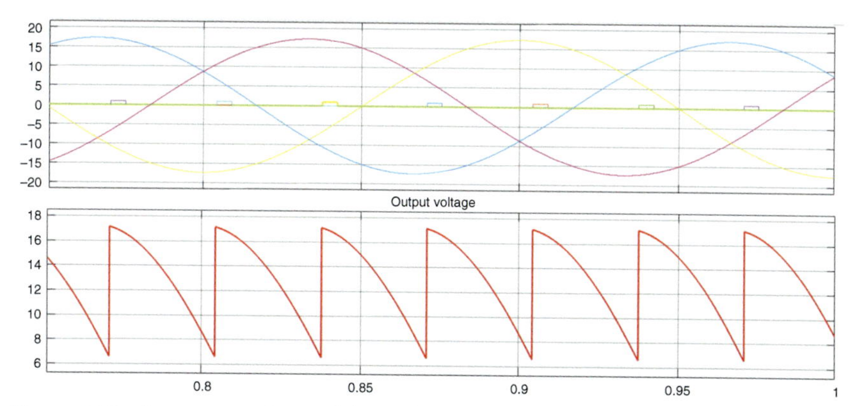

Figure 2.33 Output waveform of three-phase controlled converters fed DC motor. *DC*, direct current.

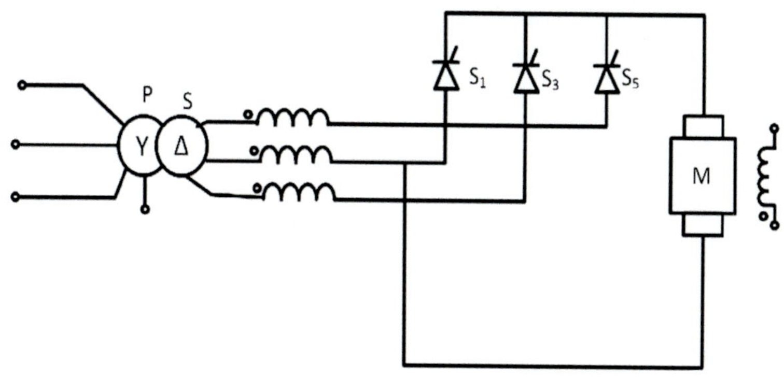

Figure 2.34 Three-phase half-wave converter.

The error amplifier receives a proportion of the DC output voltage, which causes the comparator to control the PWM ON and OFF periods. The feedback adjusts the duty cycle to keep the output voltage at the required level if the filtered output of the power MOSFET changes. The error amplifier accepts the feedback signal input and a stable voltage reference to produce a signal that is proportional to the difference between the two inputs. The comparator compares the output voltage of the error amplifier to the oscillator's ramp (sawtooth), resulting in a modified pulse width. The comparator output is applied to the switching logic, whose output is sent to the external power MOSFET's output driver. The switching logic controls whether the PWM signal applied to the power MOSFET is enabled or disabled.

To avoid instability, PWM duty cycles greater than 50% require a compensating ramp, also known as slope-compensation. Slope compensation

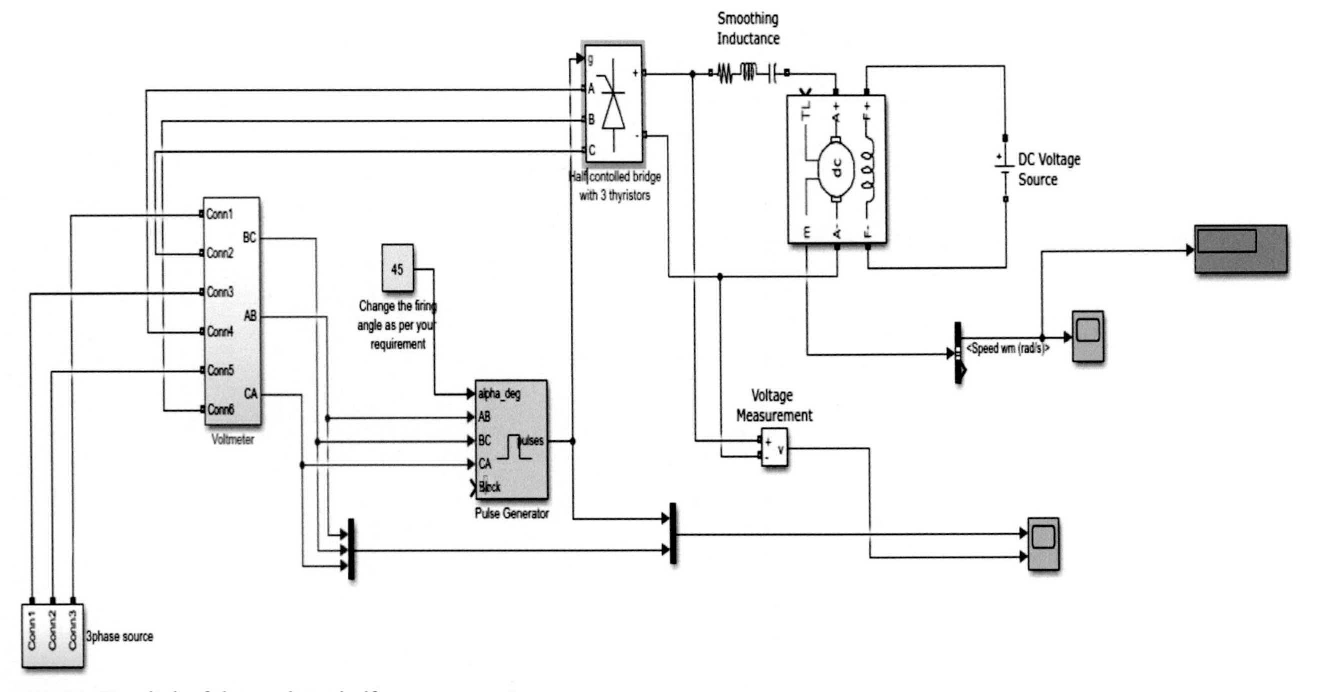

Figure 2.35 Simulink of three-phase half-wave converter.

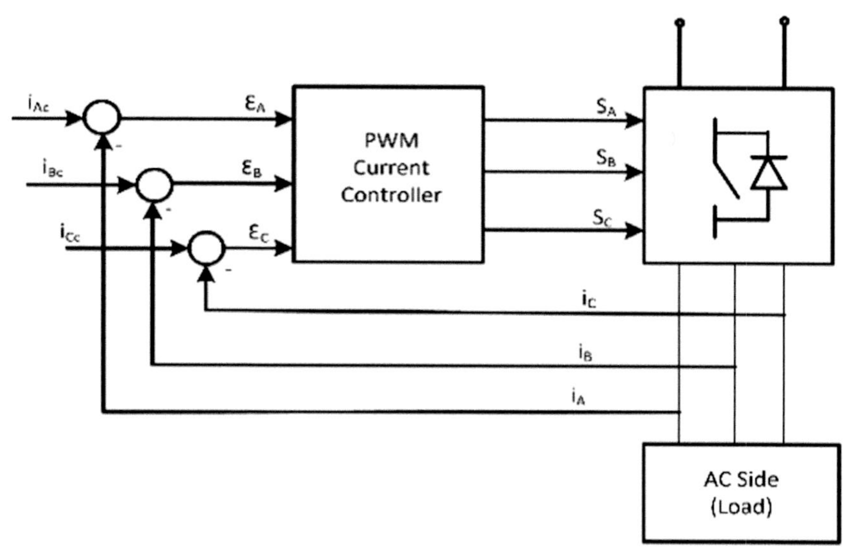

Figure 2.36 PWM fed converter block. *PWM*, pulse width modulated.

is even more important at higher duty cycles. To maintain the system stable, slope compensation must be utilized if the PWM switch is turned on for more than 50% of the switching period. Traditional slope compensation may cause the switching converter to become unstable at duty cycles nearing 100%, necessitating the use of a specific slope compensation. A PWM operating block is shown in Fig. 2.36.

Single-ended or dual-ended outputs are available on PWM controllers. MOSFETs with dual output types are designed for push–pull, bridge, or synchronous rectifier applications. The PWM controller must either precisely adjust the two outputs' dead time or prevent their overlap in certain arrangements. Allowing both outputs to be turned on at the same time would increase power dissipation and EMI. Special circuits are used in certain PWM controllers to control dead time or overlap. Figs. 2.37 and 2.38 depict PWM fed converter Simulink and waveform.

2.10 Multiquadrant operation of fully controlled converter fed DC motor

The multiquadrant operation means making the motor to operate in four quadrants such as forward motoring, forward braking, reverse motoring, reverse braking, etc. The operation is depicted in Fig. 2.39.

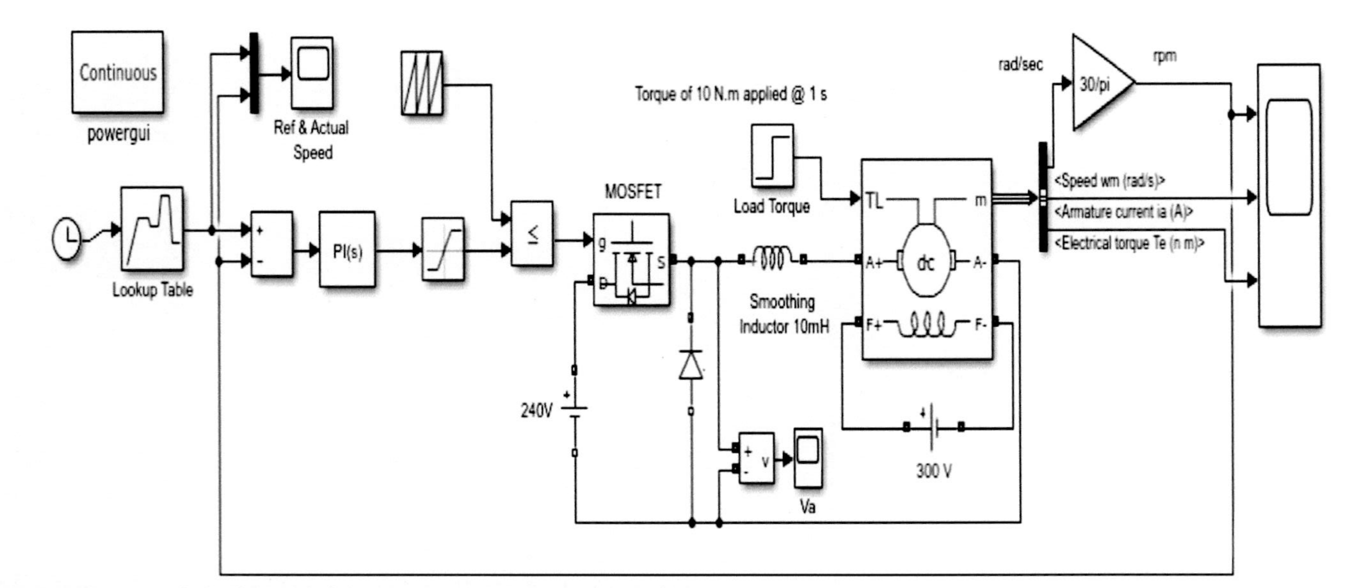

Figure 2.37 Simulink of PWM fed converter. *PWM*, pulse width modulated.

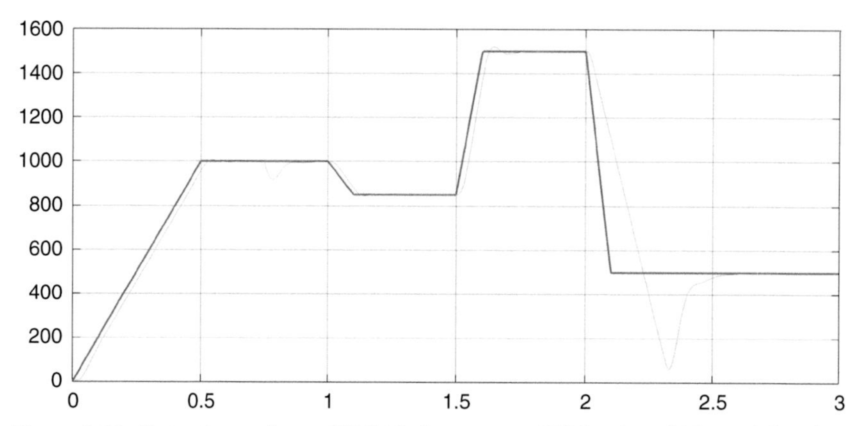

Figure 2.38 Output waveform of PWM fed converter. *PWM*, pulse width modulated.

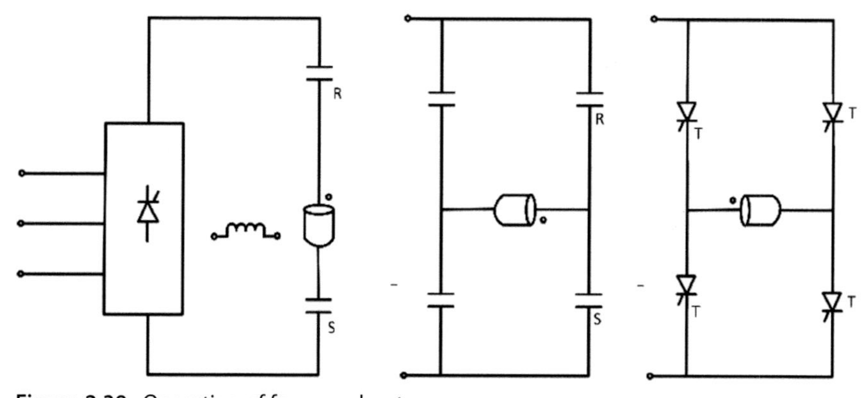

Figure 2.39 Operation of four quadrants.

The motor is fed by a completely regulated rectifier via a reversing switch RS, which is used to reverse the armature connection to the rectifier. A fully controlled rectifier can operate in the first and fourth quadrants. The armature connection is reversed, allowing functioning in quadrants third and second.

As indicated in the diagram, the reversing switch could be a relay-operated contactor with two typically open and two normally closed contacts. When the contactor's slow operation and regular maintenance are unacceptable, the reversing switch is made from a thyristor, as indicated in the diagram above.

The two-channel converter the term itself denotes the presence of two converters, as shown in Fig. 2.40. It is an electric component that is

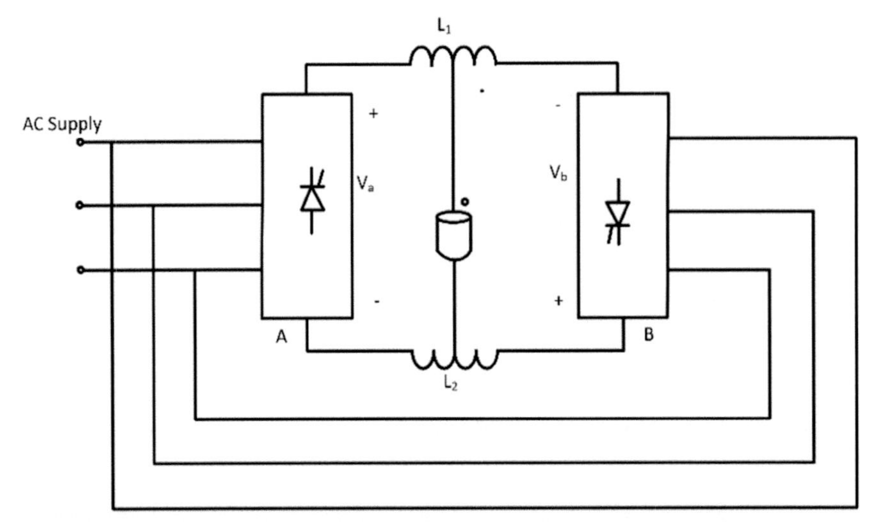

Figure 2.40 Multiquadrant operation of DC motor fed from dual converter. *DC*, direct current.

commonly seen in variable-speed drives. It is a power electronics control system that uses forward and reverse converters to obtain either polarity DC from AC rectification. Two converters are joined back to back in a dual converter.

One bridge act as a rectifier (converting AC to DC), while the other acts as an inverter (converting DC to AC) and is coupled to a DC load. A dual converter is one that performs two conversion procedures at the same time. The dual converter may operate in four quadrants.

Rectifier A facilitates motor control in quadrants one and four by providing positive motor current and voltage in either direction. Because it gives negative motor current and voltage in any direction, Rectifier B provides motor control in quadrants three and four. Figs. 2.41 and 2.42 depicts multiquadrant converter Simulink and waveform.

2.11 Closed-loop control of converter fed DC motor

Fed by a chopper DC motor control is a technique for controlling the speed of a motor by reducing the fixed DC voltage and replacing it with variable DC voltage. The four-quadrant chopper approach is used here, which means that four switching devices are utilized to control the motor's speed. The duty cycle applied to the switching devices can be used to control the motor

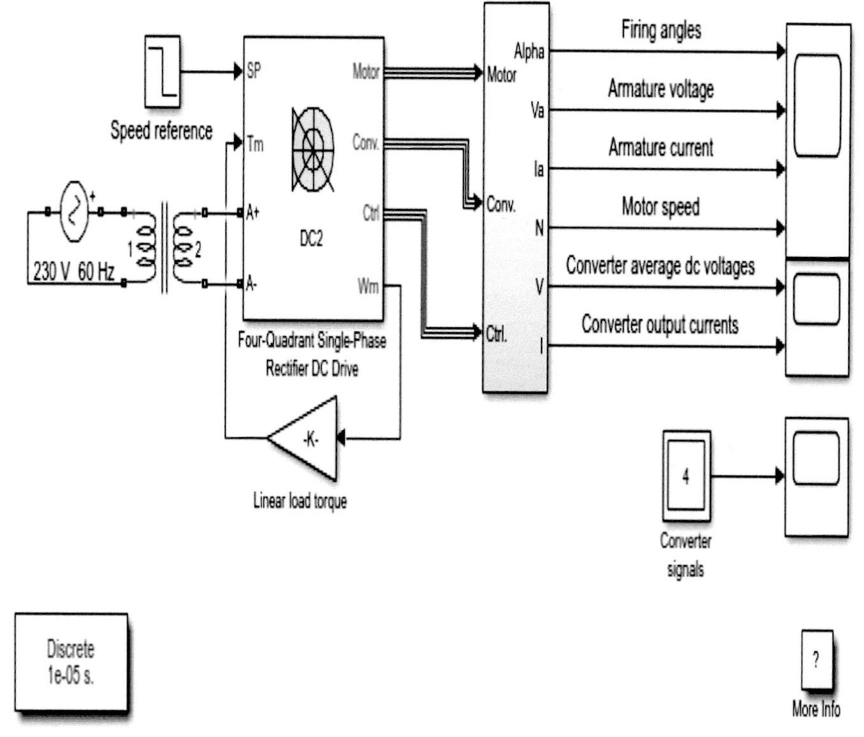

Figure 2.41 Simulink of multiquadrant converter.

speed. The output voltage is entirely dependent on the device's ON time; as the switching frequency's on time increases, so does the output voltage. Fig. 2.43 depicts the closed-loop control block.

DC motor speed control in a closed-loop — The converters (rectifiers and choppers) are made of semiconductor devices with limited thermal capacity. As a result, they have the same transient and steady-state current ratings. During transient activities of short duration, such as starting, braking, and reversing, DC motors can handle 2 to 3.5 times the rated current. The torque increases as the current increases, and the transient response accelerates. When a quick response is required during transient activities, the motor current is allowed to reach its maximum allowable amount. The converter rating is then set to the maximum motor current allowable value. The converter cost will now be greater due to the higher current rating. When a fast-transient reaction is not needed, the converter current rating is set to the same as the motor current rating to keep the converter cost down.

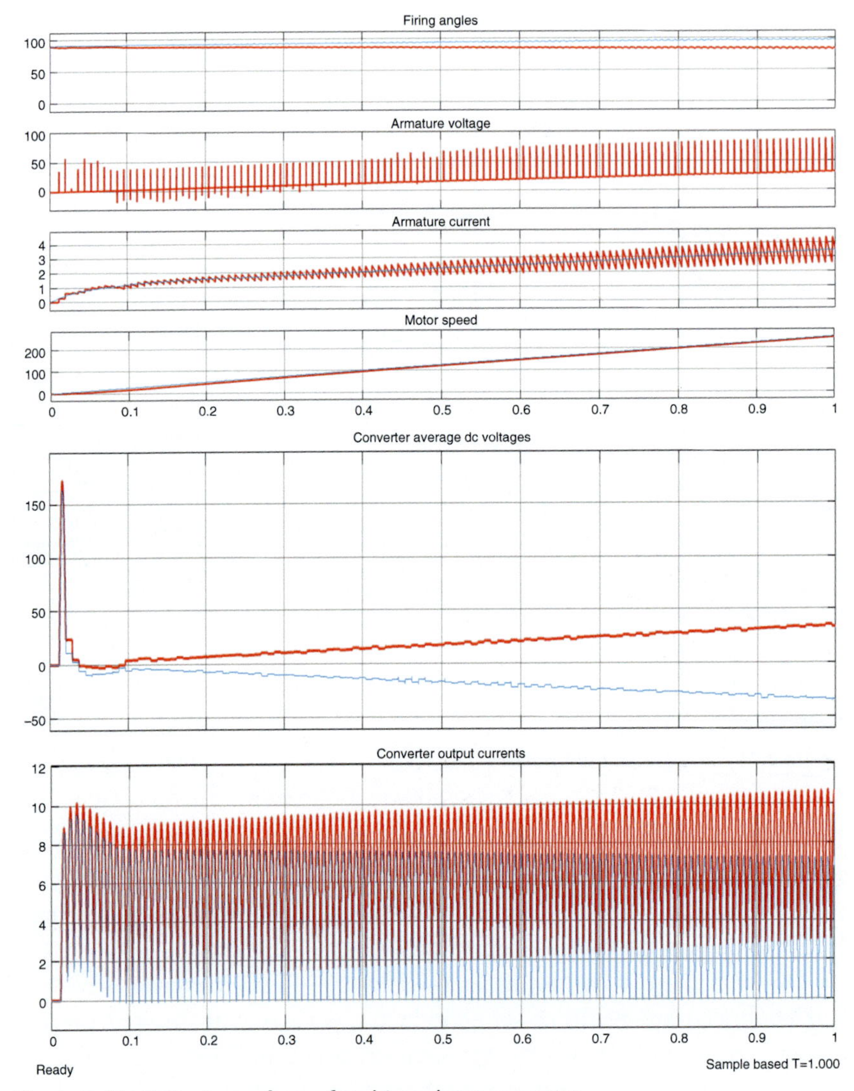

Figure 2.42 Output waveform of multiquadrant converter.

To protect the converter from current overloads, open-loop drives are equipped with current limit control. The inner current control loop in closed-loop speed control of DC motor schemes is used to keep the current within a safe range and to accelerate and decelerate the drive at the maximum permitted current and torque during transient operations. It should be emphasized, however, that deceleration at maximum current

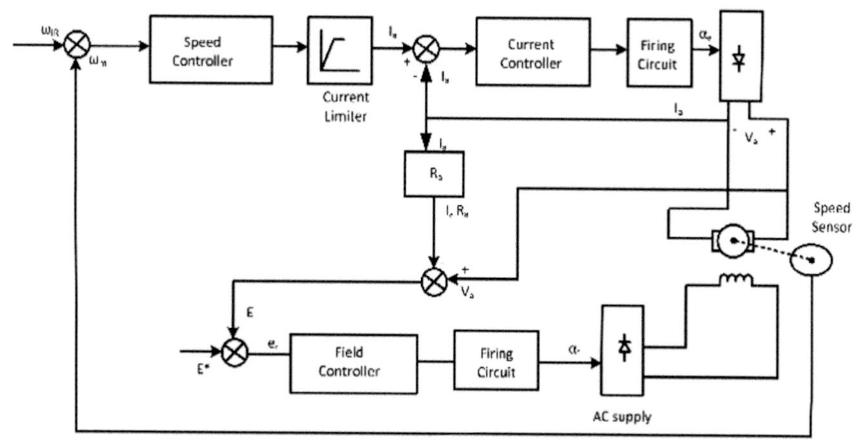

Figure 2.43 Block diagram closed-loop control of converter.

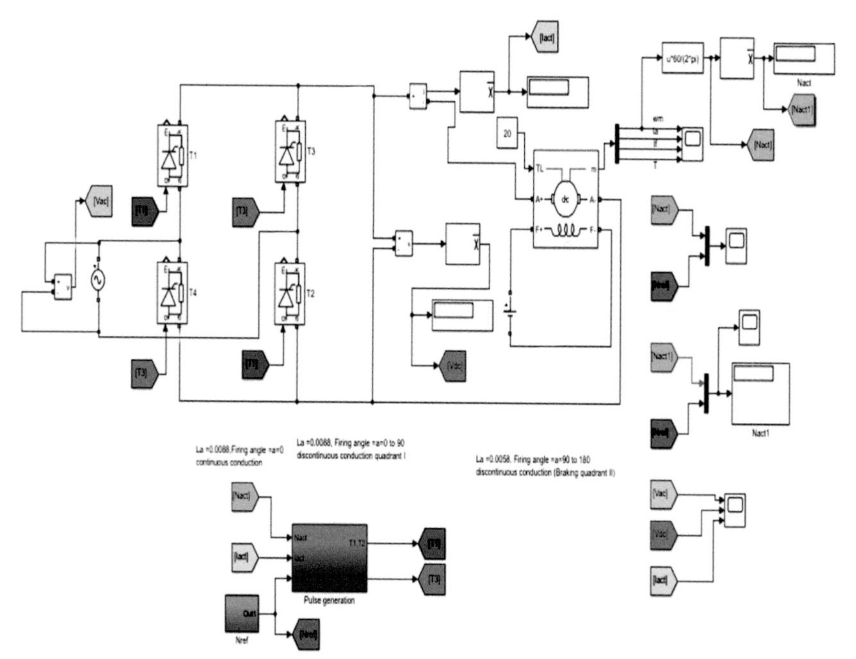

Figure 2.44 Simulink of closed-loop converter.

or torque will be achievable if the converter employed also has braking capacity. It should also be noted that when the supply is AC, a controlled rectifier will be utilized, and when the supply is DC, a chopper will be utilized. Figs. 2.44 and 2.45 depict the Simulink and waveform of closed-loop converter.

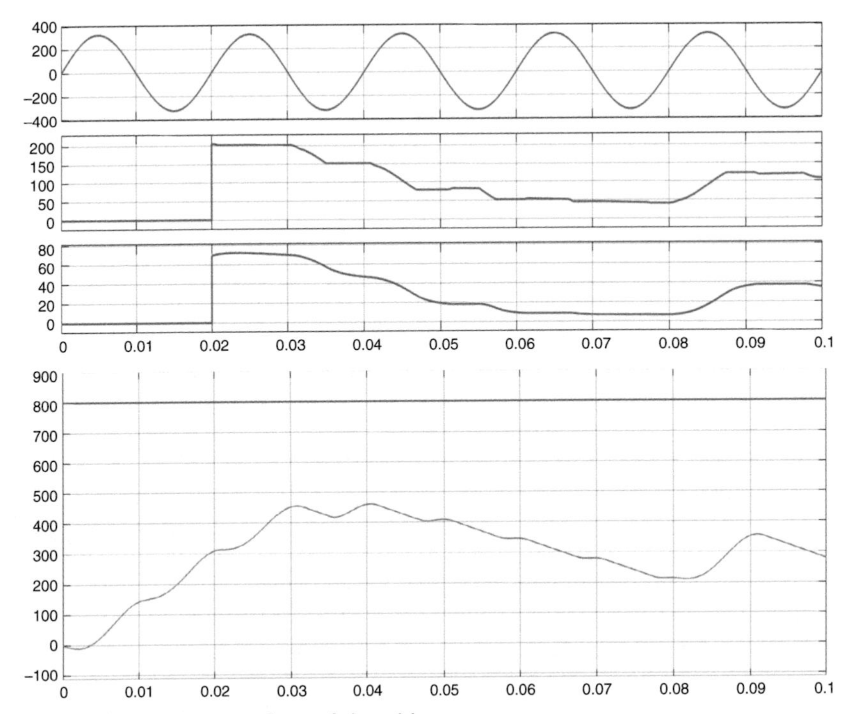

Figure 2.45 Output waveform of closed-loop converter.

2.12 Summary

This chapter provides the basic concepts of converter-fed DC drives with Simulation. The MATLAB/Simulink library details are provided with appropriate screenshots taken from MATLAB software. The concept of single- and three-phase controlled converter is detailed with simulation. The concept of PWM and closed-loop control fed DC drive is explained.

2.13 Review questions

1. Define uncontrolled converter fed DC drives.
2. What are basic types of converters?
3. What is Ripple current and Ripple Voltage?
4. Define DC chopper.
5. Draw the circuit for chopper fed DC motor.
6. Draw genetic scheme of rectifier.
7. What are the types of Rectifier and write two difference?
8. What is ZVT and ZCT models?

9. Draw and explain bridge rectifier.
10. Write in steps to select an element from MATLAB.
11. What is separately excited DC motor.
12. Create a Simulink model for single phase fully fed DC drive.
13. What is PWM pulse?
14. Differentiate open and closed loop system.
15. Draw a block diagram and explain the operation of PWM fed converter block.

CHAPTER 3

Chopper fed electric drives with simulation

Contents

3.1 Introduction to choppers and its classification

As the invention and use of technological devices need for electricity is also increasing. To meet this need for continuous electricity various methods and systems are being introduced. Among the gadgets and devices we use, some are powered by AC while some are DC powered. Not all devices require the same amount of power to operate. But the power given to the households through main power supply is AC and of a fixed amount of about 240V. Then to operate devices that work on DC some converters are required. To use only a small required amount of power from the 240V supply another type of circuit namely chopper circuit is required.

Chopper circuits are known as DC-to-DC converters. Similar to the transformers of the AC circuit, choppers are used to step up and step down the DC power. They change the fixed DC power to variable DC power. Using these, DC power supplied to the devices can be adjusted to the required amount.

Electric Motor Drives and Their Applications with Simulation Practices.
DOI: https://doi.org/10.1016/B978-0-323-91162-7.00002-3

- A chopper is a static device that converts fixed DC input voltage to a variable DC output voltage directly.
- A chopper is considered as DC equivalent of an AC transformer since it behaves in an identical manner. The other name of chopper is DC transformer.
- The choppers are more efficient as they involve in one stage conversion.
- The choppers are used in trolley cars, marine hoists, forklift trucks, and mine hauler.
- The future electric automobiles are likely to use choppers for their speed control and braking.
- The chopper systems offer smooth control, high-efficiency, fast response, and regeneration.
- Above all, the load current is very irregular and it may oblige for a smoothing or high exchanging frequency to maintain a strategic distance from undesirable impacts. In signal preparing circuits, utilization of chopper has become very important in a framework of electronic drifts. The initial standard can be enhanced using the synchronous demodulators that basically undo the "cleaving" methodology.

3.1.1 Classification of choppers

The static device that converts settled or fixed DC input voltage or power into a variable voltage or power is referred to as chopper. The chopper meets the standards of an AC transformer, where their characteristics are identical. Since the choppers involve one-stage conversion, they are more efficient. Just like the transformer, the chopper is subjected to step up or even step down the fixed DC voltage. The wide range of applications of choppers includes various electronic equipment worldwide. The chopper circuits are highly efficient, responsive with smooth control on four-quadrant operations.

Choppers may be grouped on a few bases, namely:

Based on input and output voltage levels:

- Step-down chopper.
- Step-up chopper.

On the basis of circuit operation:

- First-quadrant.
- Two-quadrant.
- Four-quadrant.

On the basis of commutated technique:

- Voltage commutated.

- Current commutated.
- Load commutated.

The effective applications of chopper circuits involve:

- Switched-mode power supplies and speed controllers.
- Electronic amplifiers of class D.
- Channel filters of switched reluctance.

For all the chopper designs working from a settled DC input voltage, the normal estimation of the output voltage is controlled by the occasional opening and shutting of the switches utilized as a part of the chopper circuit.

The normal output voltage can be controlled by distinctive systems specifically:

- Pulse width modulation system.
- Frequency modulation.

3.2 Control strategies of chopper

The average value of output voltage Vo can be controlled through the duty cycle by opening and closing the semiconductor switch periodically. The various control strategies for varying duty cycle are as follows:

1. Time ratio control, and
2. Current-limit control.

These are now explained below.

3.2.1 Time ratio control

In this control scheme, time ratio Ton/T(duty ratio) is varied. This is realized in two different ways called constant frequency system and variable frequency system as described below.

Constant frequency system: In this scheme, on-time is varied but chopping frequency f is kept constant. Variation of Ton means adjustment of pulse width, as such this scheme is also called pulse-width-modulation scheme.

Variable Frequency System: In this technique, the chopping frequency f is varied and either (1) on-time Ton is kept constant or (2) off-time Toff is kept constant. This method of controlling duty ratio is also called frequency-modulation scheme.

Current-limit control: In this control strategy, the on and off of the chopper circuit is decided by the previous set value of load current. The two set values are maximum load current and minimum load current. When the load current reaches the upper limit, chopper is switched off. When

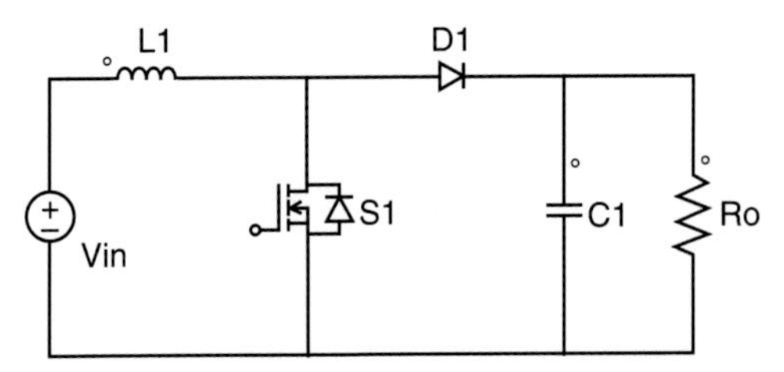

Figure 3.1 Boost converter.

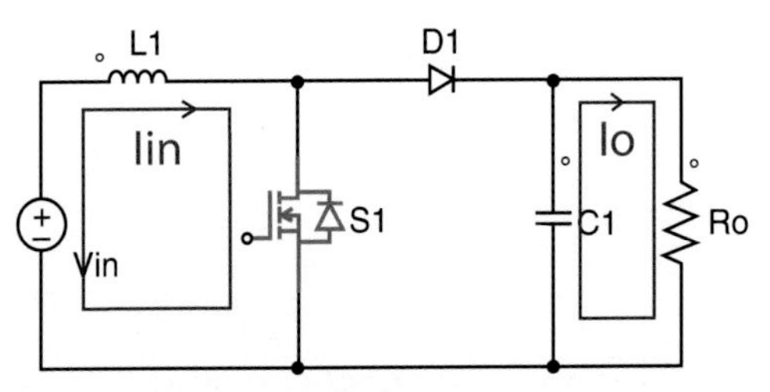

Figure 3.2 Boost converter-MOSFET ON.

the load current falls below lower limit, the chopper is switched on. Switching frequency of chopper can be controlled by setting maximum and minimum level of current. Current limit control involves feedback loop, the trigger circuit for the chopper is, therefore, more complex. Pulse width modulation (PWM) technique is the commonly chosen control strategy for the power control in chopper circuit.

3.3 Design of boost converter

Boost converters are used for the conversion of lower DC input voltage into higher DC output voltage. Conventional DC–DC boost converter using a single power switch, as shown in Fig. 3.1, in which the MOSFET is used as the power switch. There are two operating modes in the converter

During the first operating mode, the switch is turned ON and current completes its path via inductor, as shown in Fig. 3.2 At this mode, inductor stores the energy and load current is maintained by the load capacitor. During the time t1, inductor current increases from minimum to maximum. During the time t2, inductor current falls from the maximum value to minimum value. ΔI is change in input current.

During the second operating mode, the power switch is turned OFF, and current completes its path via inductor, as shown in Fig. 3.3. At this

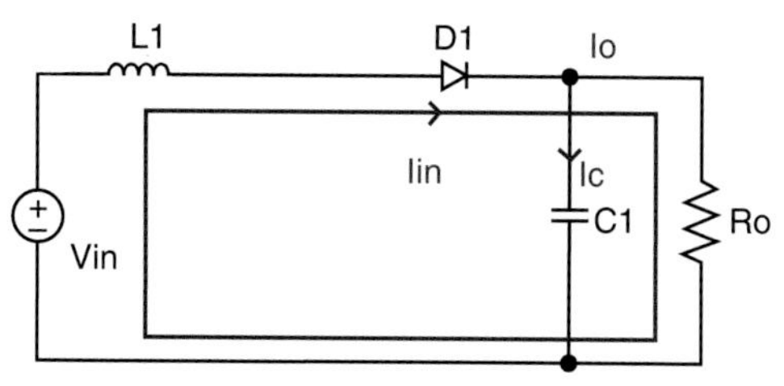

Figure 3.3 Boost converter-MOSFET OFF.

mode, inductor releases the stored energy, and the load receives the load current. The diode and MOSFET switch is operating in a complementary manner. The current flows through the inductor, diode, capacitor, and load. The output side capacitor across the load is used as a filter in the converter circuit (Rashid, 2009). For high switching frequencies the switching device can be replaced by IGBT.

Similarly, average current in the output is calculated by assuming $V_{in} I_{in} = V_o I_o$. The output voltage V_o is calculated as follows and it shows that V_o is the function of the duty cycle.

$$V_0 = V_{in} \frac{T}{t_2} V_0 = \frac{V_{in}}{(1-D)} I_{in} = \frac{I_o}{(1-D)}$$

3.3.1 Simulation of boost converter using MATLAB

Before getting into the simulation diagram, we will see how to select the components in MATLAB simulation.

Now will see the step-by-step process of MATLAB simulation.

>>Type Simulink on MATLAB Command Window>>File>> New>>Model after that start selecting the components from library by using the command.

>>Libraries>>Simscape>>SpecializedPowerSystems>>Fundamentalblocks>>Electrical Sources>>DC Voltage Source.

Now you can see the window like Fig. 3.4. After selecting the DC voltage source, if you double click on that parameter dialogue box will open, in that you can set the amplitude value of the DC voltage source.

>>Libraries>>Simscape>>SpecializedPowerSystems>>Fundamentalblocks>> PowerElectronics>>Thyristor/MOSFET.

Using the above command, you can select the MOSFET switch, then click on that parameter dialogue box will open and set the required values (Fig. 3.5).

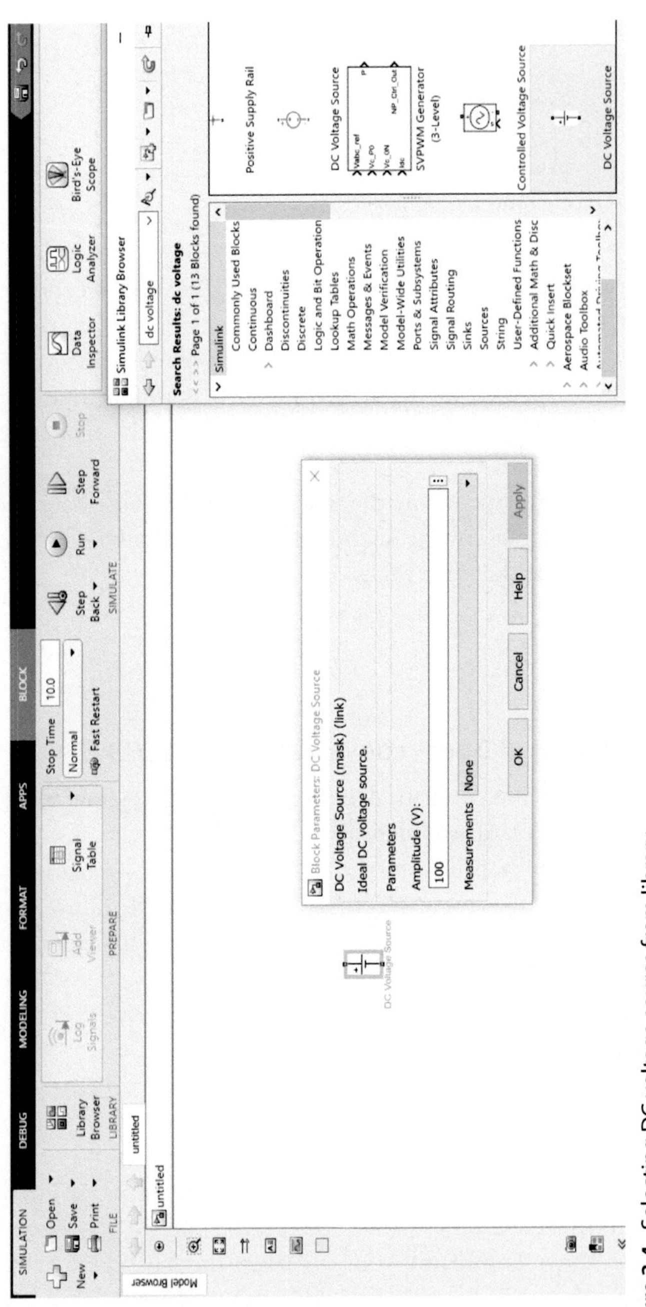

Figure 3.4 Selecting DC voltage source from library.

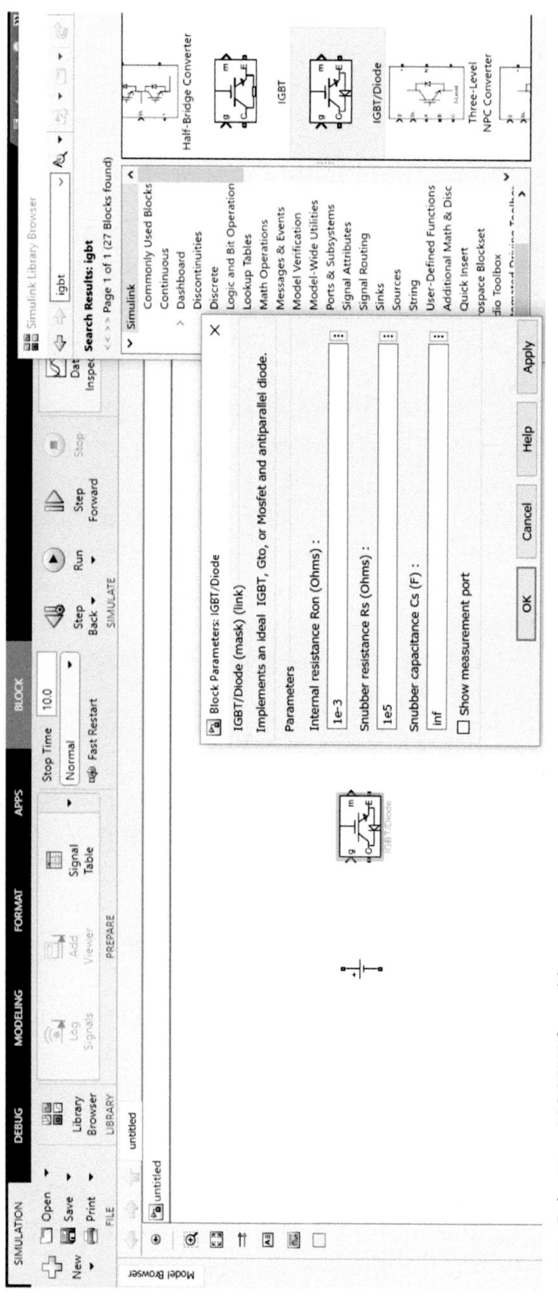

Figure 3.5 Selecting MOSFET from library.

>>Libraries>>Simscape>>SpecializedPowerSystems>>Fundamentalblocks>> PowerElectronics>>Diode.

Select the diode by using the above command, then click on that parameter dialogue box will open and set the required values (Fig. 3.6).

>>Libraries>>Simscape>>SpecializedPowerSystems>>Fundamentalblocks>> Elements>>Series RLC Branch.

Select the series RLC branch using the above command. Even though you do not have RLC series branch in your circuit you have select it for the purpose of resistance, inductance, and capacitance. In MATLAB, we do not have separate R, L, and C, so you have to select series RLC branch and change that into R, L, & C by providing the required value in the parameter setting.

To get R, L, & C, here you need to use the copy command for series RLC branch. Use the copy and paste command, get one more series RLC branch.

Now to get inductance, you have to change the branch type as L instead of RLC and set the inductance value in the parameter dialogue box (Fig. 3.7).

Now to get capacitance, you have to change the branch type as C instead of RLC and set the capacitance value in the parameter dialogue box (Fig. 3.8).

>>Libraries>>Simscape>>Electrical>>Specialized power system >>Fundamental Blocks>> Machines >> DC Machines (Fig. 3.9).

>>Libraries>>Simscape>>SpecializedPowerSystems>>Fundamentalblocks>>Electrical Sources>>DC Voltage Source (Fig. 3.10).

>>Libraries>>Simulink>>Quickinsert>>Logicandbitoperations>> Greaterthanorequal (Fig. 3.11).

>>Libraries>>Simulink>>Sources>>Repeating Sequence (Fig. 3.12).

>>Libraries>>Simulink>>Continuous>>PID Controller (Fig. 3.13).

>>Libraries>>Simulink>>Commonly used blocks>>Gain (Figs. 3.14 and 3.15).

>>Libraries>>Simulink>>Commonly used blocks>>Sum (Fig. 3.16).

>>Libraries>>Simulink>>Sources>>Step (Figs. 3.17 and 3.18).

>>Libraries>>Simulink>>Commonlyusedblocks>>Busselector (Figs. 3.19 and 3.20).

>>Libraries>>Simulink>>Commonlyusedblocks>>Scope.

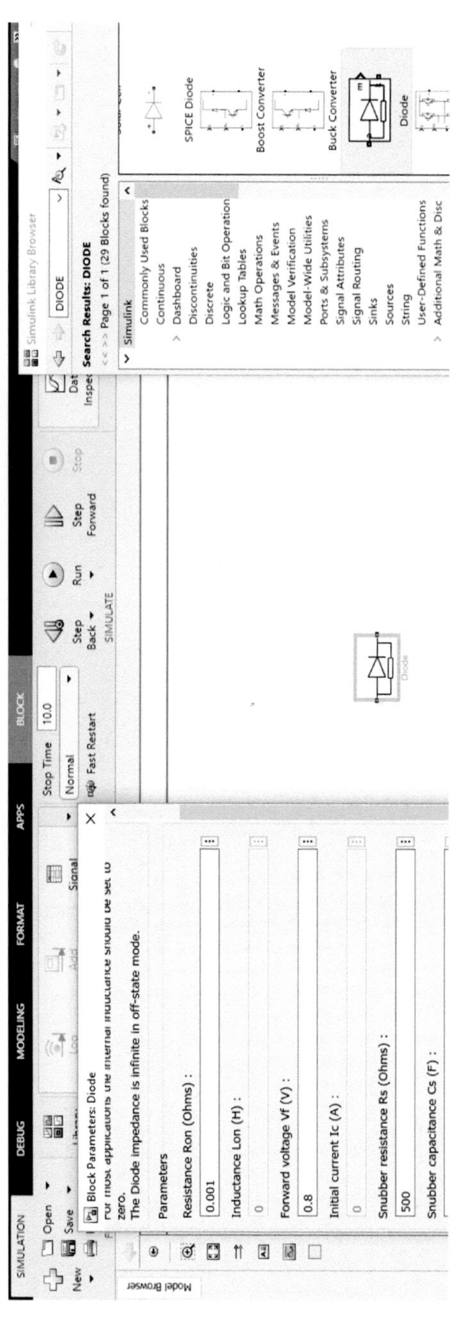

Figure 3.6 Selecting diode from library.

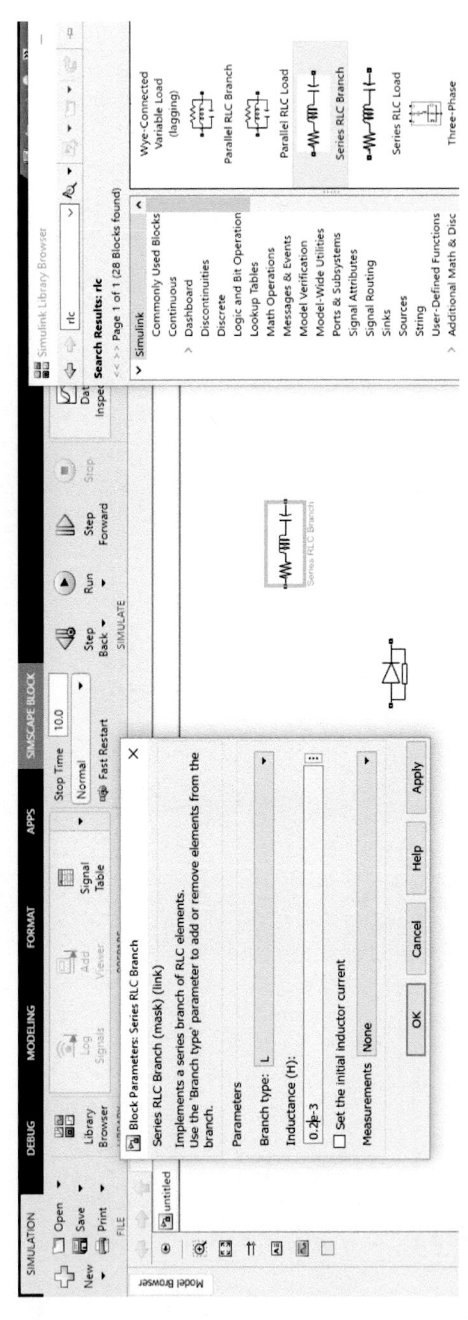

Figure 3.7 Selecting series RLC branch from library.

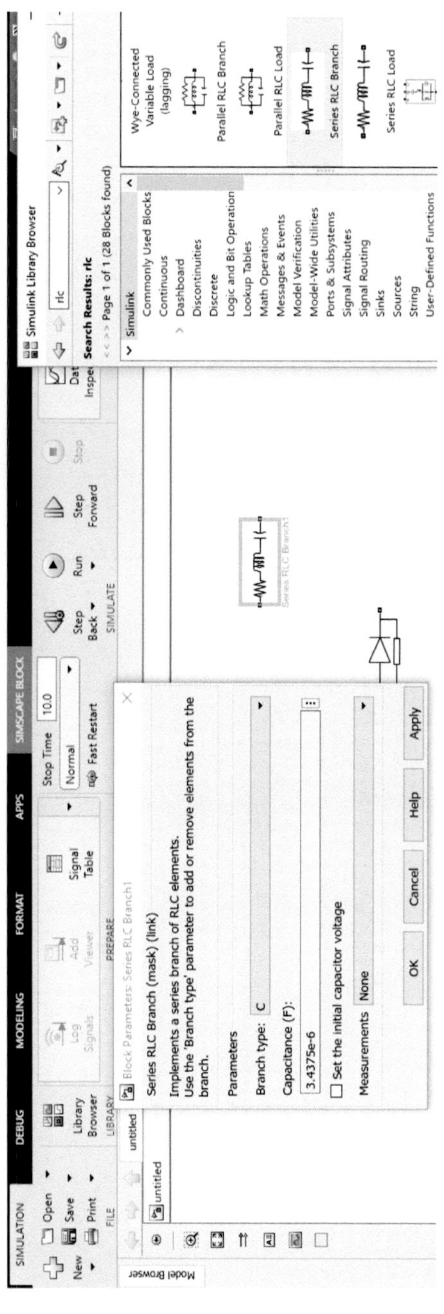

Figure 3.8 Getting capacitance from series RLC branch.

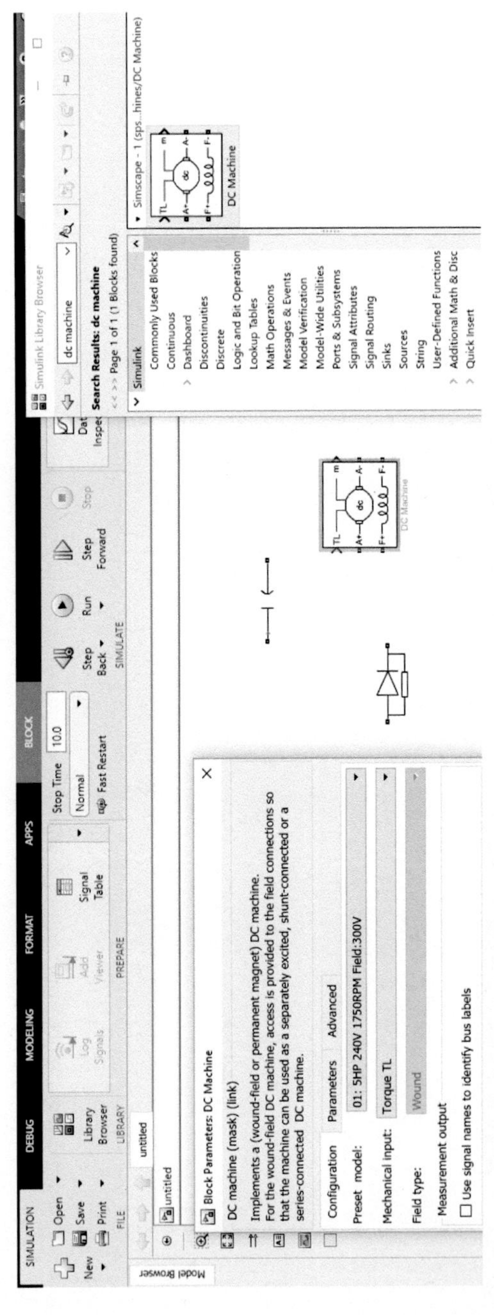

Figure 3.9 Selecting DC machines from library.

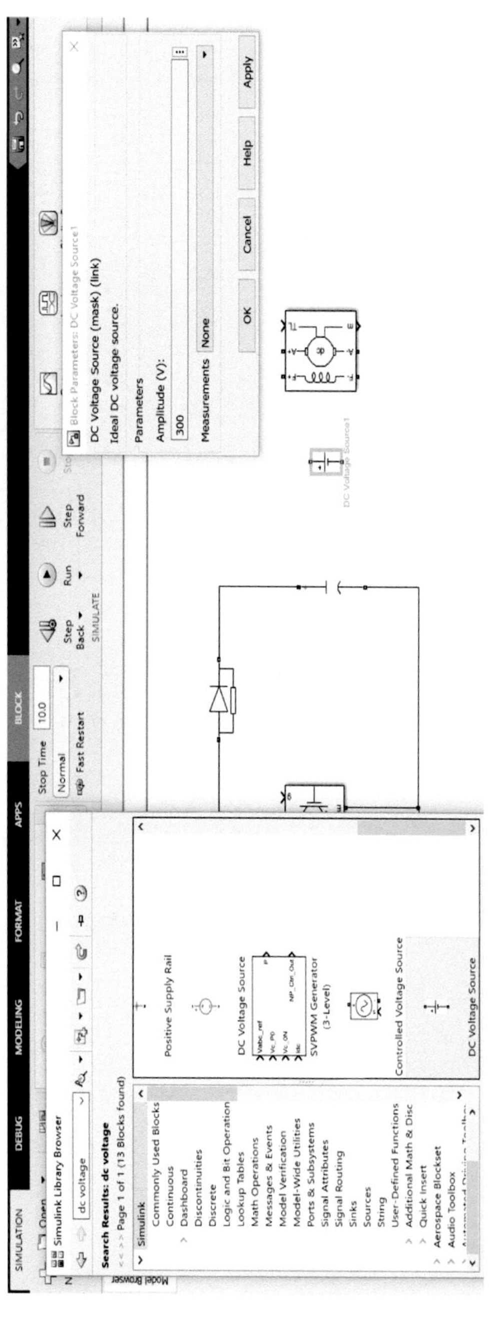

Figure 3.10 Selecting DC voltage source from library.

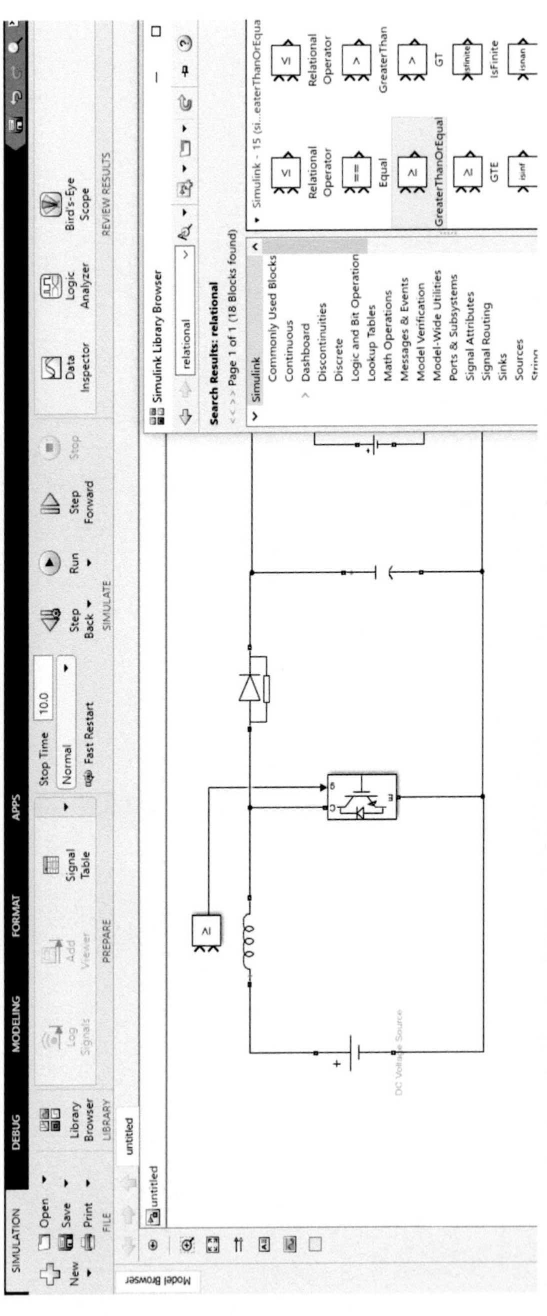

Figure 3.11 Selecting greater than or equal from library.

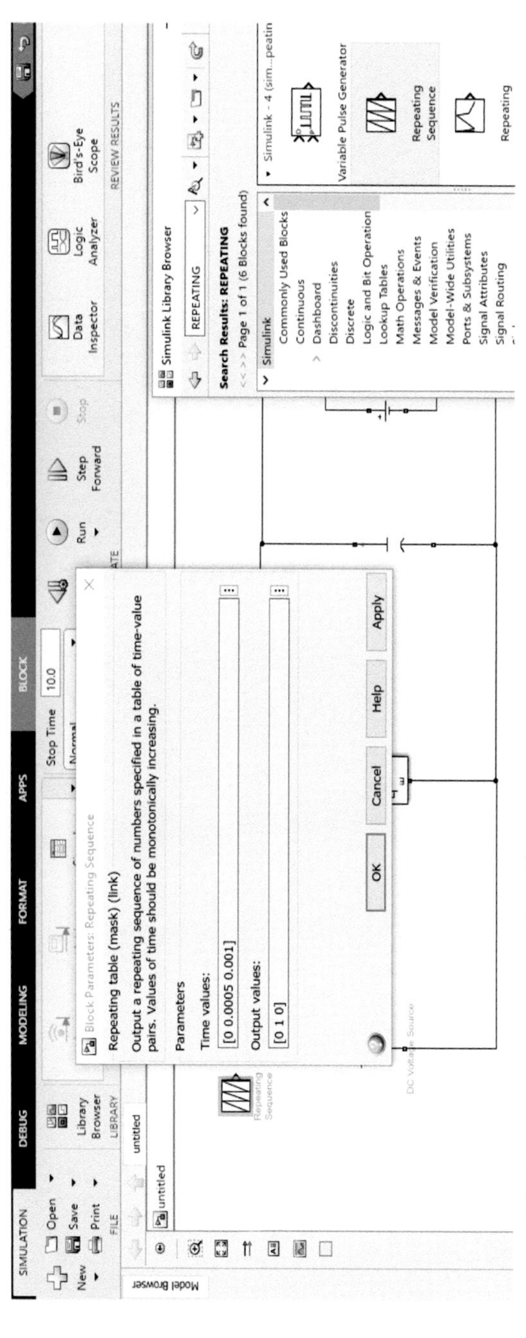

Figure 3.12 Selecting repeating sequence from library.

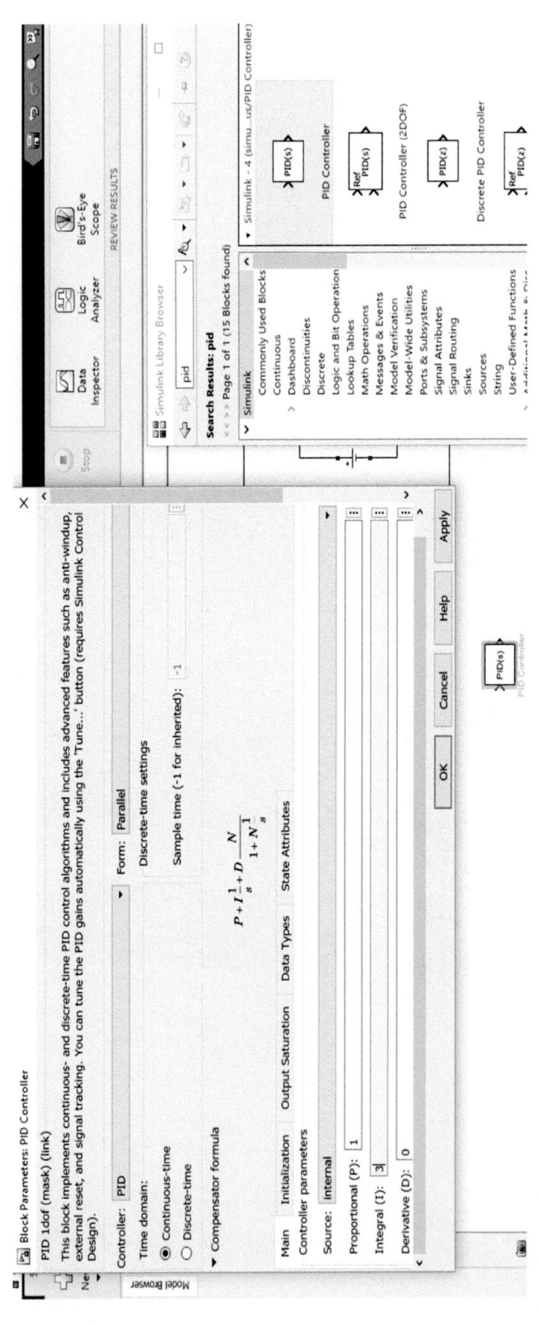

Figure 3.13 Selecting PID controller from library.

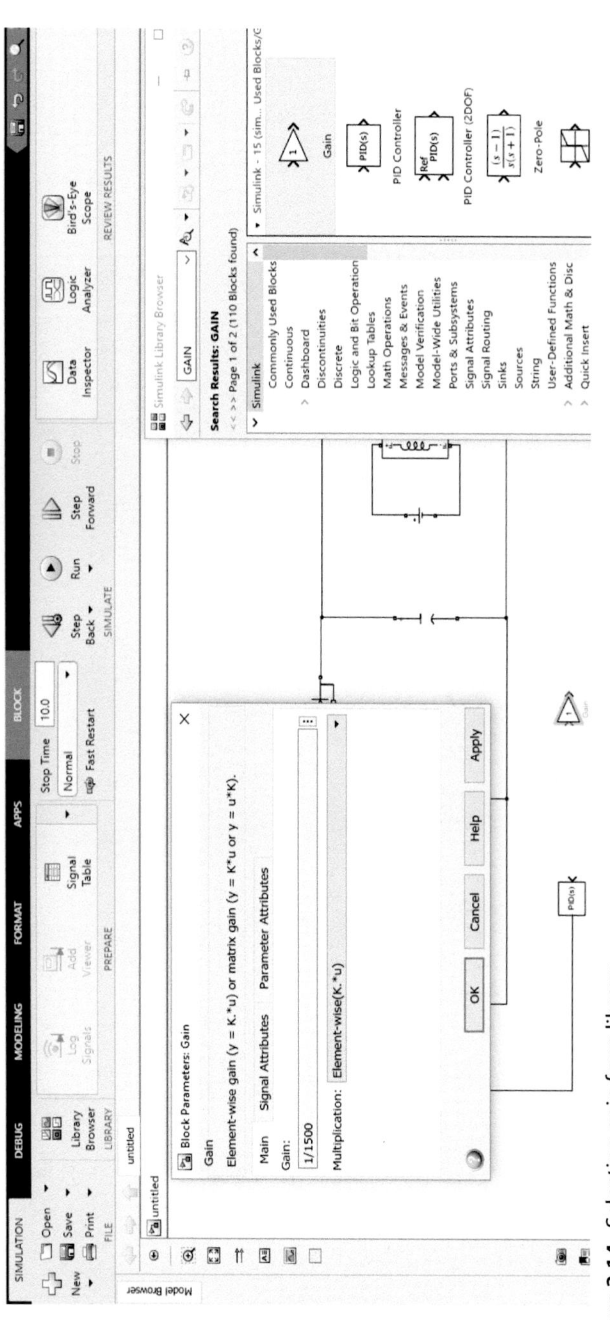

Figure 3.14 Selecting gain from library.

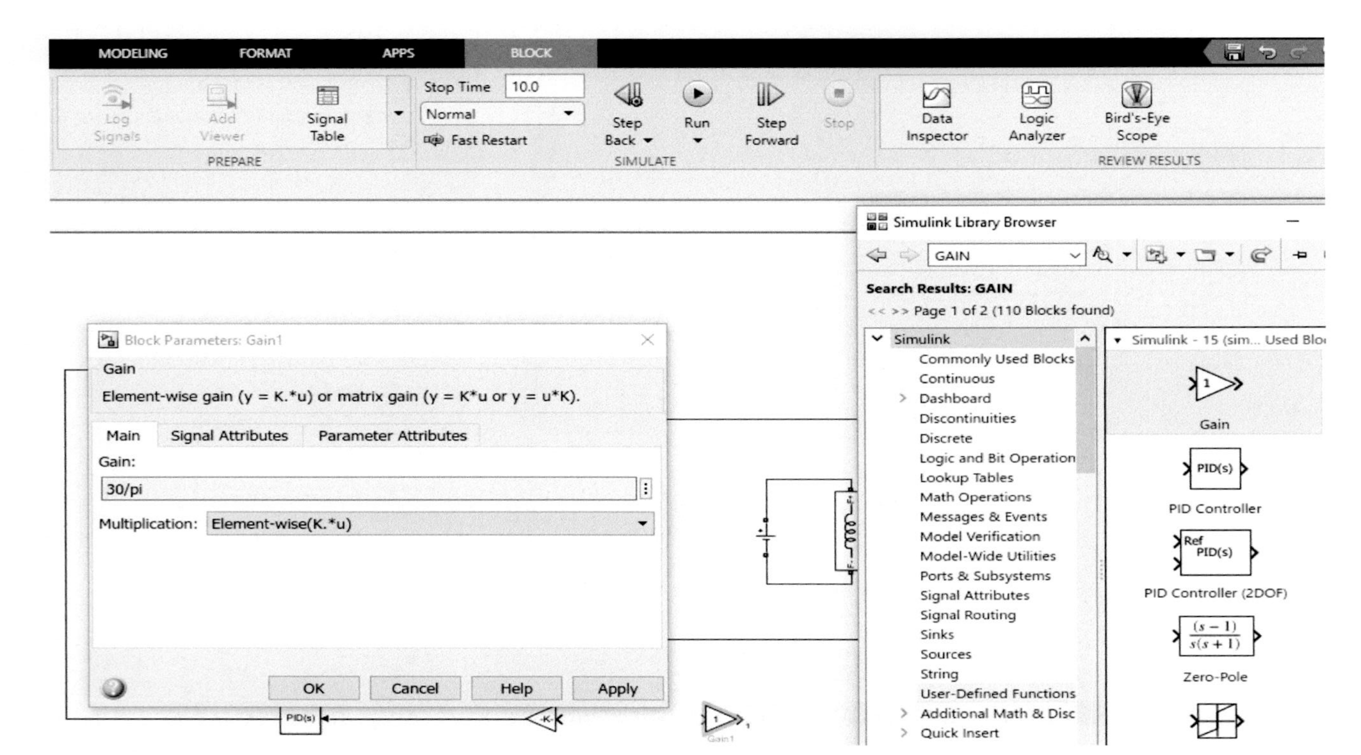

Figure 3.15 Setting value for gain.

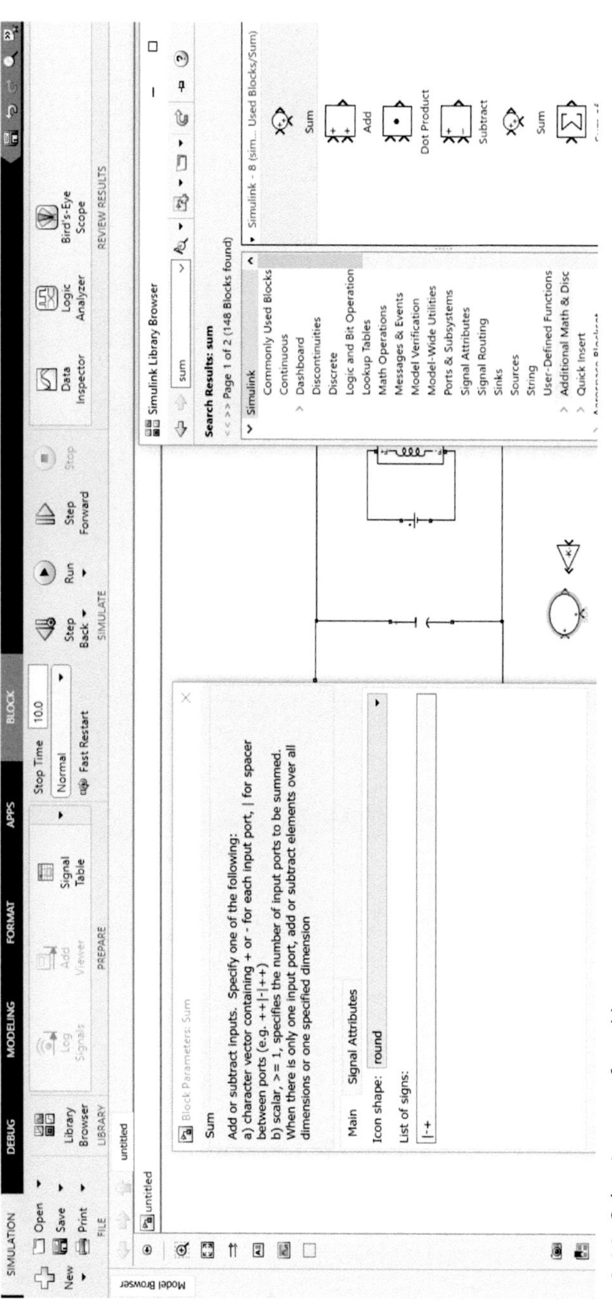

Figure 3.16 Selecting sum from library.

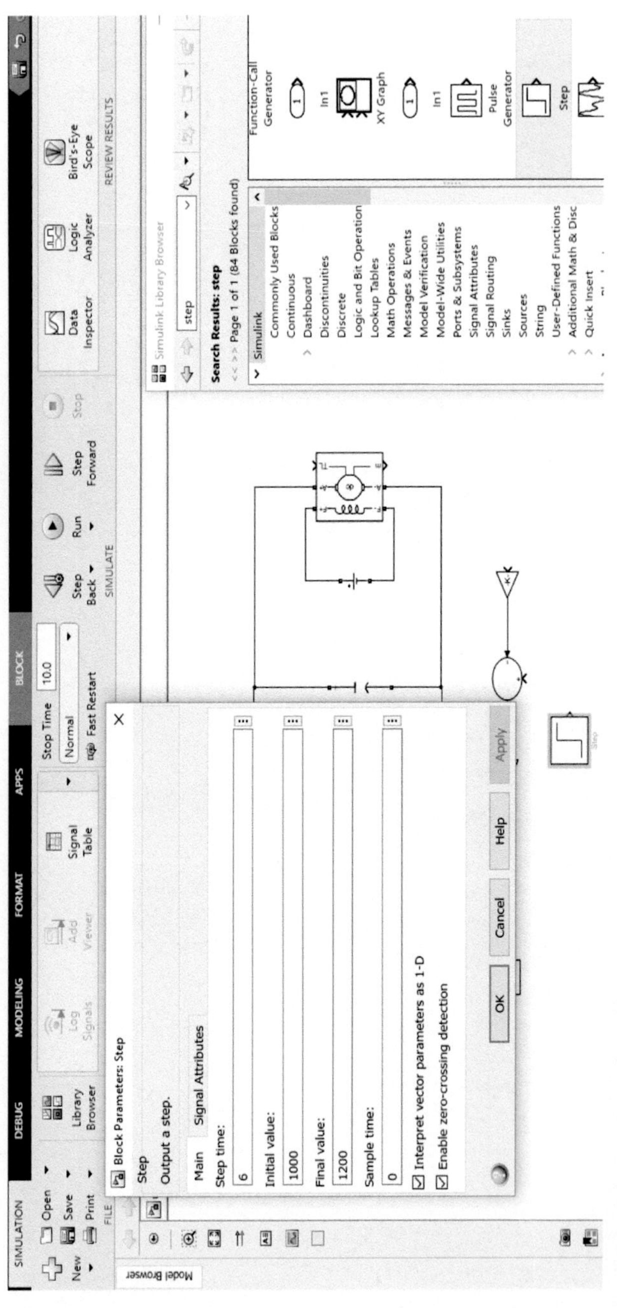

Figure 3.17 Selecting step from library.

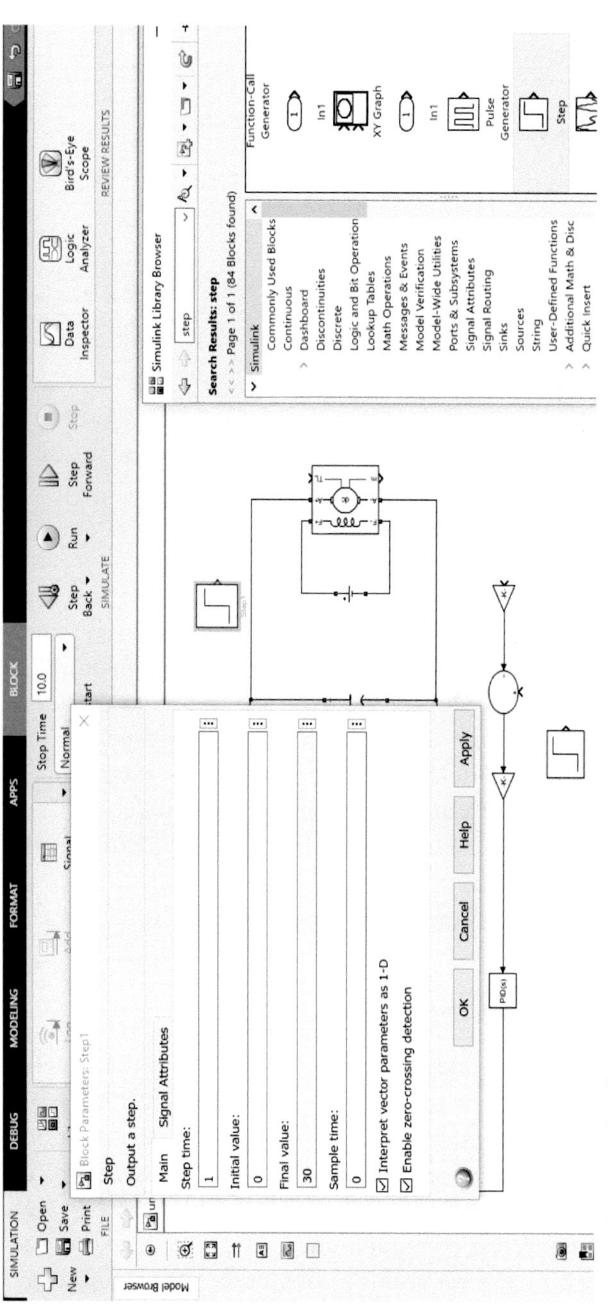

Figure 3.18 Setting value for step.

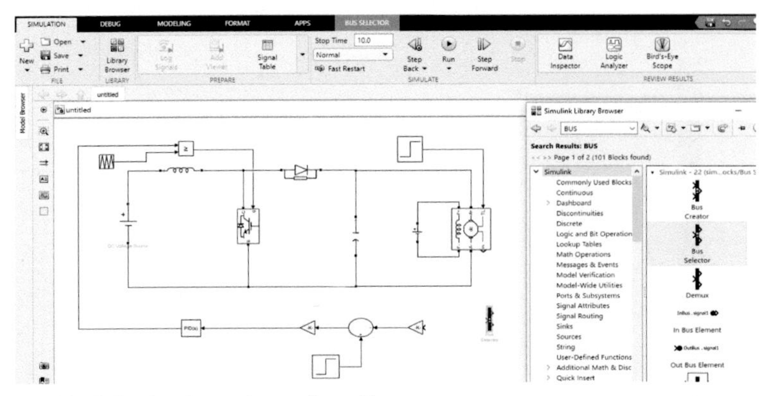

Figure 3.19 Selecting bus selector from library.

To see the waveform you need CRO, select the scope by using the above command (Fig. 3.21).

>>Libraries>>Simspace>>Powerlib>>Powergui

By dragging the selected blocks into the Simulink editor, connect all the blocks using the pointer to make it a closed circuit. The Simulink editor provides a finished control over what we see and use inside the model. We will get the simulation circuit of boost converter, as shown in Fig. 3.22.

To simulate the behavior of the structure and point of view the results, focus run-time decisions, including the sort and properties of the solver, model start and stop times, and whether to load or extra propagation data. Different combinations could be saved with the model. You can run your change spontaneously from the Simulink Editor or proficiently from the MATLAB summon line.

After simulation, you can see the graph in scope by clicking it (Fig. 3.23).

3.4 Design of buck converter

Due to the nonlinear and discontinuous nature of the power electronic converters, they should be linearized in order to analyze the circuit. Standard linear circuit theory cannot be used to analyze the power electronic converters due to their inherent large-signal nature. So, it is essential to linearize the circuit so that the designer can apply the control theory. Fig. 3.24 illustrates a typical buck converter configuration.

To analyze how the variations in the source current, output current, or the duty ratio affect the load voltage, a dynamic switching converter model is essential. Normally, the small signal-averaged equations of PWM switching

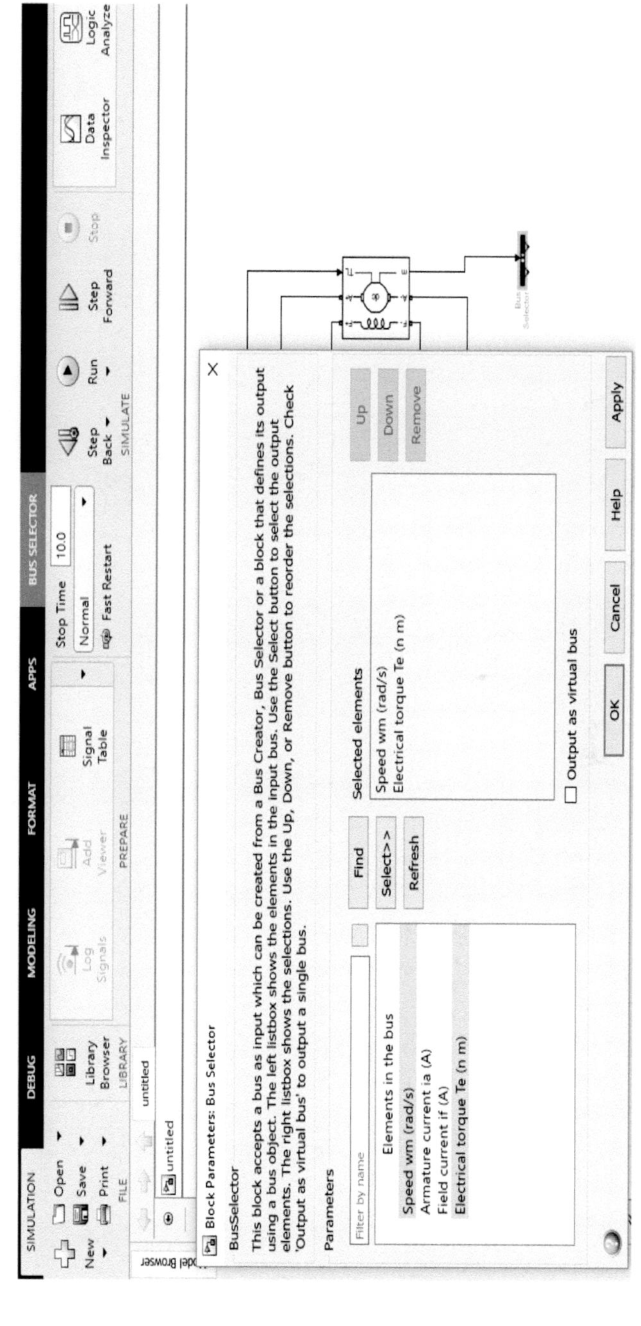

Figure 3.20 Setting parameters for bus selector.

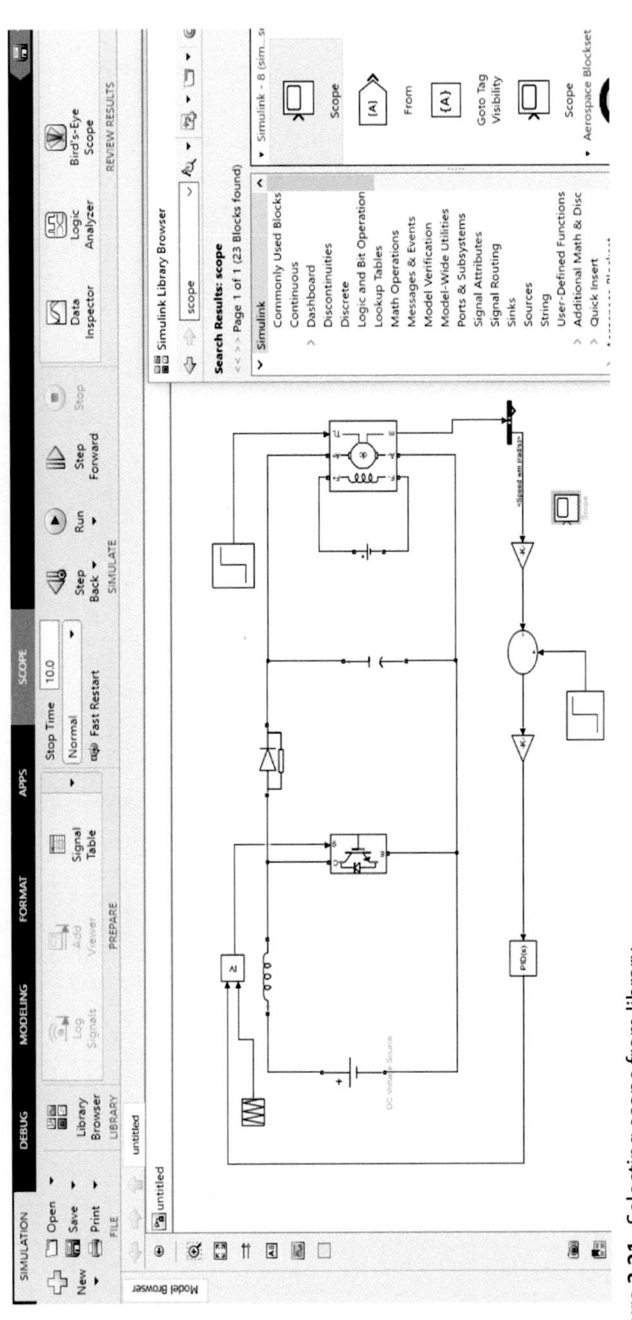

Figure 3.21 Selecting scope from library.

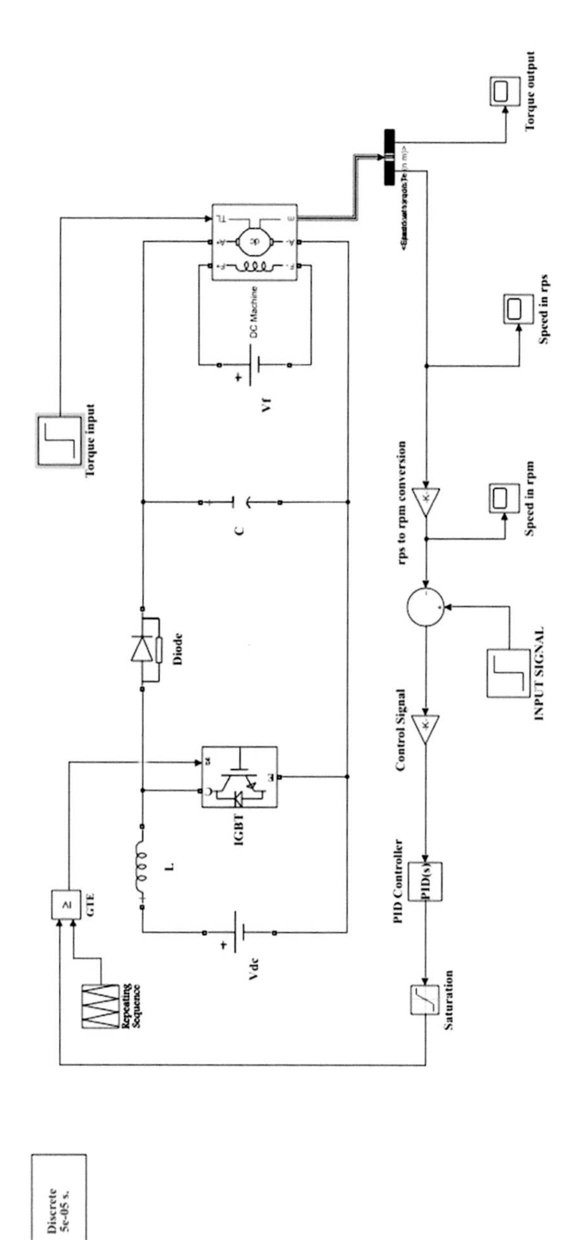

Figure 3.22 MATLAB/Simulink model of boost converter.

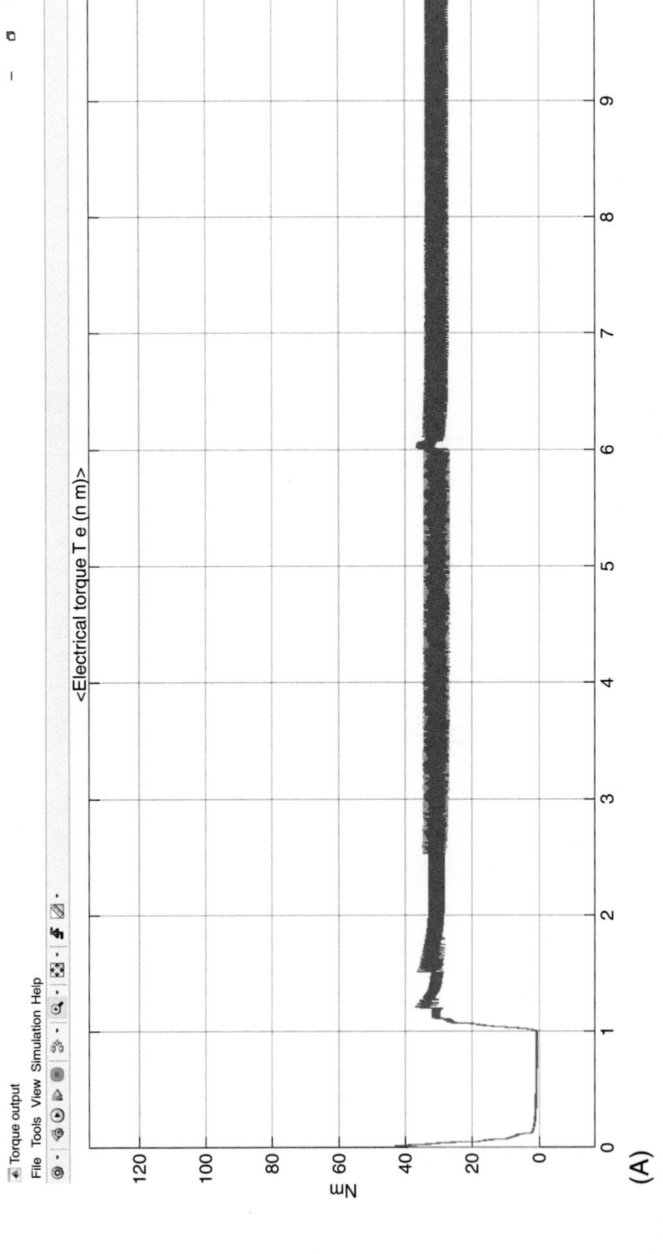

(A)

Figure 3.23 Simulation results of boost converter. (A) Torque output, (B) Speed in rps, and (C) speed in rpm.

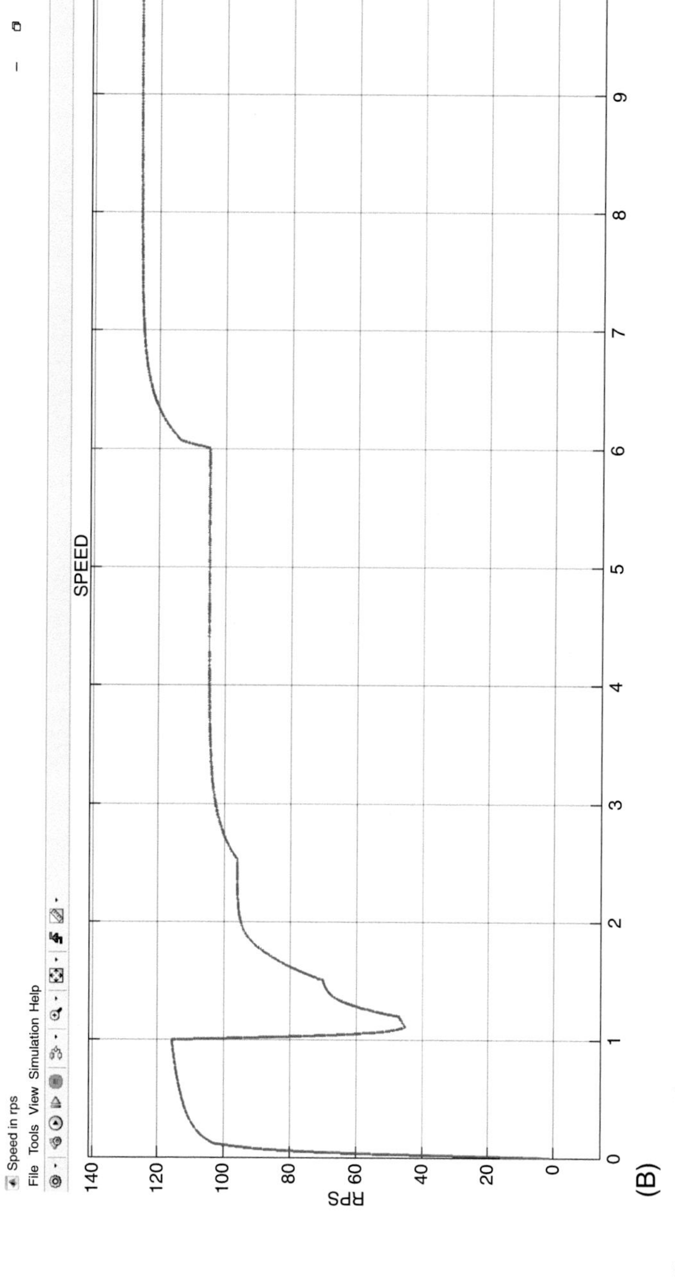

Figure 3.23, cont'd.

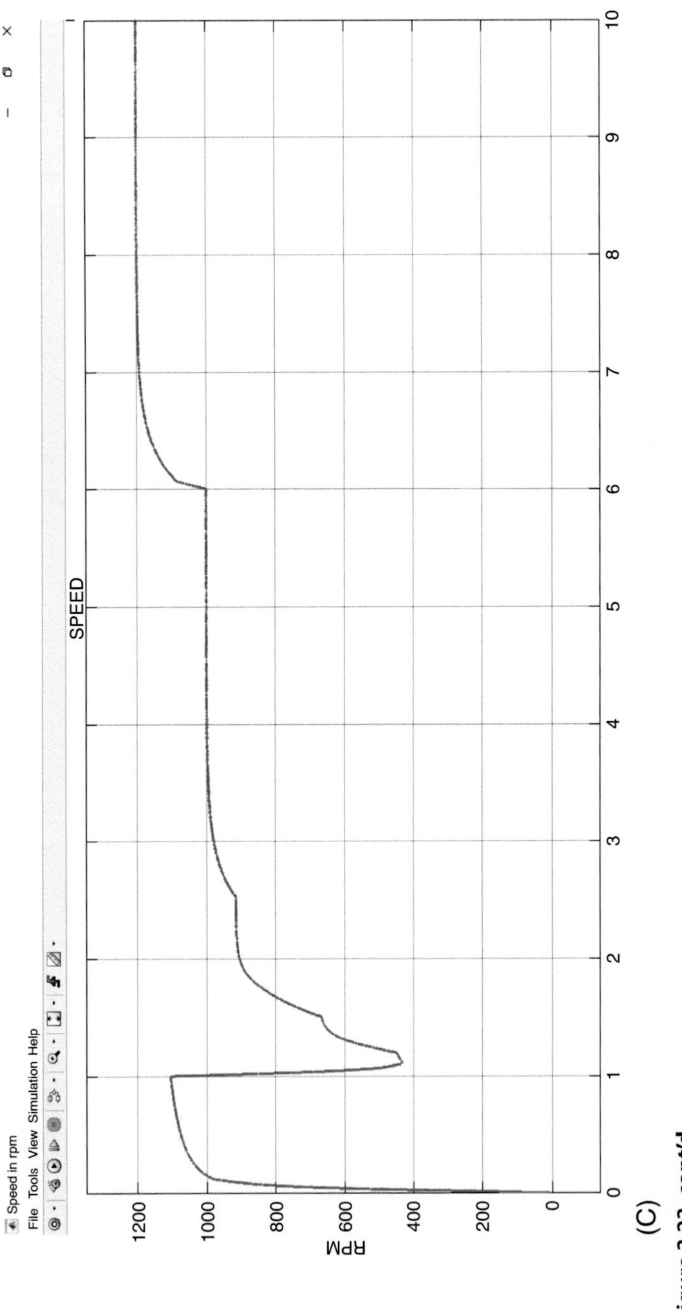

(C)

Figure 3.23, cont'd.

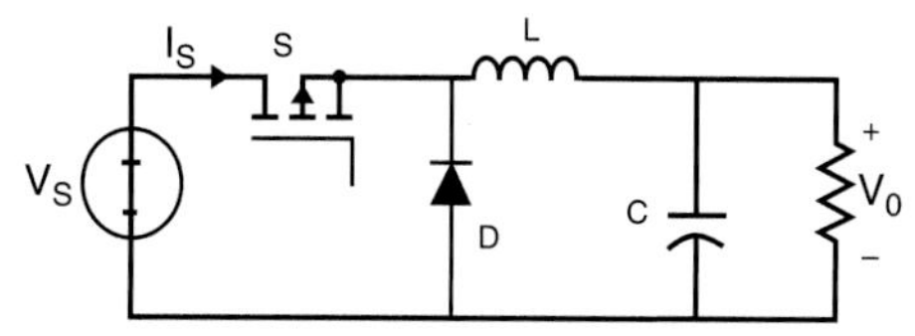

Figure 3.24 Buck converter.

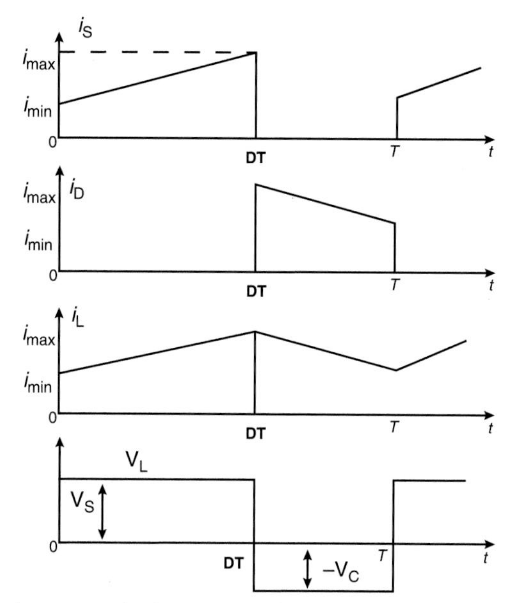

Figure 3.25 Supply current, diode current, inductor current, and inductor voltage waveforms of Buck converter.

converters are derived using state space representation of dynamic systems. The nonlinear switching action of the circuit can be modeled F as a three-terminal circuit element. Any converter topology can be easily analyzed using this model. Fig. 3.25 shows supply current, diode current, inductor current, and inductor voltage waveforms of the buck converter.

3.4.1 Simulation of buck converter using MATLAB

Now will see the step-by-step process of MATLAB simulation.

>>Type Simulink on Matlab Command Window>>File>>New>> Model after that start selecting the components from library by using the command.

>>Libraries>>Simscape>>SpecializedPowerSystems>>Fundamentalblocks>>Electrical Sources>>DC Voltage Source.

Now you can see the window like in Fig. 3.26. After selecting the DC voltage source, if you double click on that parameter dialogue box will open, in that you can set the amplitude value of the DC voltage source.

>>Libraries>>Simscape>>SpecializedPowerSystems>>Fundamentalblocks>> PowerElectronics>>IGBT.

Using the above command you can select the IGBT switch, then click on that parameter dialogue box will open and set the required values (Fig. 3.27).

>>Libraries>>Simscape>>SpecializedPowerSystems>>Fundamentalblocks>> PowerElectronics>>Diode (Fig. 3.28).

Select the diode by using the above command, then click on that parameter dialogue box will open and set the required values.

>>Libraries>>Simscape>>SpecializedPowerSystems>>Fundamentalblocks>> Elements>>Series RLC Branch.

Select series RLC branch using the above command. Even though you do not have RLC series branch in your circuit you have select it for the purpose of resistance, inductance, and capacitance. In MATLAB, we do not have separate R, L, and C, so you have to select series RLC branch and change that into R, L, & C by providing the required value in the parameter setting.

To get R, L, & C, here you need to use the copy command for the series RLC branch. Use the copy and paste command, get one more series RLC branch.

Now to get inductance, you have to change the branch type as L instead of RLC and set the inductance value in the parameter dialogue box (Fig. 3.29).

Now to get capacitance, you have to change the branch type as C instead of RLC and set the capacitance value in the parameter dialogue box.

>>Libraries>>Simscape>>Electrical>>Specialized power system >>Fundamental Blocks>>Machines>>DC Machines (Fig. 3.30).

>>Libraries>>Simscape>>SpecializedPowerSystems>>Fundamentalblocks>>Electrical Sources>>DC Voltage Source (Fig. 3.31).

>>Libraries>>Simulink>>Quickinsert>>Logicandbitoperations>> Greaterthanorequal (Fig. 3.32).

>>Libraries>>Simulink>>Sources>>Repeating Sequence (Fig. 3.33).

>>Libraries>>Simulink>>Continuous>>PIDController (Fig. 3.34).

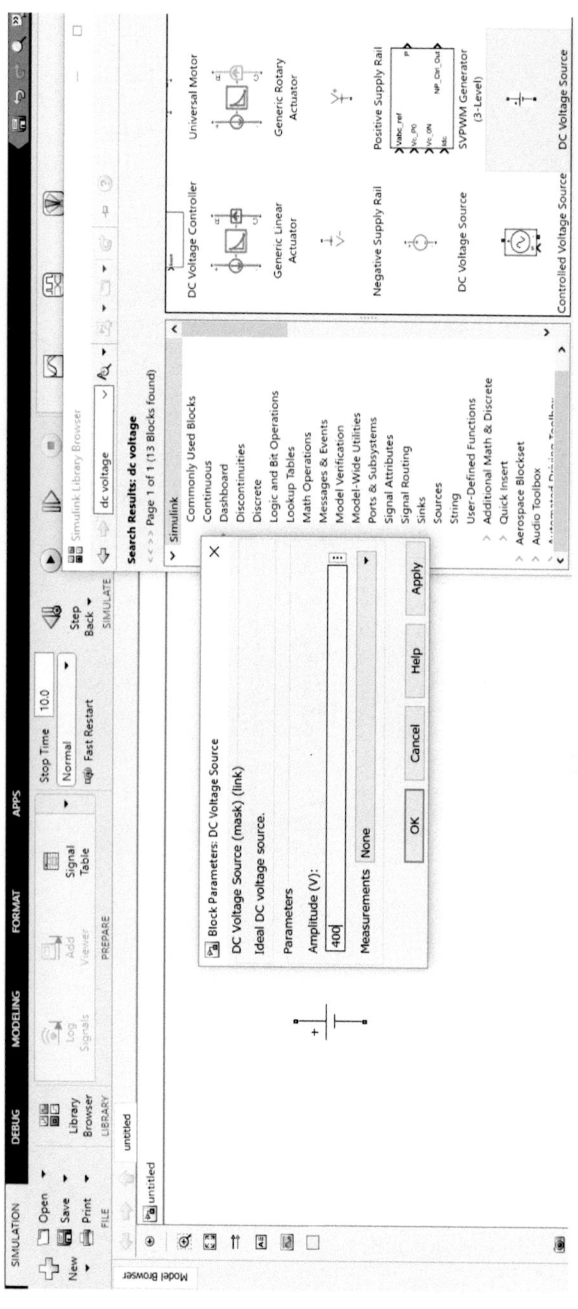

Figure 3.26 Selecting DC voltage source from Library.

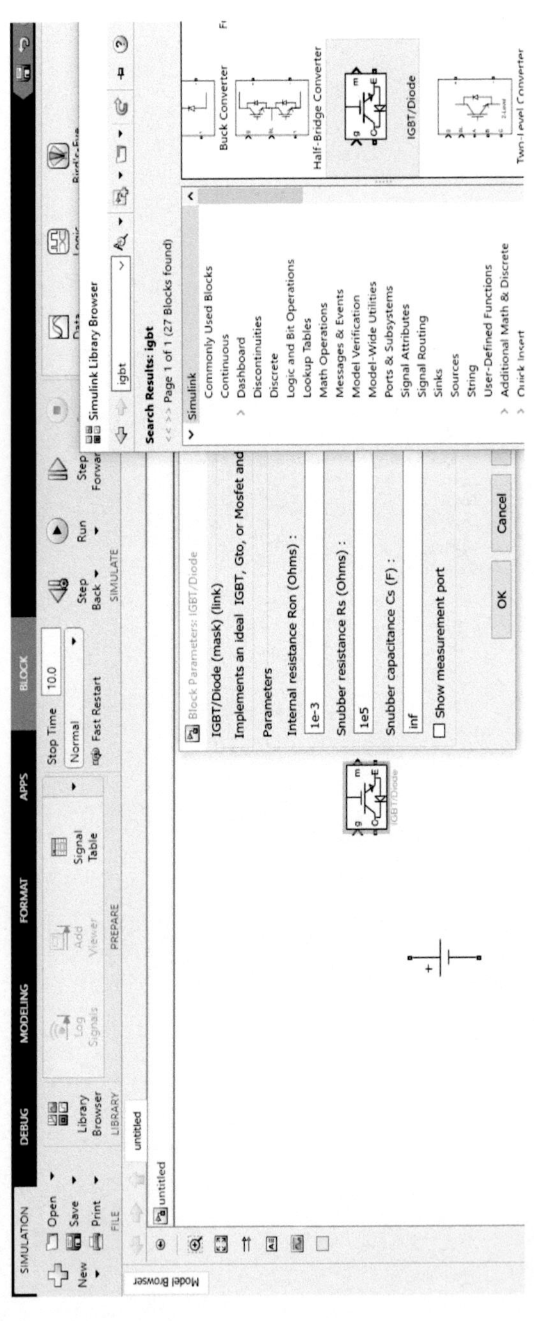

Figure 3.27 Selecting IGBT from library.

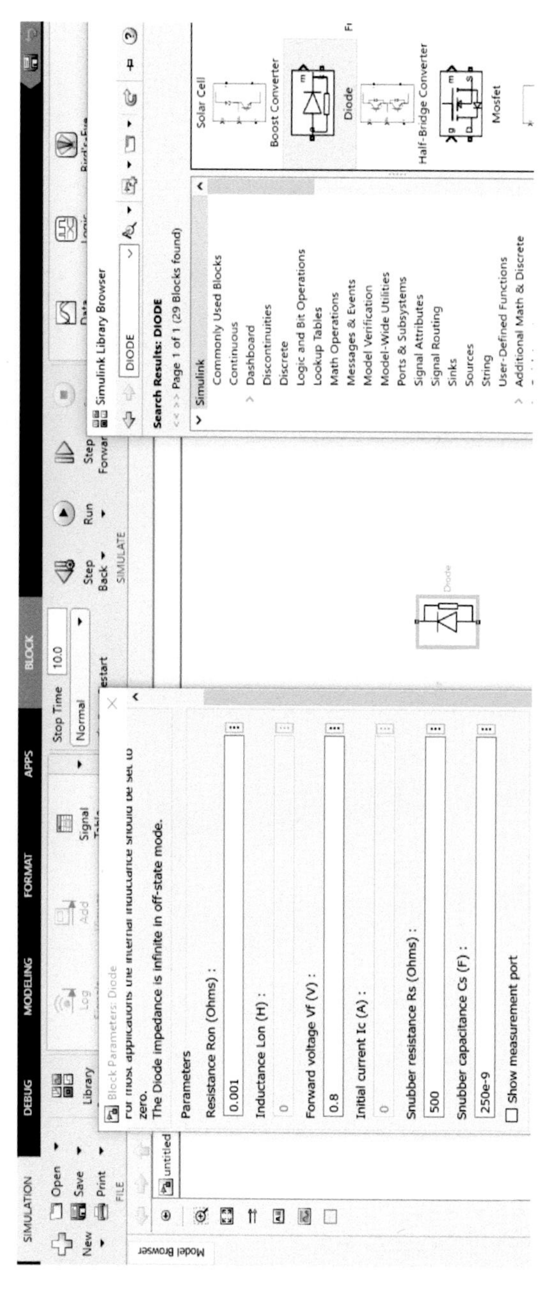

Figure 3.28 Selecting diode from library.

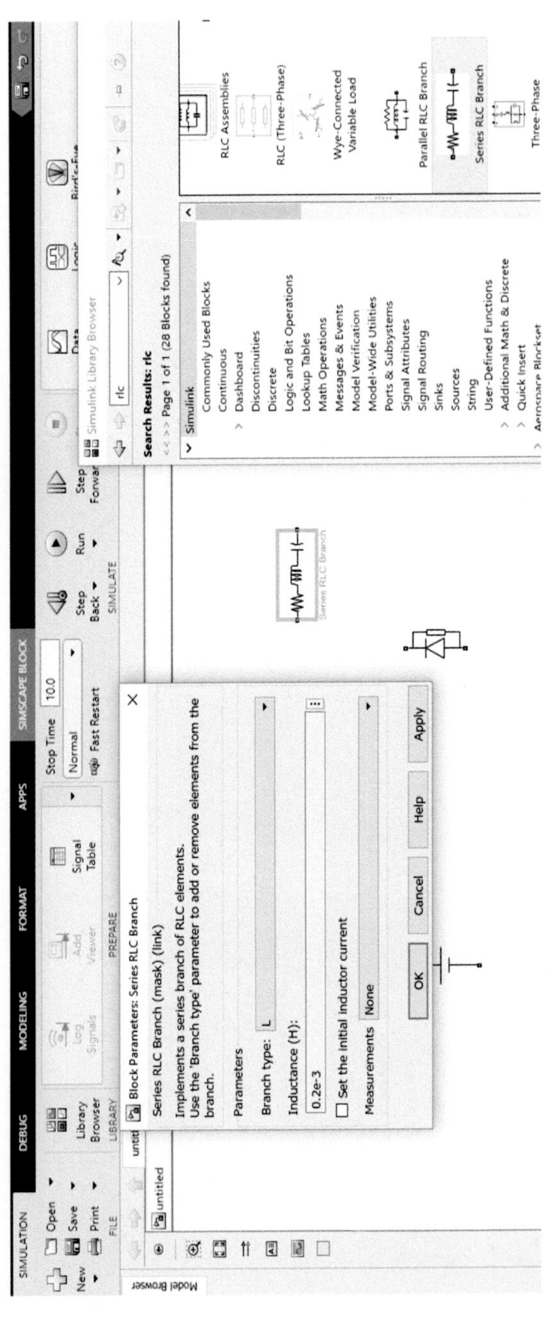

Figure 3.29 Selecting series RLC branch from library.

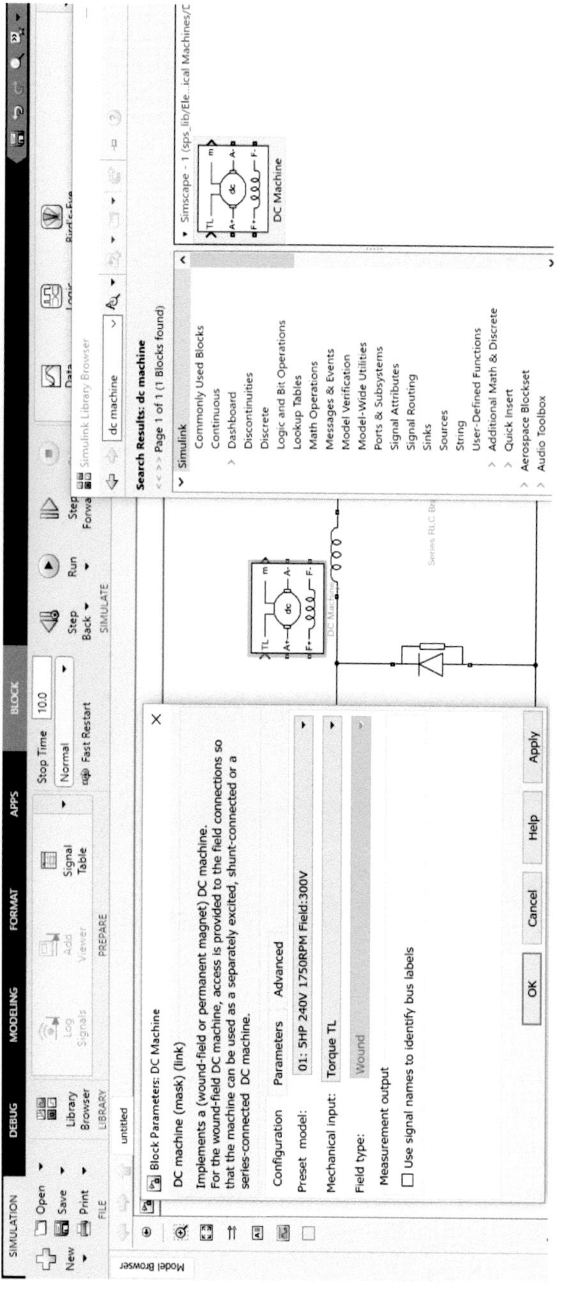

Figure 3.30 Selecting DC machines from library.

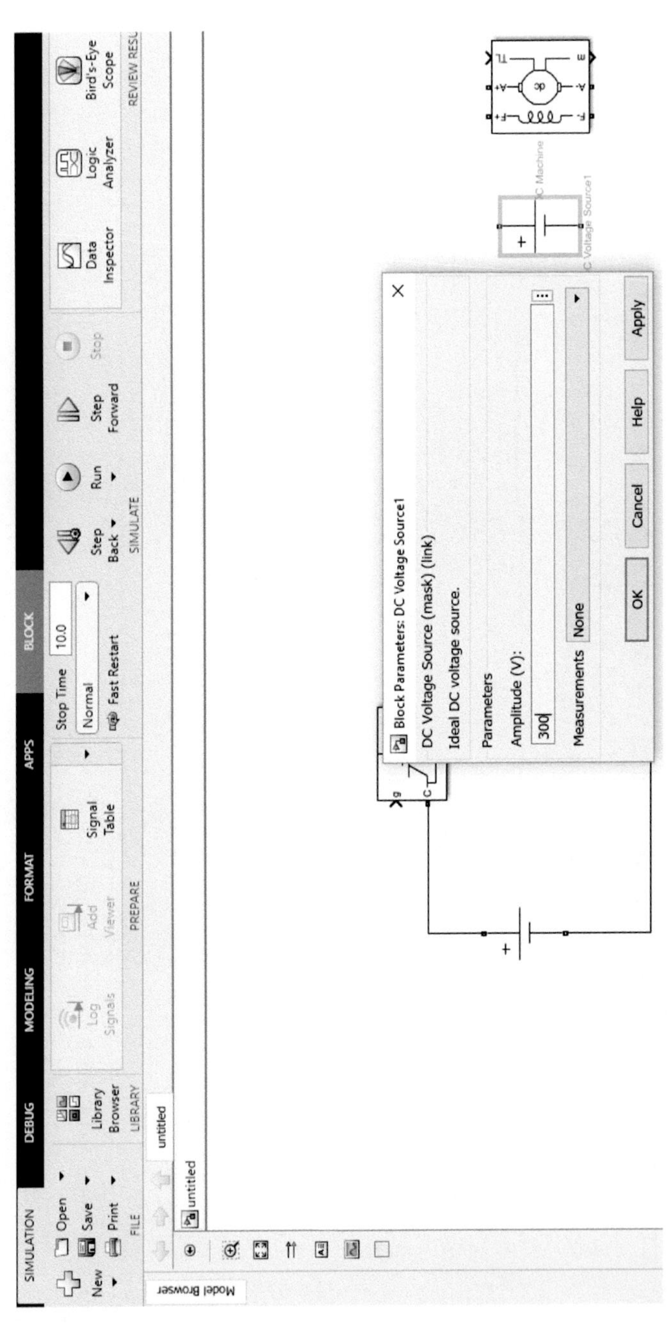

Figure 3.31 Selecting DC voltage source from library.

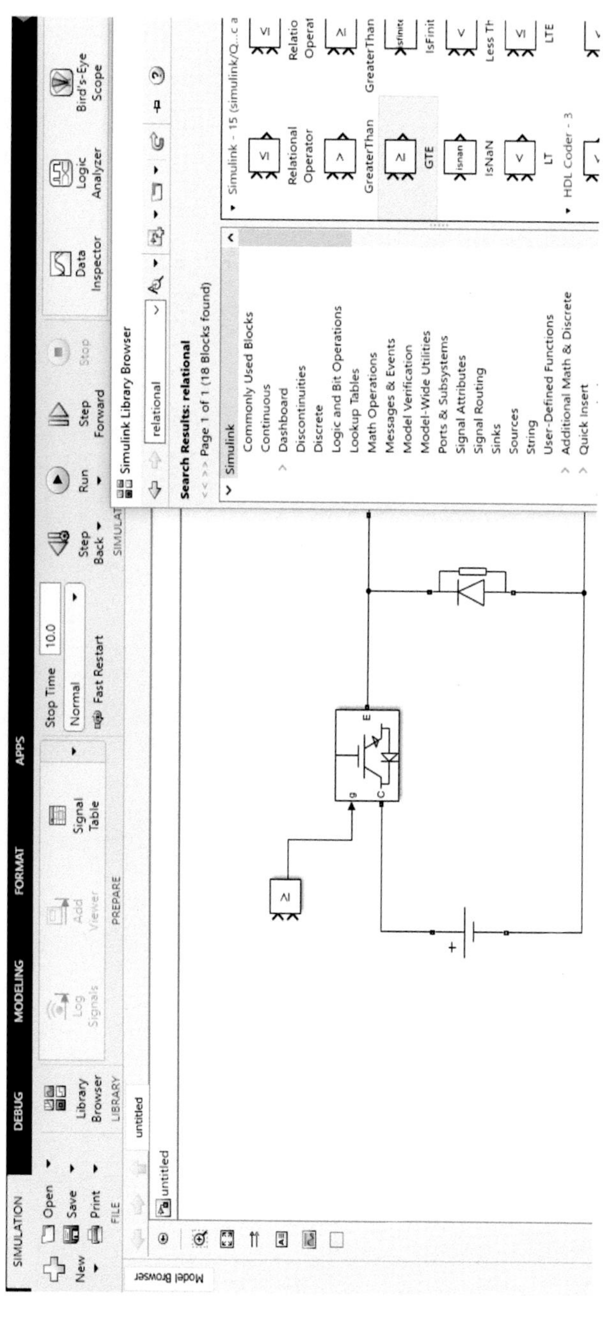

Figure 3.32 Selecting greater than or equal from library.

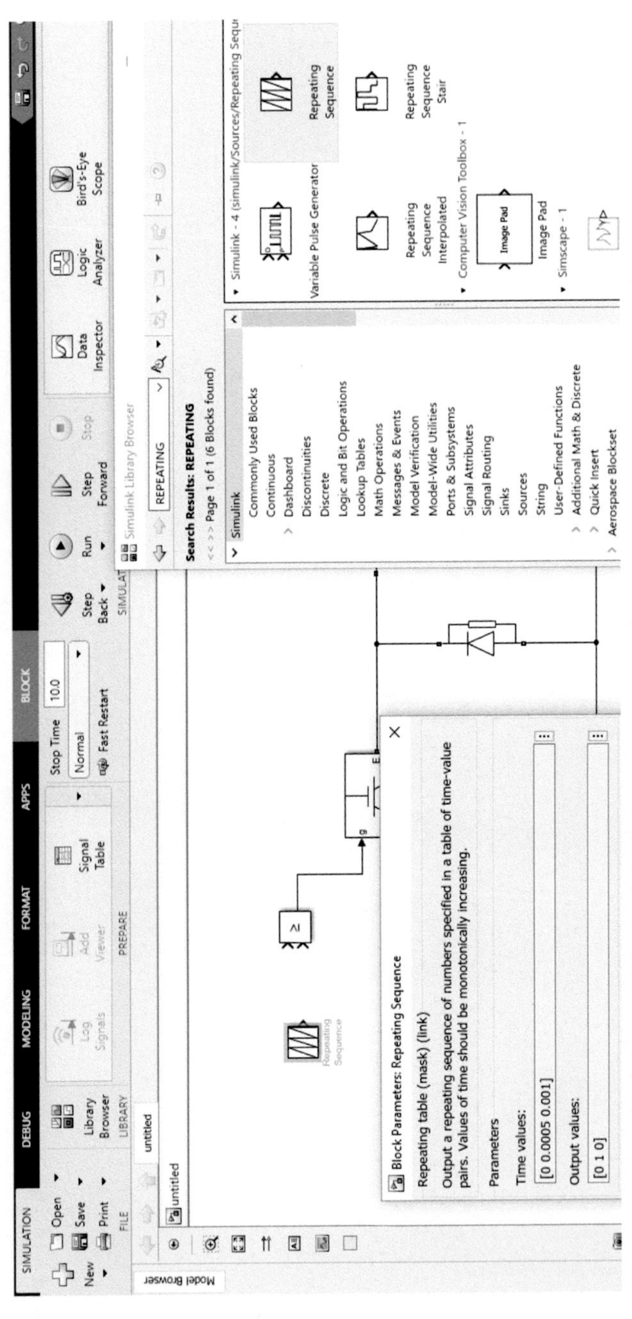

Figure 3.33 Selecting repeating sequence from library.

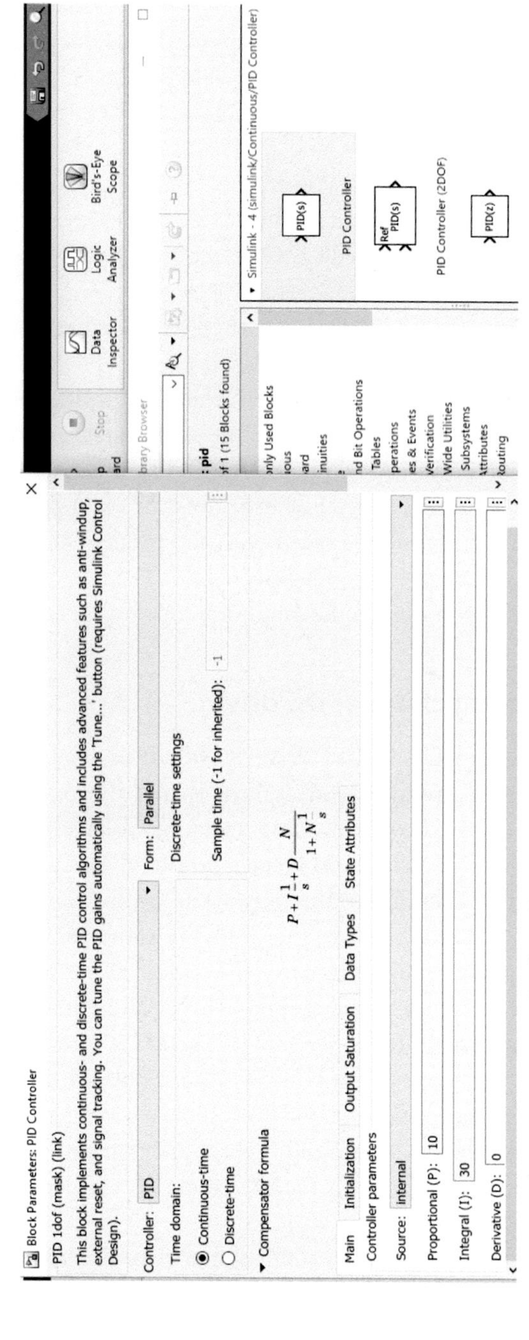

Figure 3.34 Selecting PID controller from library.

>>Libraries>>Simulink>>Commonlyusedblocks>>Gain (Figs. 3.35 and 3.36).

>>Libraries>>Simulink>>Commonlyusedblocks>>Sum (Fig. 3.37).

>>Libraries>>Simulink>>Sources>>Step (Figs. 3.38 and 3.39).

>>Libraries>>Simulink>>Commonlyusedblocks>>Busselector (Figs. 3.40 and 3.41).

>>Libraries>>Simspace>>Powerlib>>Powergui (Fig. 3.42).

>>Libraries>>Simulink>>Commonlyusedblocks>>Scope.

To see the waveform you need CRO, select the scope by using the above command (Fig. 3.43).

By dragging the selected blocks into the Simulink editor, connect all the blocks using the pointer to make it a closed circuit. The Simulink Editor provides a finished control over what we see and use inside the model.

To simulate the behavior of the structure and point of view the results, focus run-time decisions, including the sort and properties of the solver, model start and stop times, and whether to load or extra propagation data. Different combinations could be saved with the model. Simulink model of the buck converter is seen in (Fig. 3.44).

After simulation, you can see the graph in scope by clicking it (Fig. 3.45).

3.5 One-quadrant chopper DC drive

A simple chopper-fed DC motor drive is shown in Fig. 3.46. The basic principle behind DC motor speed control is that the output speed of a DC motor can be varied by controlling armature voltage for speed below and up to rated value keeping field voltage constant. The armature voltage can be controlled by controlling the duty cycle of the converter (here the converter used is a DC chopper). In Fig. 3.46, the converter output gives the DC output voltage Va required to drive the motor at the desired speed. In this diagram, the DC motor is represented by its equivalent circuit consisting of inductor La and resistor Ra in series with the ideal source back emf (Ea). The thyristor T1 is triggered by a pulse width modulated (PWM) signal to control the average motor voltage.

Theoretical waveforms illustrating the chopper operation are shown in Fig. 3.47. In this case, the average armature voltage is a direct function of the chopper duty cycle γ, that is, Vav $= \gamma$ Vd.

Two modes of operation may result when the chopper operates with DC motor load, namely, continuous armature current operation, and discontinuous armature current operation modes, as illustrated in Fig. 3.47.

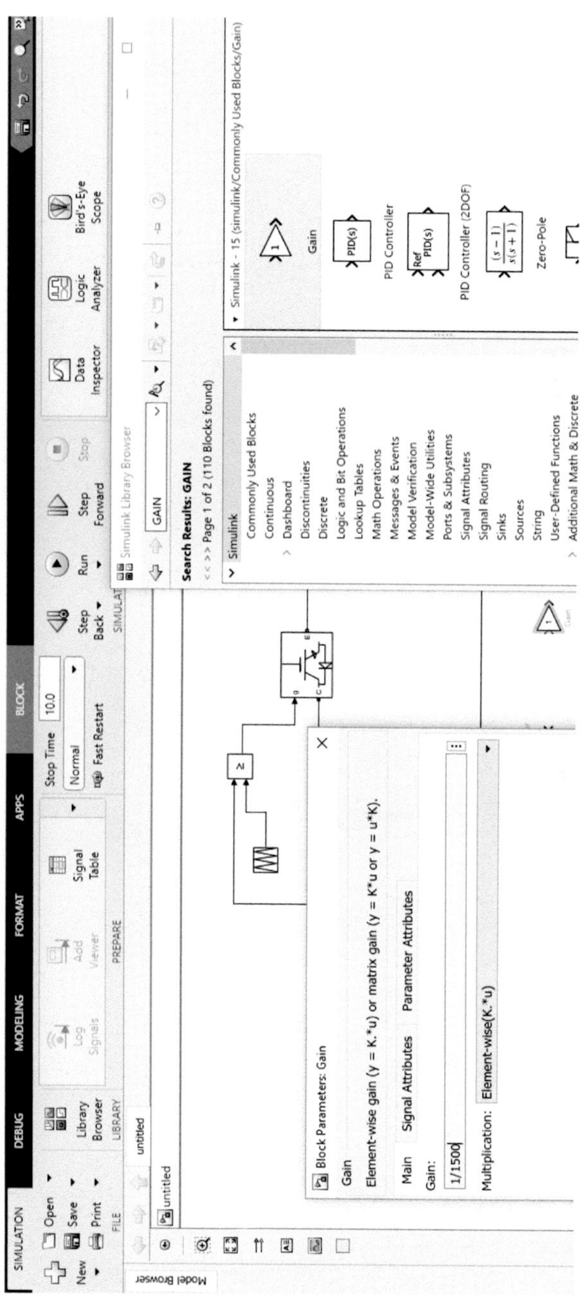

Figure 3.35 Selecting gain from library.

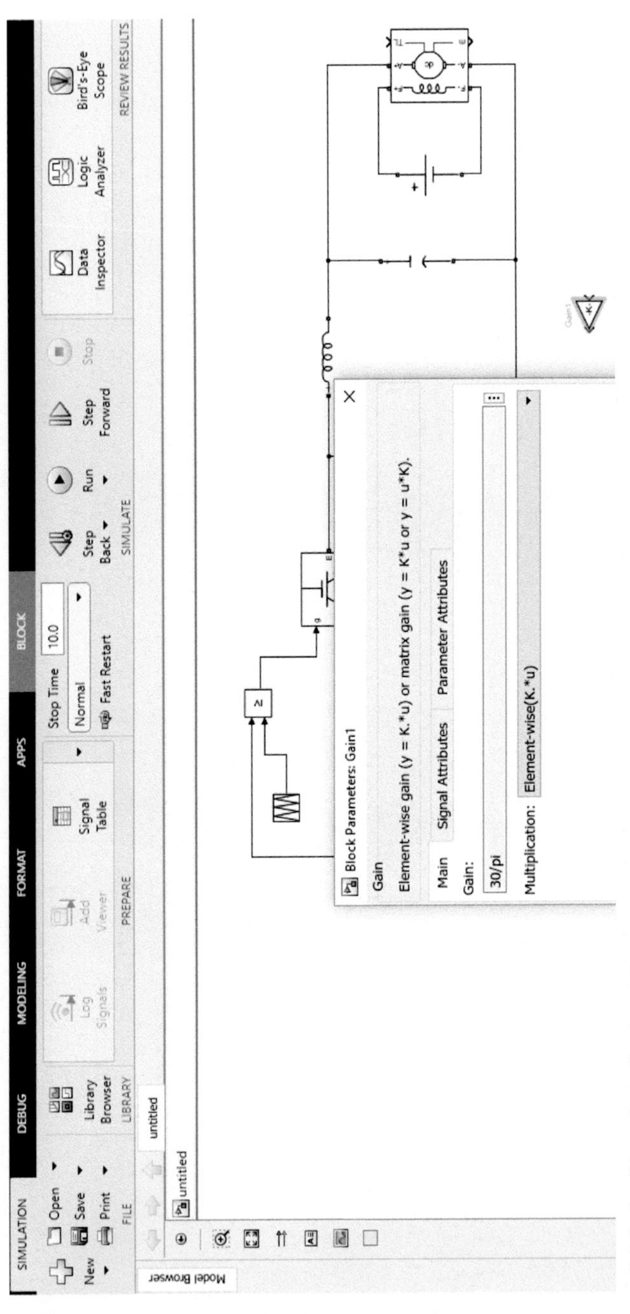

Figure 3.36 Setting value of gain.

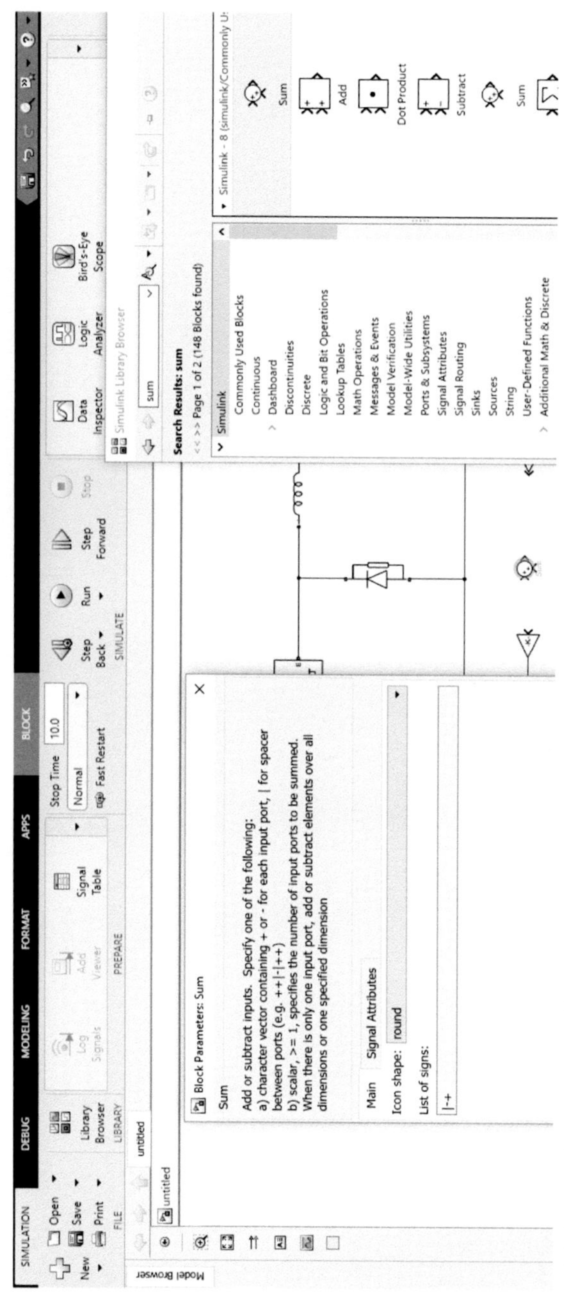

Figure 3.37 Selecting sum from library.

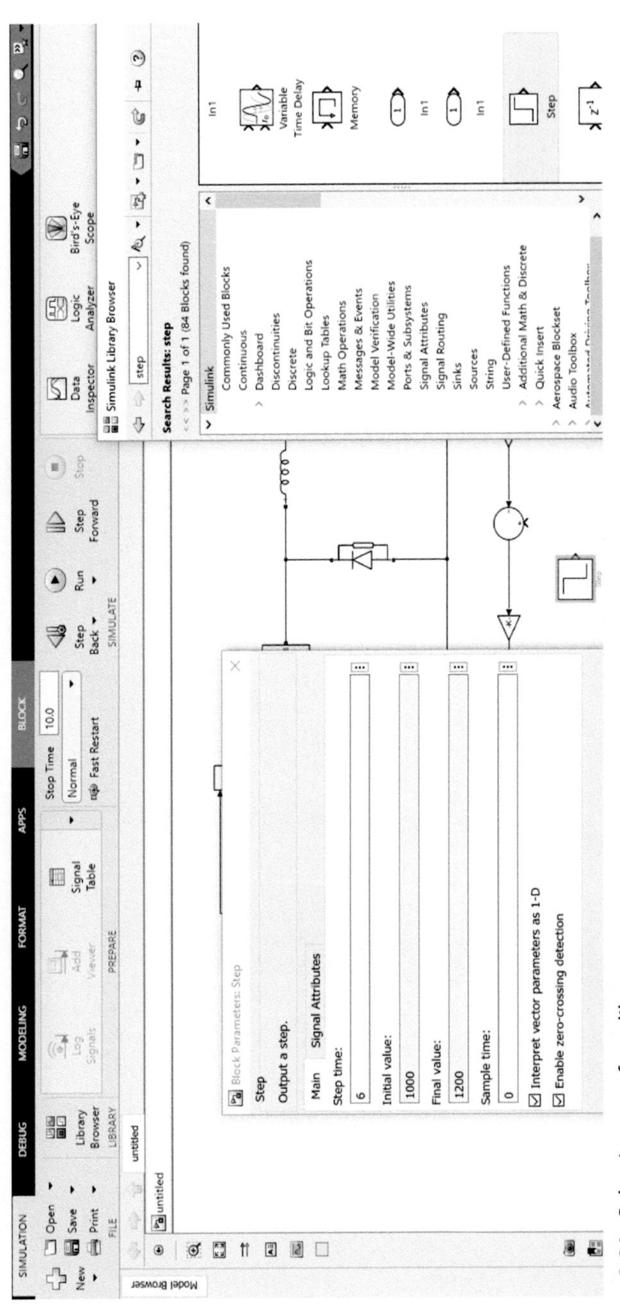

Figure 3.38 Selecting step from library.

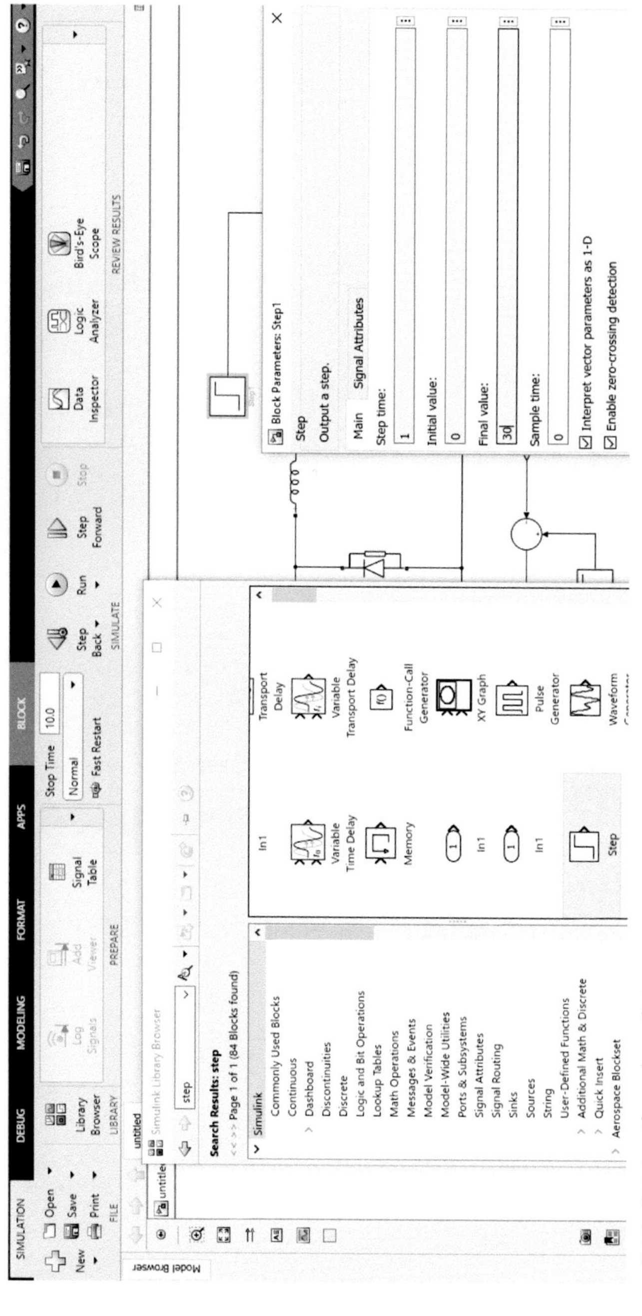

Figure 3.39 Setting the value of step.

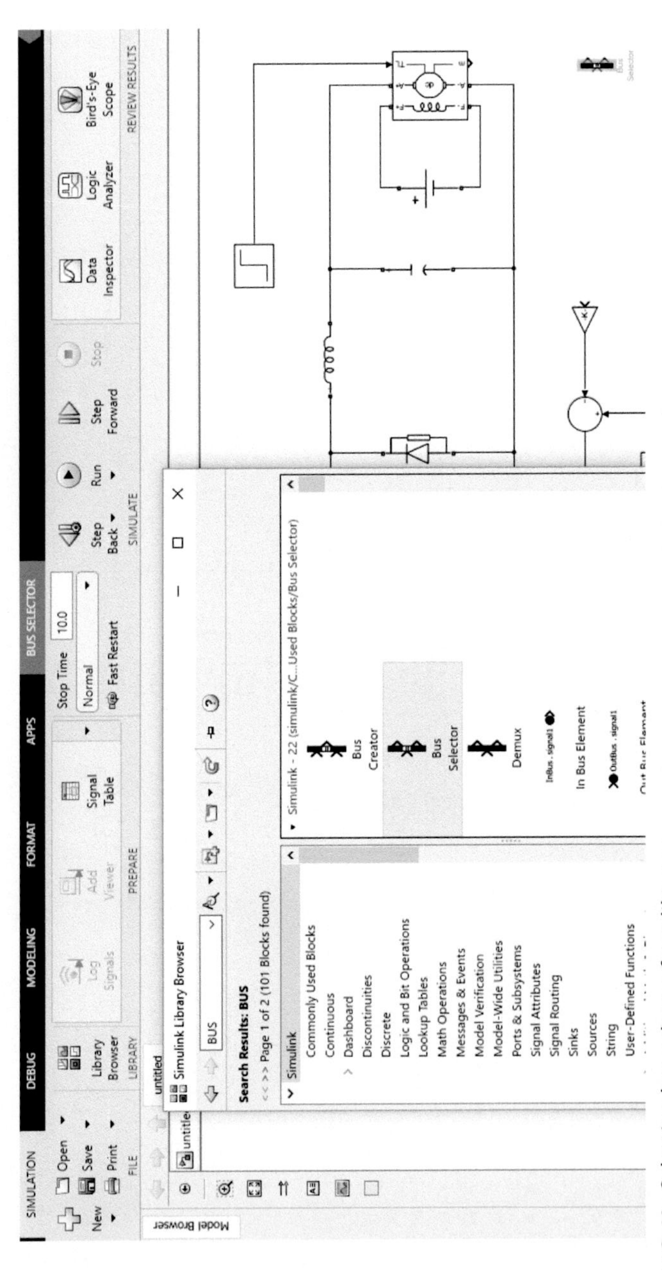

Figure 3.40 Selecting bus selector from library.

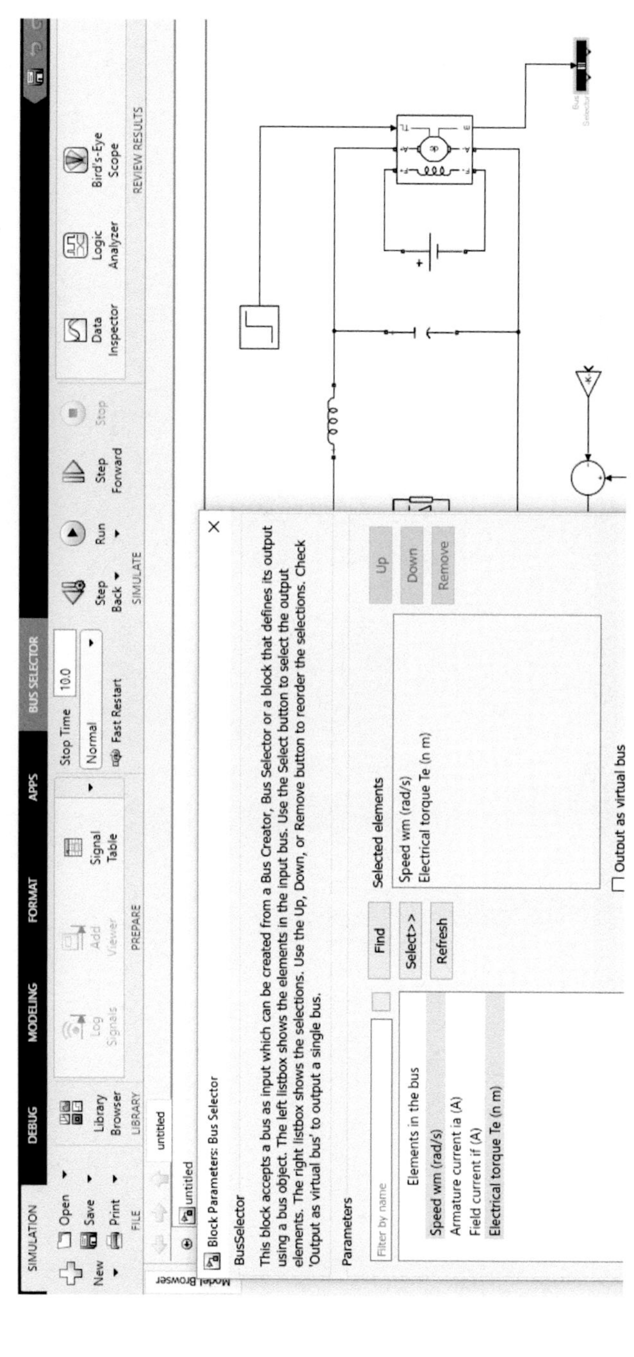

Figure 3.41 Setting parameters for bus selector.

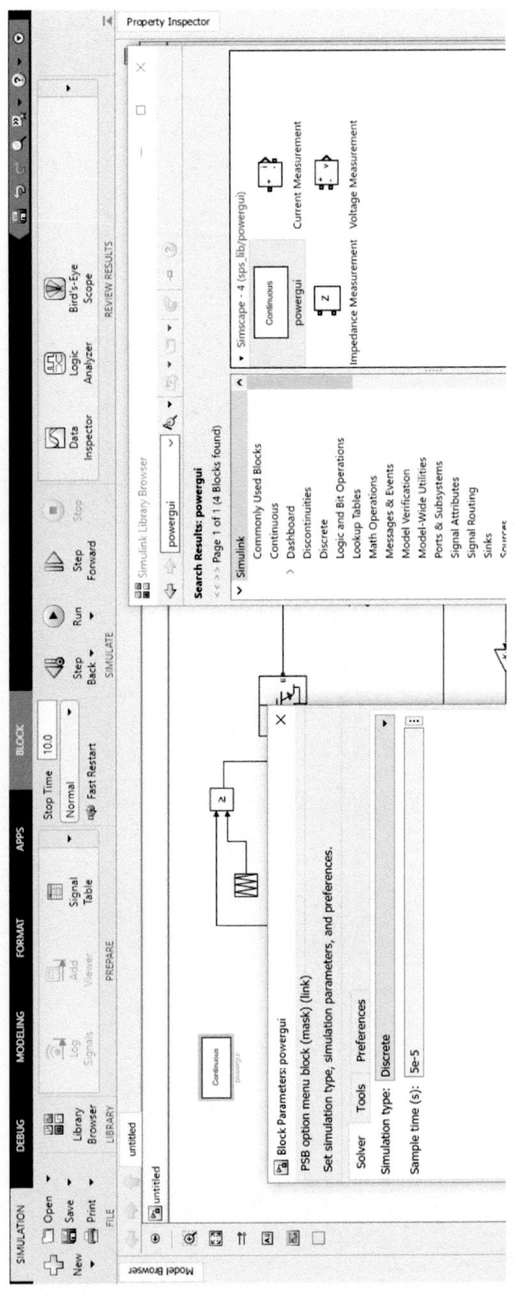

Figure 3.42 Selecting Powergui from library.

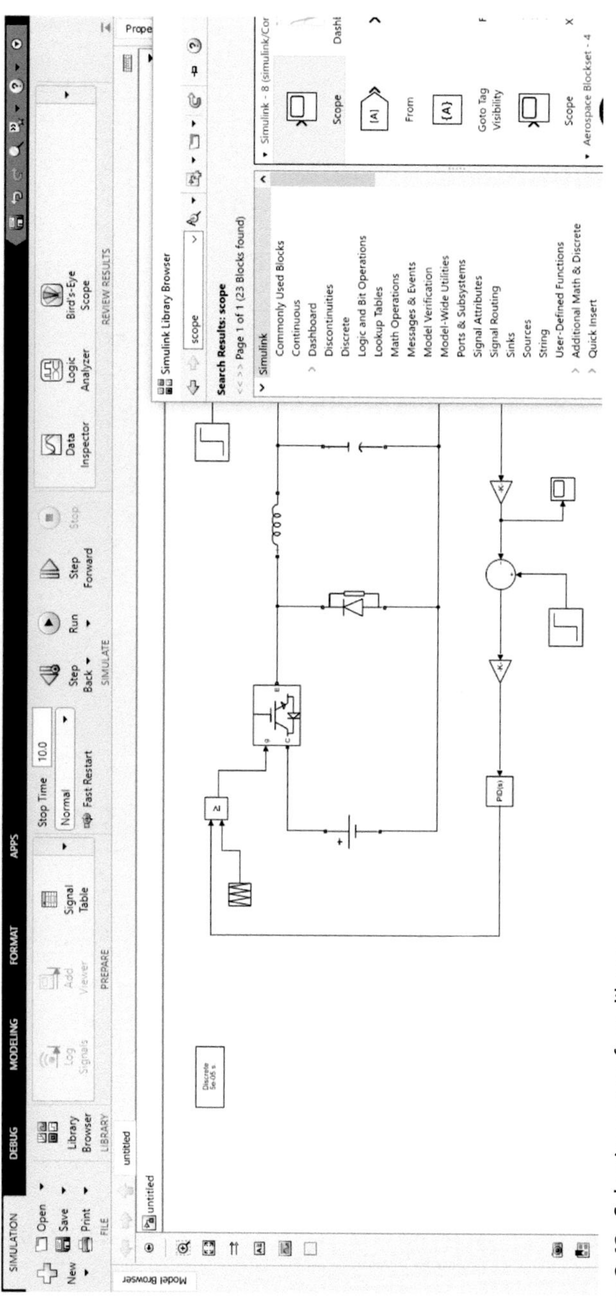

Figure 3.43 Selecting scope from library.

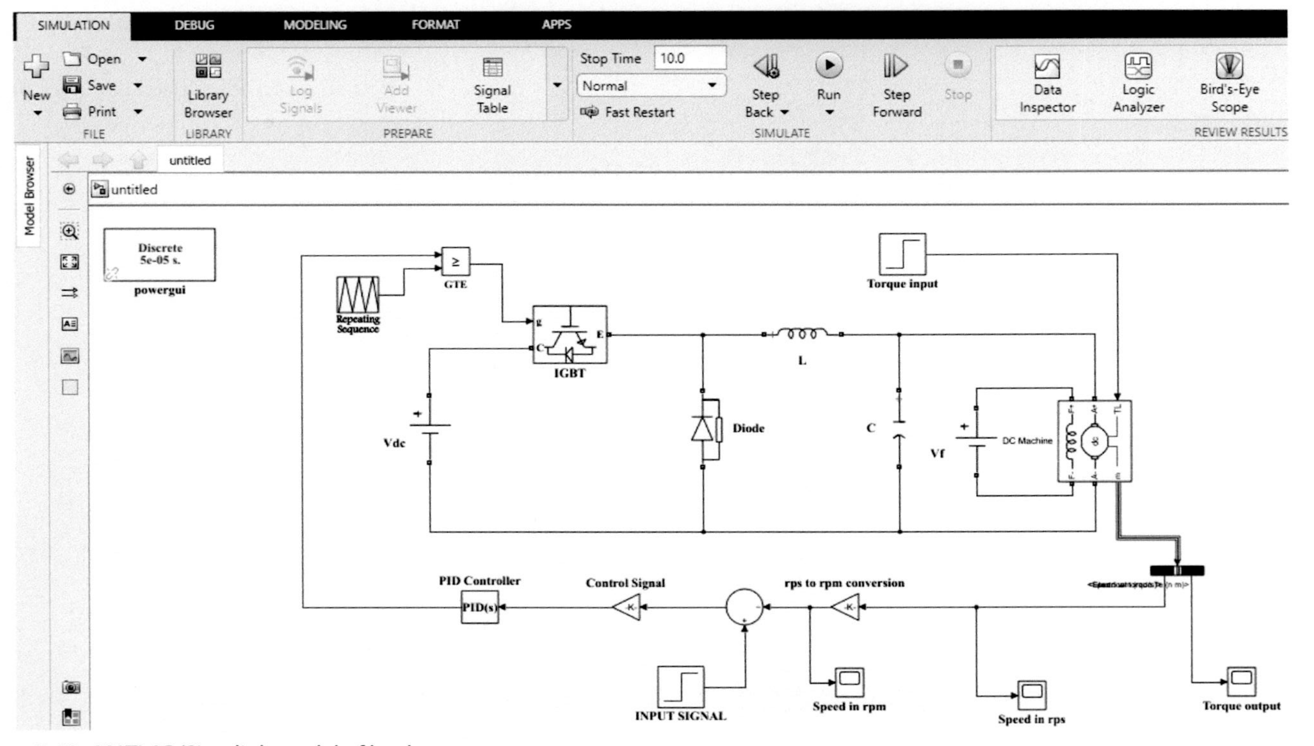

Figure 3.44 MATLAB/Simulink model of buck converter.

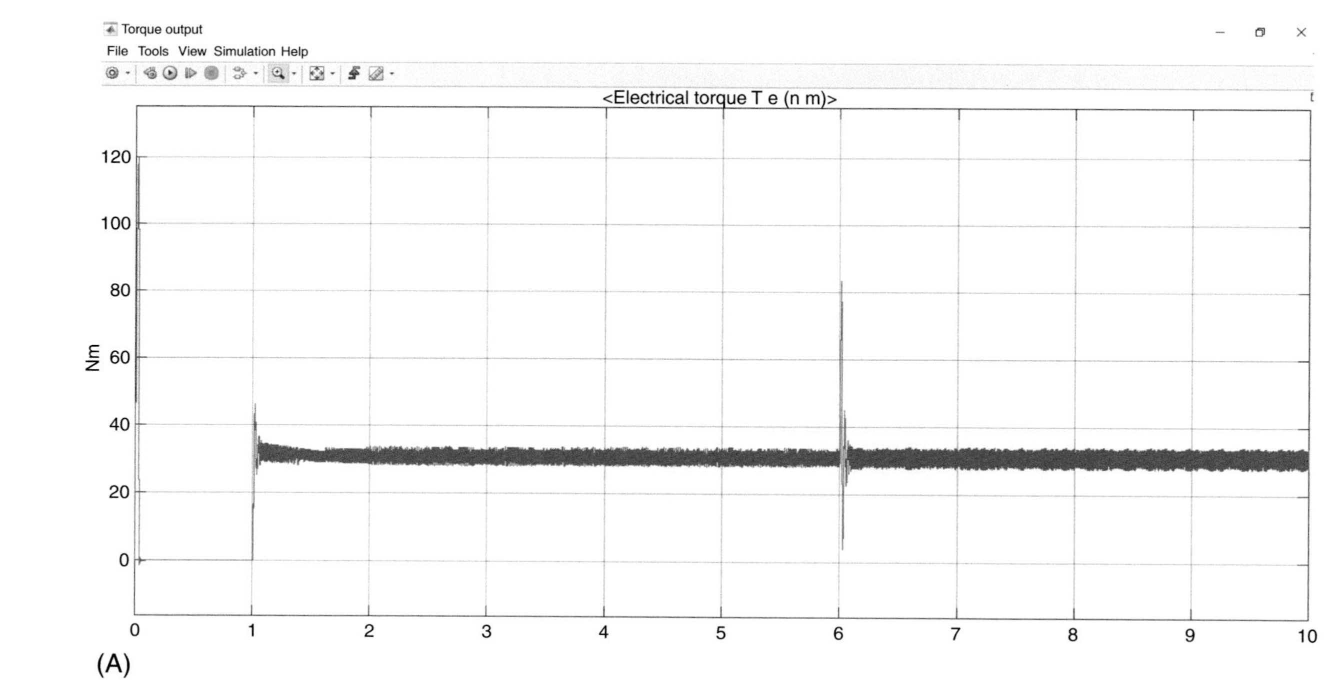

Figure 3.45 Simulation results of buck converter. (A) Torque output, (B) speed in rps, and (C) Speed in rpm.

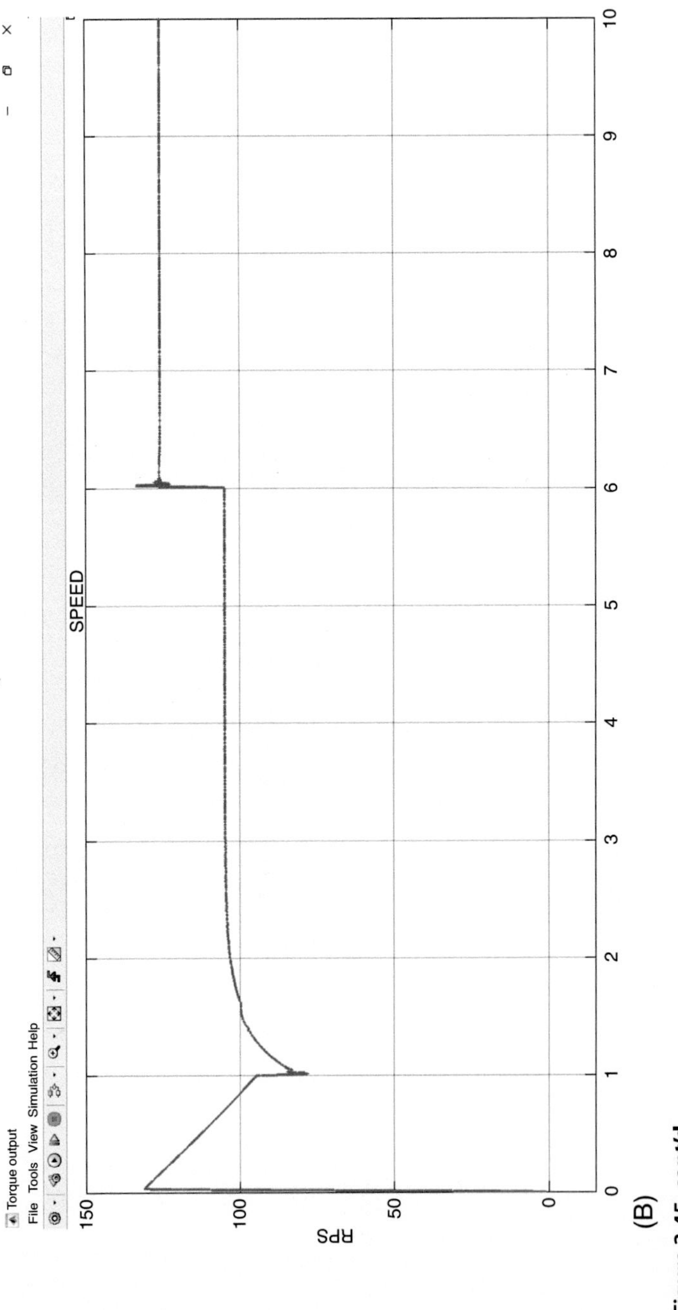

(B)

Figure 3.45, cont'd.

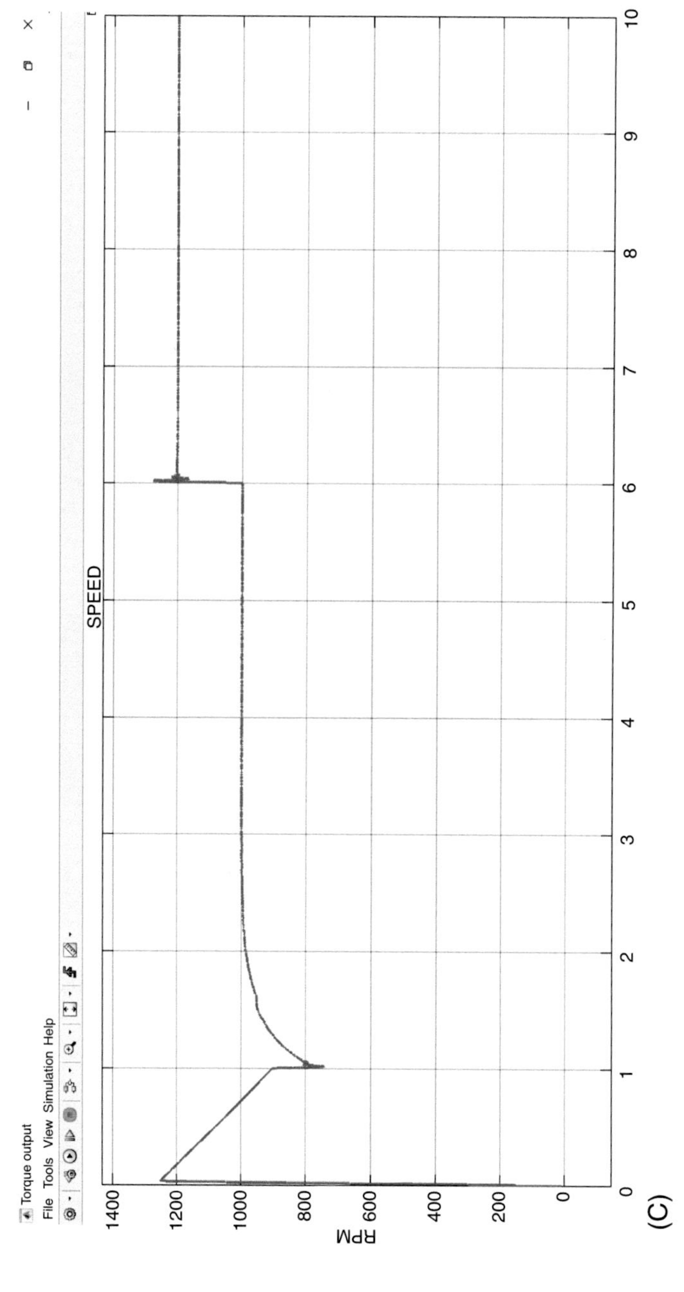

(C)

Figure 3.45, cont'd.

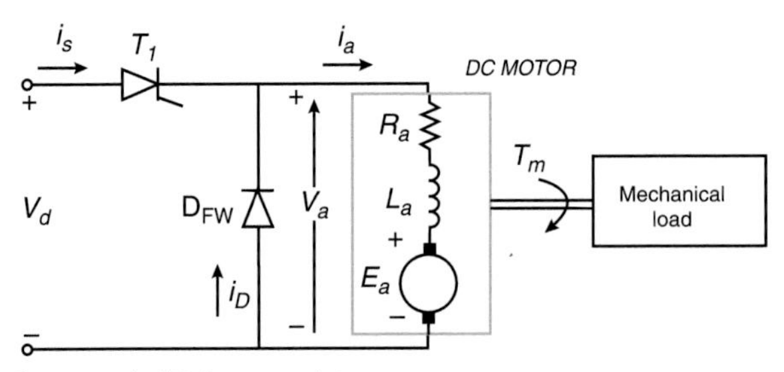

Figure 3.46 Chopper-fed DC motor drive.

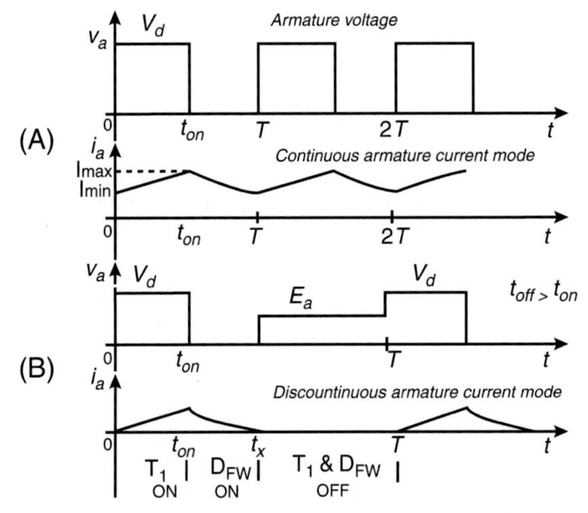

Figure 3.47 Load current and voltage waveforms with motor load: (A) continuous armature current, (B) discontinuous armature current.

In both cases, the armature voltage and current waveforms are different in shape and each has its own analytical properties.

3.5.1 Simulation of one-quadrant chopper DC drive using MATLAB

>>Libraries>>Simscape>>SpecializedPowerSystems>>Fundamentalblocks>>Electrical Sources>>DC Voltage Source (Fig. 3.48).

>>Libraries>>Simscape>>SpecializedPowerSystems>>Fundamentalblocks>> PowerElectronics>>Ideal Switch (Fig. 3.49)

>>Libraries>>Simscape>>SpecializedPowerSystems>>Fundamentalblocks>> PowerElectronics>>Diode (Fig. 3.50)

>>Libraries>>Simscape>>SpecializedPowerSystems>>Fundamentalblocks>> Elements>>Series RLC Branch

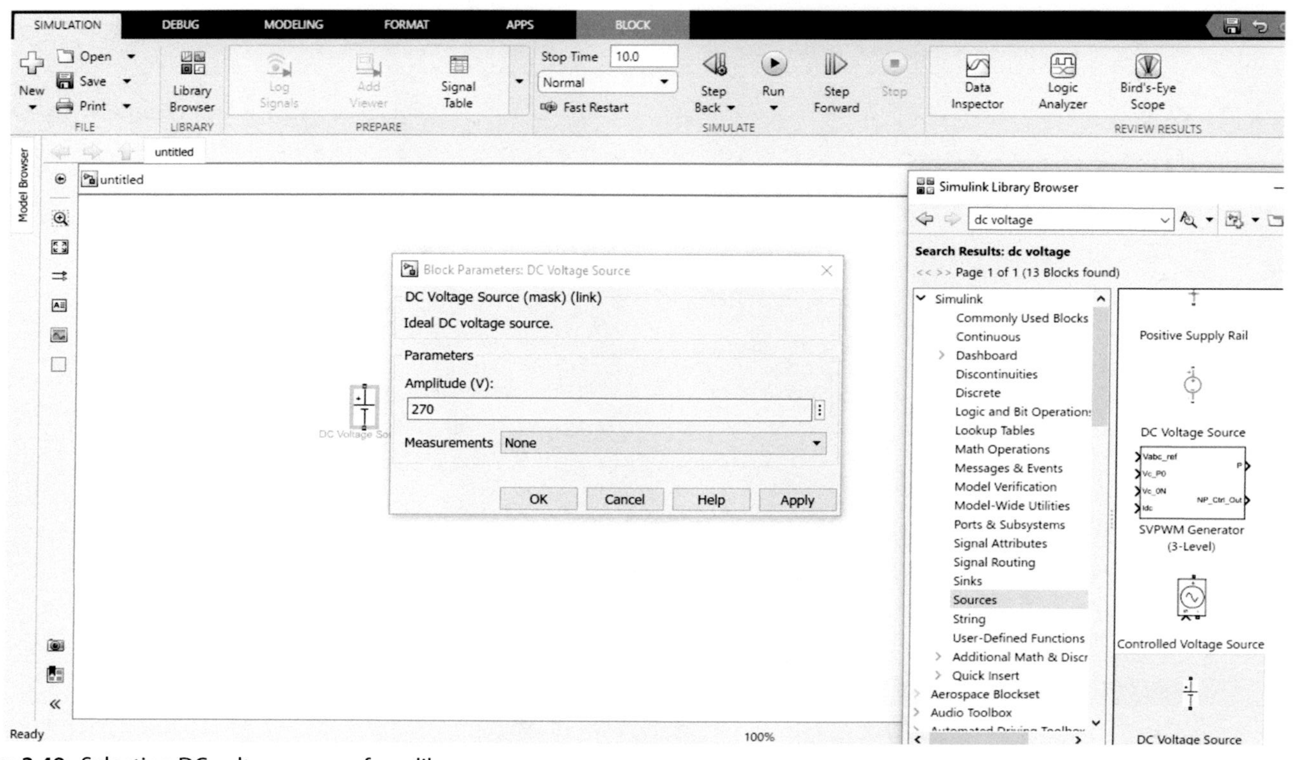

Figure 3.48 Selecting DC voltage source from library.

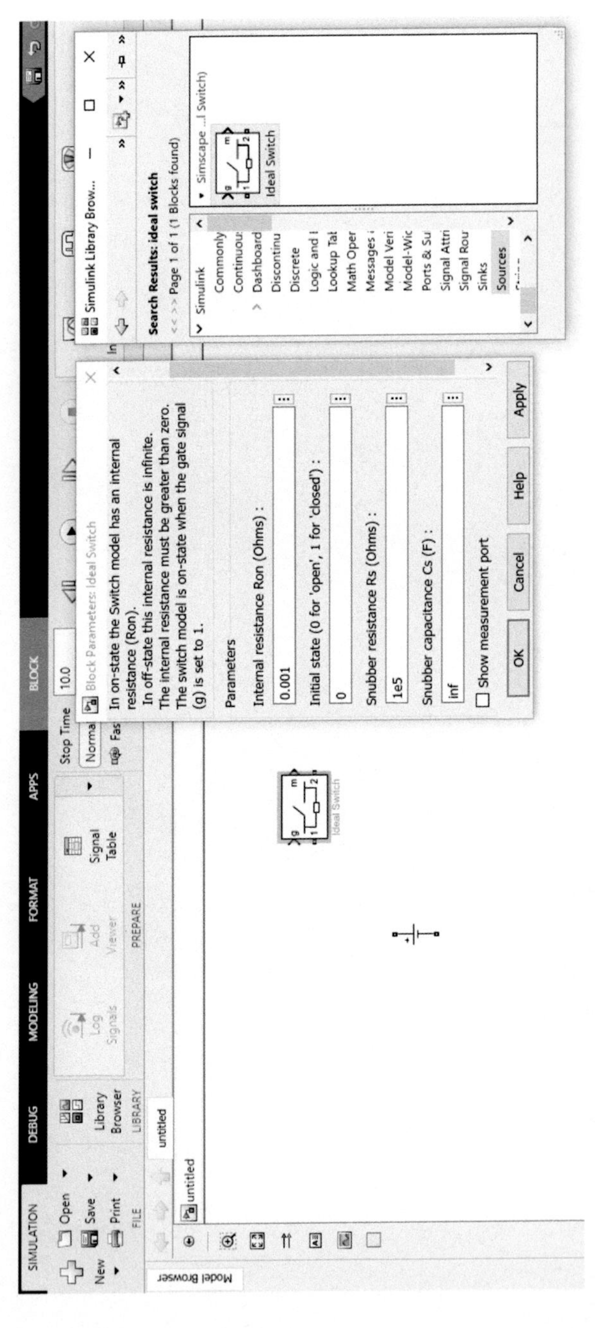

Figure 3.49 Selecting ideal switch from library.

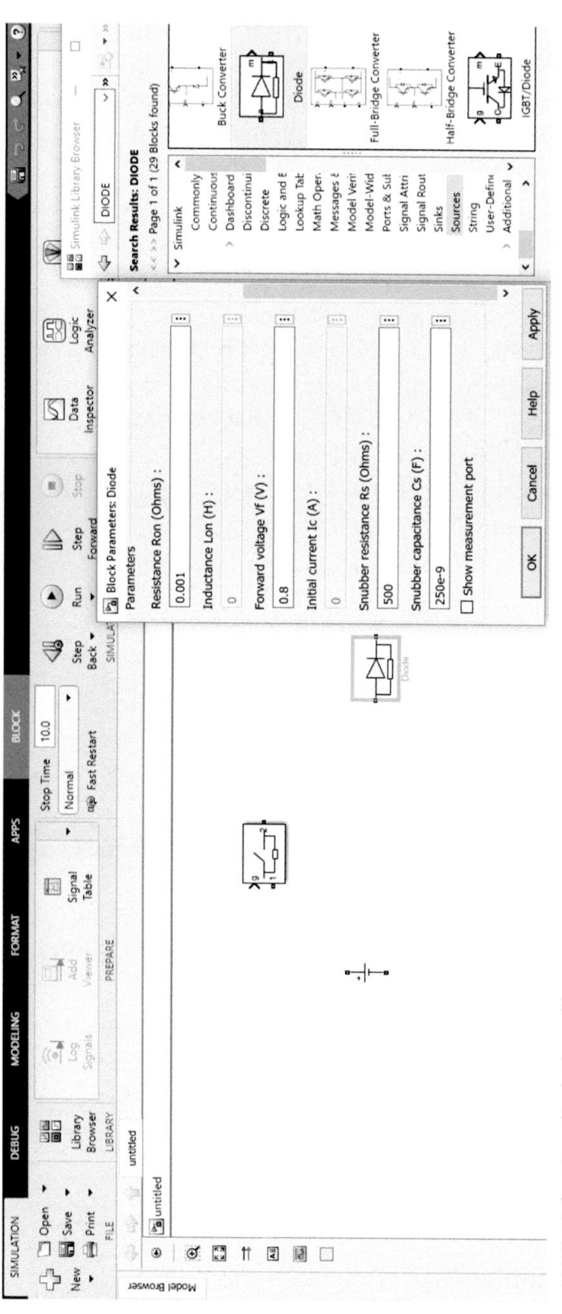

Figure 3.50 Selecting diode from library.

Select series RLC branch using the above command. Even though you do not have an RLC series branch in your circuit you have select it for the purpose of resistance, inductance, and capacitance. In MATLAB, we do not have separate R, L, and C. So you have to select series RLC branch and change that into R, L, & C by providing the required value in the parameter setting.

Now to get inductance, you have to change the branch type as L instead of RLC and set the inductance value in the parameter dialogue box (Fig. 3.51).

>>Libraries>>Simscape>>Electrical>>Specialized power system >>Fundamental Blocks>>Machines>>DC Machines (Fig. 3.52)

>>Libraries>>Simscape>>SpecializedPowerSystems>>Fundamentalblocks>>Electrical Sources>>DC Voltage Source (Fig. 3.53).

>>Libraries>>Simulink>>Sources>>Step (Fig. 3.54).

>>Libraries>>Simulink>>Quickinsert>>Logicandbitoperations>> Relational Operator (Fig. 3.55).

>>Libraries>>Simulink>>Sources>>Repeating Sequence (Fig. 3.56).

>>Libraries>>Simulink>>Commonlyusedblocks>>Gain (Figs. 3.57 and 3.58).

>>Libraries>>Simulink>>Commonlyusedblocks>>Busselector (Fig. 3.59).

>>Libraries>>Simulink>>Commonlyusedblocks>>Scope (Fig. 3.60).

To see the waveform you need CRO, select the scope by using the above command.

The Simulink model of one-quadrant chopper 5 HP DC is given in Fig. 3.61. Through the above Simulink model will get the waveforms as given in Fig. 3.62.

Simulation of one-quadrant chopper 5 HP DC drive

To simulate one quadrant chopper, in MATLAB separate block is available as one-quadrant chopper DC drive (Fig. 3.63).

The one-quadrant chopper DC drive block uses these blocks from the electric drives/fundamental drive blocks library:

- Speed controller (DC).
- Regulation switch.
- Current controller (DC).
- Chopper.

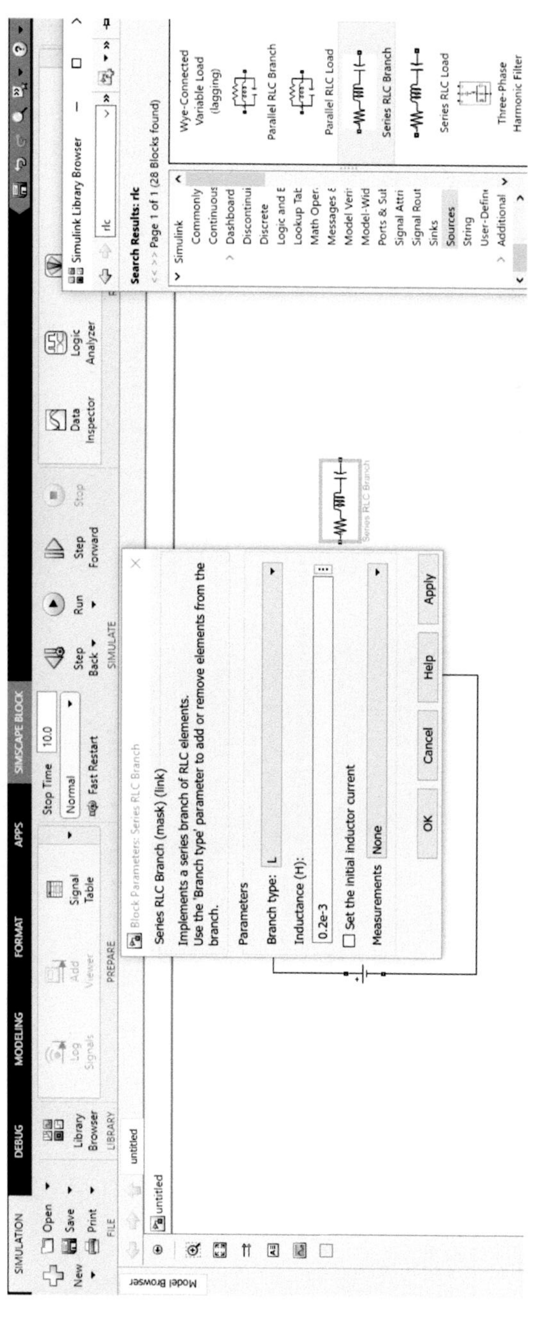

Figure 3.51 Selecting series RLC branch from library.

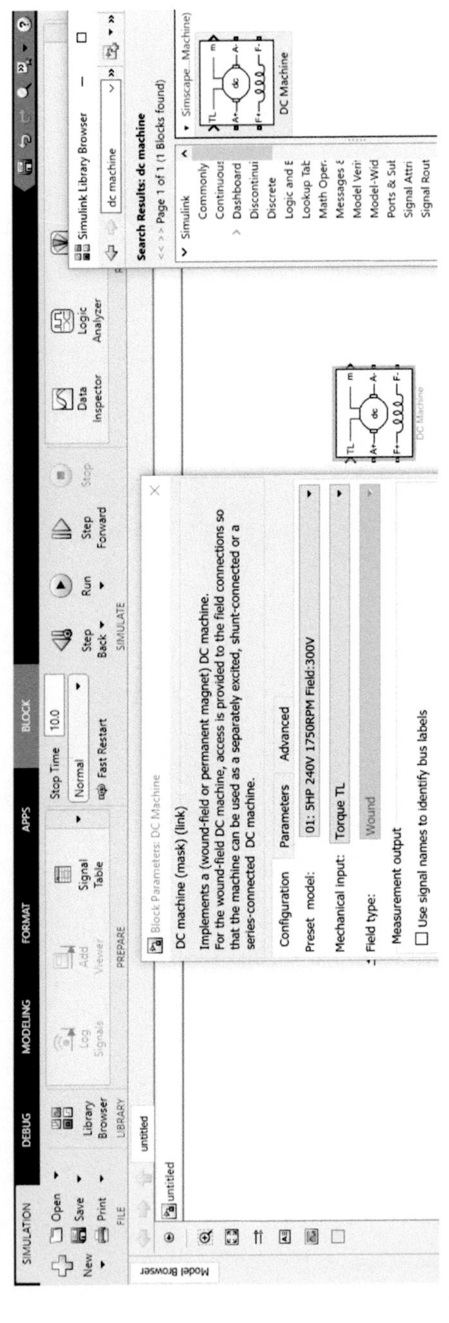

Figure 3.52 Selecting DC machines from library.

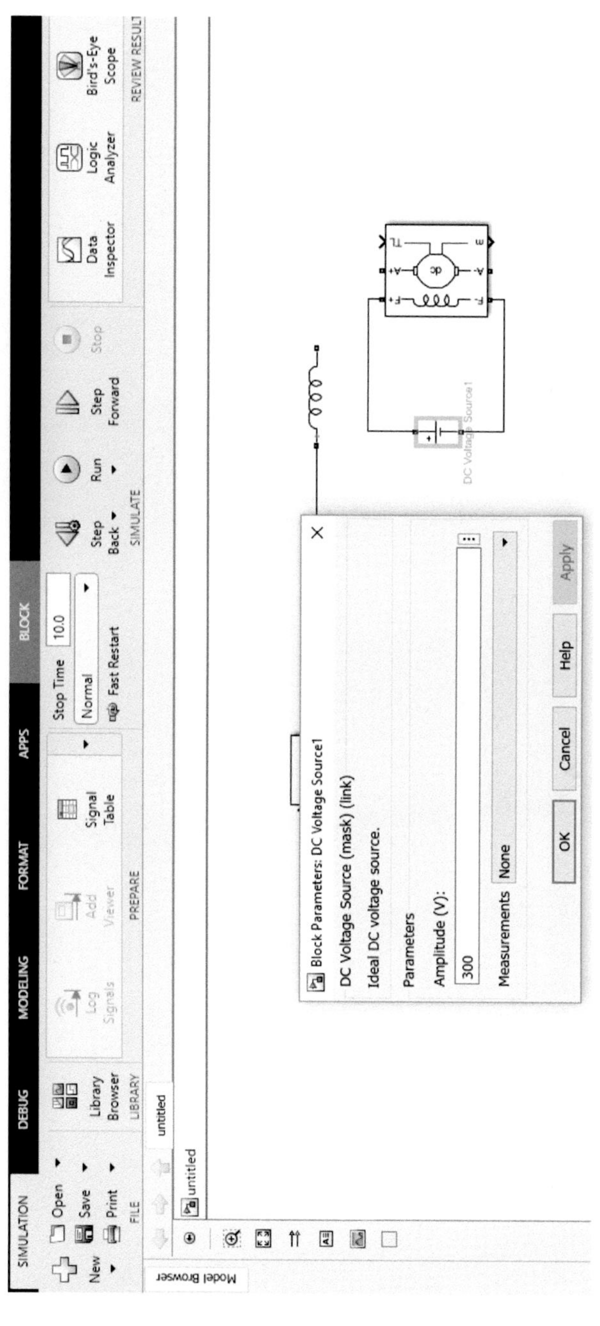

Figure 3.53 Selecting DC voltage source from library.

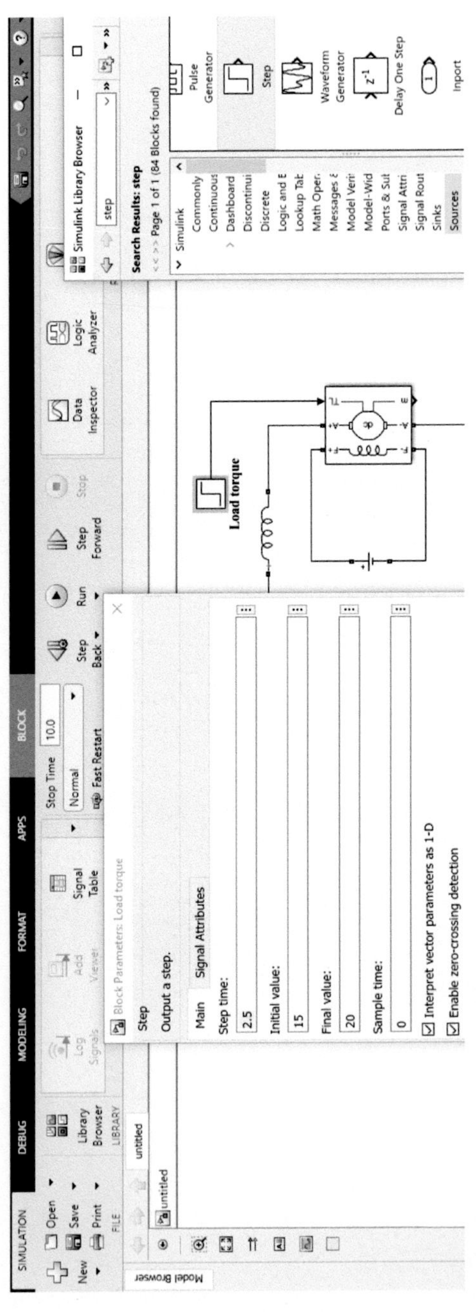

Figure 3.54 Selecting step from library.

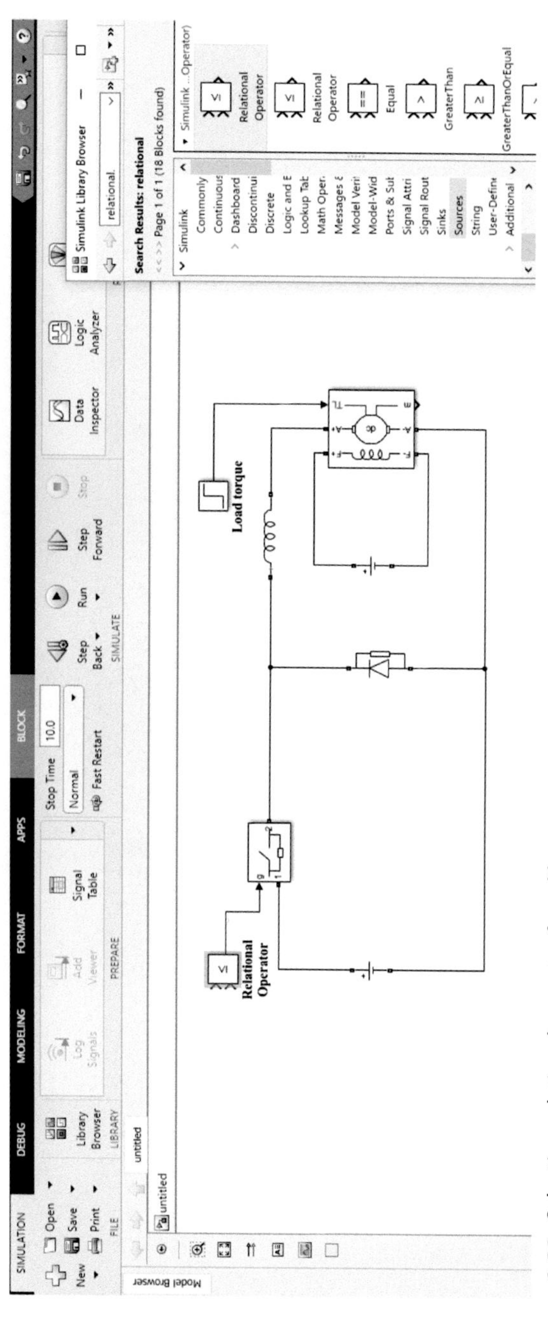

Figure 3.55 Selecting relational operator from library.

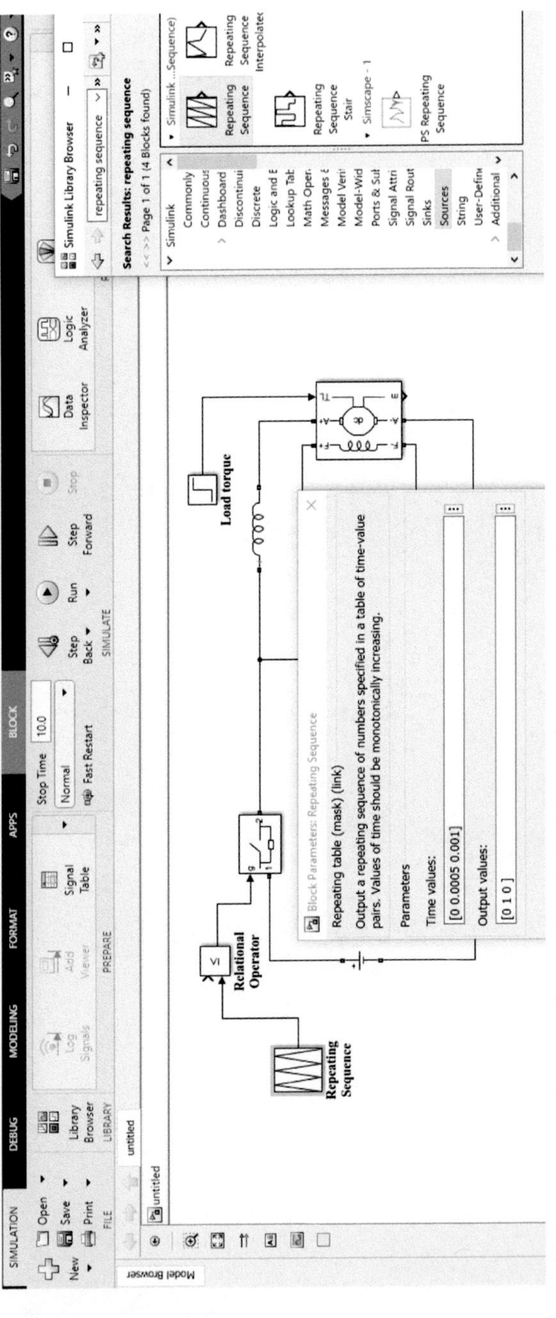

Figure 3.56 Selecting repeating sequence from library.

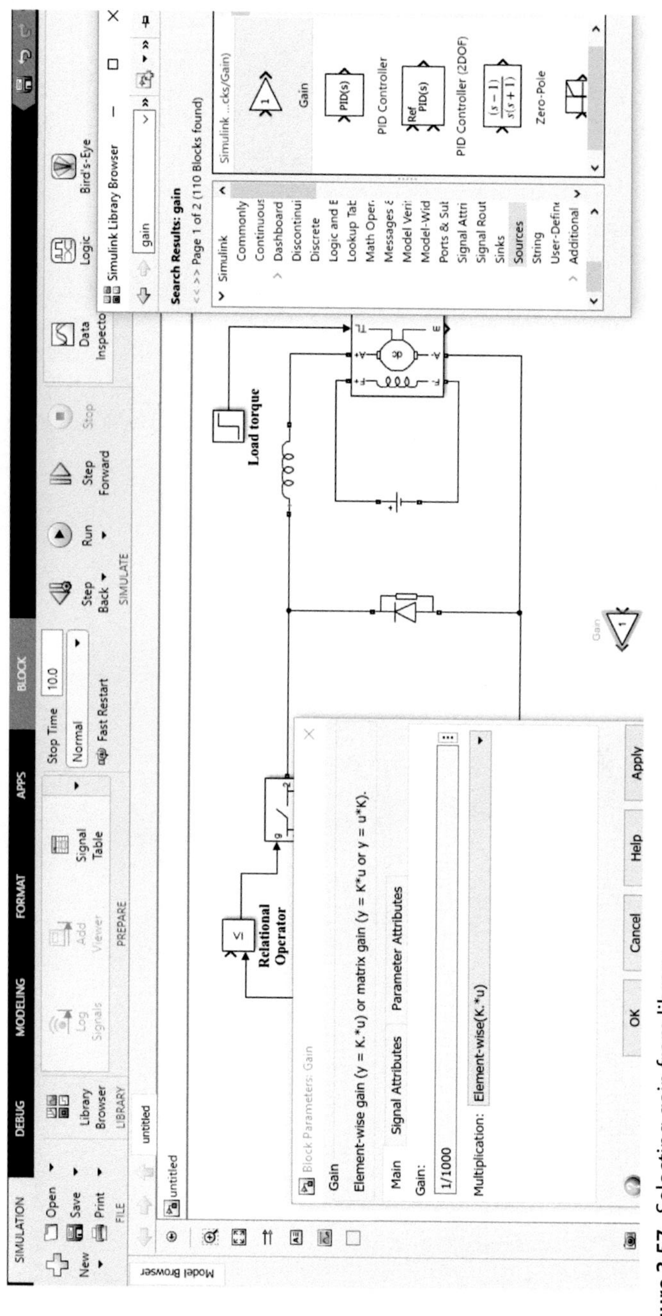

Figure 3.57 Selecting gain from library.

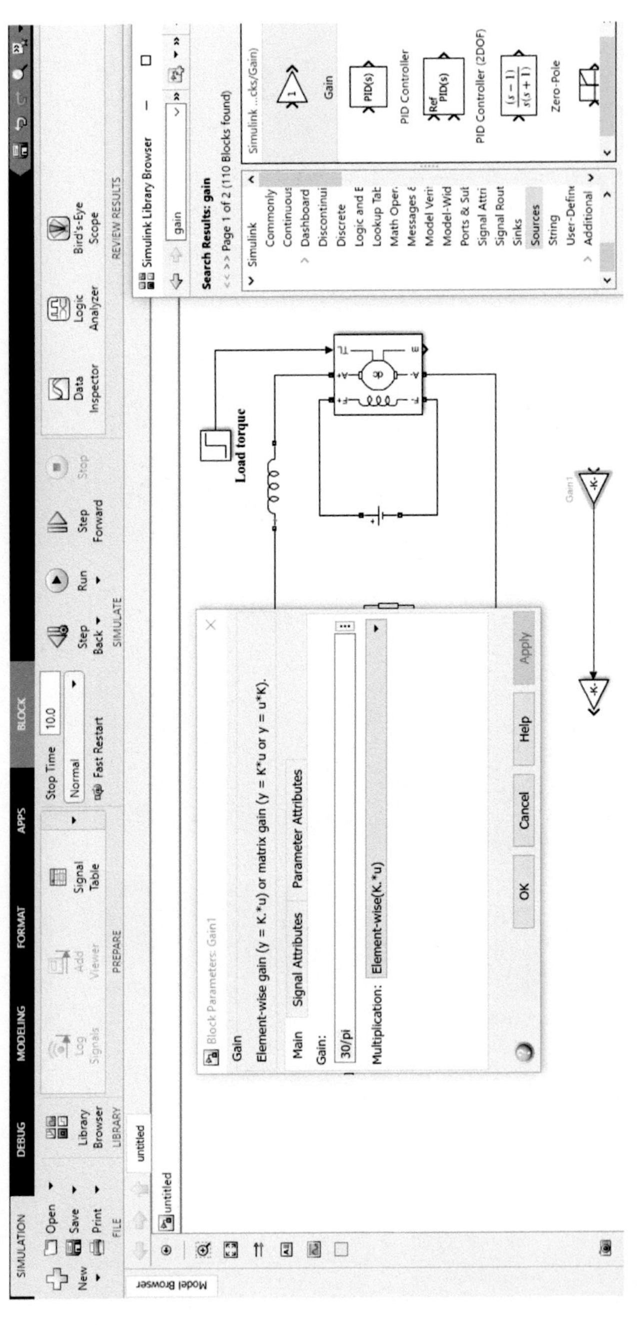

Figure 3.58 Setting value for gain.

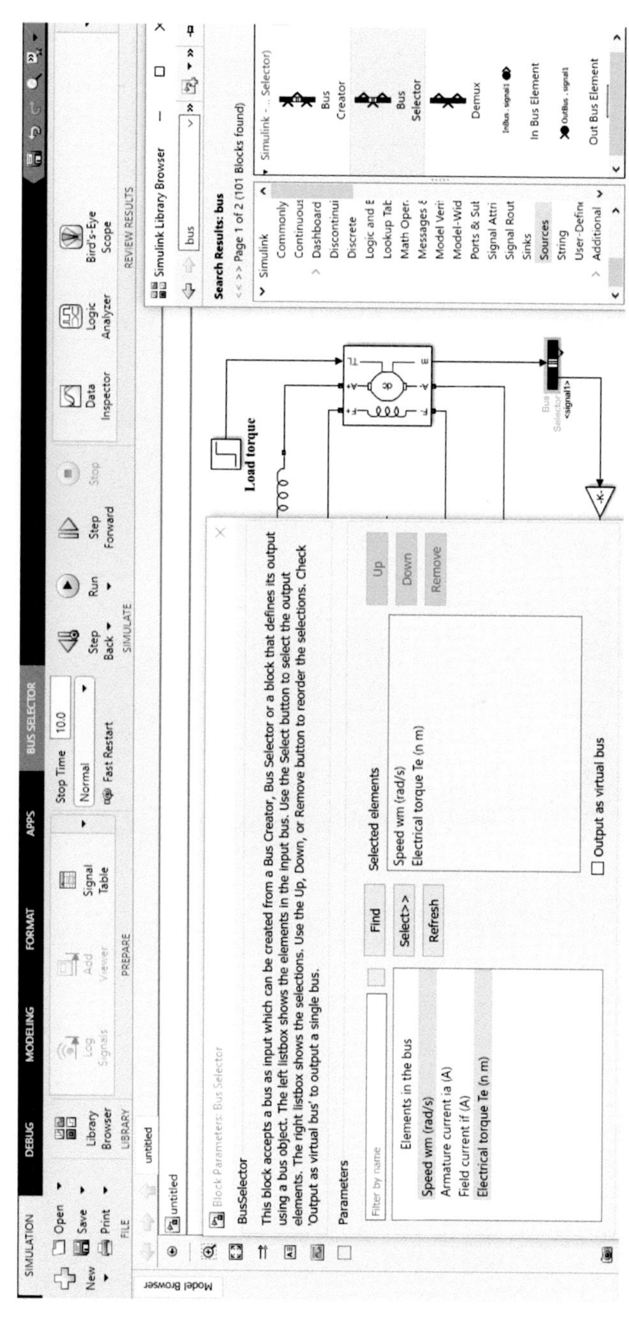

Figure 3.59 Selecting bus selector from library.

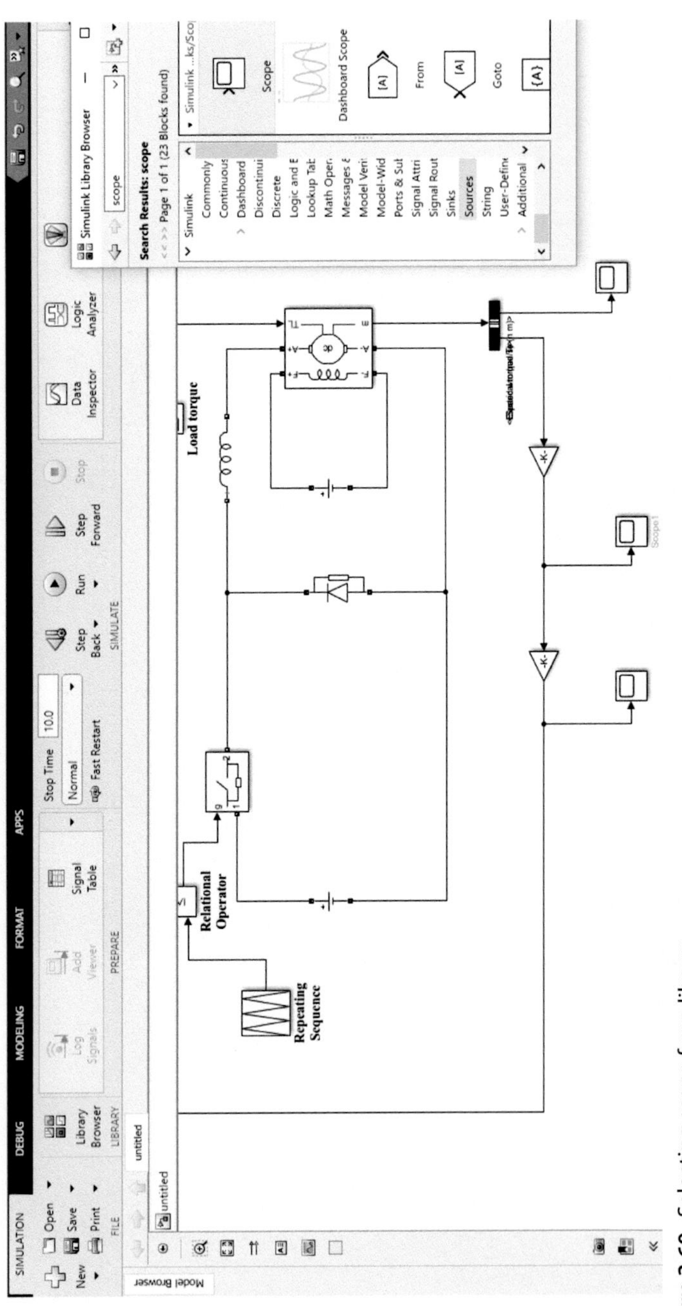

Figure 3.60 Selecting scope from library.

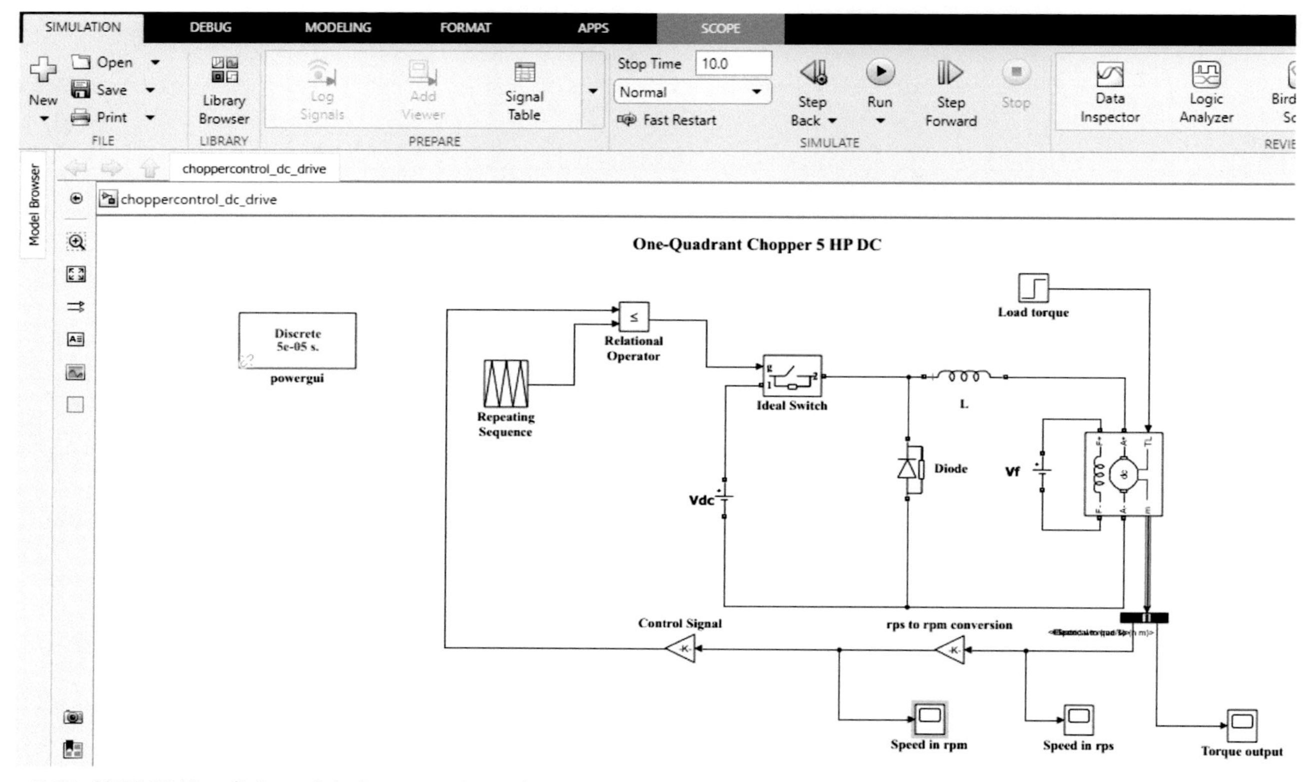

Figure 3.61 MATLAB/Simulink model of one-quadrant chopper 5 HP DC.

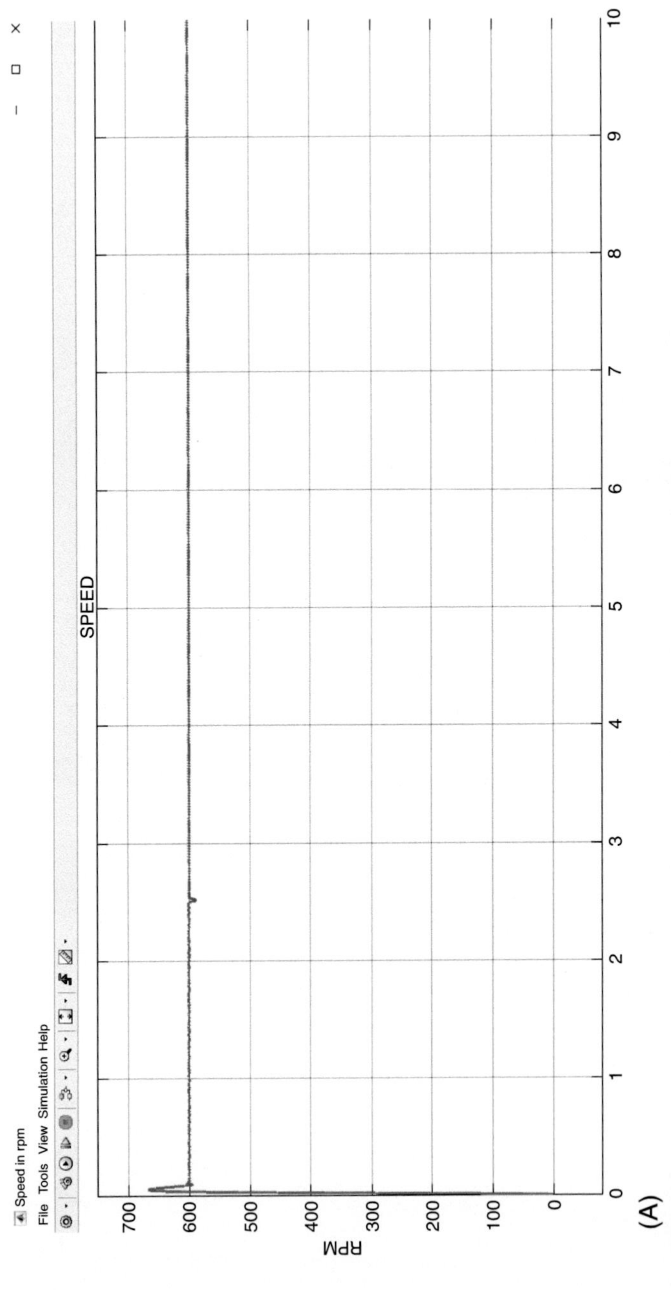

Figure 3.62 Simulation results of one-quadrant chopper. (A) Speed in rpm, (B) speed in rps, and (C) torque output.

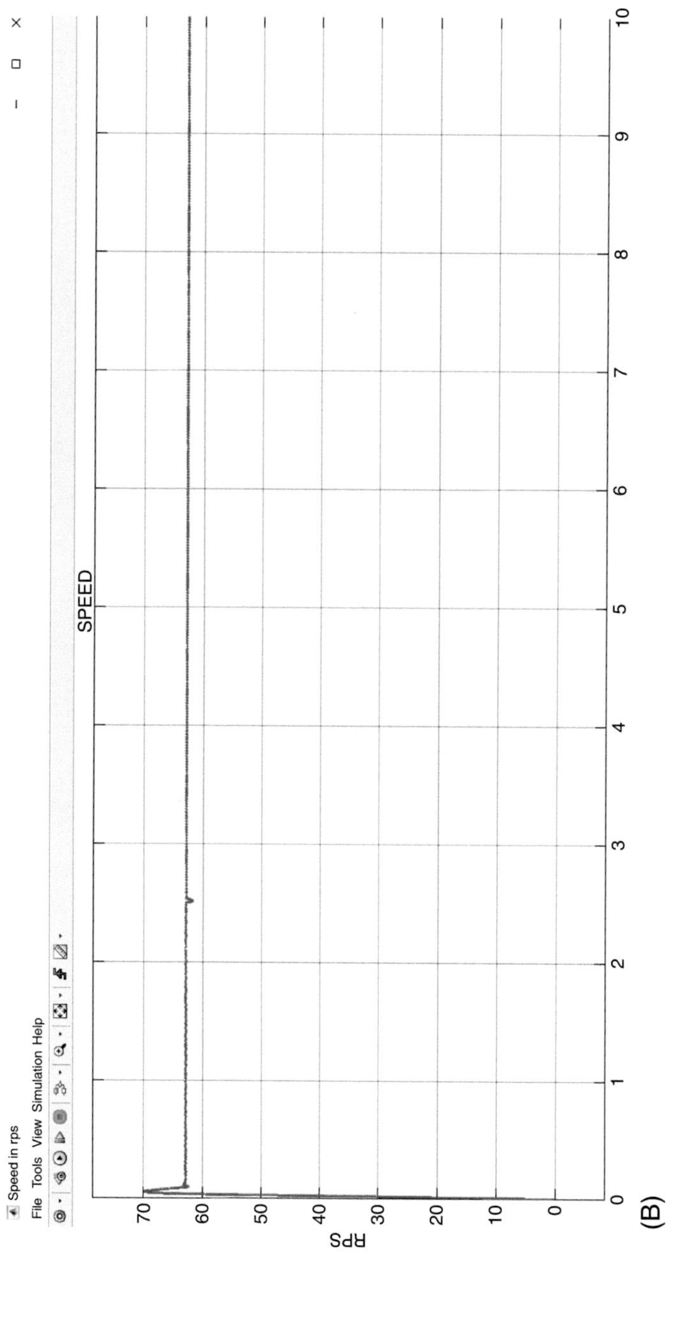

(B)

Figure 3.62, cont'd.

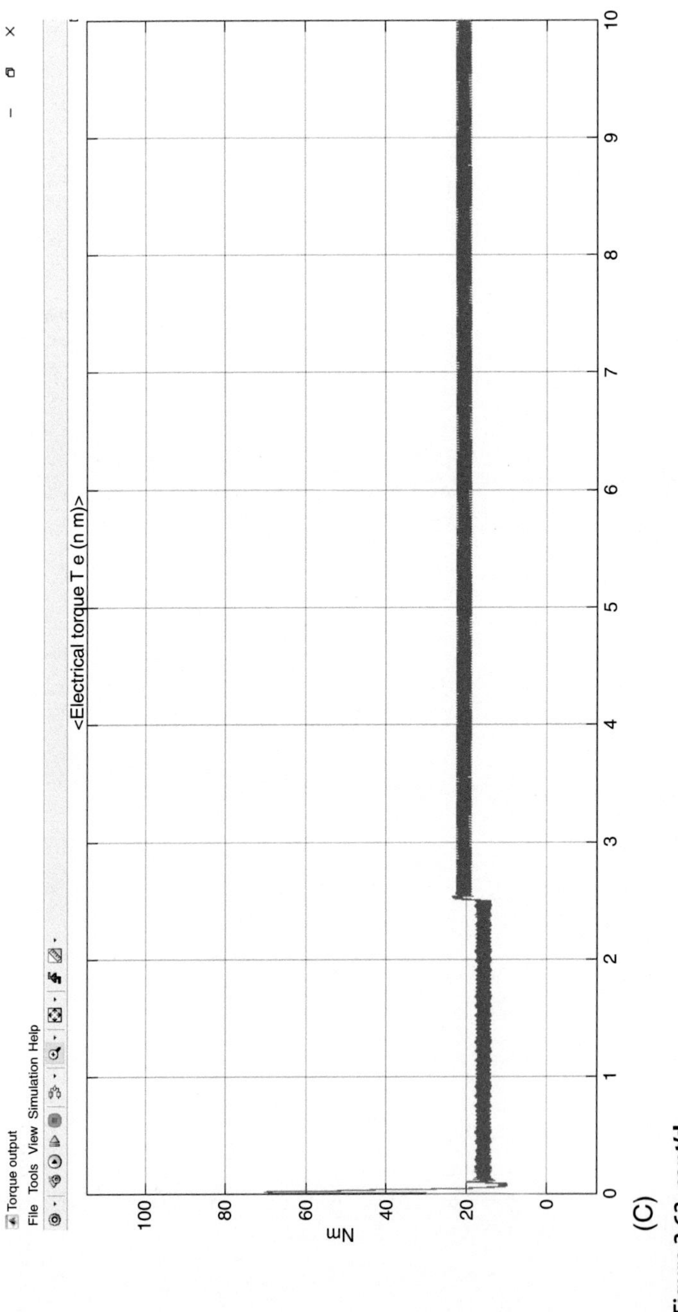

(C)

Figure 3.62, cont'd.

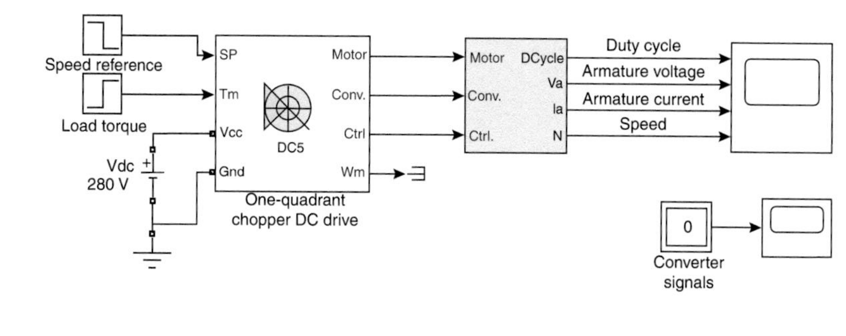

Figure 3.63 MATLAB/Simulink model of one-quadrant chopper 5 HP DC.

The machine is separately excited with a constant DC field voltage source. There is thus no field voltage control. By default, the field current is set to its steady-state value when a simulation is started.

The armature voltage is provided by an IGBT buck converter controlled by two PI regulators. The buck converter is fed by a constant DC voltage source. Armature current oscillations are reduced by a smoothing inductance connected in series with the armature circuit.

The model is discrete. Good simulation results have been obtained with a 1-μs time step (Fig. 3.64). In order to simulate a digital controller device, the control system has two different sampling times:

- The speed controller sampling time.
- The current controller sampling time.

The speed controller sampling time has to be a multiple of the current sampling time. The latter sampling time has to be a multiple of the simulation time step.

3.6 One-quadrant chopper DC drive with hysteresis current control

This circuit is based on the DC5 block of specialized power systems. It models a one-quadrant chopper (buck converter) drive for a 5 HP DC motor.

The 5 HP DC motor is separately excited with a constant 150 V DC field voltage source. The armature voltage is provided by an IGBT buck converter controlled by two regulators. The buck converter is fed by a 280 V DC voltage source.

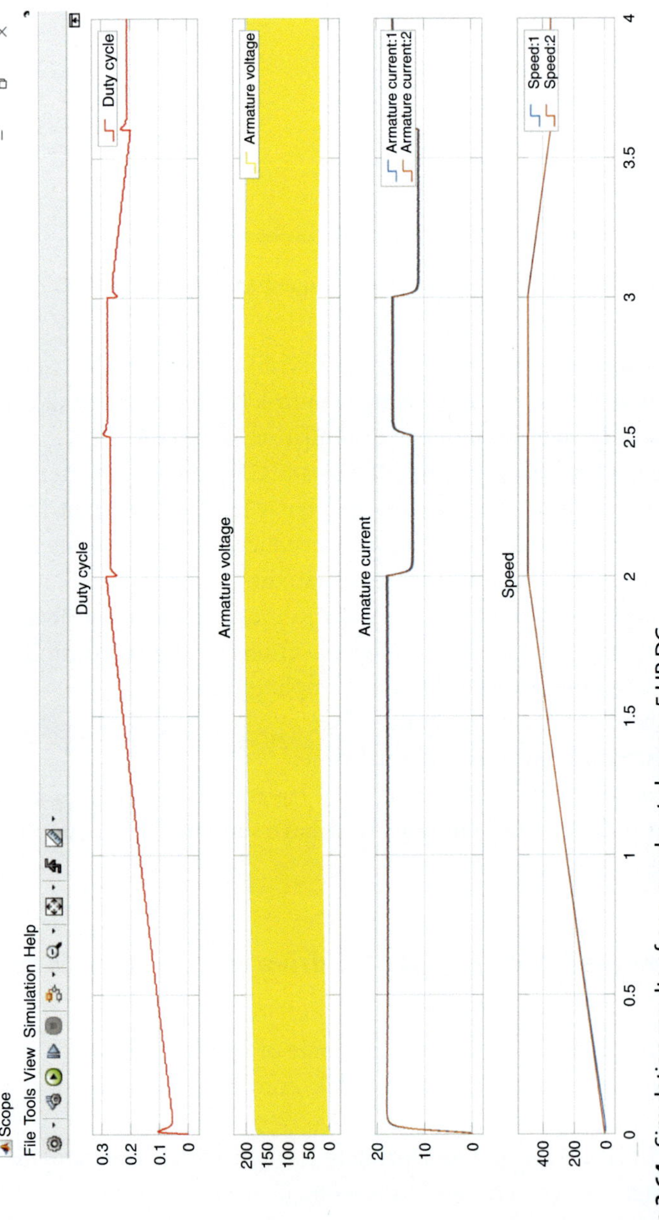

Figure 3.64 Simulation results of one-quadrant chopper 5 HP DC.

The first regulator is a PI speed regulator, followed by a hysteresis current regulator. The speed regulator outputs the armature current reference (in p.u.) used by the current controller in order to obtain the electromagnetic torque needed to reach the desired speed. The speed reference change rate follows acceleration and deceleration ramps in order to avoid sudden reference changes that could cause armature over-current and destabilize the system. The current regulator controls the armature current by delivering the correct pulses to the IGBT device in order to keep the armature current inside a user-defined hysteresis band. The switching frequency of the IGBT device is limited by the motor inductance and an external inductance placed in series with the armature circuit.

>>Libraries>>Simscape>>SpecializedPowerSystems>>Electrical-Drives>>DCdrives>>onequadrantchopperdcdrive (Fig. 3.65).

>>Libraries>>Simulink>>Sources>>Step (Fig. 3.66).

Now subsystem need to include with the help of library.

>>Libraries>>Simulink>>commonlyusedblocks>>Subsystems (Fig. 3.67A).

Inside the subsystem, need to include bus selector, Mux, Input, and output ports. Three input ports are included from the library. The subsystem input ports are connected to the one quadrant chopper DC drive block. So that we can select the parameters for the bus selectors.

>>Libraries>>Simulink>>commonlyusedblocks>>In1 (Fig. 3.67B).

Two bus selectors are included from the library using>> Libraries>>Simulink>>commonlyusedblocks>>bus selectors (Fig. 3.68).

Four output ports are included in the subsystem from the library. >>Libraries>>Simulink>>commonlyusedblocks>>Out1 (Fig. 3.69).

Include two multiplers in the subsystem using >>Libraries>> Simulink>>commonlyusedblocks>>Mux (Fig. 3.70).

The bus parameters are selected as in the given Figs. 3.71 and 3.72.

Now do the connection as per the given Fig. 3.73 with the help of wire.

To display all the output, we need a scope. So select the scope by the command >>Libraries>>Simulink>>Commonlyusedblocks>>Scope (Fig. 3.74).

After including all the elements from the library. Do the connection as per the diagram given below (Fig. 3.75).

After giving run command we will get the simulation results of the circuit (Fig. 3.76).

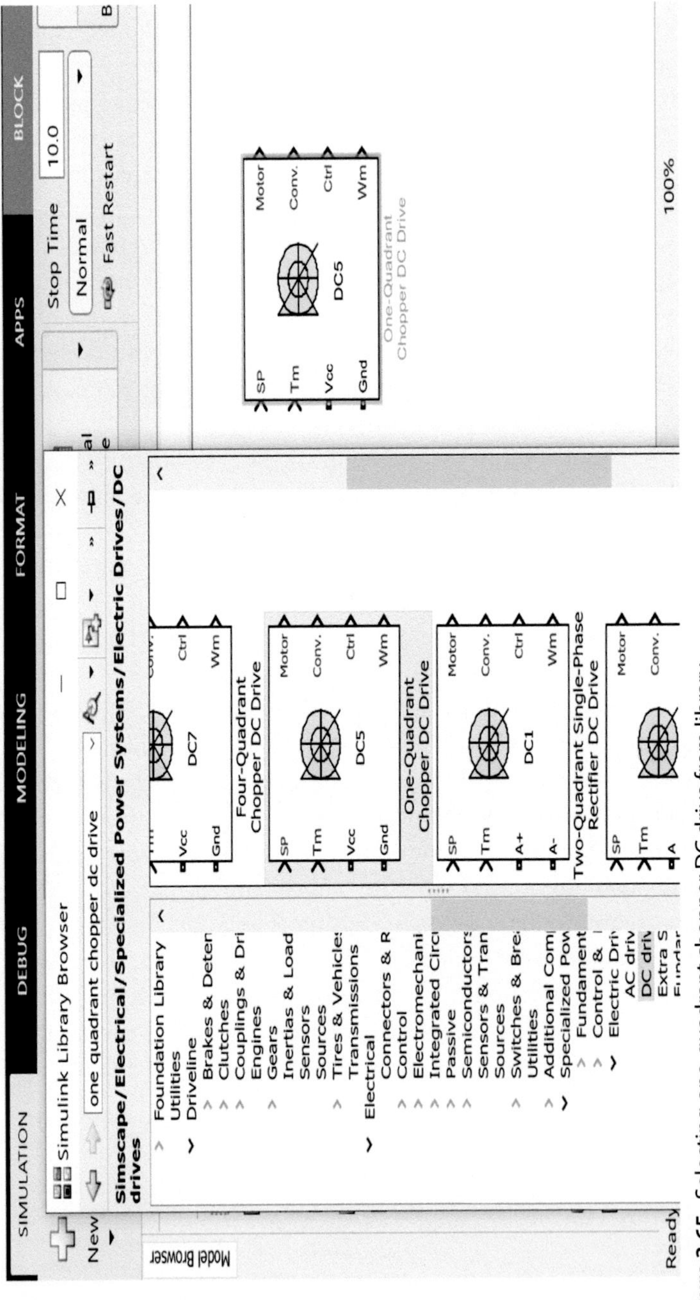

Figure 3.65 Selecting one-quadrant chopper DC drive from library.

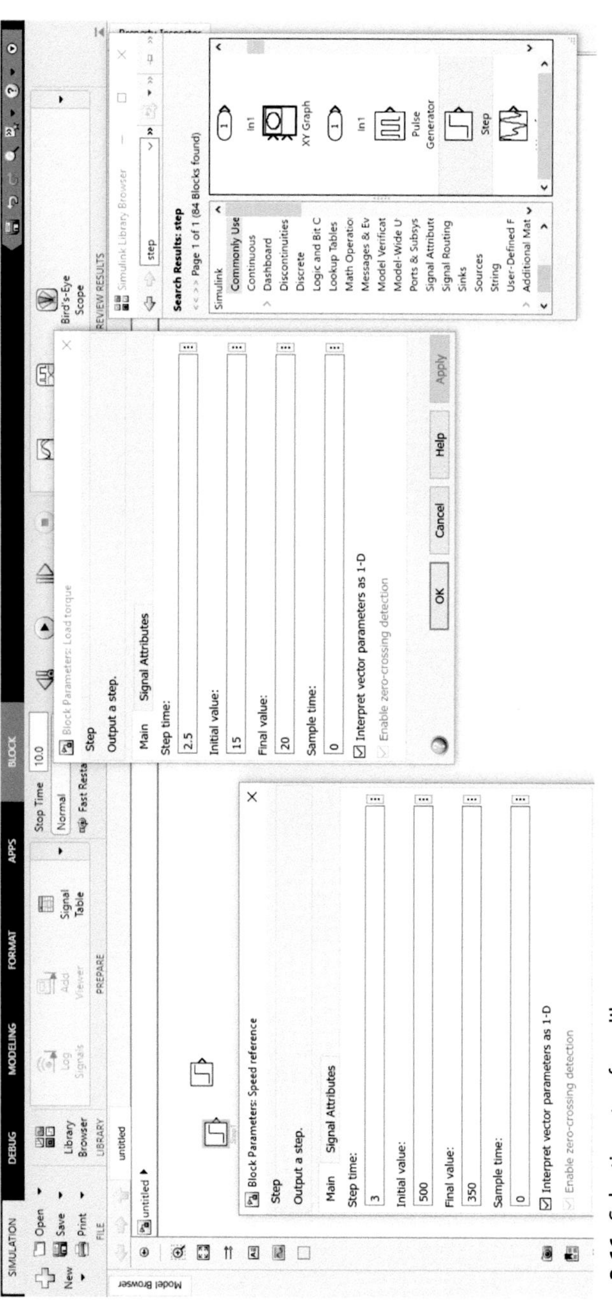

Figure 3.66 Selecting step from library.

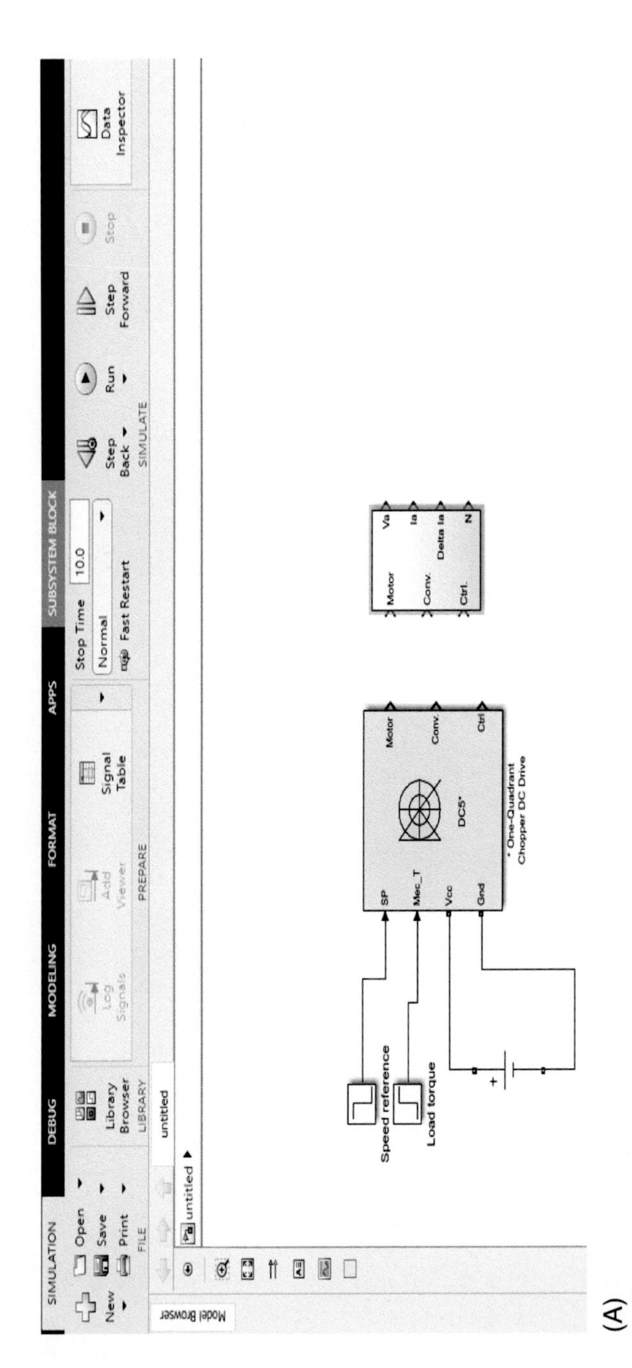

Figure 3.67A Selecting subsystem from library.

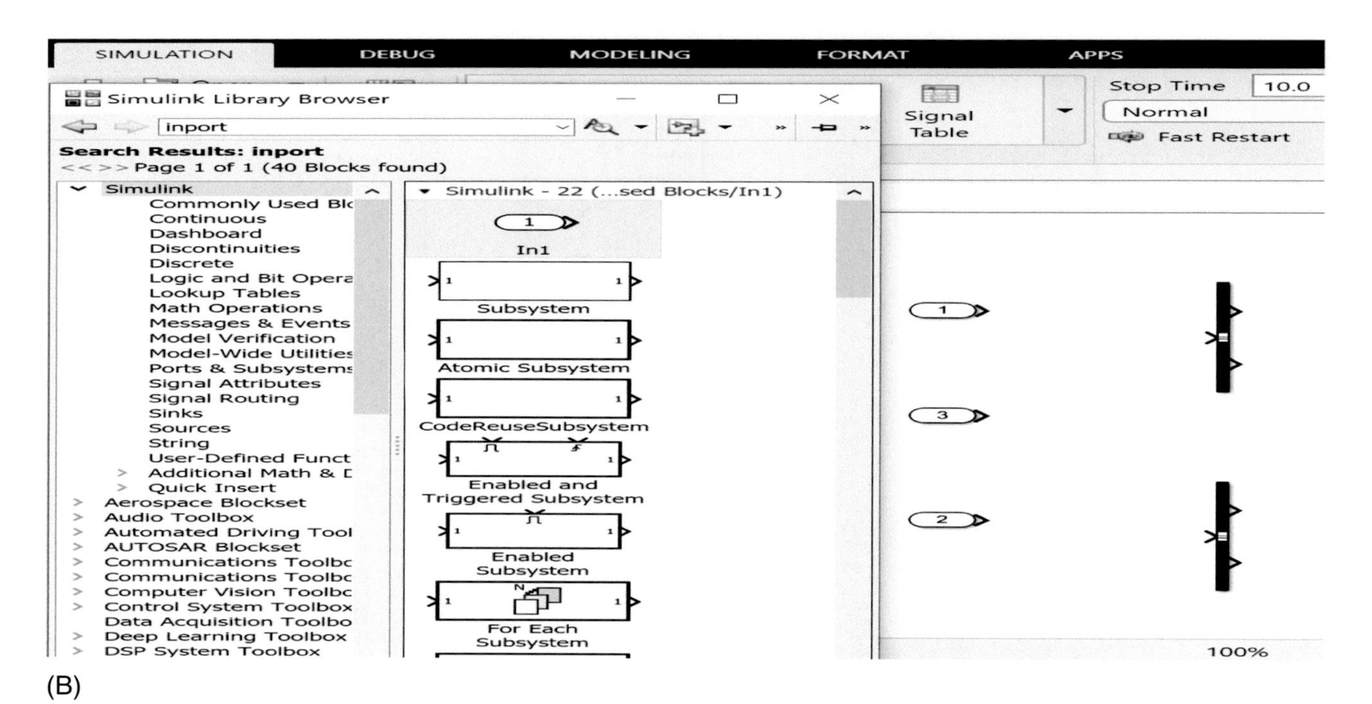

Figure 3.67B Selecting input ports from library.

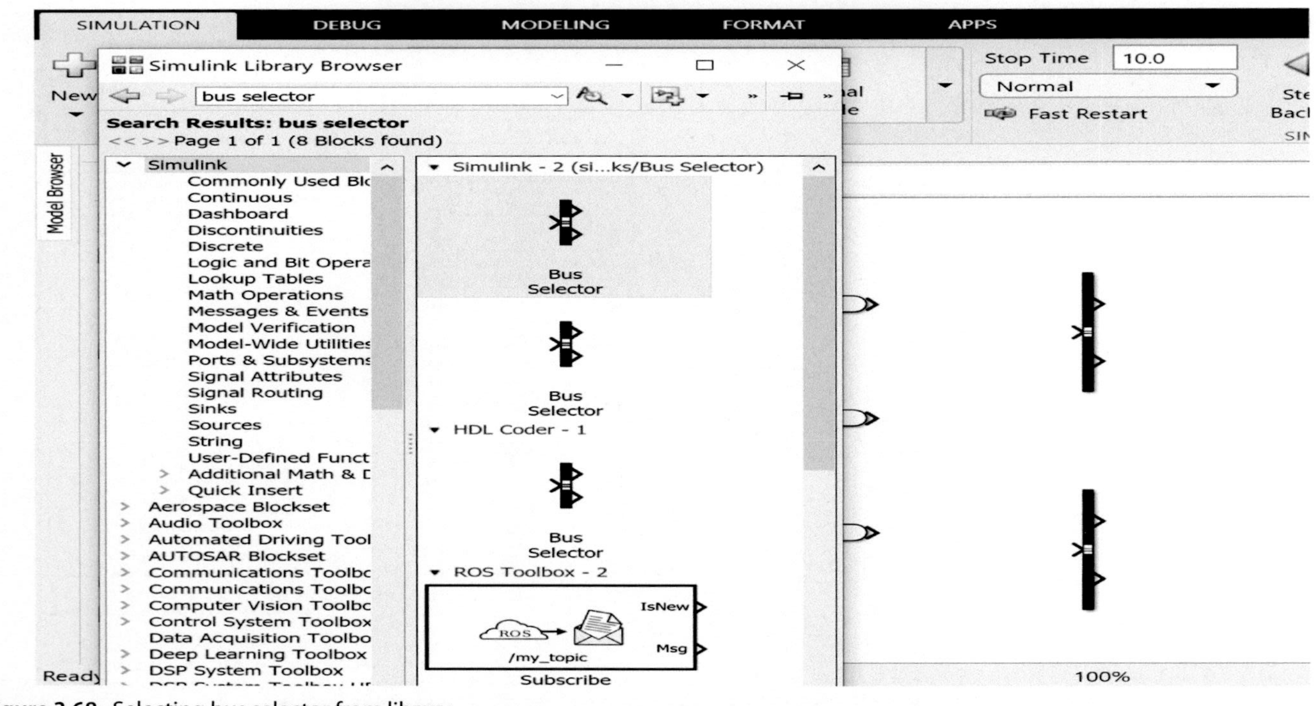

Figure 3.68 Selecting bus selector from library.

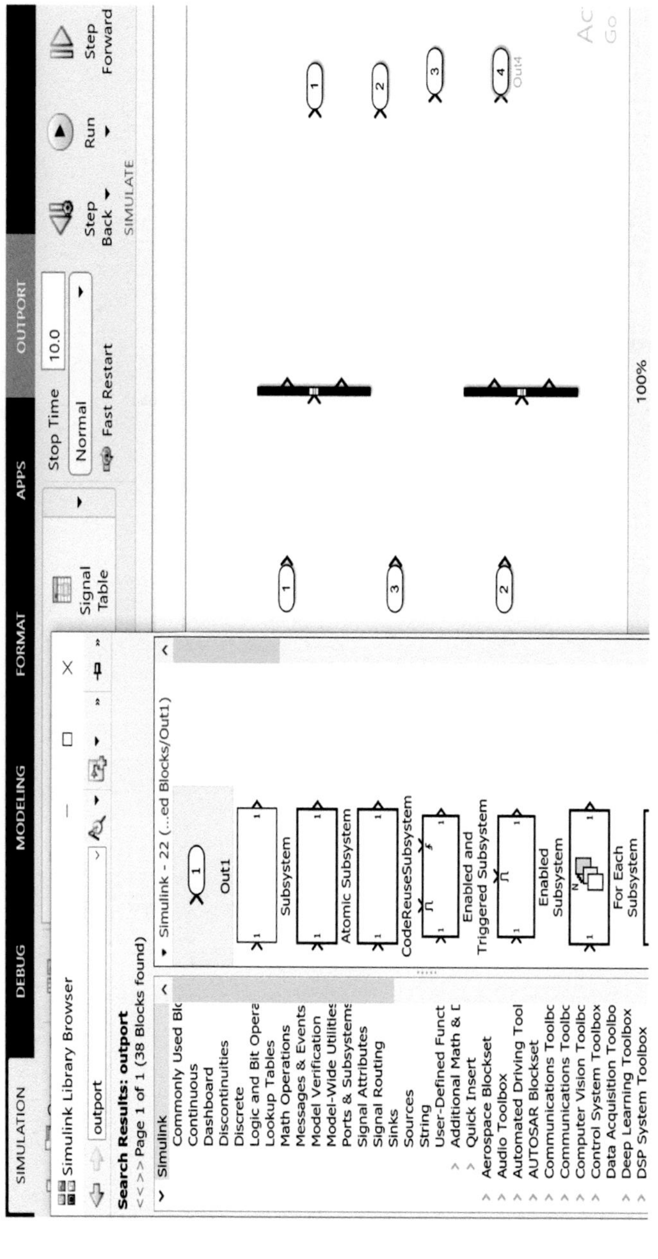

Figure 3.69 Selecting outport from library.

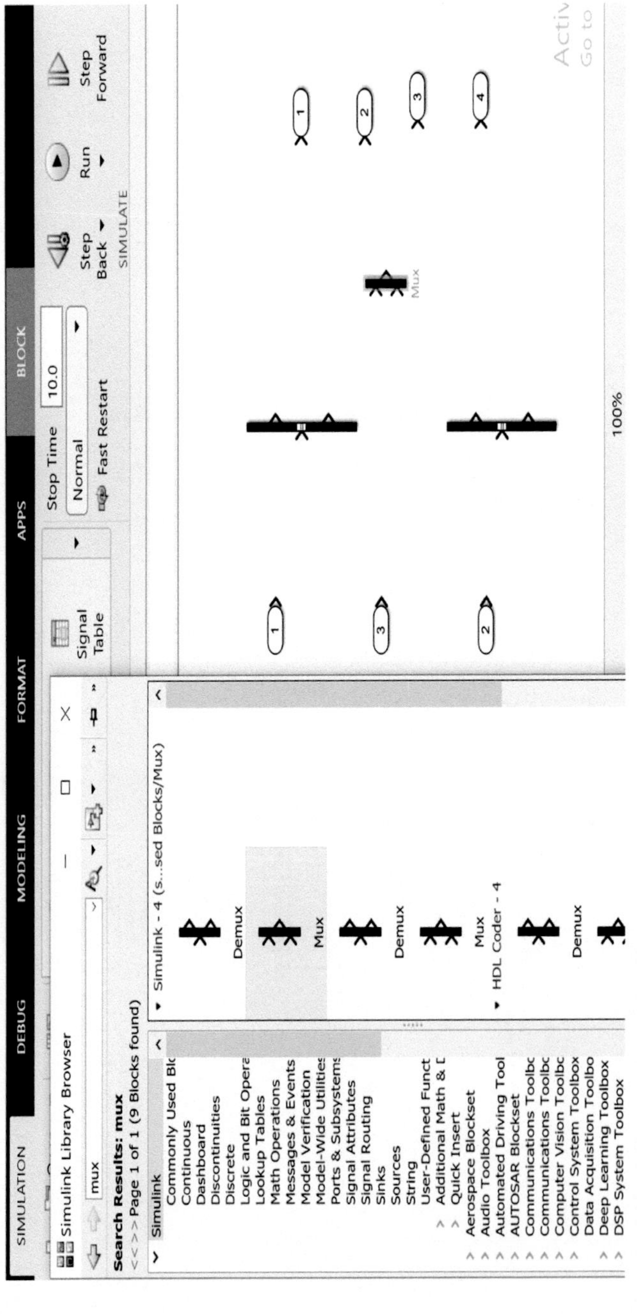

Figure 3.70 Selecting Mux from library.

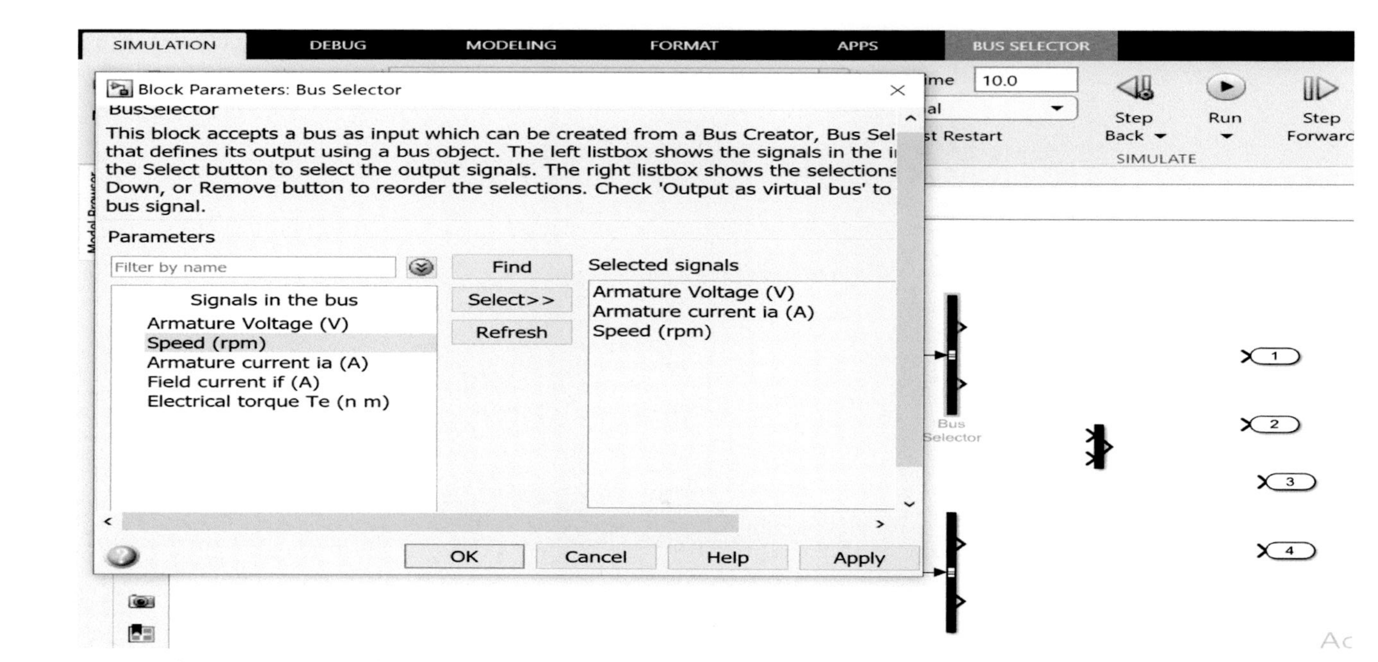

Figure 3.71 Selecting parameters for bus1.

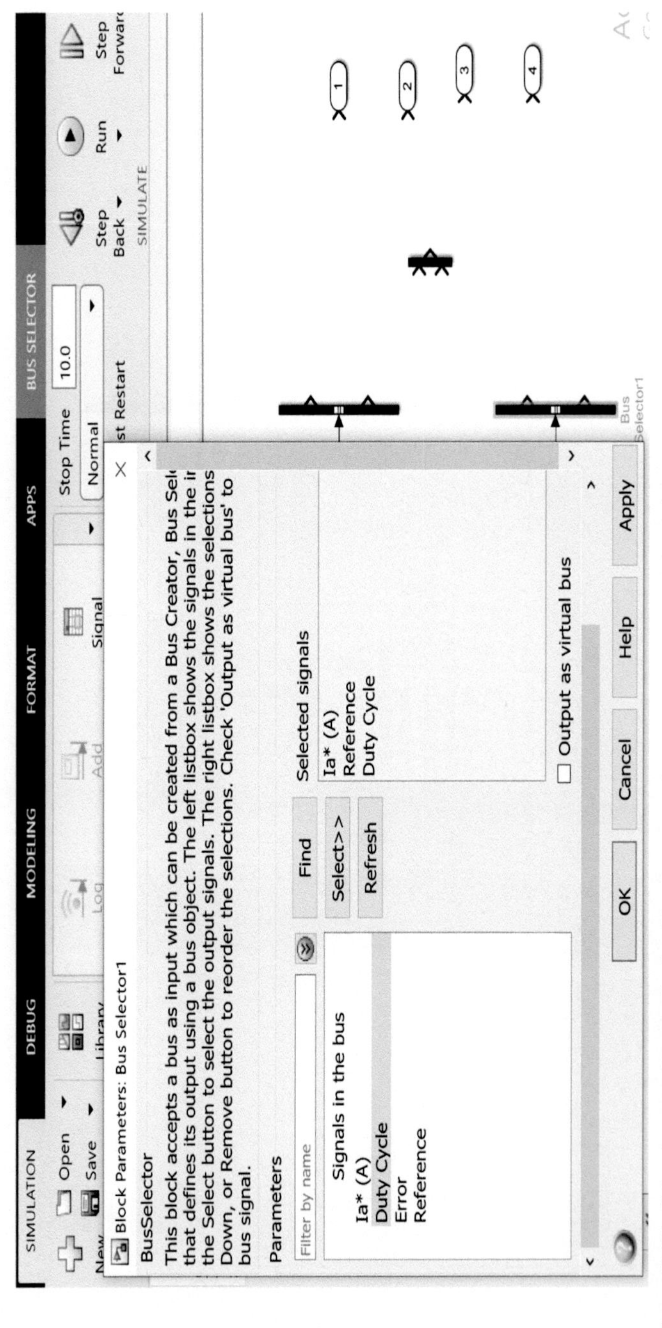

Figure 3.72 Selecting parameters for bus2.

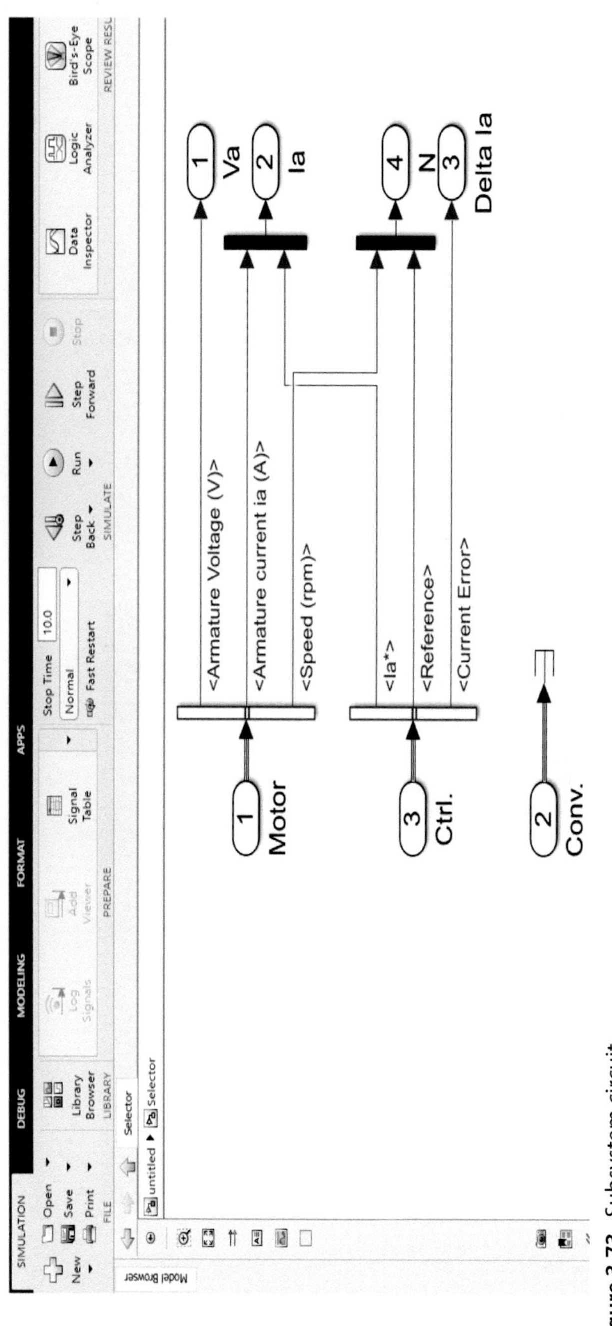

Figure 3.73 Subsystem circuit.

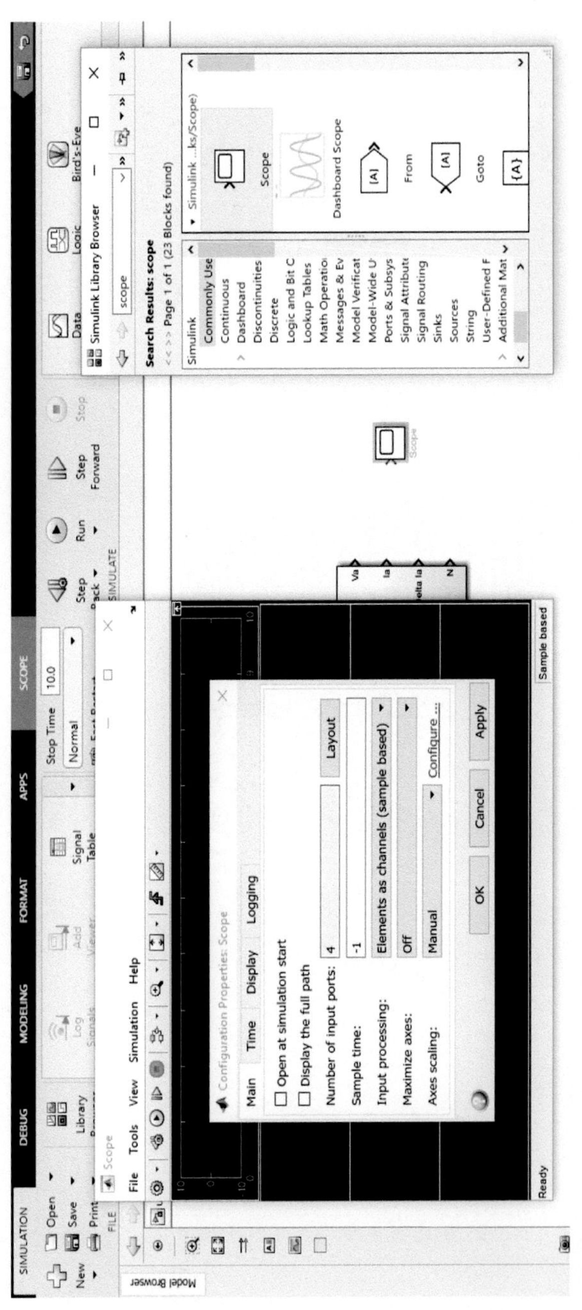

Figure 3.74 Selecting scope from library.

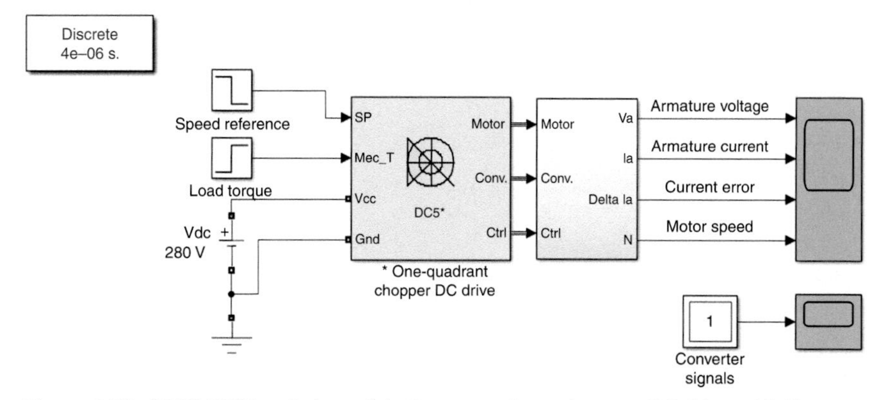

Figure 3.75 MATLAB/Simulink model of one-quadrant chopper DC drive with Hysteresis current control.

3.7 Two-quadrant chopper DC drive

3.7.1 Type-C chopper or two-quadrant type-A chopper [3]

Type C chopper is obtained by connecting type-A and type-B choppers in parallel. We will always get a positive output voltage V_0 as the freewheeling diode FD is present across the load. When the chopper is on the freewheeling diode starts conducting and the output voltage v_0 will be equal to V_s. The direction of the load current i_0 will be reversed. The current i_0 will be flowing toward the source and it will be positive regardless the chopper is on or the FD conducts. The load current will be negative if the chopper is or the diode D2 conducts. We can say the chopper and FD operate together as type-A chopper in first quadrant. In the second quadrant, the chopper and D2 will operate together as type-B chopper (Fig. 3.77).

The average voltage will be always positive but the average load current might be positive or negative. The power flow may be life the first quadrant operation, that is, from source to load or from load to source like the second quadrant operation. The two choppers should not be turned on simultaneously as the combined action my cause a short circuit in supply lines. For regenerative braking and motoring this type of chopper configuration is used.

3.7.2 Type-D chopper or two-quadrant type-B chopper

The circuit diagram of the type D chopper is shown in Fig. 3.78. When the two choppers are on the output voltage V_0 will be equal to V_s. When $V_0 = -V_s$ the two choppers will be off but both the diodes D1 and D2 will

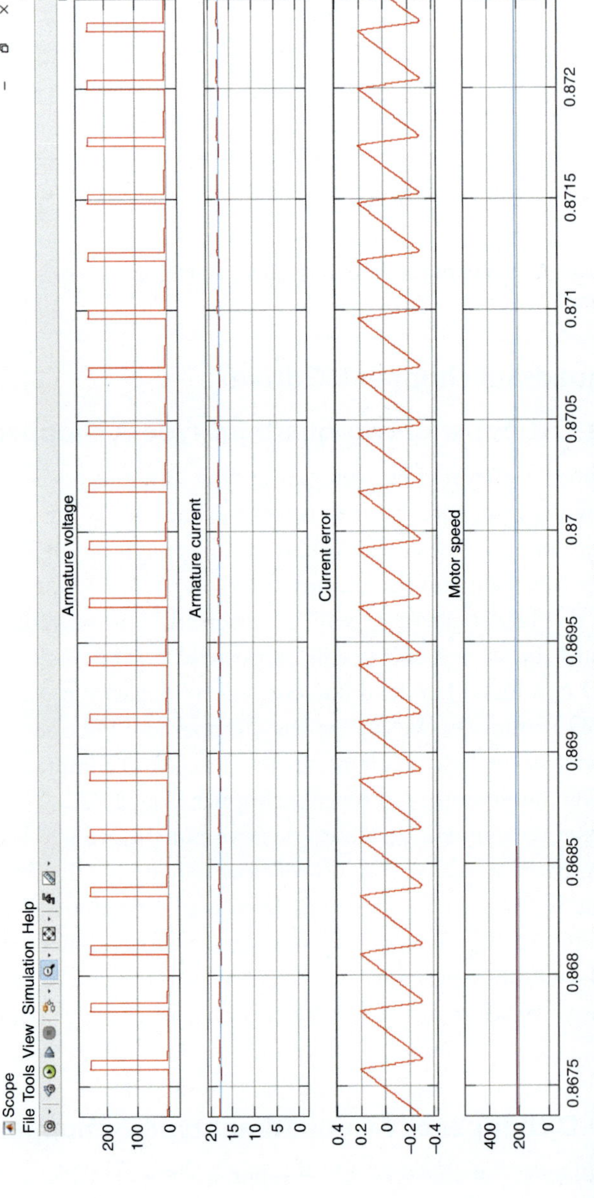

Figure 3.76 Simulation results of one-quadrant chopper DC drive with hysteresis current control.

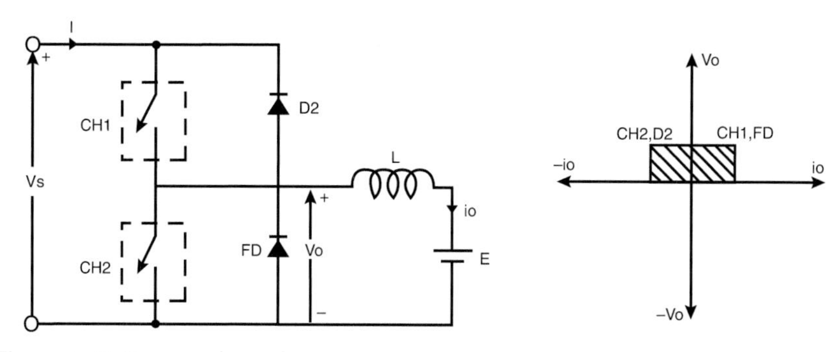

Figure 3.77 Two-quadrant chopper.

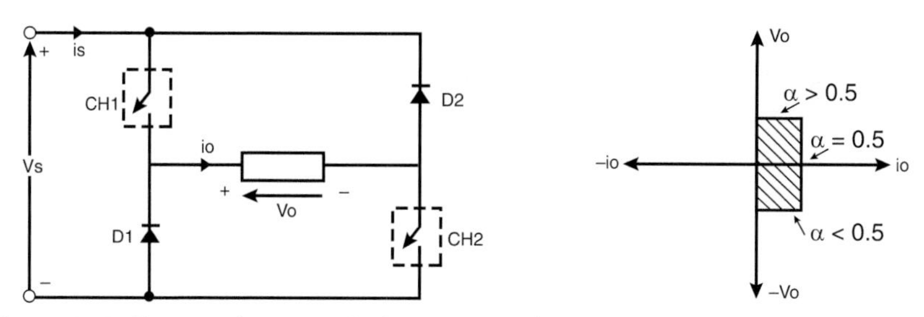

Figure 3.78 Two-quadrant type B chopper or D chopper circuit.

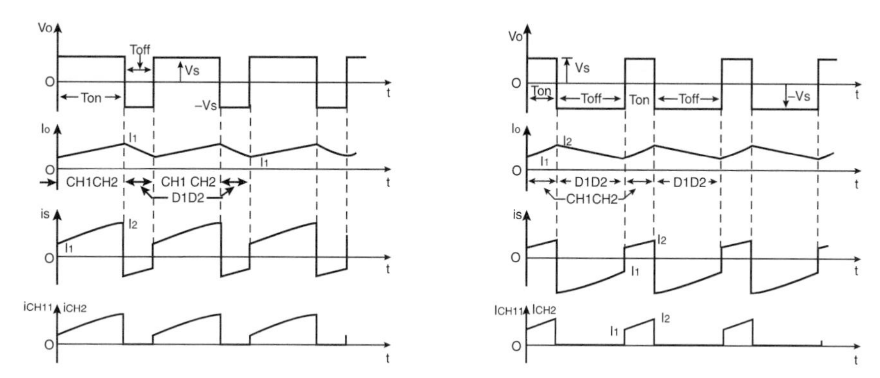

Figure 3.79 Positive first-quadrant operation and negative fourth-quadrant operation.

start conducting. V_0 the average output voltage will be positive when the choppers turn-on the time T_{on} will be more than the turn off time T_{off} its shown in the waveform below. As the diodes and choppers conduct current only in one direction the direction of load current will be always positive.

The power flows from source to load as the average values of both v_0 and i_0 is positive. From the waveform (Fig. 3.79), it is seen that the average value of V_0 is positive thus the fourth quadrant operation of type D chopper is obtained.

From the waveforms the Average value of output voltage is given by

$$V_0 = (V_sT_{on} - V_sT_{off})/T = V_s.(T_{on} - T_{off})/T$$

To do the simulation of two-quadrant chopper DC drive, we need to select the components. First, we select the IGBT. To activate these semiconductor devices we need DC voltage source. We can select the DC voltage sources as previously.

>>Libraries>>Simscape>>SpecializedPowerSystems>>Fundamentalblocks>> PowerElectronics>>IGBT

The parameters need to be given as per Fig. 3.80.

Pulse generator has to be selected with the help of library, then change the parameters of pulse generator, as given in Fig. 3.81.

Logic operator is selected with the help of library (Fig. 3.82).

DC machines are selected from library as previously. Parameters are given as per Fig. 3.83.

Select constant with the help of library and give the value, as mentioned in Fig. 3.84.

Select the bus selector and parameters as previous. Here, armature current, speed, and electrical torques are selected (Fig. 3.85).

To display the outputs, CRO or scope is needed. So select scope as previous, then so to settings and arrange the layout as required. From the bus selector, armature current, speed, and electrical torques are given to scope (Fig. 3.86). These three outputs can been seen in scope.

To find the voltage across the IGBT, voltage measurement is needed. So select voltage measurement from library (Fig. 3.87).

Select mean by using command.

>>Libraries>>Simscape>>Electrical>>SpecializedPowerSystems>> Fundamentalblocks>> Measurements>>Additional Measurements>> Mean (Fig. 3.88)

Select display by using library. Libraries>>Simulink>>Sinks>>Display (Fig. 3.89).

After selecting all the elements, connect it as per Fig. 3.90.

After completion of circuit diagram, simulate it to see the simulation results (Fig. 3.91).

3.8 Four-quadrant chopper DC drive

Type E or the fourth quadrant chopper consists of four semiconductor switches and four diodes arranged in antiparallel (Fig. 3.92). The four

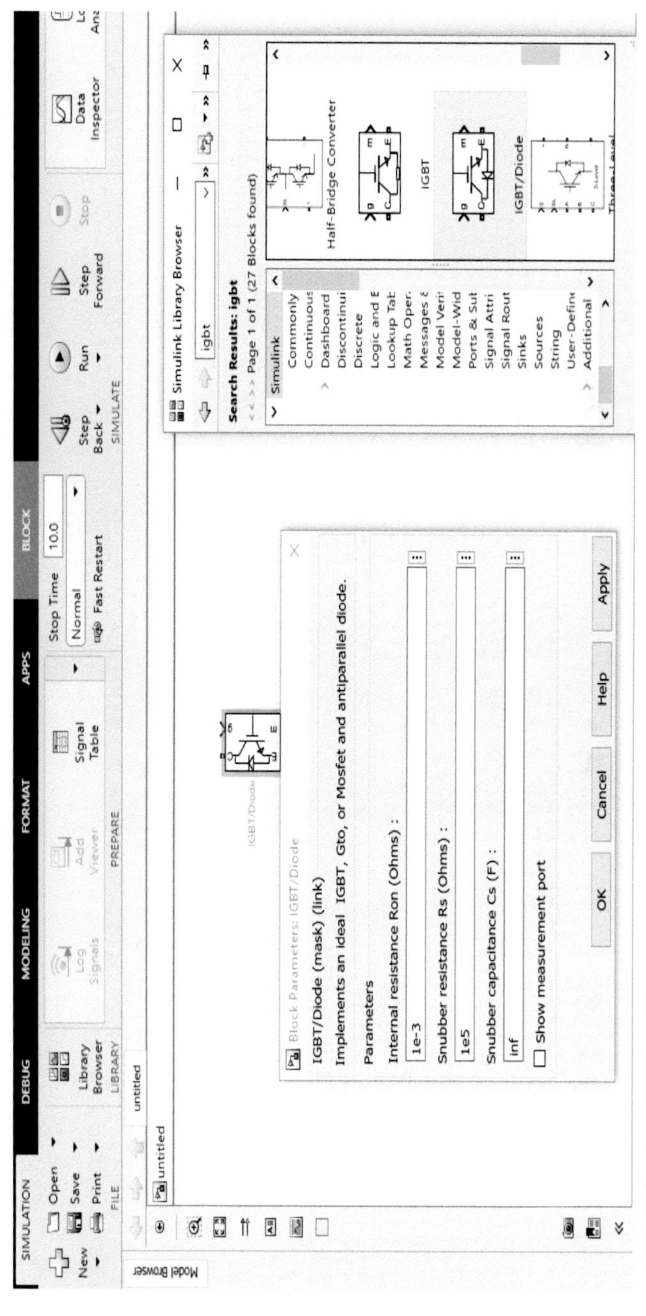

Figure 3.80 Selecting IGBT from library.

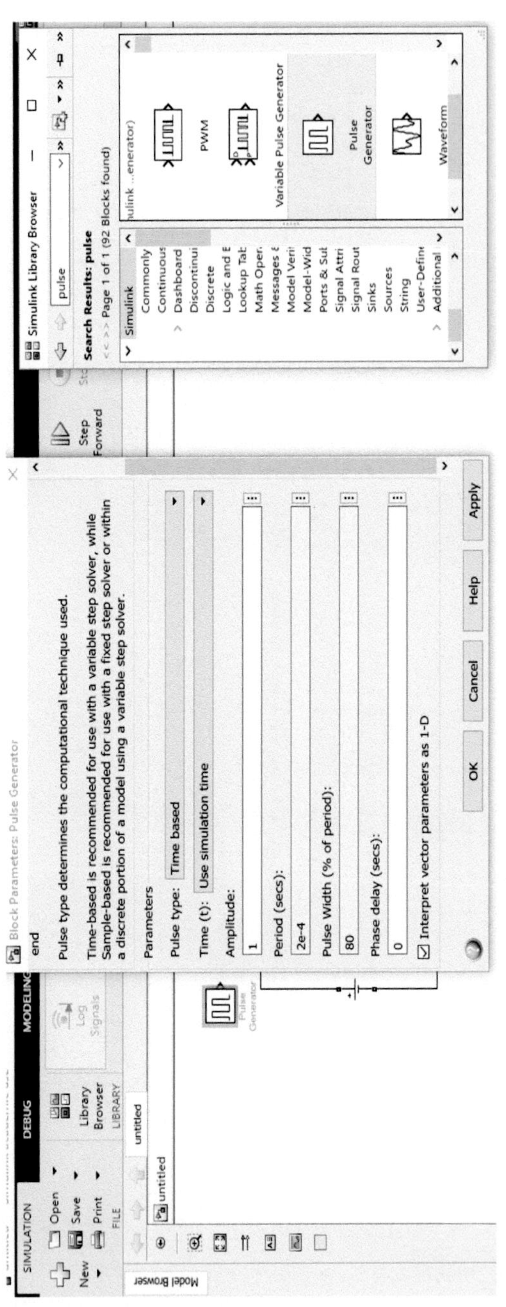

Figure 3.81 Selecting pulse generator from library.

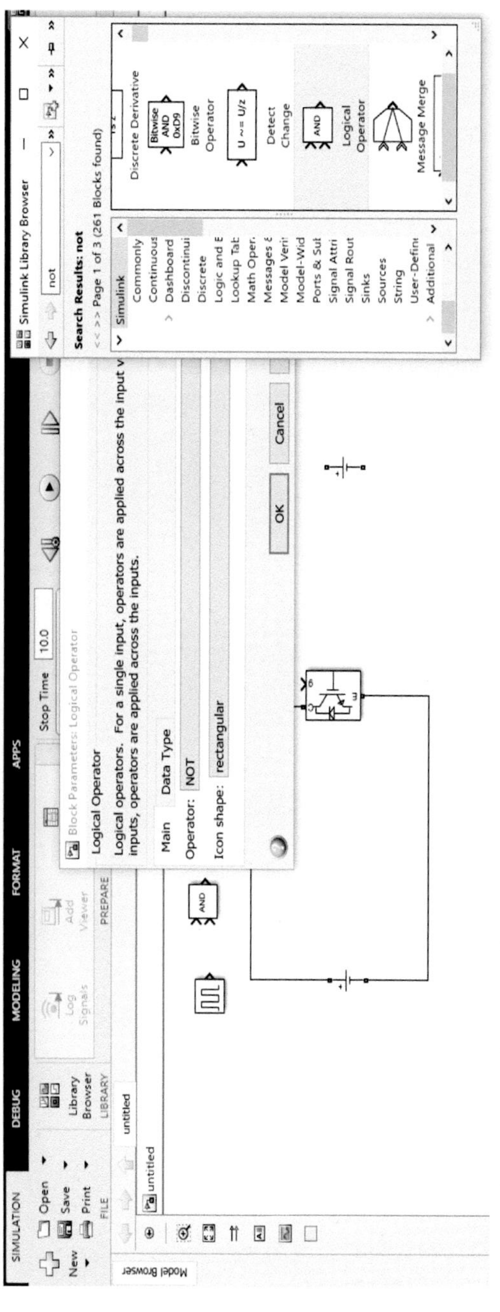

Figure 3.82 Selecting logical operator from library.

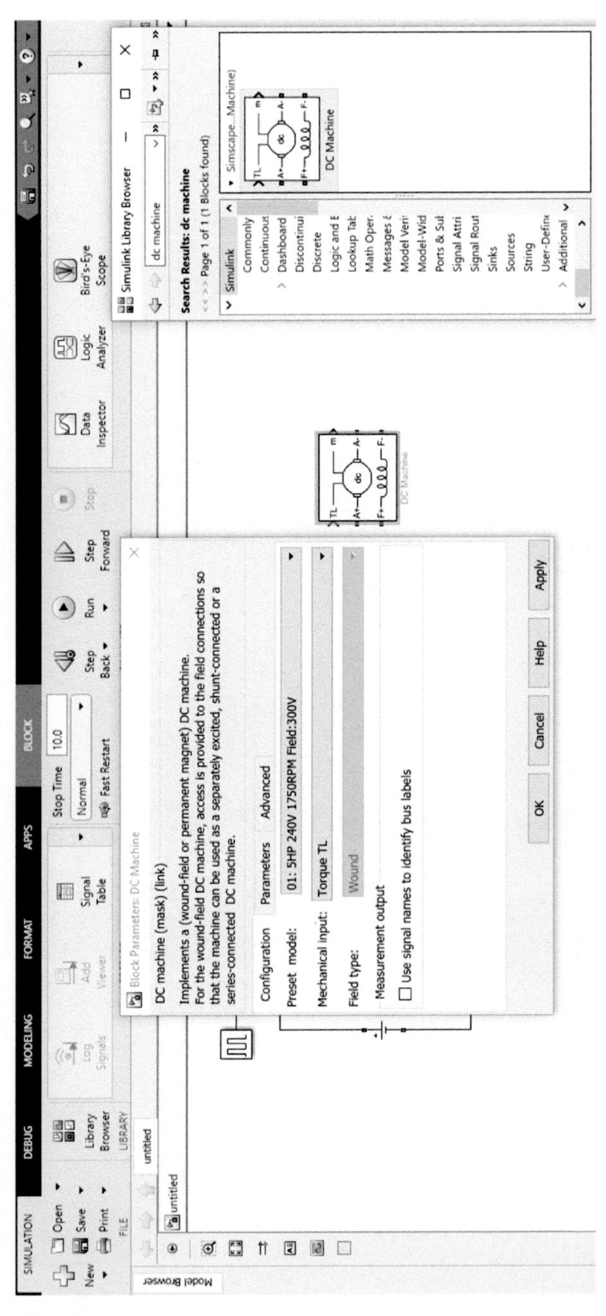

Figure 3.83 Selecting DC machines from library.

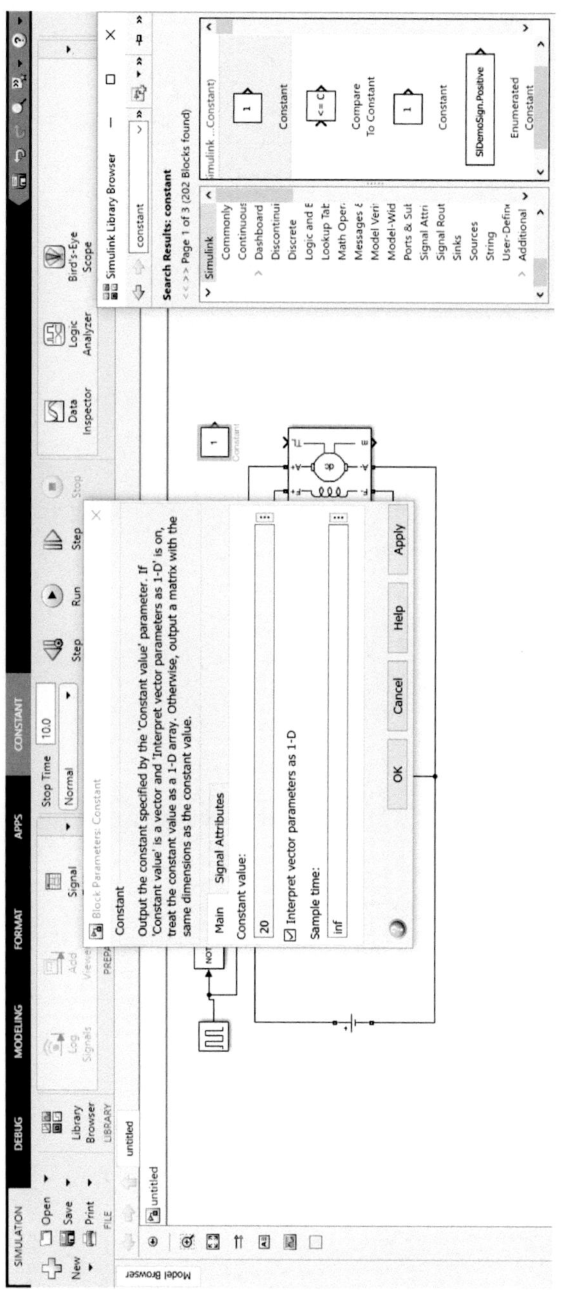

Figure 3.84 Selecting constant from library.

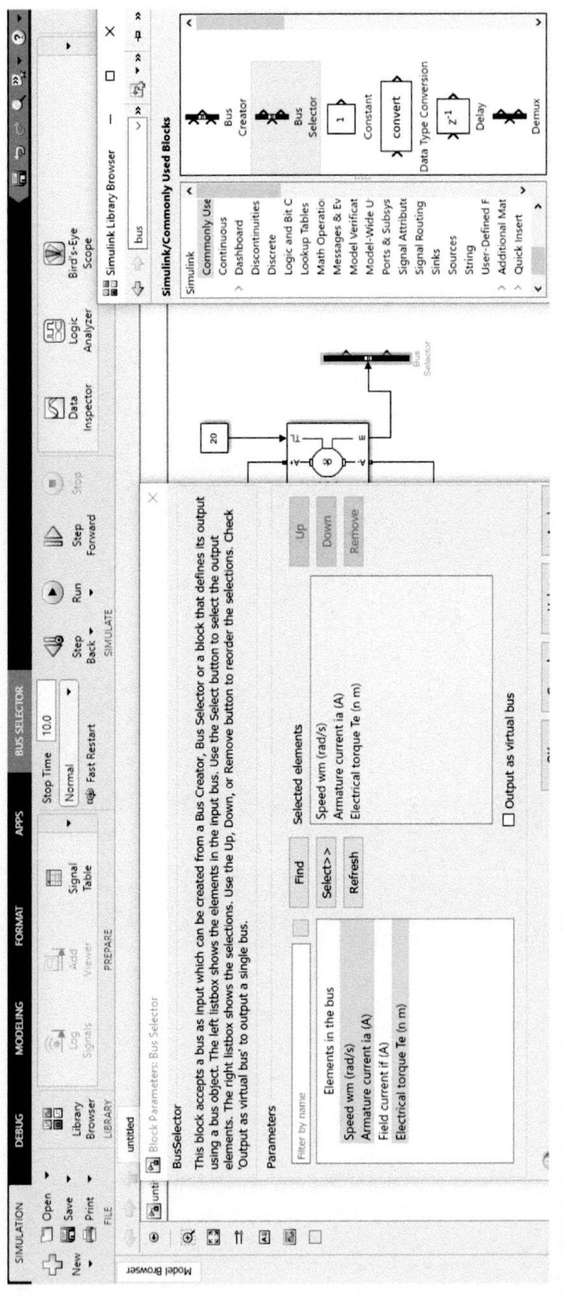

Figure 3.85 Selecting bus selector from library.

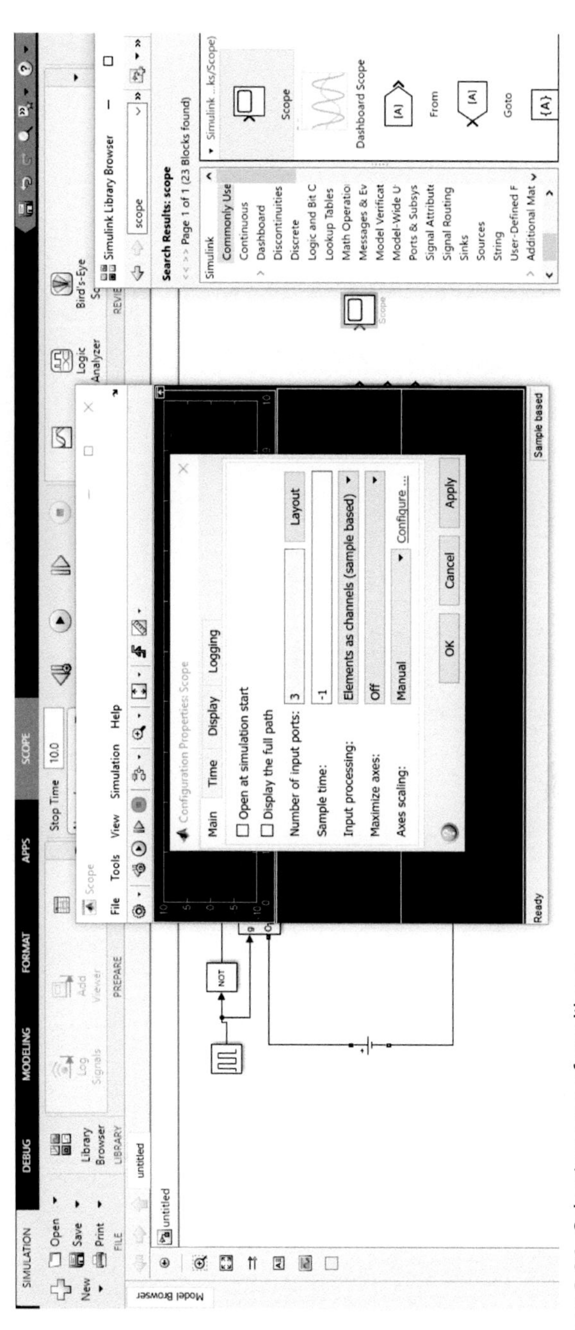

Figure 3.86 Selecting scope from library.

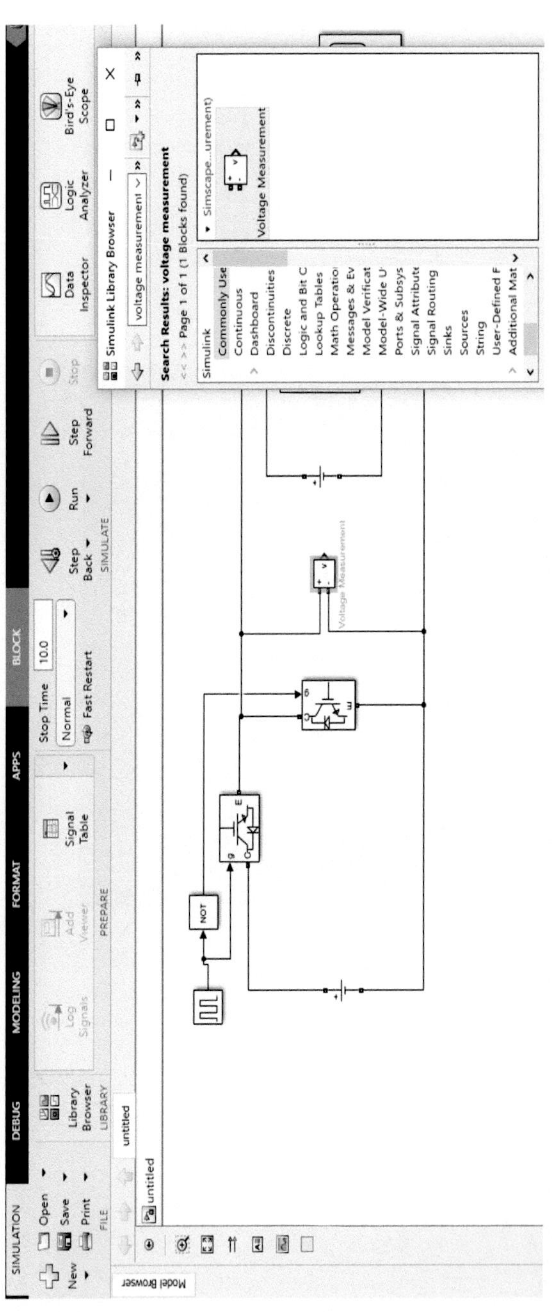

Figure 3.87 Selecting voltage measurement from library.

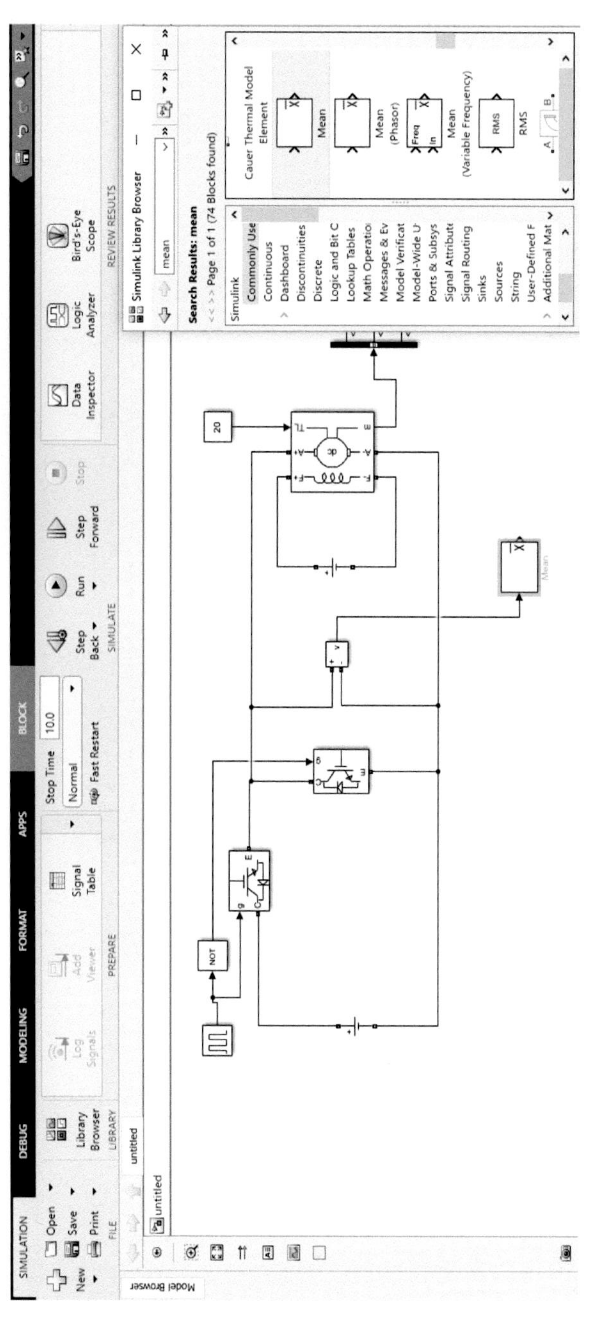

Figure 3.88 Selecting mean from library.

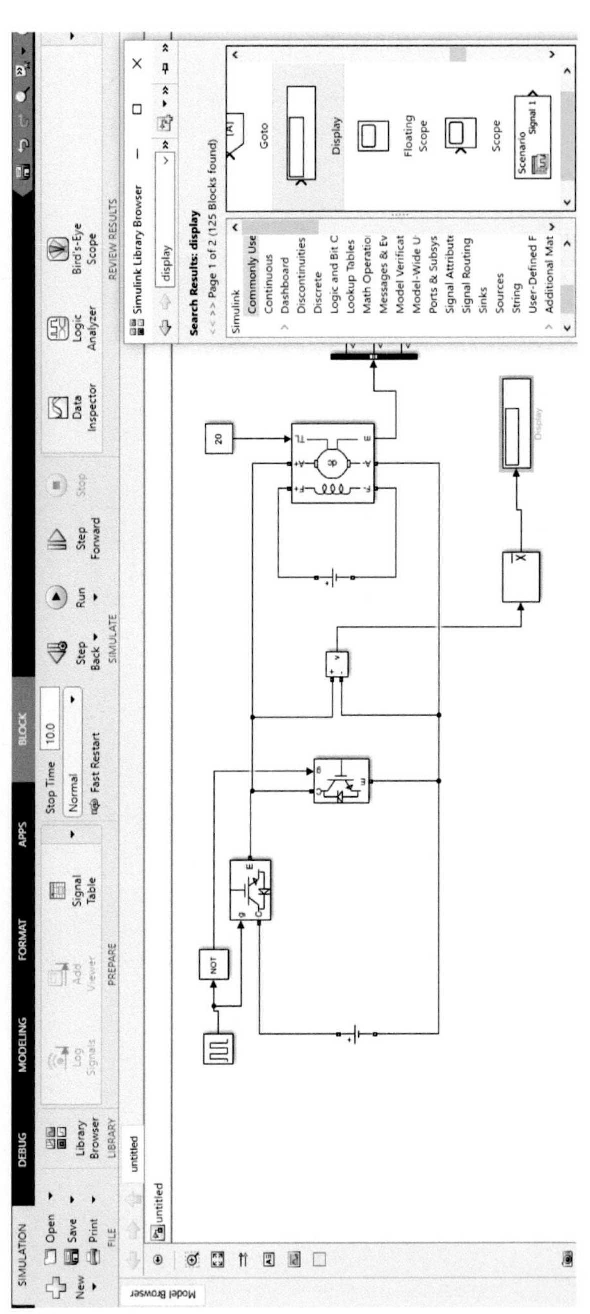

Figure 3.89 Selecting display from library.

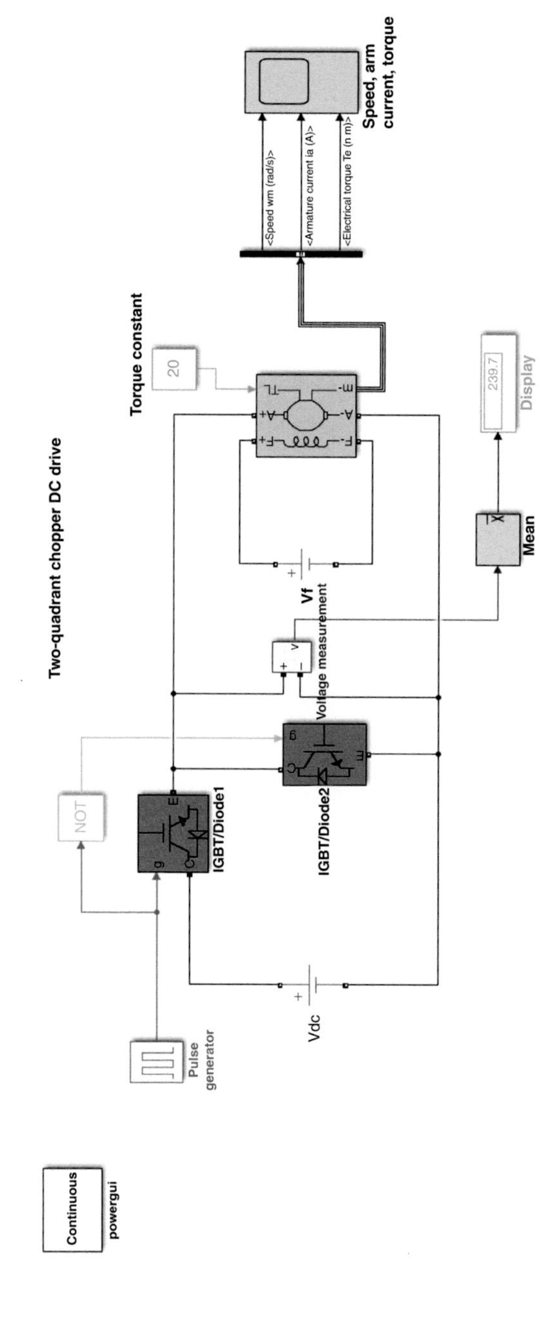

Figure 3.90 MATLAB/Simulink model of two-quadrant chopper DC drive.

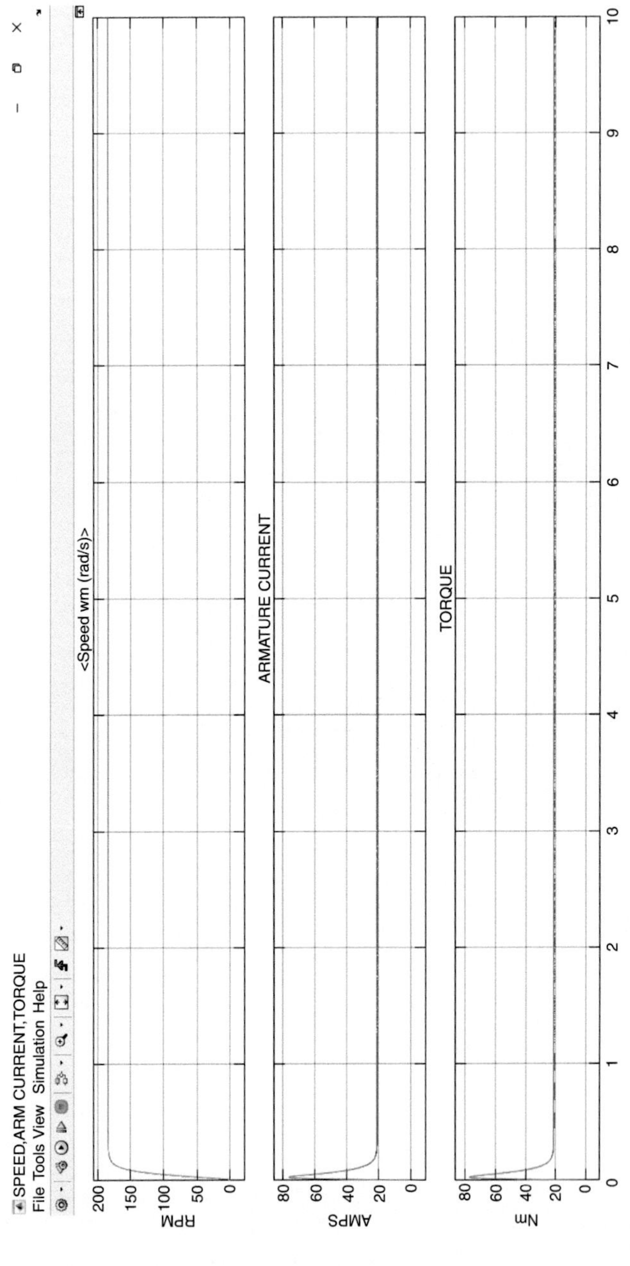

Figure 3.91 Simulation results of two-quadrant chopper DC drive.

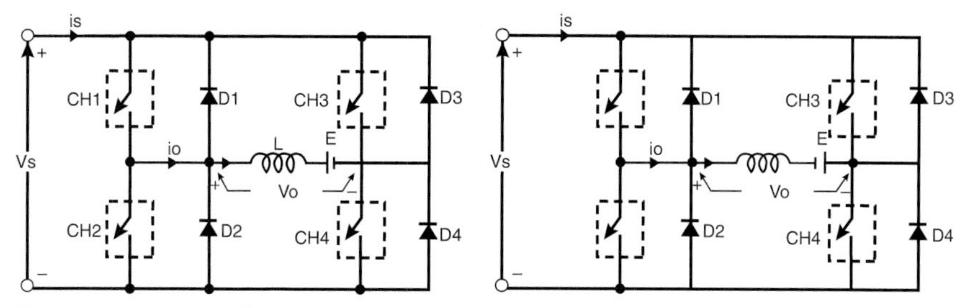

Figure 3.92 E-type chopper circuit diagram with load emf E and E reversed.

choppers are numbered according to which quadrant they belong. Their operation will be in each quadrant and the corresponding chopper only be active in its quadrant.

- *First quadrant*

During the first quadrant operation the chopper CH4 will be on. Chopper CH3 will be off and CH1 will be operated. AS the CH1 and CH4 is on the load voltage v_0 will be equal to the source voltage V_s and the load current i_0 will begin to flow. v_0 and i_0 will be positive as the first quadrant operation is taking place. As soon as the chopper CH1 is turned off, the positive current freewheels through CH4 and the diode D2. The type E chopper acts as a step-down chopper in the first quadrant.

- *Second quadrant*

In this case, the chopper CH2 will be operational and the other three are kept off. As CH2 is on negative current will start flowing through the inductor L. CH2, E, and D4. Energy is stored in the inductor L as the chopper CH2 is on. When CH2 is off the current will be fed back to the source through the diodes D1 and D4. Here (E+L.di/dt) will be more than the source voltage V_s. In the second quadrant, the chopper will act as a step-up chopper as the power is fed back from load to source

- *Third quadrant*

In the third quadrant operation CH1 will be kept off, CH2 will be on and CH3 is operated. For this quadrant working the polarity of the load should be reversed. As the chopper CH3 is on, the load gets connected to the source V_s and v_0 and i_0 will be negative and the third quadrant operation will takes place. This chopper acts as a step-down chopper

- *Fourth quadrant*

CH4 will be operated and CH1, CH2, and CH3 will be off. When the chopper CH4 is turned on positive current starts to flow through CH4, D2, E and the inductor L will store energy. As the CH4 is turned off the current is feedback to the source through the diodes D2 and D3, the operation will be in the fourth quadrant as the load voltage is negative but the load current is positive. The chopper acts as a step up chopper as the power is fed back from load to source.

To get the simulation of four-quadrant chopper, select the components as previous (Figs. 3.93 and 3.94).

For four-quadrant chopper, four MOSFETs are required.

Here 5HP DC machines are selected. The voltage rating of the DC machines is 240V. Excitation voltage needs to be given as 300V to the field winding. The speed rating of the machine is 1750rpm (Fig. 3.95).

Connect the bus selector to the DC machine and select three parameters as per Fig. 3.96.

Like previous select gain, scope, and other elements from library. Select add from the library using

Libraries>>Simulink>>mathoperations>>Add (Fig. 3.97)

Libraries>>Simulink>>audiotoolbox>>Sinks>>Spectrum Analyser (Fig. 3.98)

Select PID Controller from library. Libraries>>Simulink>>Discrete>> DiscretePIDcontroller (Fig. 3.99)

Select Abs from library. Libraries>>Simulink>>Mathoperation>>Abs (Fig. 3.100)

Select Signal generator from library. Libraries>>Simulink>>Sources>> Signal generator (Fig. 3.101)

Select Compare to Zero from library. Libraries>>Simulink>>Logic and bit operation>>Compare to zero (Fig. 3.102)

Select Goto from library. Libraries>>Simulink>>Signal routing>>Goto (Fig. 3.103)

Select From from library. Libraries>>Simulink>>Signal routing>>From (Fig. 3.104)

After selecting all the elements, need to connect the circuits as per Fig. 3.105

Simulate the connected circuit to get the simulation results (Fig. 3.106).

3.9 Closed-loop control of chopper fed DC drive

In closed-loop system, the output of the system is feedback to the input. The closed-loop system controls the electrical drive, and the system is self-adjusted. Feedback loops in an electrical drive may be provided to satisfy the following requirements.

1. Enhancement of speed of torque.
2. To improve steady-state accuracy.

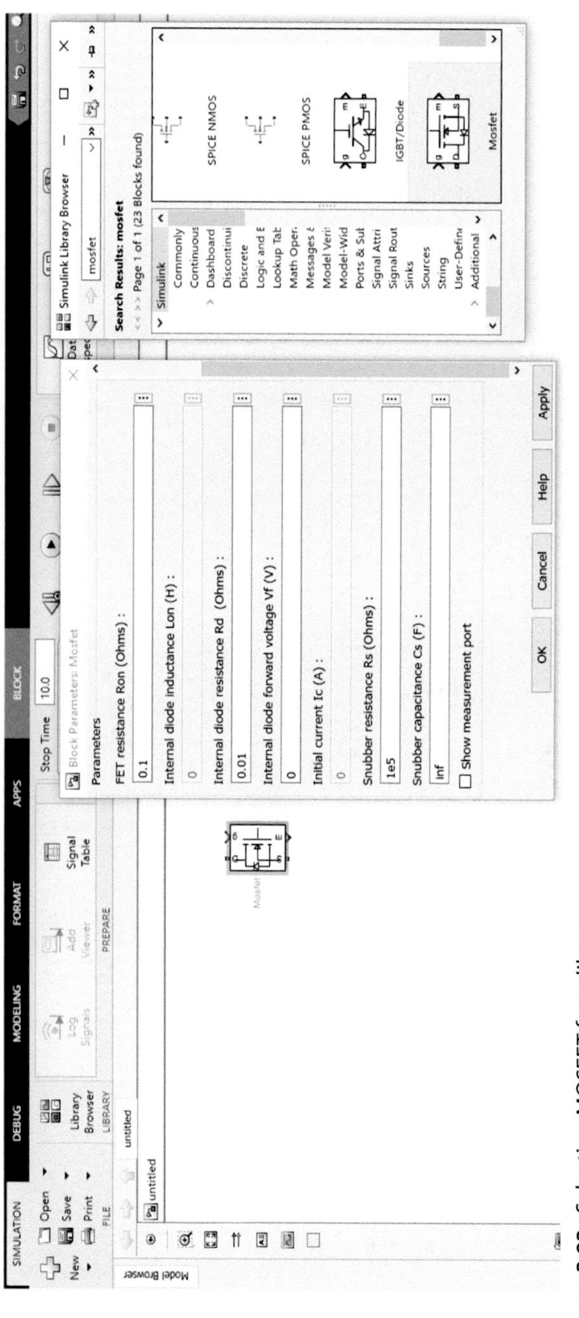

Figure 3.93 Selecting MOSFET from library.

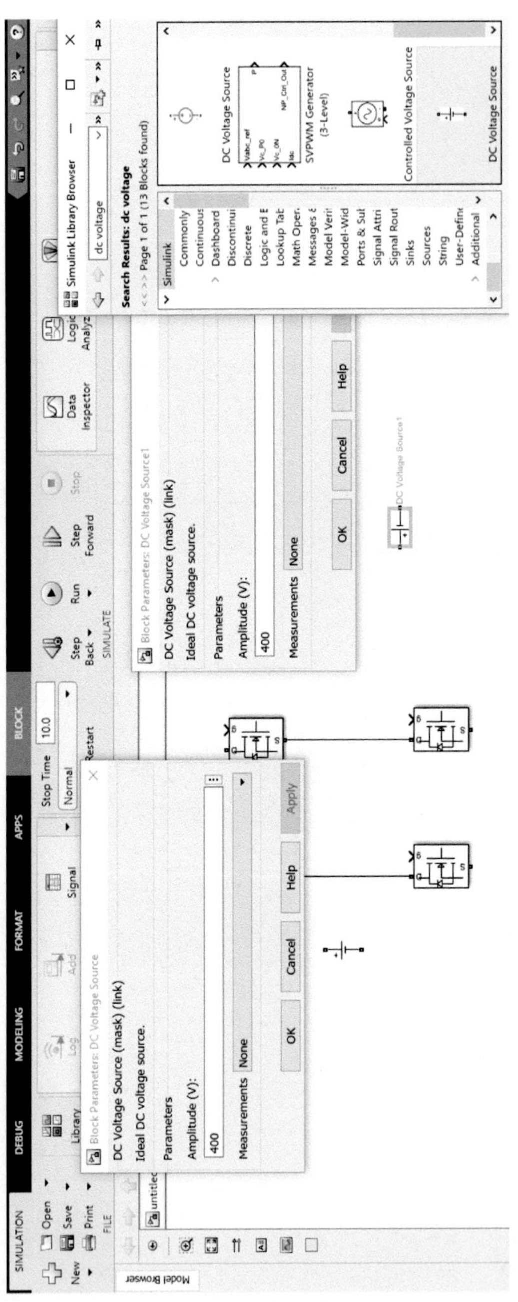

Figure 3.94 Selecting DC voltage source from library.

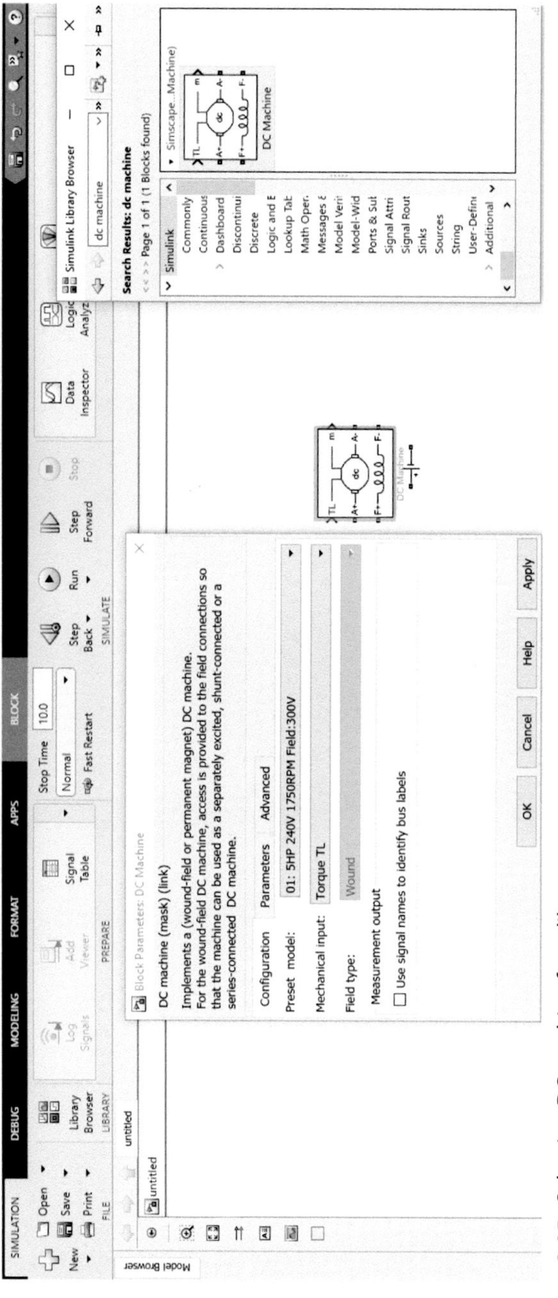

Figure 3.95 Selecting DC machine from library.

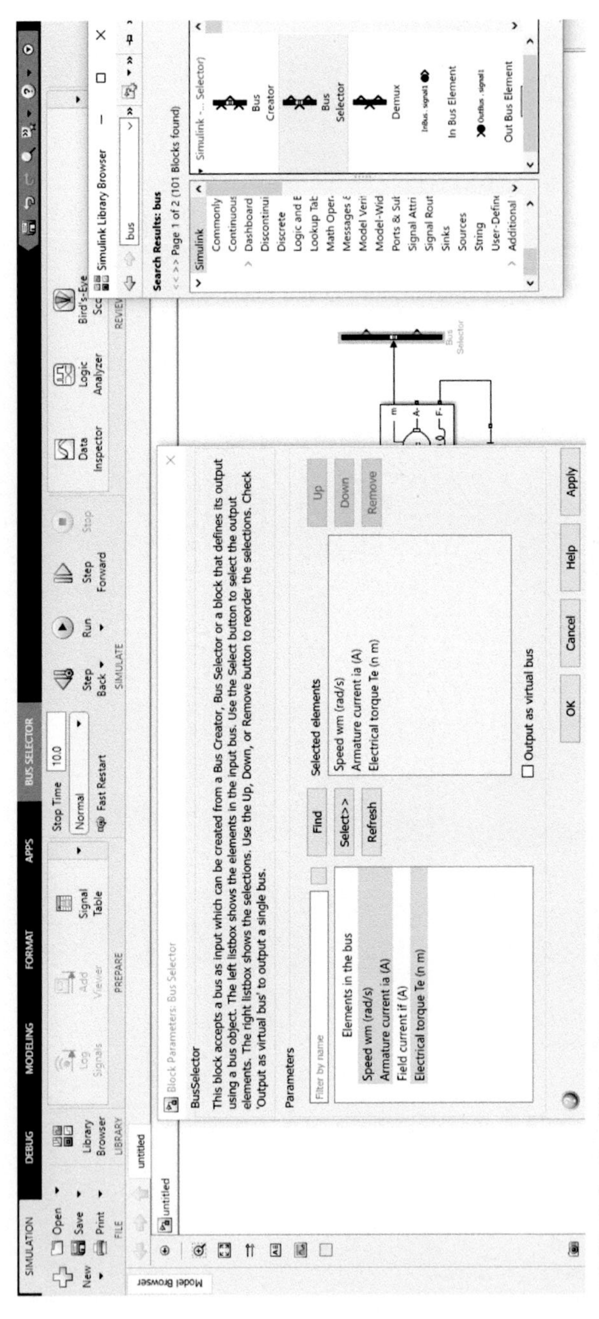

Figure 3.96 Selecting DC machine from library.

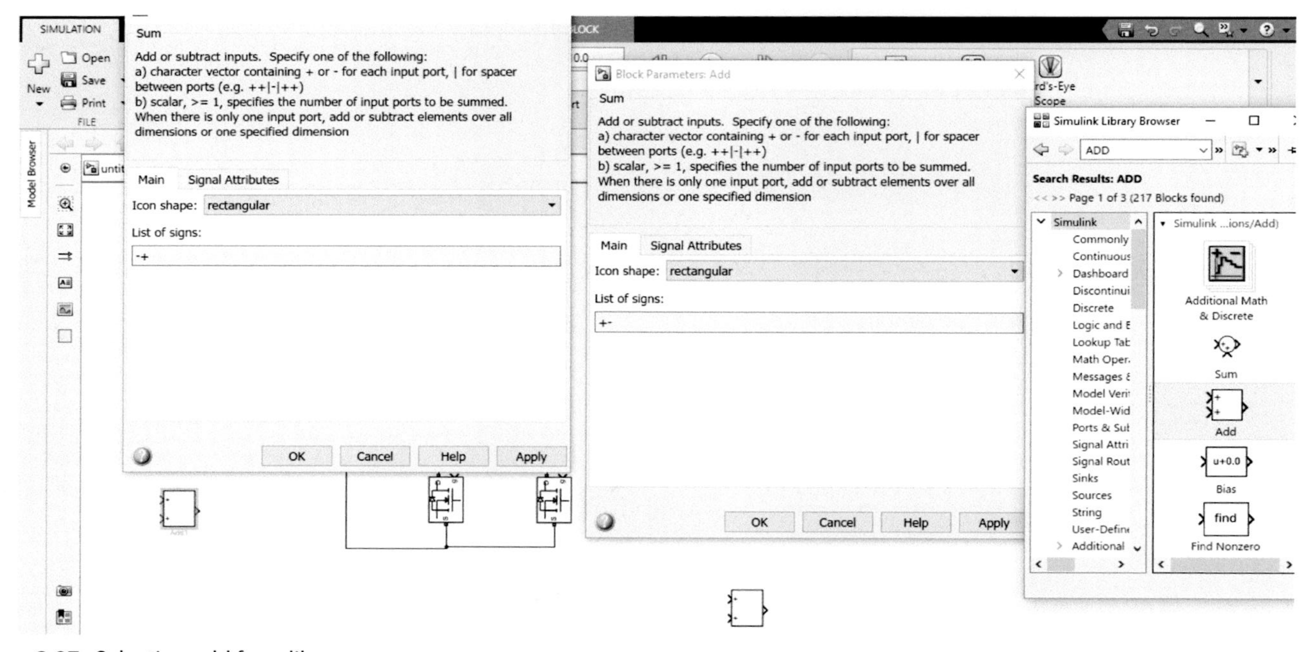

Figure 3.97 Selecting add from library.

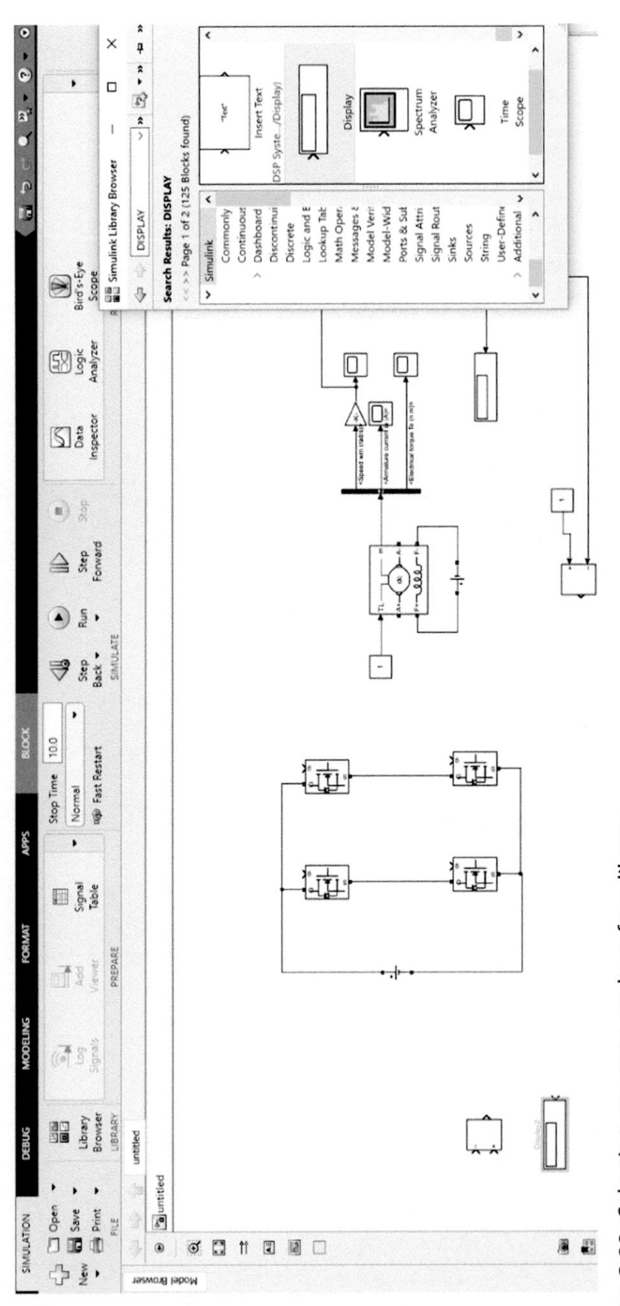

Figure 3.98 Selecting spectrum analyzer from library.

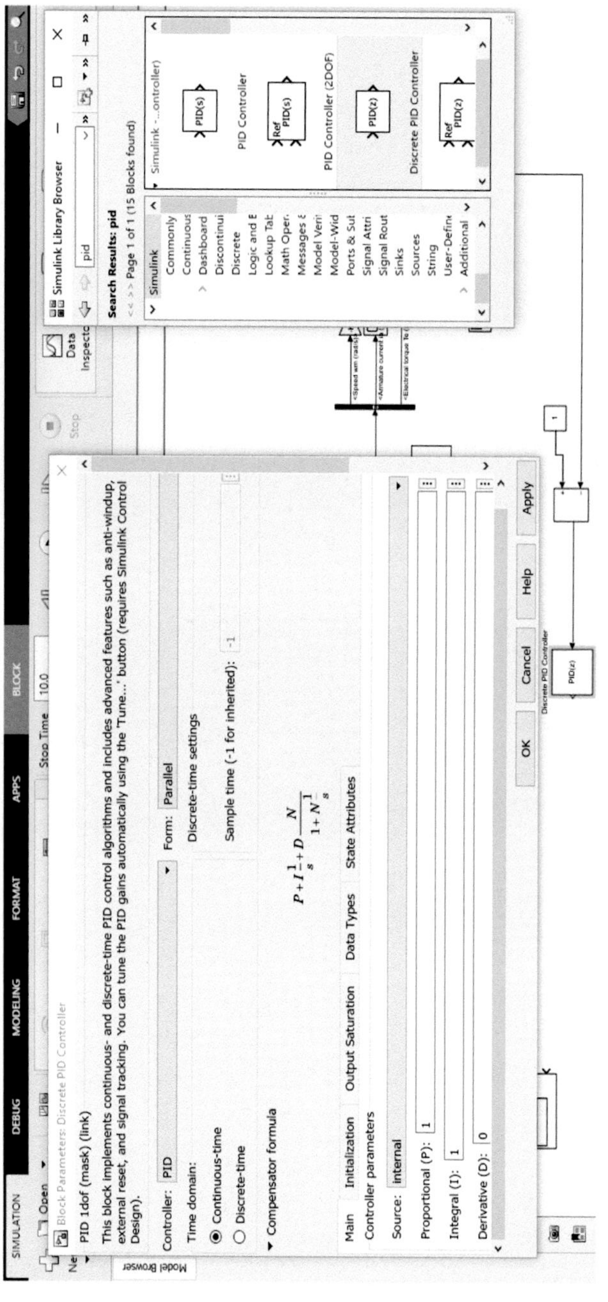

$$P + I\frac{1}{s} + D\frac{N}{1 + N\frac{1}{s}}$$

Figure 3.99 Selecting discrete PID controller from library.

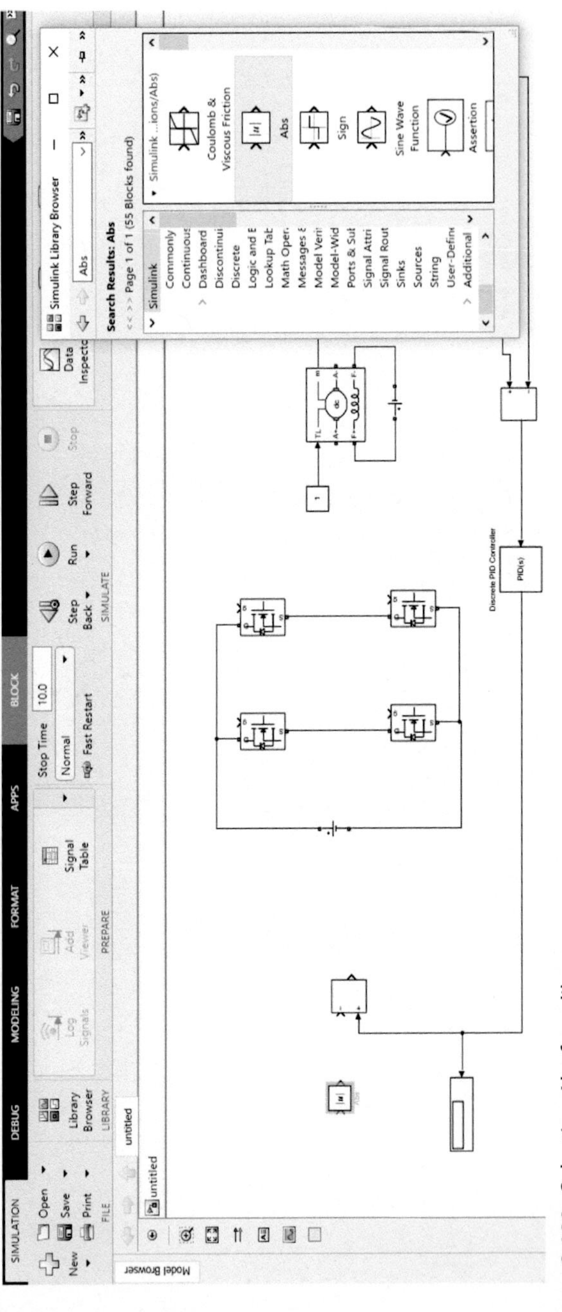

Figure 3.100 Selecting Abs from library.

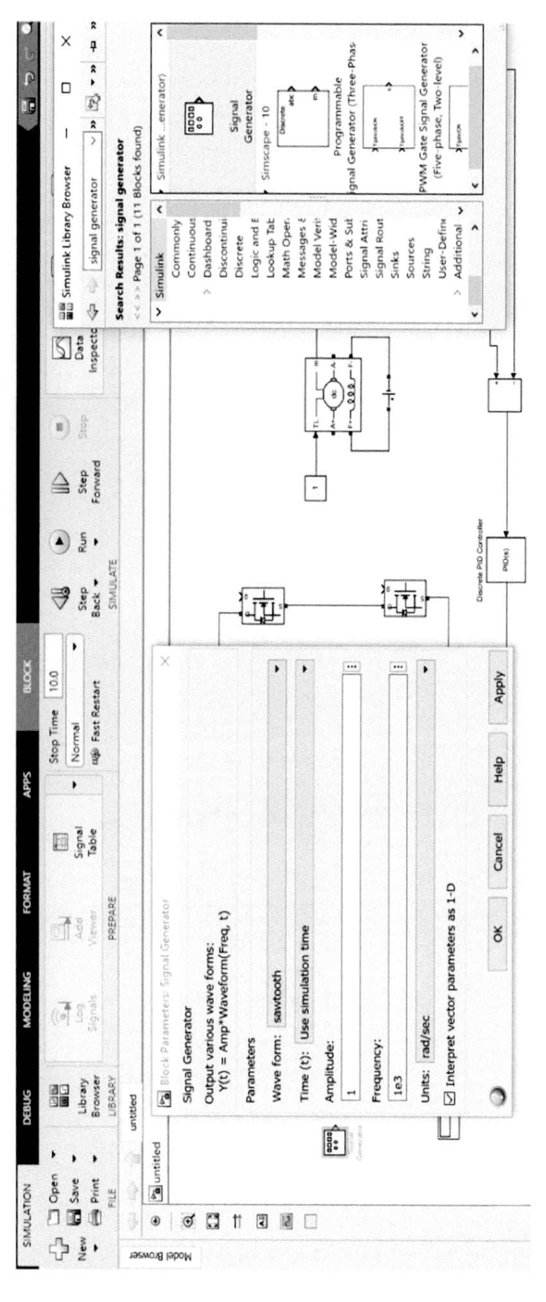

Figure 3.101 Selecting signal generator from library.

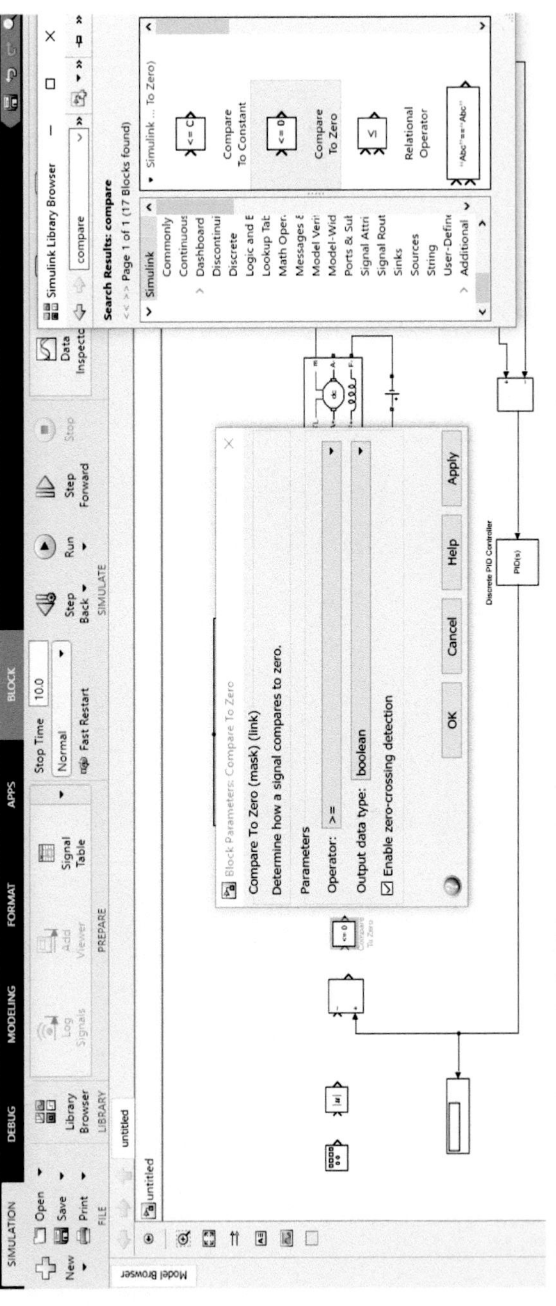

Figure 3.102 Selecting compare to zero from library.

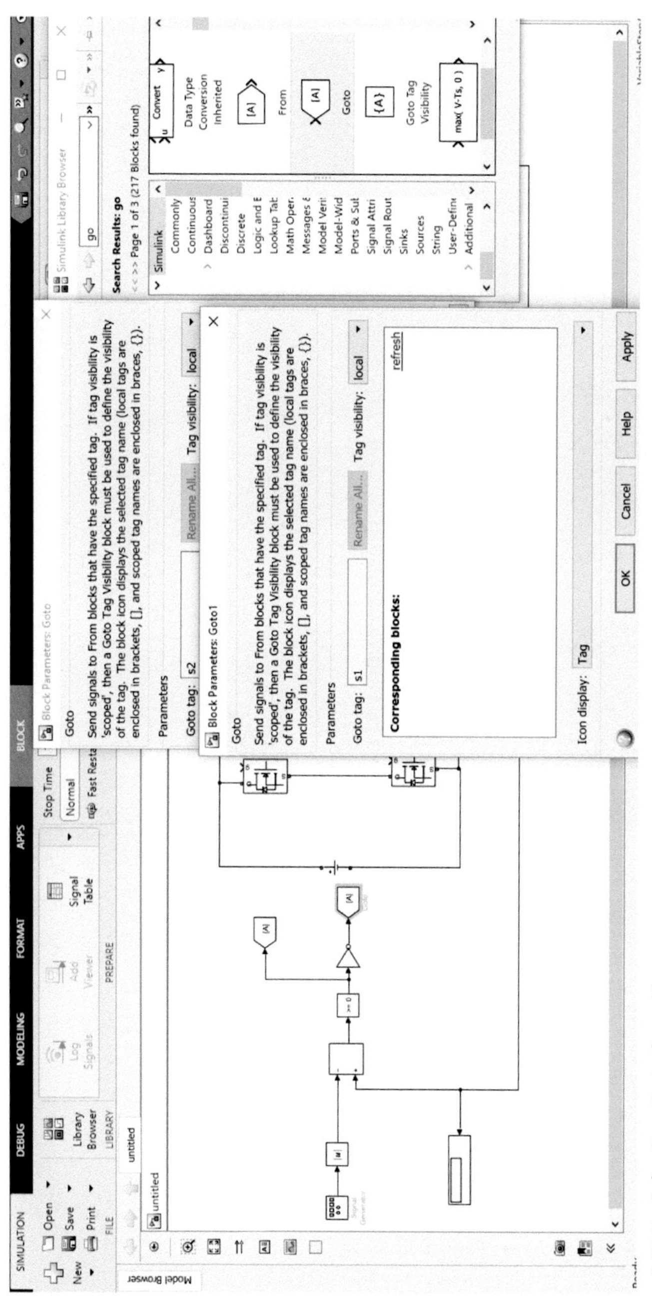

Figure 3.103 Selecting Goto from library.

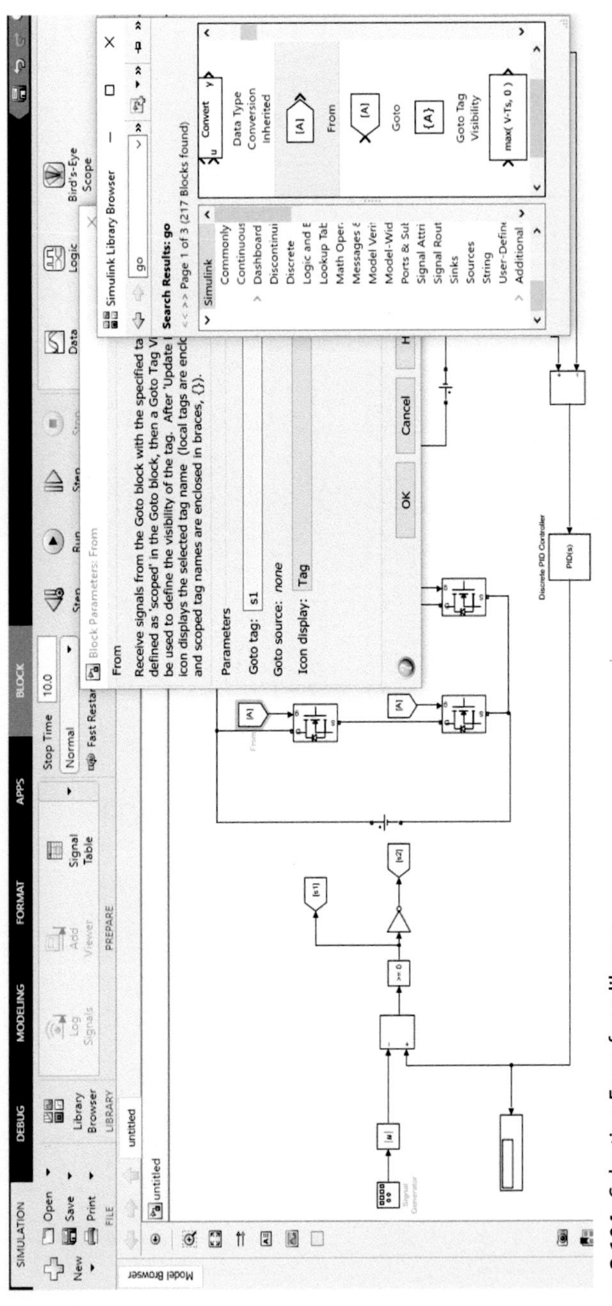

Figure 3.104 Selecting From from library.

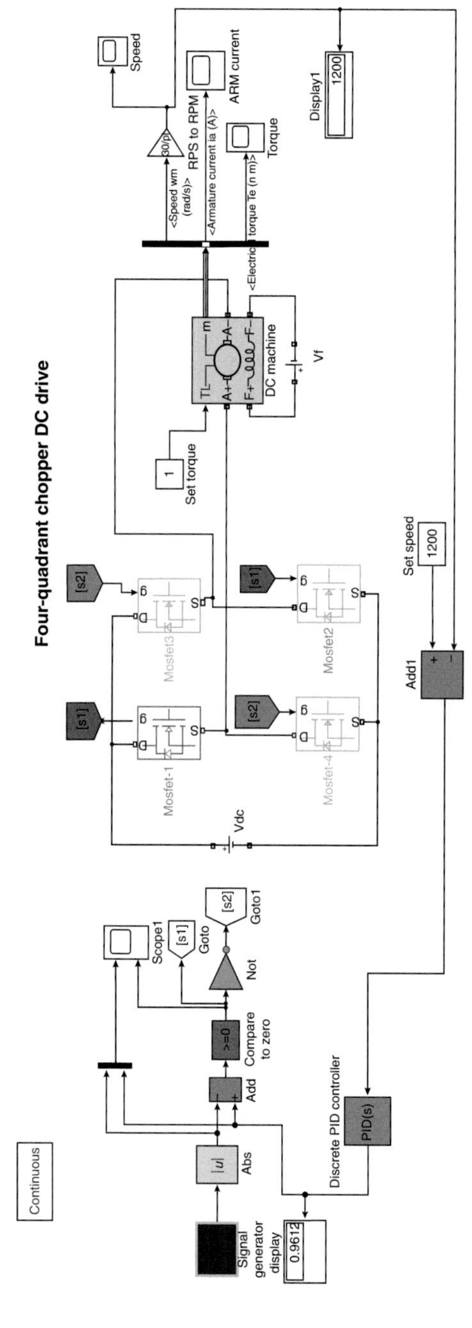

Figure 3.105 MATLAB/Simulink model of four-quadrant chopper DC drive.

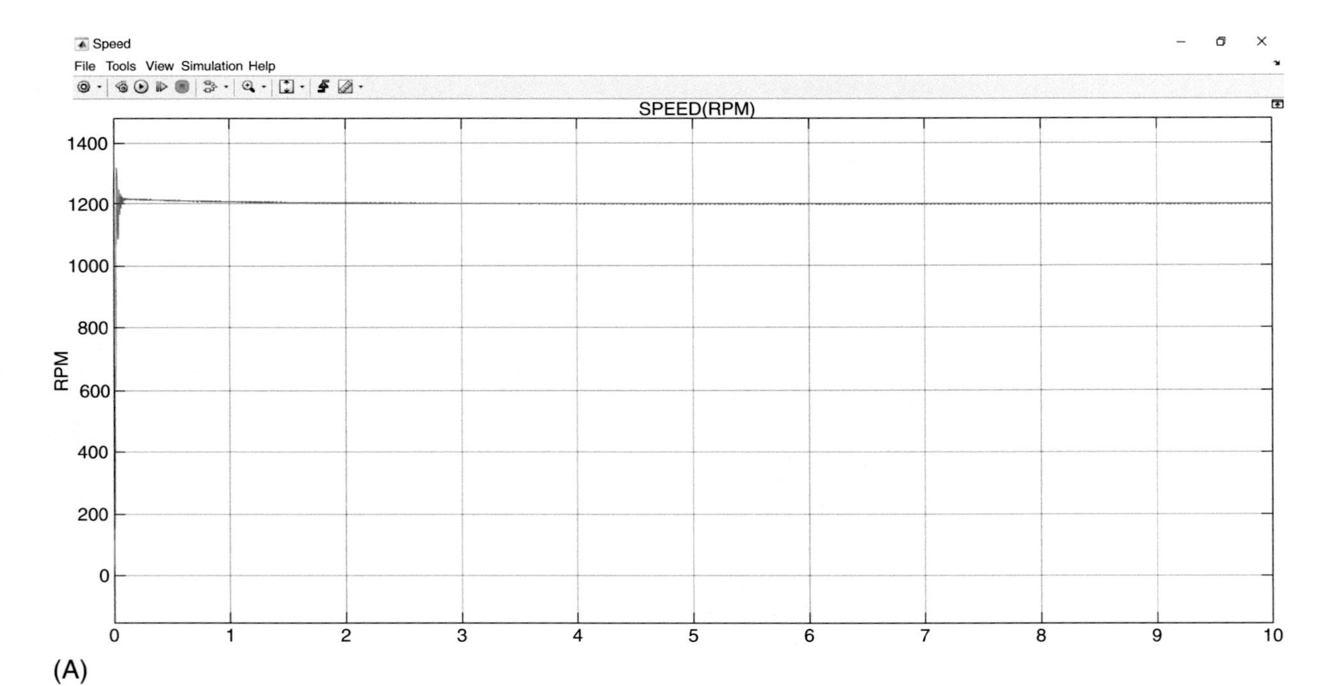

(A)

Figure 3.106 Simulation results of four-quadrant chopper DC drive. (A) Speed, (B) torque output, and (C) Abs, discrete PID controller and Compare to zero.

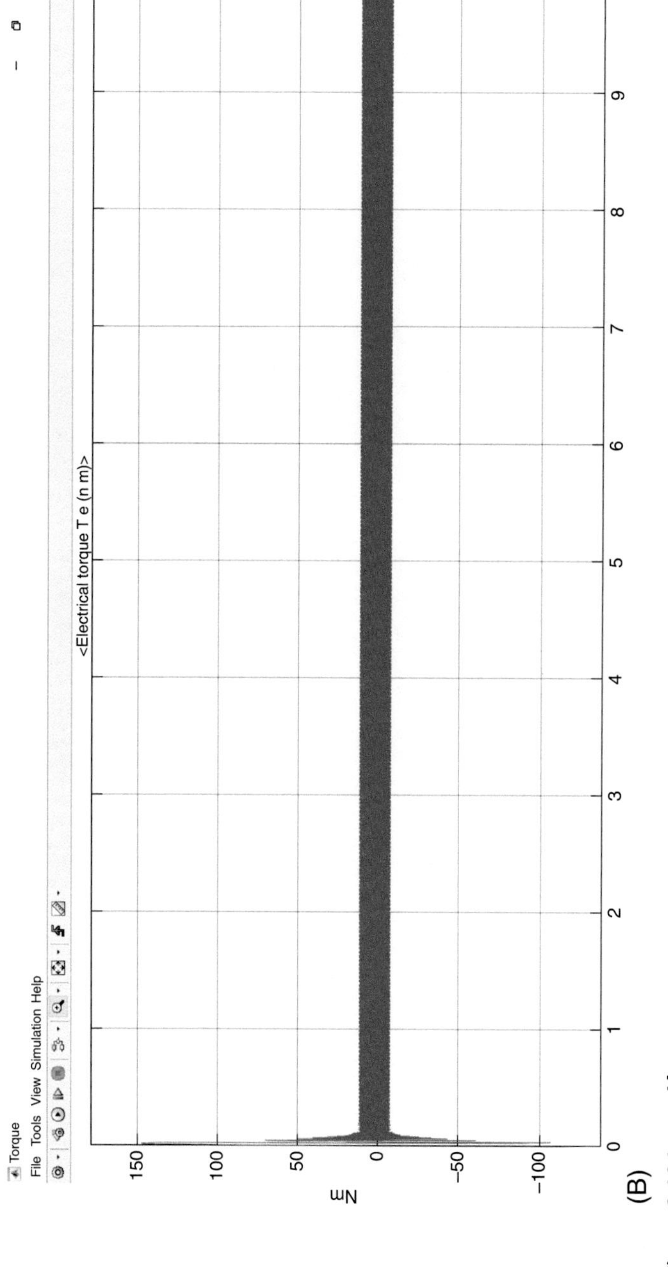

(B)

Figure 3.106, cont'd.

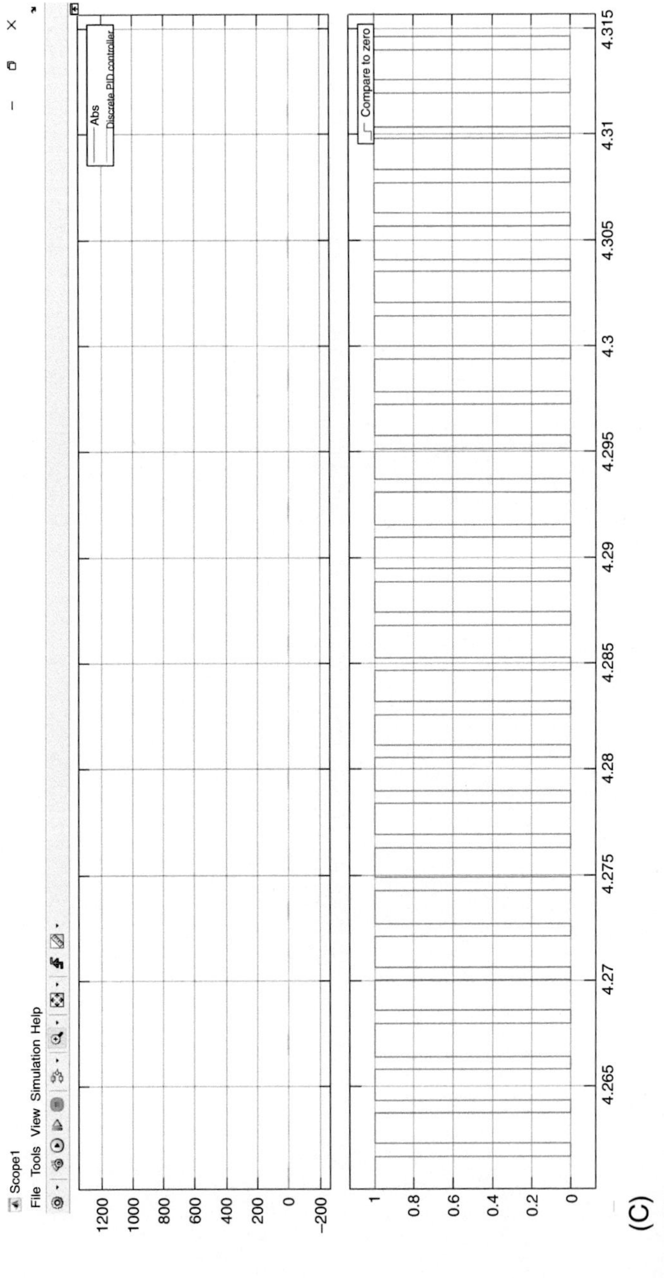

(C)

Figure 3.106, cont'd.

3. Protection.

The main parts of the closed-loop system are the controller, converter, current limiter, current sensor, etc. The converter converts the variable frequency into fixed frequency and vice-versa. The current limiter limits the current to rise above the maximum set value.

To simulate the closed-loop control of chopper fed DC drive, the required elements need to be selected from library as previous. The elements which are not used in the previous circuits are shown here.

Select GTO from library. >>Libraries>>Simscape>>Specialized Power Systems>>Fundamental blocks>> Power Electronics>>GTO (Fig. 3.107)

Select transfer function from library. >>Libraries>>Simulink>> Continuous>>Transfer fcn.

The parameters need to be given as per Fig. 3.108.

Select Relay from library. >>Libraries>>Simulink>>Discontinuous>> Relay (Fig. 3.109).

Select Saturation from library. >>Libraries>>Simulink>>Commonly used blocks >>Saturation (Fig. 3.110).

Select Integrator limited from library. >>Libraries>>Simulink>> Continuous >>Integrator Limited (Fig. 3.111).

Once all the elements are selected from library, connect them as per Fig. 3.112.

Simulate the connected circuit to get the simulation results (Fig. 3.113).

3.10 Case studies

3.10.1 Speed regulation of DC motor by buck converter

The MATLAB/Simulink Model and results of speed regulation of DC motor by buck converter [1] are given in Figs. 3.114 and 3.115.

3.11 Numerical solutions with simulation

1. Construct two-quadrant chopper 200Hp DC drive using MATLAB\ SIMULINK

The two-quadrant chopper DC drive block represents a two-quadrant, DC-supplied, chopper (or DC-DC PWM converter) drive for DC motors (Figs. 3.116 and 3.117). This drive features closed-loop speed control with

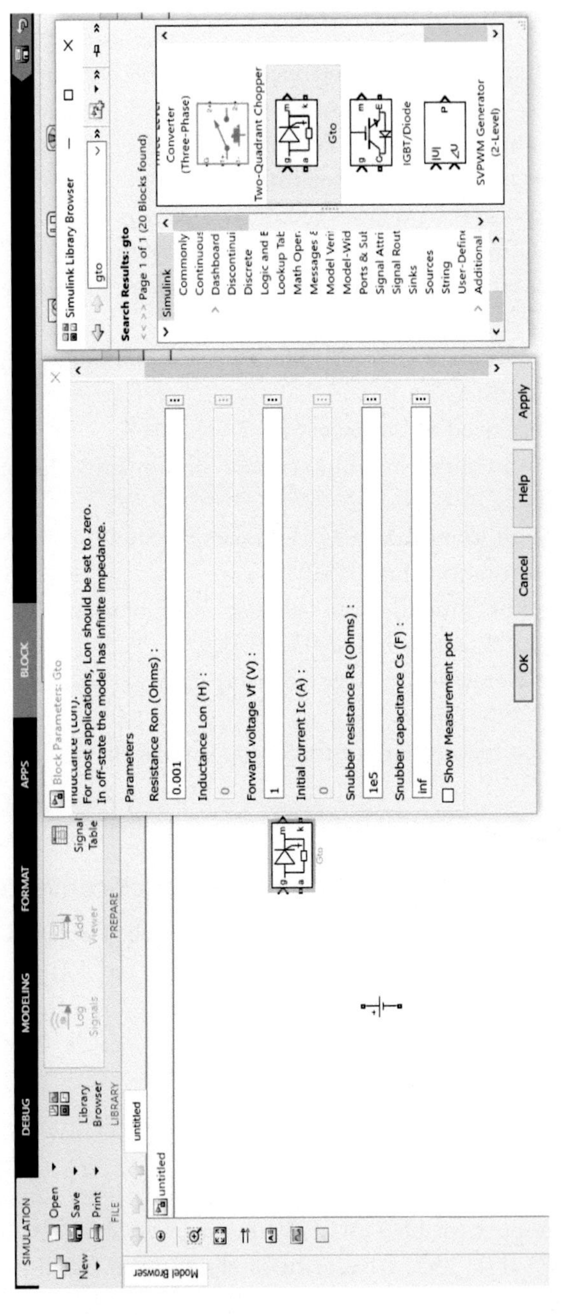

Figure 3.107 Selecting GTO from library.

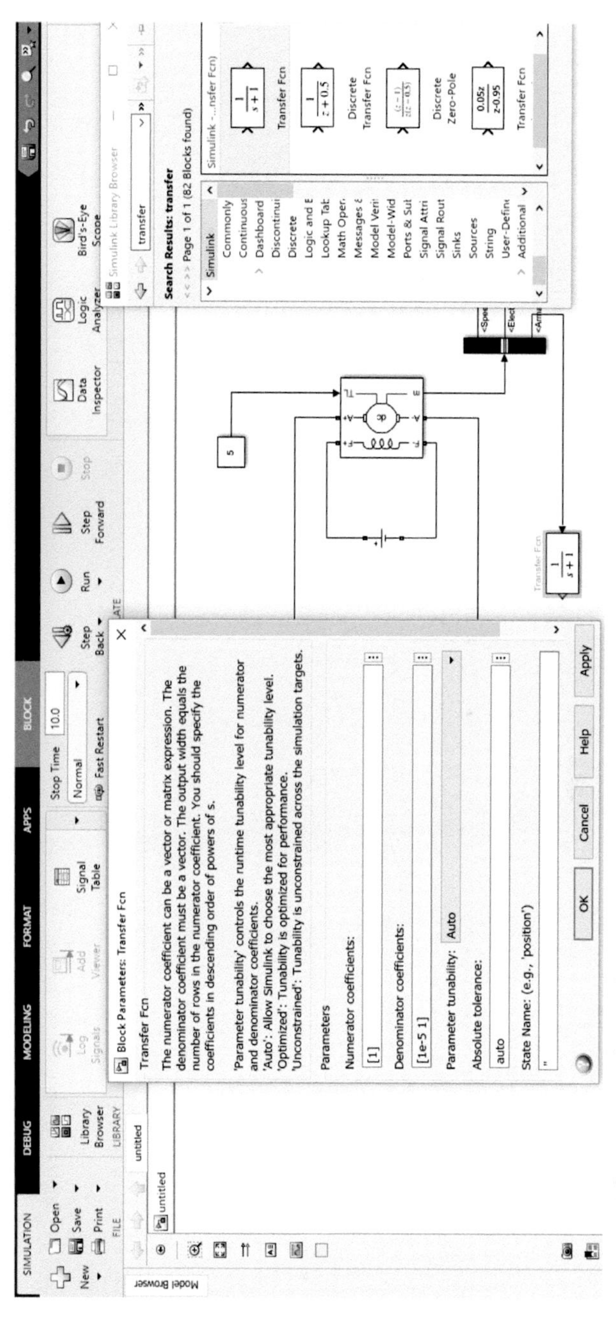

Figure 3.108 Selecting Transfer Fcn from library.

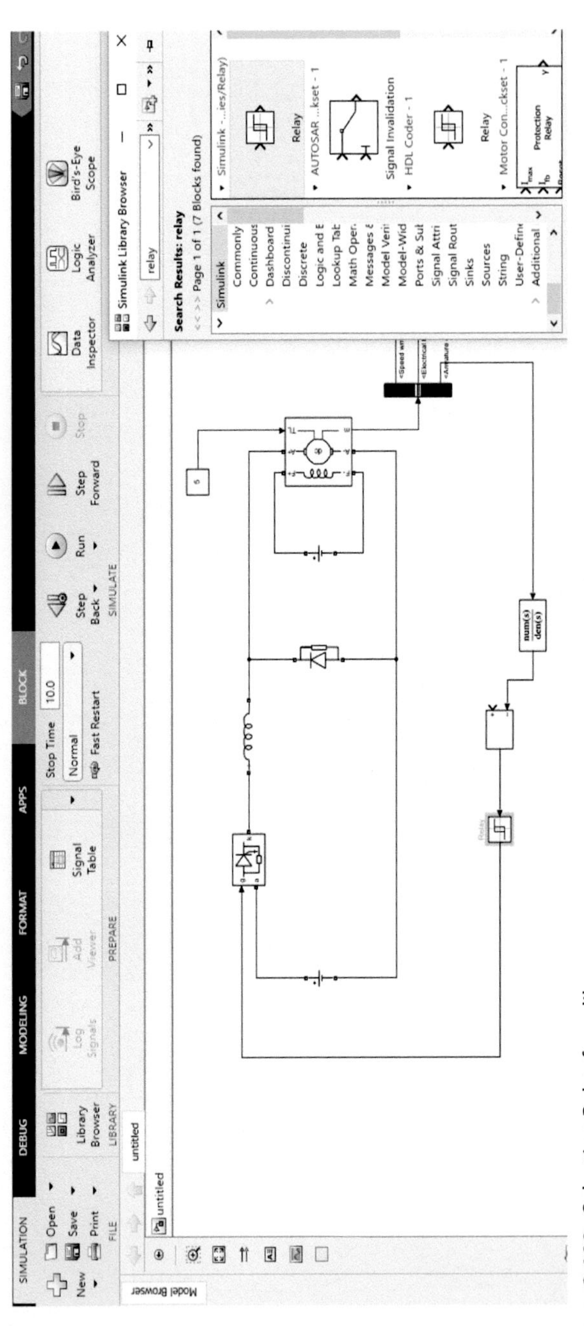

Figure 3.109 Selecting Relay from library.

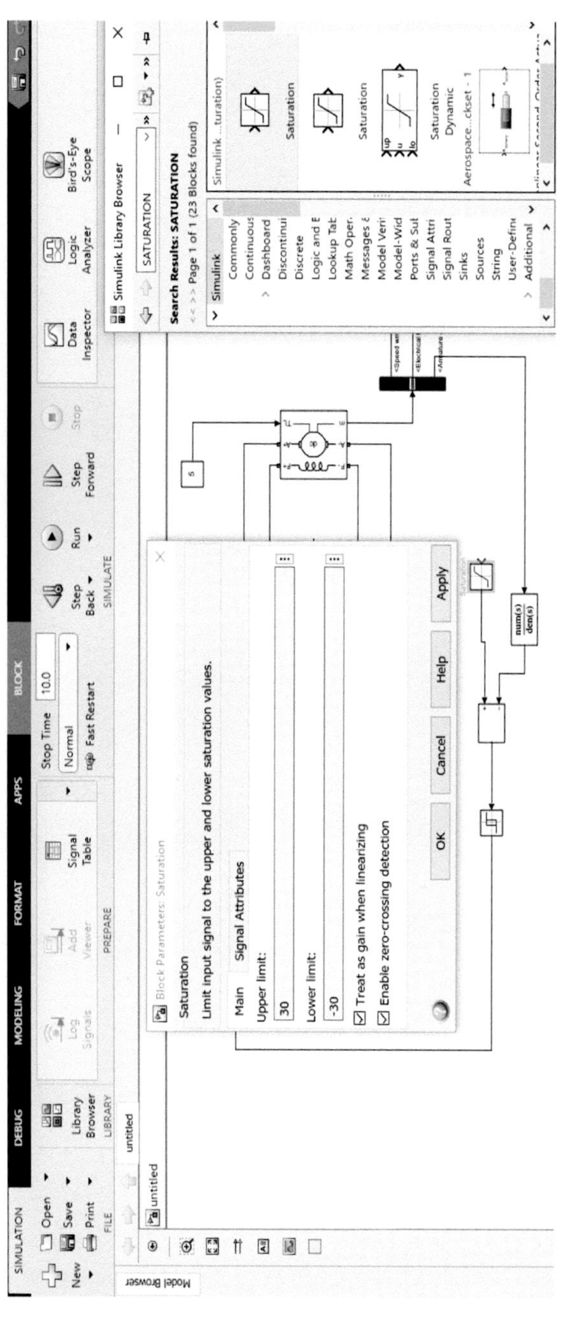

Figure 3.110 Selecting saturation from library.

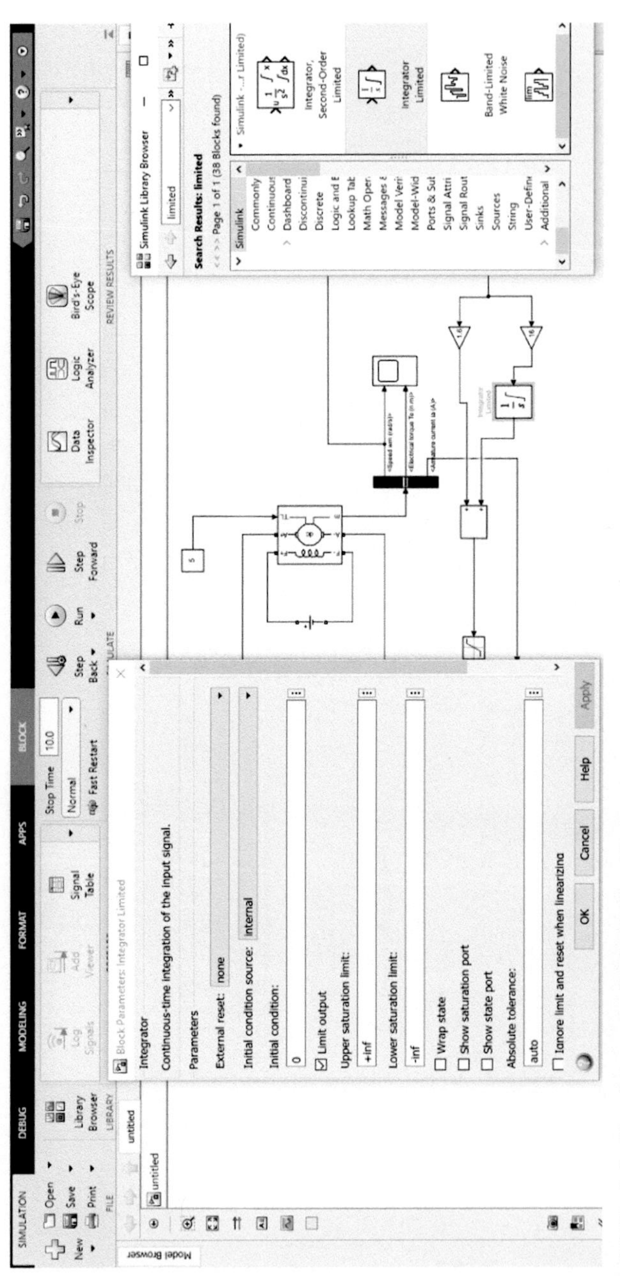

Figure 3.111 Selecting integrator limited from library.

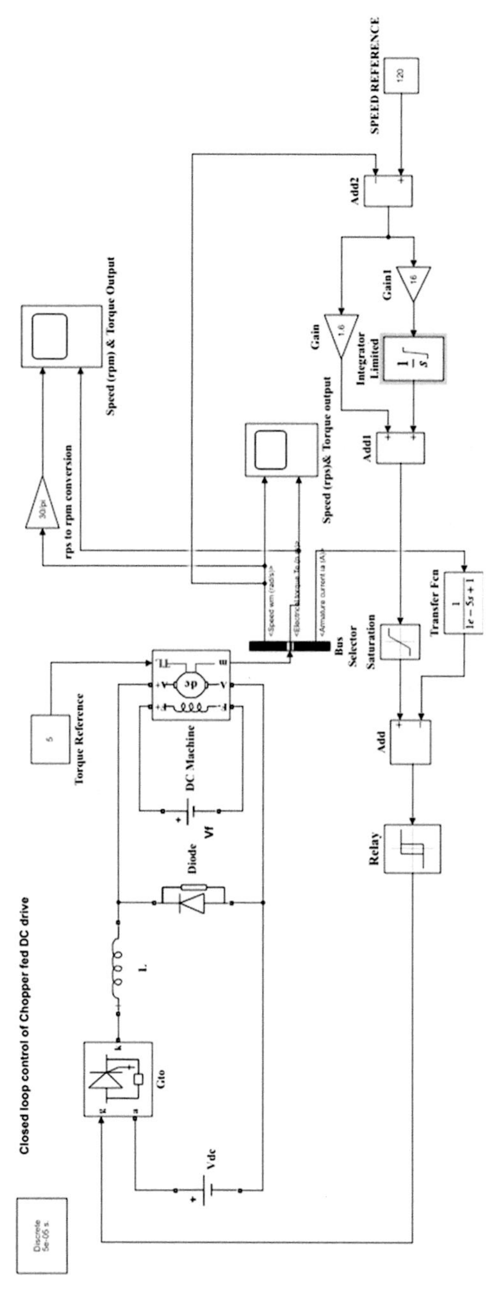

Figure 3.112 MATLAB/Simulink model of closed-loop control of chopper fed DC drive.

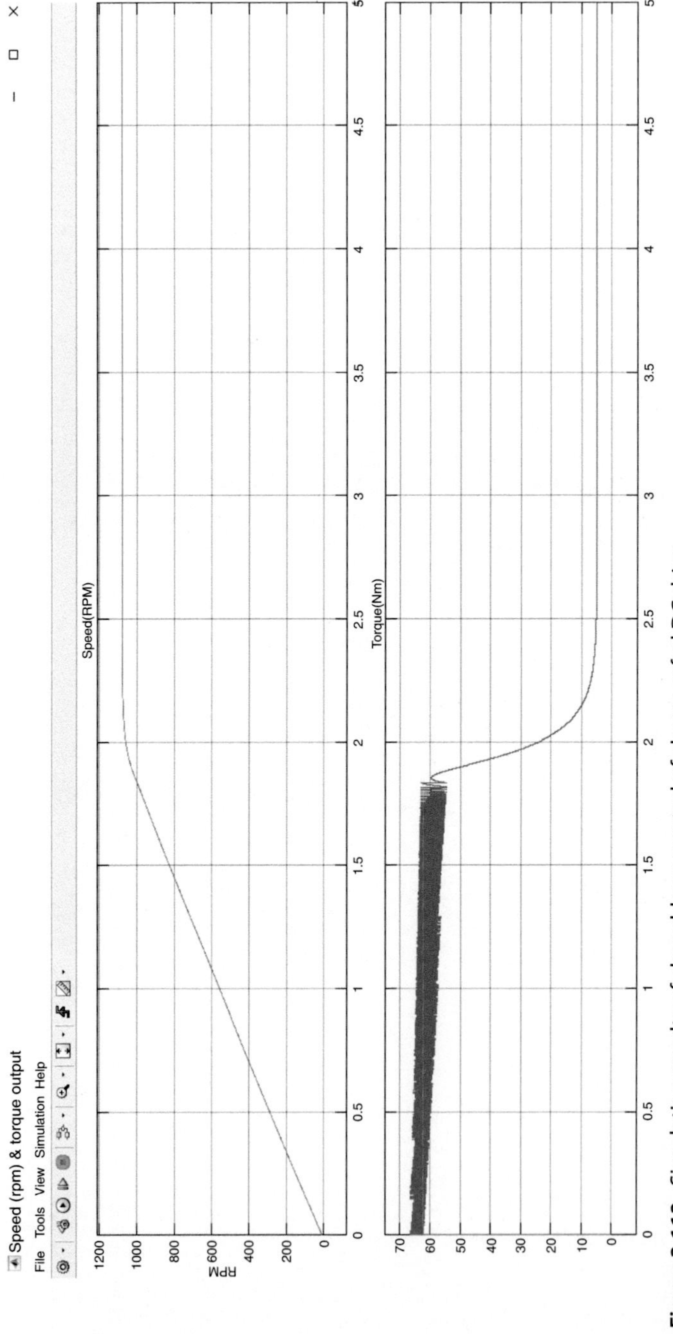

Figure 3.113 Simulation results of closed-loop control of chopper fed DC drive.

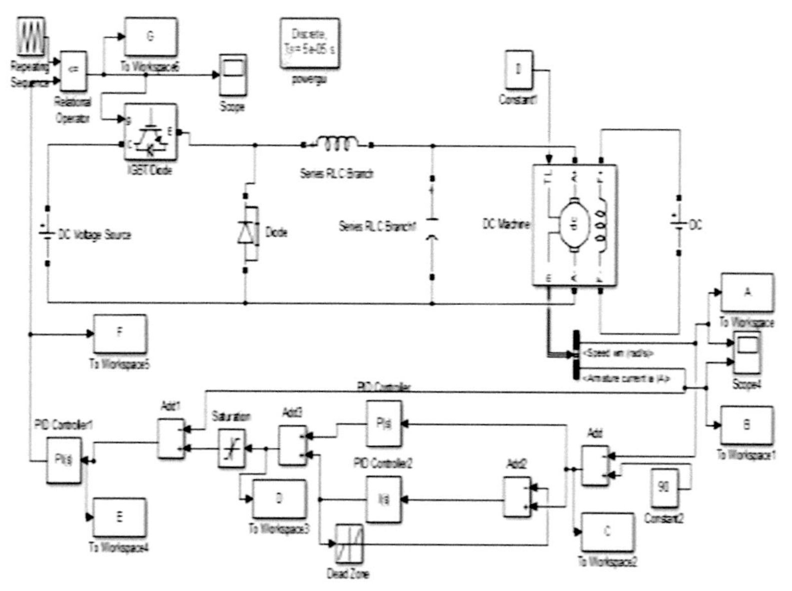

Figure 3.114 MATLAB/Simulink model of speed regulation of DC motor by buck converter.

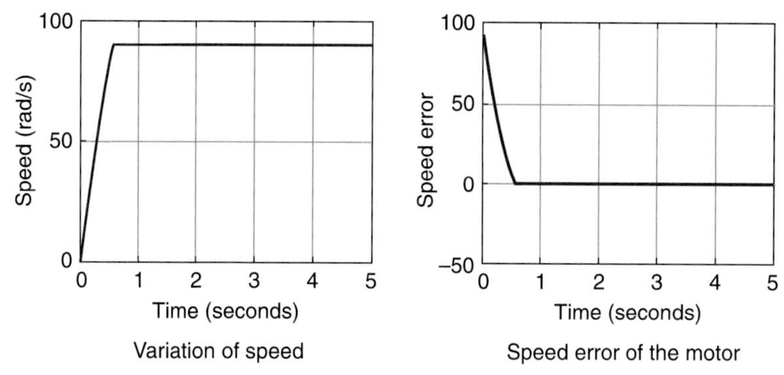

Figure 3.115 MATLAB/Simulink mode simulation results of speed regulation of DC motor by buck converter.

two-quadrant operation. The speed control loop outputs the reference armature current of the machine. Using a PI current controller, the chopper duty cycle corresponding to the commanded armature current is derived. This duty cycle is then compared with a saw tooth carrier signal to obtain the required PWM signals for the chopper.

The main advantages of this drive, compared with other DC drives, are its implementation simplicity and that it can operate in two quadrants

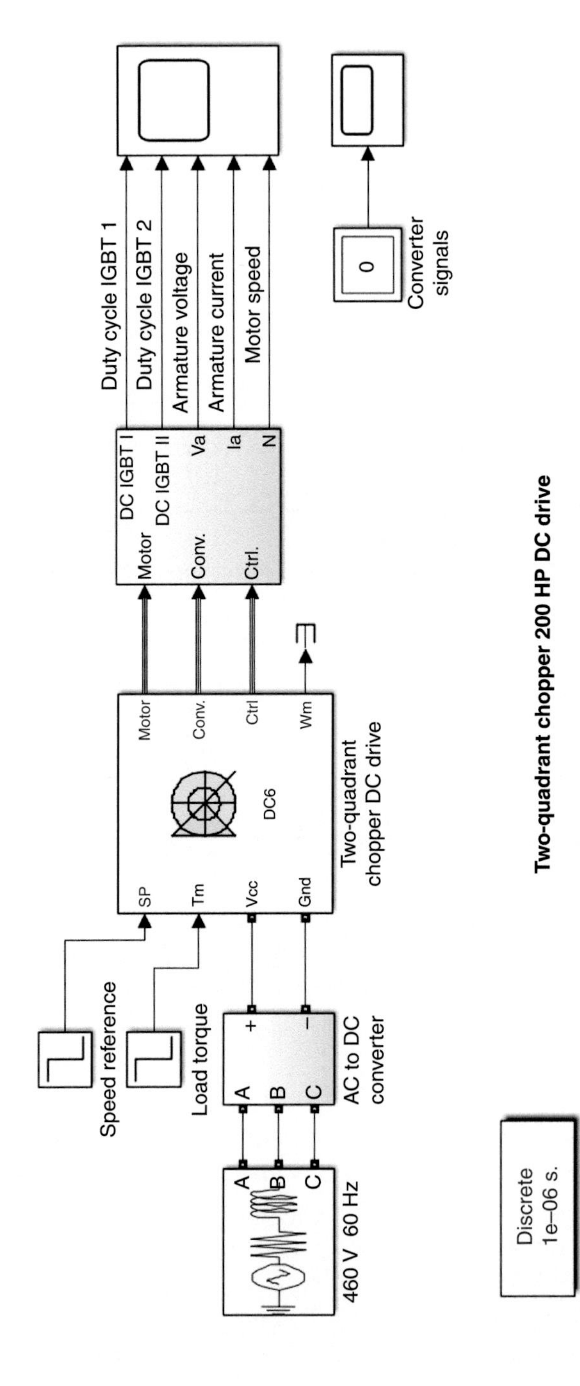

Figure 3.116 MATLAB/Simulink model of two-quadrant chopper [2].

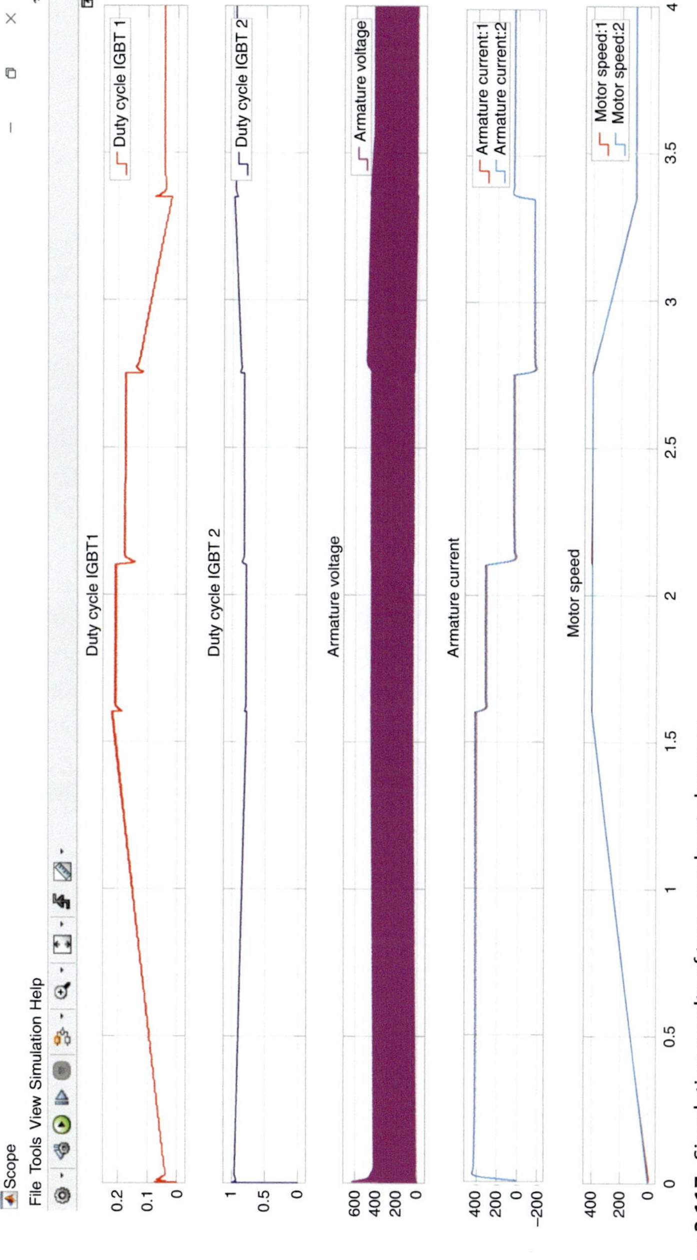

Figure 3.117 Simulation results of two-quadrant chopper.

(forward motoring and reverse regeneration). In addition, due to the use of high switching frequency DC–DC converters, a lower armature current ripple (compared with thyristor-based DC drives) is obtained. However, for all two-quadrant DC drives, reversible and regenerative operations (reverse motoring and forward regeneration), which are required in most DC drives, cannot be obtained.

The two-quadrant chopper DC drive block uses these blocks from the electric drives/fundamental drive blocks library:

- Speed controller (DC).
- Regulation switch.
- Chopper.

3.11.1 Simulation results

2. Construct four-quadrant chopper 200Hp DC drive using MATLAB\ SIMULINK

The four-quadrant chopper DC drive (DC7) block represents a four-quadrant, DC-supplied, chopper (or DC-DC PWM converter) drive for DC motors (Figs. 3.118 and 3.119). This drive features closed-loop speed control with four-quadrant operation. The speed control loop outputs the reference armature current of the machine. Using a PI current controller, the chopper duty cycle corresponding to the commanded armature current is derived. This duty cycle is then compared with a sawtooth carrier signal to obtain the required PWM signals for the chopper.

The main advantage of this drive, compared with other DC drives, is that it can operate in all four quadrants (forward motoring, reverse regeneration, reverse motoring, and forward regeneration). In addition, due to the use of high switching frequency DC–DC converters, a lower armature current ripple (compared with thyristor-based DC drives) is obtained. However, four switching devices are required, which increases the complexity of the drive system.

The four-quadrant chopper DC drive block uses these blocks from the electric drives/fundamental drive blocks library:

- Speed controller (DC).
- Regulation switch.
- Current controller (DC).
- Chopper.

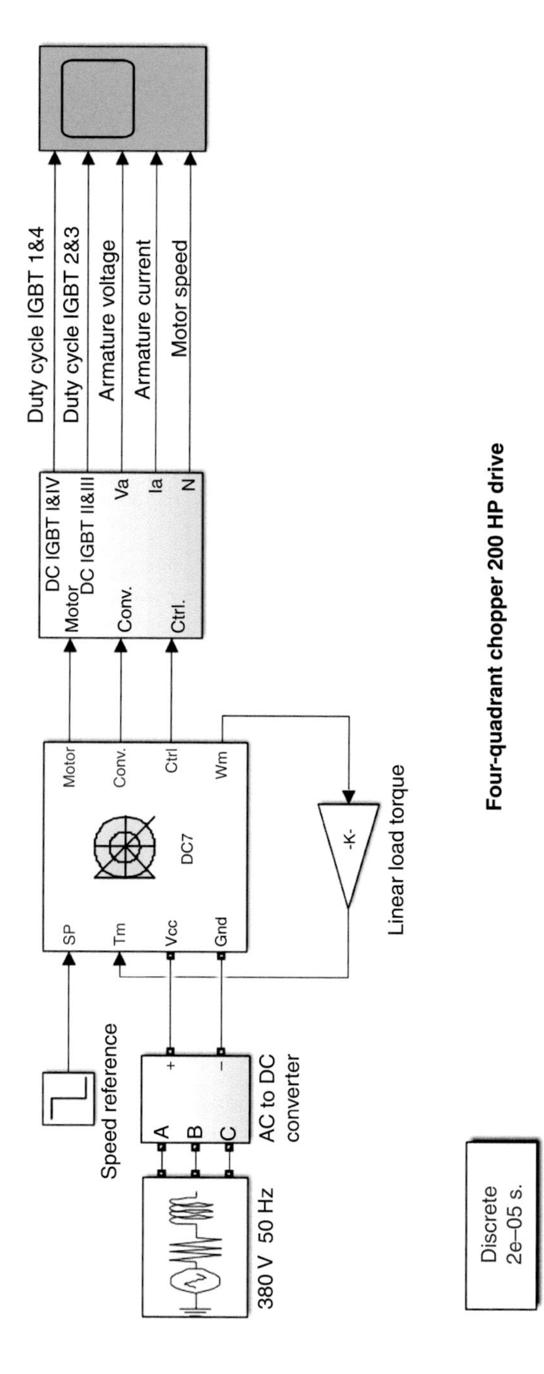

Figure 3.118 MATLAB/Simulink model of four-quadrant chopper.

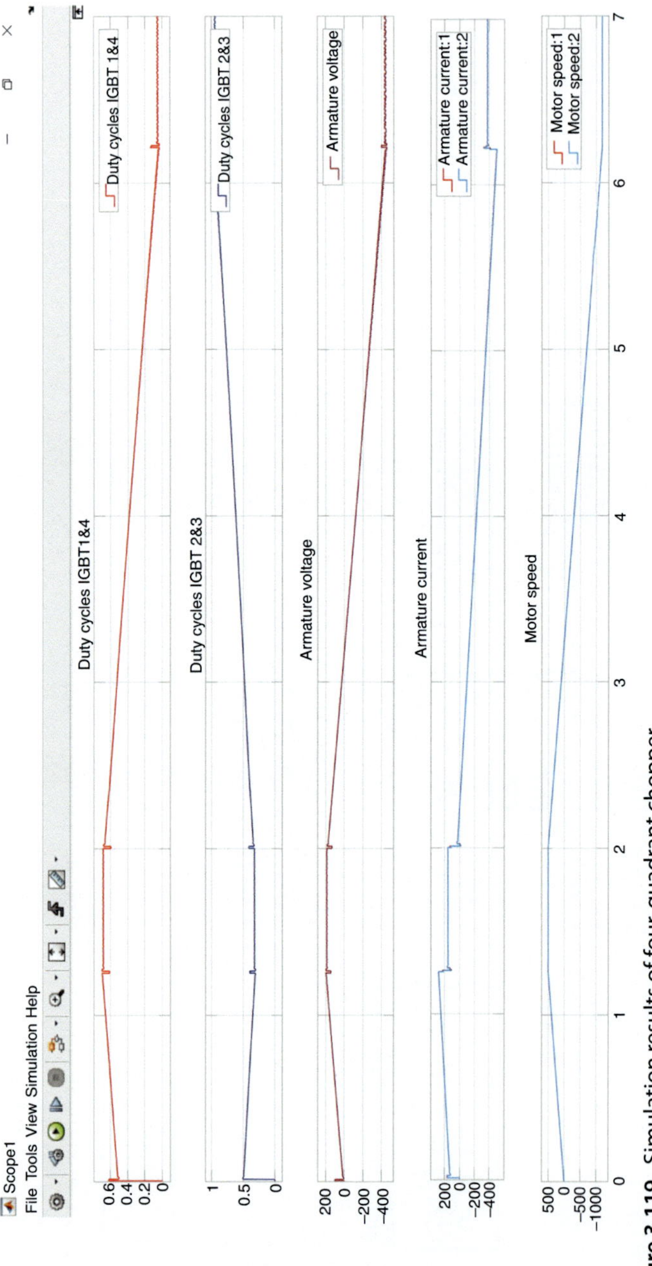

Figure 3.119 Simulation results of four-quadrant chopper.

3.12 Summary

This chapter provides the step-by-step process of chopper fed dc drives simulations using MATLAB/Simulink. The circuit diagrams and results are taken as screenshots from MATLAB/Simulink. On the onset of this chapter, the beginner will be able to execute chopper fed DC drives using MATLAB software.

Practice Questions

1.

Multiple Choice Questions

1. Choppers is a_______
   (a) AC – DC Converter
   (b) AC – AC Converter
   (c) DC – AC Converter
   (d) DC – DC Converter
   Answer: Option (d)
2. The principle of step-up chopper can be employed for the__________
   (a) Motoring mode
   (b) Regenerative mode
   (c) Plugging
   (d) Reverse motoring mode
   Answer: Option (b)
3. Duty cycle (D) is _______
   (a) Ton÷Toff
   (b) Ton÷(Ton+ Toff)
   (c) Ton÷2×(Ton+ Toff)
   (d) Ton÷2×Toff
   Answer: Option (b)
4. A step - down choppers can be used in
   (a) Electric Traction
   (b) Electric Vehicle
   (c) Machine Tools
   (d) All of these
   Answer: Option (d)

5. In DC chopper, the waveform for input and output voltages is respectively_________
   - **(a)** Discontinuous and Continuous
   - **(b)** Continuous and Discontinuous
   - **(c)** Both Continuous
   - **(d)** Both Discontinuous

   Answer: Option (b)
6. A four quadrant operation requires
   - **(a)** two full converters in series
   - **(b)** two full converters connected in parallel
   - **(c)** two full converter connected in back to back
   - **(d)** two semi converters connected in back to back

   Answer: Option (c)
7. Which braking is not possible in series motor?
   - **(a)** Regenerative braking
   - **(b)** Dynamic braking
   - **(c)** Counter electric current braking
   - **(d)** Rheostat braking

   Answer: Option (a)
8. An elevator drive is required to operate in
   - **(a)** one quadrant only
   - **(b)** two quadrants
   - **(c)** three quadrants
   - **(d)** four quadrants

   Answer: Option (d)
9. In industries which electrical braking is preferred?
   - **(a)** Regenerative braking
   - **(b)** Dynamic braking
   - **(c)** Plugging
   - **(d)** None of the above

   Answer: Option (a)
10. High braking torque produced in
    - **(a)** plugging
    - **(b)** dynamic braking
    - **(c)** regenerative braking
    - **(d)** none of above

    Answer: Option (a)

References

Sinha, R., Kasari, P.R., Chakrabarti, A., et al., 2018. Speed regulation of DC motor by buck converter. In: IEEE International Conference on Power Electronics, Drives and Energy Systems (PEDES).

Mathworks.com.

https://www.circuitstoday.com/types-of-chopper-circuits.

CHAPTER 4

Induction motor drives and its simulation

Contents

4.1 Introduction

For variable-speed applications, DC motor drives are used in the past. But this motor has several disadvantages like the presence of commutator and brushes due to which frequent maintenance is required. This problem is overcome by the variable speed induction motor drive. The induction motor drive is cheaper, lighter, smaller, more efficient, and requires low maintenance. The only disadvantage of an induction motor drive is its higher cost.

The induction motor drive has many applications like it is used in fans, blowers, mill run-out tables, cranes conveyers, traction, etc. The induction motor drive is self-starting, or we can say when the supply is given to the motor, it starts rotating without any external supply.

The initial resistance of the supply is zero, and hence large current flows through the motor, which damages the windings of the motor. For reducing the flow of starting current, different starting methods are used. These methods keep the magnitude of starting current within a prescribed limit such that it does not cause overheating.

Electric Motor Drives and Their Applications with Simulation Practices.
DOI: https://doi.org/10.1016/B978-0-323-91162-7.00010-2

In a 3-phase AC induction motor, there are three stator windings, each usually in two halves, with the rotor winding short-circuited by end rings. As the current passes through the coils on opposite sides of the stator, a two-pole electromagnet is established, creating a two-pole motor. Applying a phase to each of the electromagnets in turn creates the rotating magnetic field that is strong enough to start moving the rotor. More winding can create more poles in the motor, with more complex control required but more accuracy in positioning the rotor. A four-pole motor is regarded as optimum for the torque and responsiveness needed for the motor drives of electric cars, for example. But higher pole counts are only possible with more sophisticated control schemes.

The typical drive has three half-bridges, each delivering a sine-wave voltage to the stator. This uses power MOSFETs or IGBTs with high-voltage gate drivers, or power modules that combine the three half-bridges and related gate drives. These can use scalar algorithms that vary the voltage to determine the frequency of the phases, or volts/hertz. More sophisticated algorithms such as vector control or field-oriented control (FOC) are used to control the frequency of multiple phases in high-end motors are now increasingly popular across the range of three-phase induction motors.

Indirect field oriented controlled (IFOC) induction motor drives are being increasingly used in high-performance drive systems, as induction motors are more reliable because of their construction and less expensive materials used, than any other motors available in the market today. As indirect field orientation utilizes an inherent slip relation, it is essentially a feed-forward scheme and hence depends greatly on the accuracy of the motor parameters used in the vector controller—particularly to the rotor resistance. It changes widely with the rotor temperature, resulting in various harmful effects—such as over (or under) excitation, the destruction of the decoupled condition of the flux and torque, etc.

Recently, attention has been given to the identification of the instantaneous value of the rotor resistance while the drive is in normal operation. So far, several approaches have been presented. A new sliding mode current observer for an induction motor is developed. Sliding mode functions are chosen to determine speed and rotor resistance of an induction motor in which the speed and rotor resistance are assumed to be unknown constant parameters. In a method using a programmable cascaded low pass filter for the estimation of rotor flux of an induction motor, with a view to estimate the rotor time constant of an indirect field orientation-controlled induction motor drive is investigated. The estimated rotor flux data has also been used for the on-line rotor resistance identification with the artificial neural network. Despite all these effects, rotor resistance estimation remains a difficult problem.

AC motor drives are extensively used in industrial application requiring high performance. In high-performance systems, the motor speed should closely follow a specified reference trajectory regardless of any load disturbance, parameter variations, and model uncertainties. In order to achieve high performance, FOC of induction motor drive is employed. However, the control design of such a system plays a role in system performance. The decoupling characteristics of vector-controlled induction motor have adversely affected the parameter changes in the motor. The speed control of IM issues is traditionally handled by fixed gain PI and PID controllers. However, the fixed-gain controllers are very sensitive to parameter variations, load disturbances, etc. Thus, the controller parameters have to be continuously adapted.

The problem can be solved by several adaptive control techniques such as model reference adaptive control, sliding mode control, variable structure control, and self-tuning PI controller. The design of the entire above controller depends on the exact system mathematical model. However, it is often difficult to develop an accurate mathematical model due to unknown load variation and unavoidable parameter variations due to saturation, temperature variations, and system disturbance. To overcome the above problems, fuzzy logic controller (FLC) is being used for motor control purposes. There is some advantage of FLC as compared to conventional PI, PID, and adaptive controller; it does not require any mathematical model—it is based on linguistic rules within "if then" general structure, which is the basic of human logic. In this chapter, the configuration and design of FLC of indirect vector control of induction motor has been investigated. The performance of FLC has been successfully compared with conventional PI controllers. The induction motor drives are simulated using software like PSIM and MATLAB. All the possible MATLAB/Simulink library, as shown in Fig. 4.1, double click on the drive for the power, voltage, and frequency parameter settings. The parameter settings tab is shown in Figs. 4.2–4.4. The measurements are done in MATLAB by using bus selector block. All the rotor parameters, stator parameters, and mechanical parameters are possible to visualize after the simulation, as shown in Figs. 4.5–4.7.

Terminals of induction motor drives

SP: The speed or torque set point. The speed setpoint can be a step function, but the speed change rate will follow the acceleration/deceleration

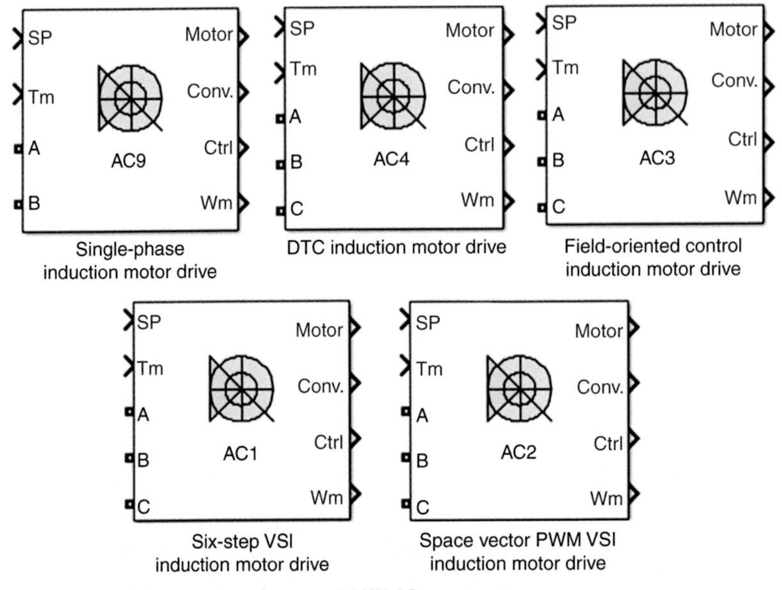

Figure 4.1 Induction motor drives in MATLAB.

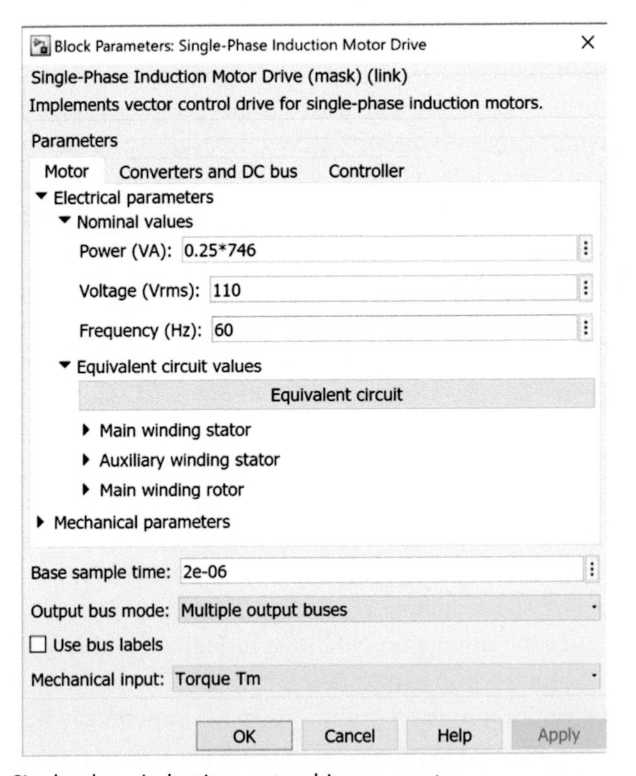

Figure 4.2 Single-phase induction motor drive parameters.

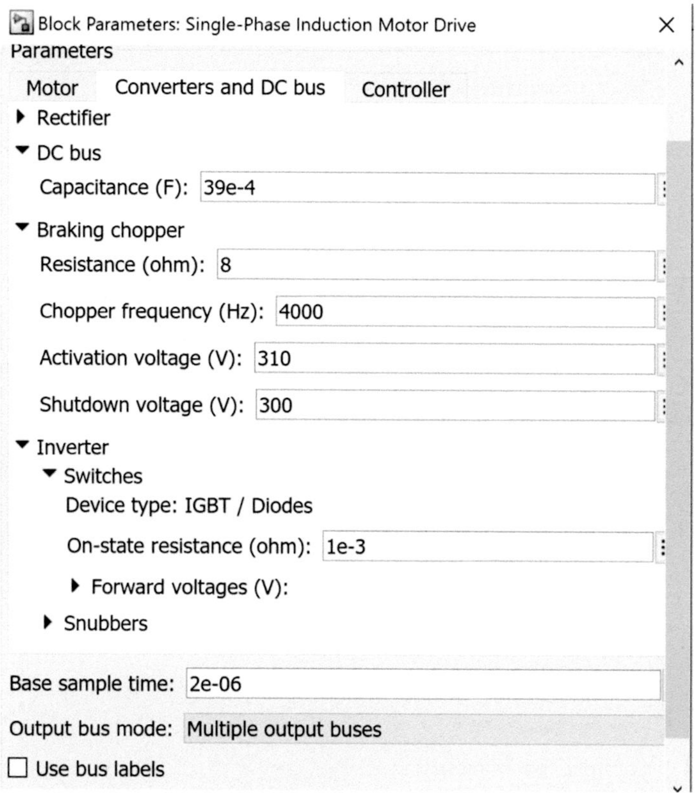

Block Parameters: Single-Phase Induction Motor Drive
Parameters
Motor | Converters and DC bus | Controller
▸ Rectifier
▾ DC bus
Capacitance (F): 39e-4
▾ Braking chopper
Resistance (ohm): 8
Chopper frequency (Hz): 4000
Activation voltage (V): 310
Shutdown voltage (V): 300
▾ Inverter
▾ Switches
Device type: IGBT / Diodes
On-state resistance (ohm): 1e-3
▸ Forward voltages (V):
▸ Snubbers
Base sample time: 2e-06
Output bus mode: Multiple output buses
☐ Use bus labels

Figure 4.3 Converter and DC bus parameters.

ramps. If the load torque and the speed have opposite signs, the accelerating torque will be the sum of the electromagnetic and load torques.
Tm or Wm: The mechanical input: load torque (Tm) or motor speed (Wm). For the mechanical rotational port (S), this input is deleted.
A, B, C: The three-phase terminals of the motor drive.
Wm, Te, or S: The mechanical output: motor speed (Wm), electromagnetic torque (Te), or mechanical rotational port (S).

When the output bus mode parameter is set to multiple output buses, the block has the following three output buses:
Motor: The motor measurement vector. This vector allows you to observe the motor's variables using the bus selector block.
Conv: The three-phase converters measurement vector.
This vector contains the DC bus voltage.
The rectifier output current.

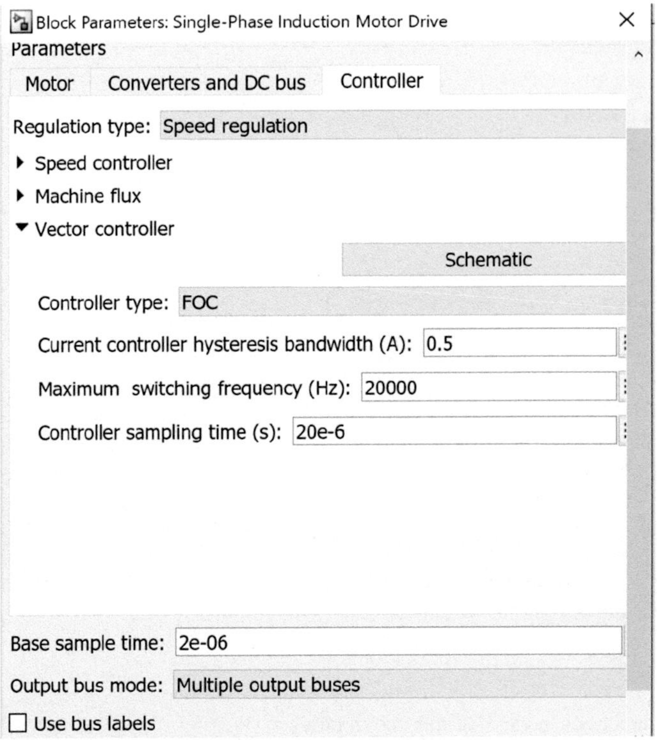

Figure 4.4 Controller parameters.

The inverter input current.

Note that all current and voltage values of the bridges can be visualized with the Multimeter block.

Ctrl: The controller measurement vector.

This vector contains the torque references.

The speed error (difference between the speed reference ramp and actual speed).

The speed reference ramp or torque reference.

4.2 Simulation of three-phase induction motor at different load conditions

The simulation diagram of induction motor fed from constant 3-phase supply designed in MATLAB/Simulink is shown in Fig. 4.8. The induction motor drive with variable 3-phase supply obtained from voltage source inverter (VSI) is presented in Fig. 4.9. The gate signal presented to the IGBT

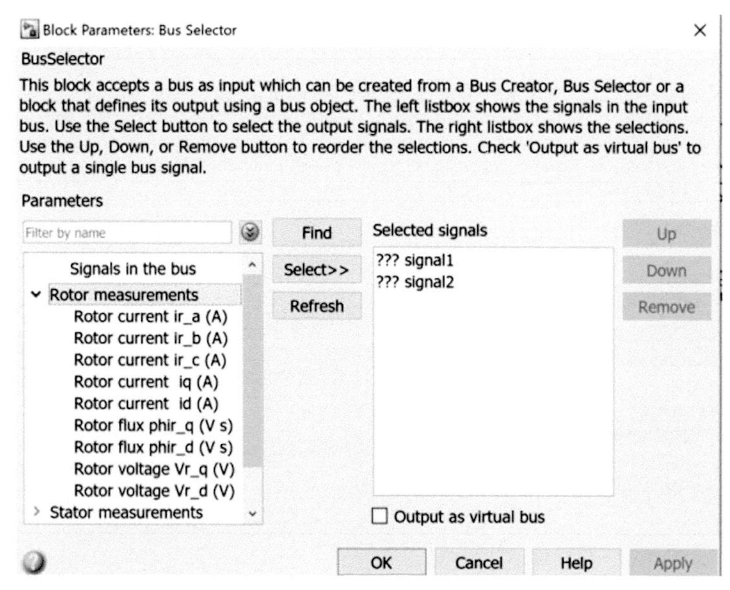

Figure 4.5 Rotor measurements.

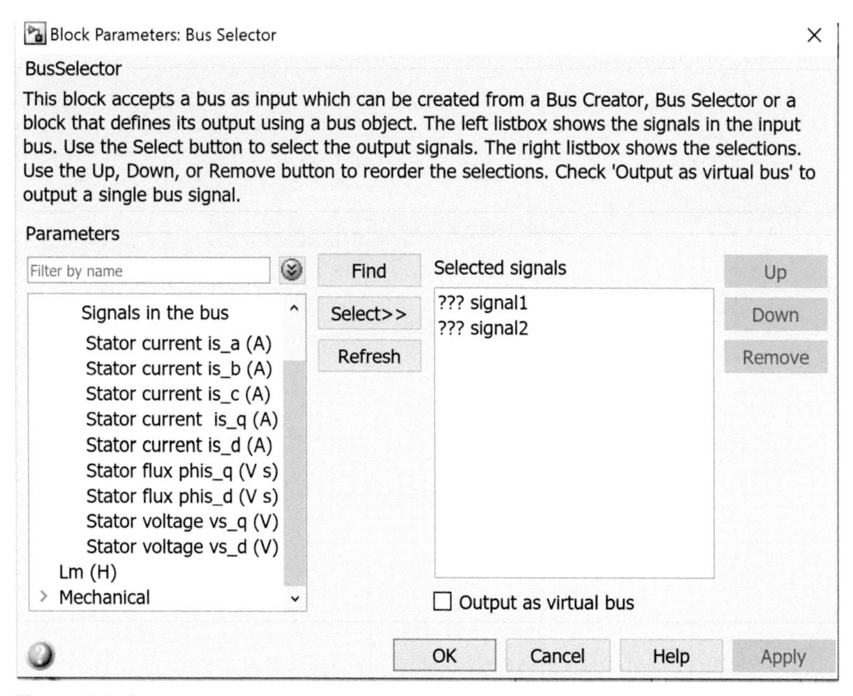

Figure 4.6 Stator measurements.

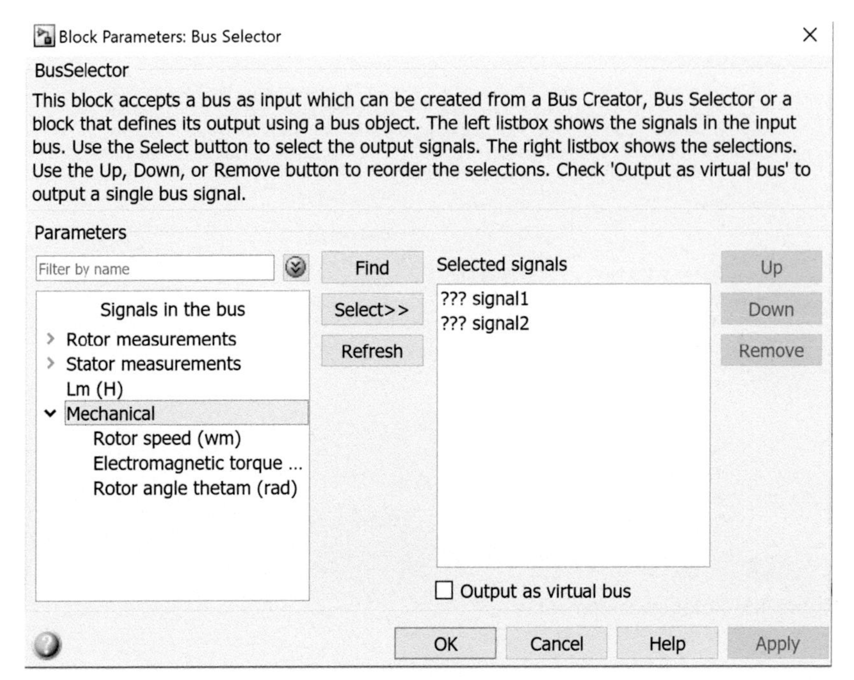

Figure 4.7 Mechanical parameter measurements.

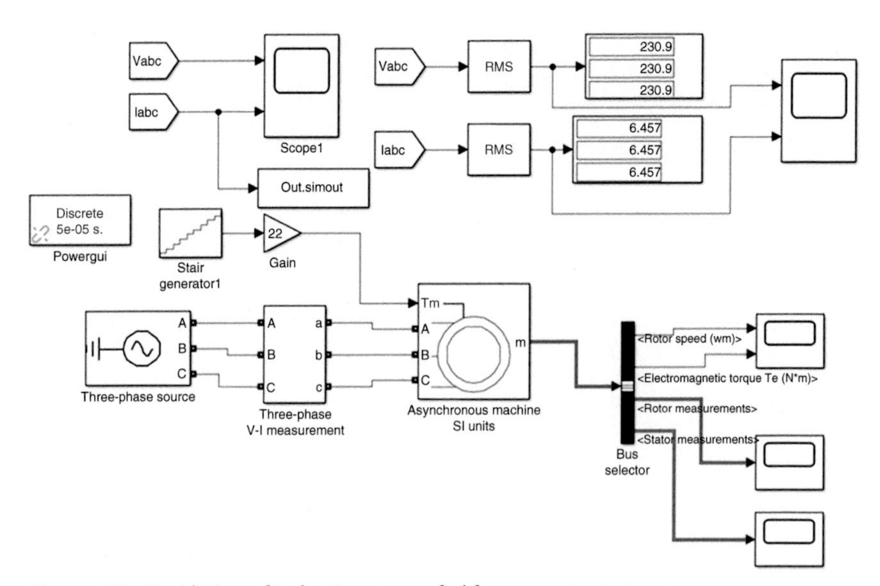

Figure 4.8 Simulation of induction motor fed from constant source.

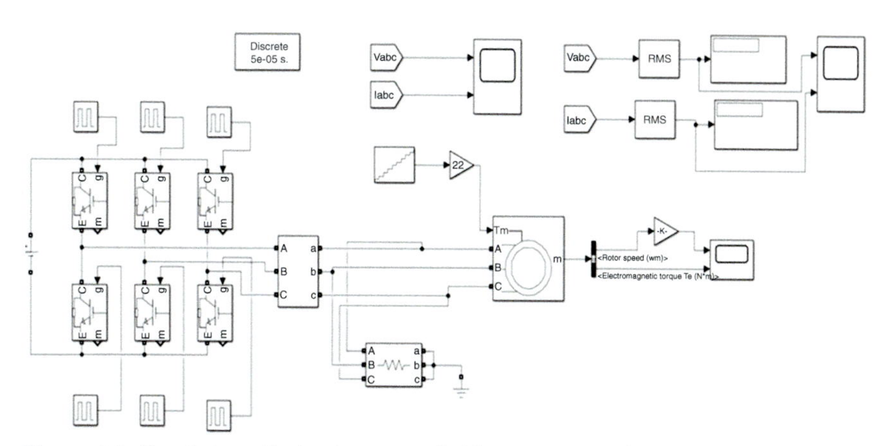

Figure 4.9 Simulation of induction motor fed from VSI. *VSI*, voltage source inverter.

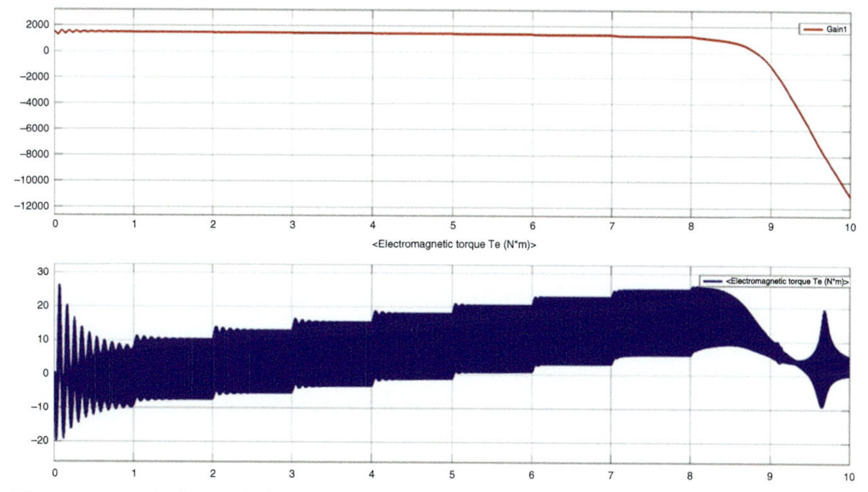

Figure 4.10 Mechanical characteristics of drive.

switches is varied to get the variable input voltage to the induction motor. It helps to control the speed of the induction motor. The results and variations of current at different load conditions are presented in Figs. 4.10–4.12. The gate pulses generated are shown in Fig. 4.13.

4.3 PWM inverter fed variable frequency drive simulation

The pulse width modulation (PWM) inverter fed variable frequency drive simulation diagram is presented in Fig. 4.14. The reference and carrier signals are presented in Fig. 4.15. The gate pulses generated after the PWM

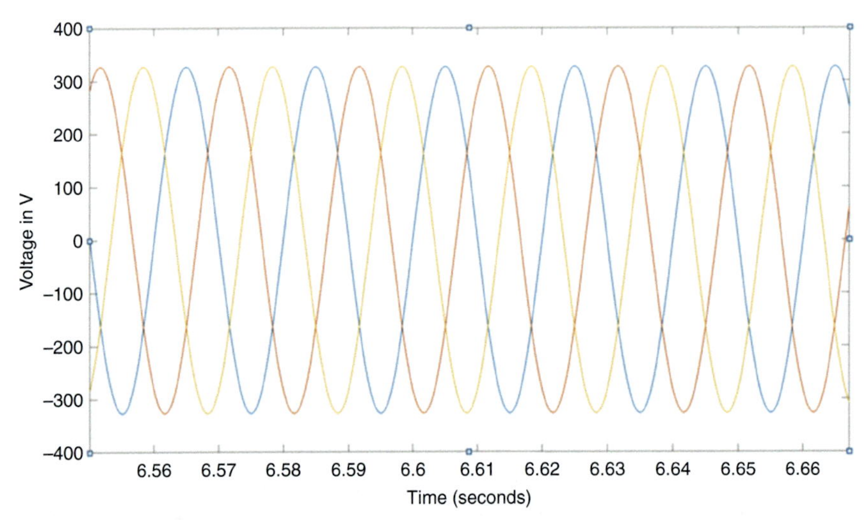

Figure 4.11 Three-phase voltage.

are presented in Fig. 4.16. The mechanical characteristics of the drive are presented in Fig. 4.17.

4.4 Simulation of the single-phase induction motor

The single-phase induction motor drive block models a vector-controlled single-phase machine drive(Fig. 4.18). The drive configuration consists of a half-bridge rectifier, a divided DC bus with two filter capacitors, and a two-leg inverter that supplies the motor windings. The single-phase induction machine, without its start-up and running capacitors, is treated as an asymmetric two-phase machine. The auxiliary and main windings are accessible and are in quadrature. This configuration provides good performances and operation in regenerating mode.

The FOC induction motor drive block represents a standard vector or rotor FOC drive for induction motors (Fig. 4.19). This drive features closed-loop speed control based on the indirect or feedforward vector control method. The speed control loop outputs the reference electromagnetic torque and rotor flux of the machine. The reference direct and quadrature (dq) components of the stator current, corresponding to the commanded rotor flux and torque, are derived based on the indirect vector control strategy. The reference dq components of the stator current are then used to obtain the required gate signals for the inverter through a hysteresis-band or PWM current controller.

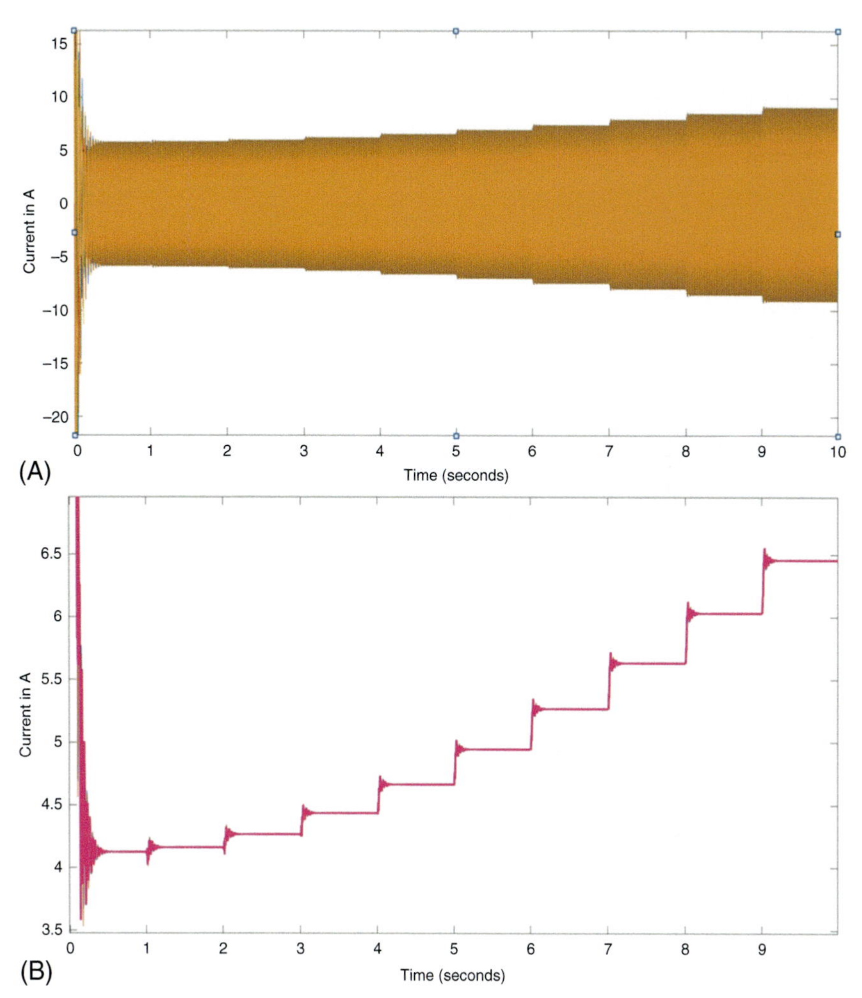

Figure 4.12 (A) Current at different load conditions, (B) Current response to step torque input.

The simulation diagram designed in Simulink environment is presented in Fig. 4.20. The measurement block, current, voltage, speed, and torque waveforms after running the simulation are shown in Figs. 4.21 and 4.22.

4.5 Speed estimated direct torque control

Stator flux linkage is estimated by integrating the stator voltages. Torque is estimated as a cross-product of estimated stator flux linkage vector and measured motor current vector. The estimated flux magnitude and torque

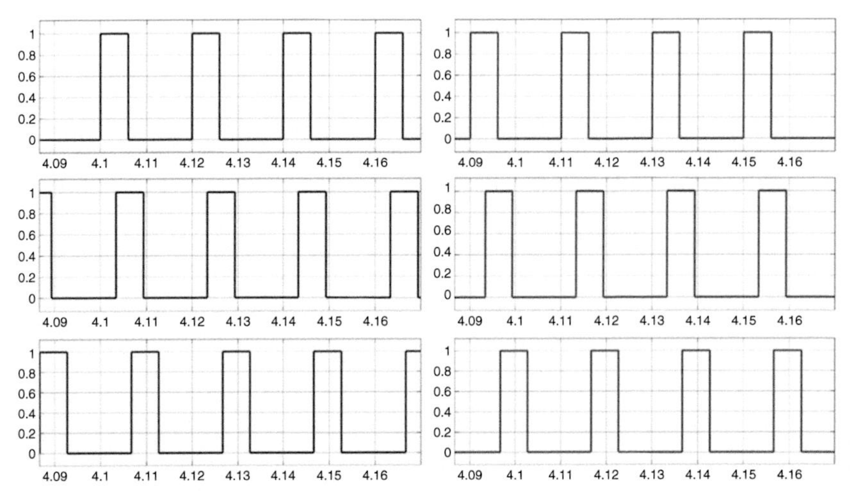

Figure 4.13 The gate pulses to IGBT's.

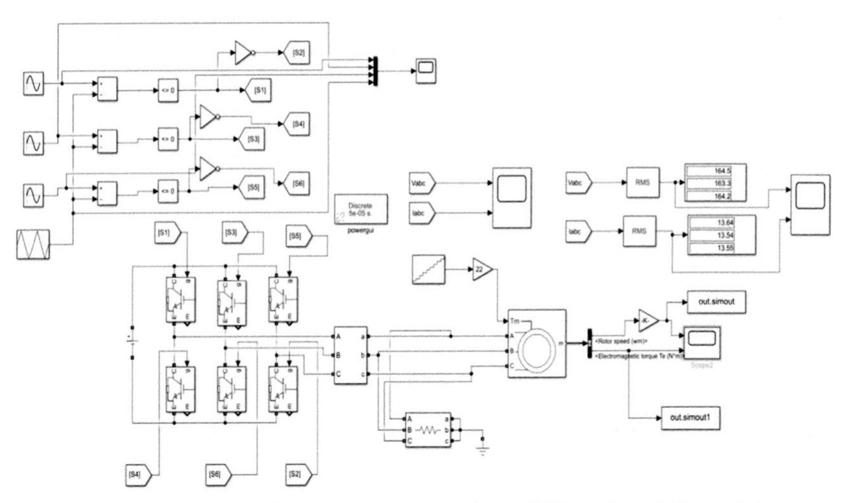

Figure 4.14 PWM inverter fed induction motor drive. *PWM*, pulse width modulation.

are then compared with their reference values. If either the estimated flux or torque deviates too far from the reference tolerance, the transistors of the variable frequency drive are turned off and on in such a way that the flux and torque errors will return in their tolerant bands as fast as possible. Thus, direct torque control (DTC) is one form of the hysteresis or bang-bang control (Fig. 4.23).

The properties of DTC can be characterized as follows:

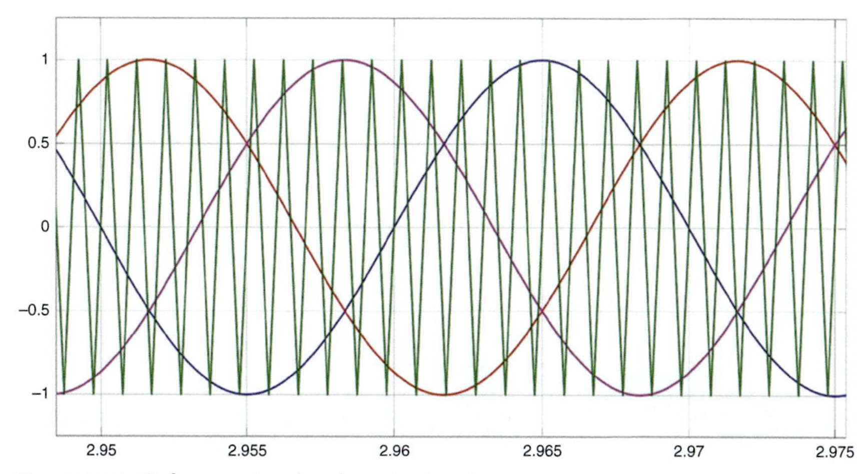

Figure 4.15 Reference signal and carrier signal.

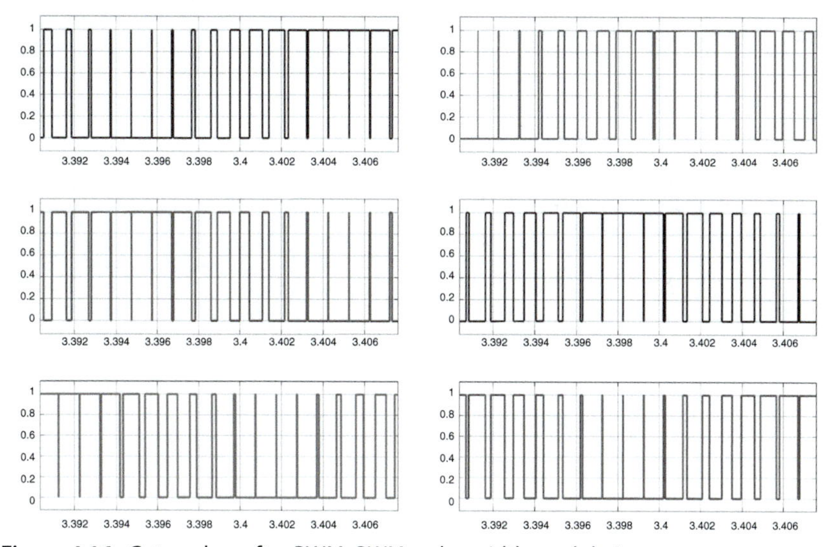

Figure 4.16 Gate pulses after PWM. *PWM*, pulse width modulation.

- Torque and flux can be changed very fast by changing the references.
- High efficiency & low losses—switching losses are minimized because the transistors are switched only when it is needed to keep torque and flux within their hysteresis bands.
- The step response has no overshoot.
- No dynamic coordinate transforms are needed, all calculations are done in stationary coordinate system.

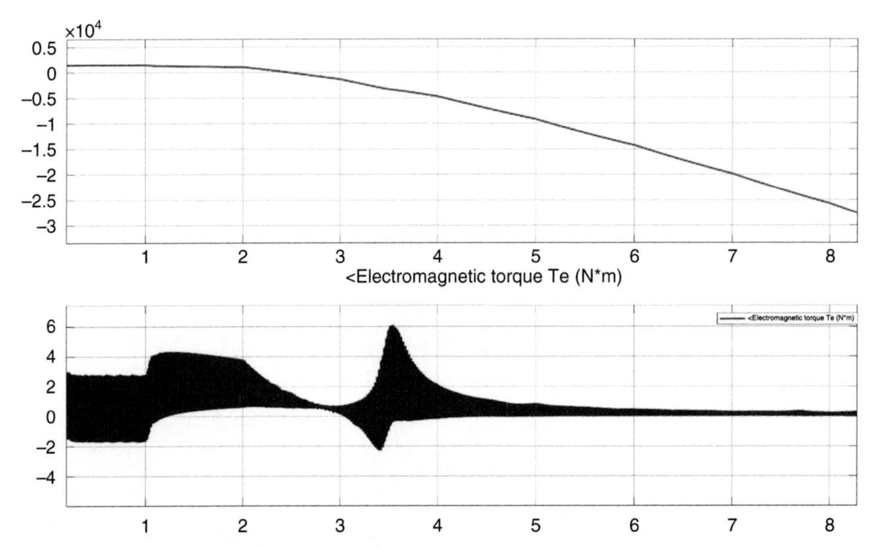

Figure 4.17 Mechanical characteristics of the drive.

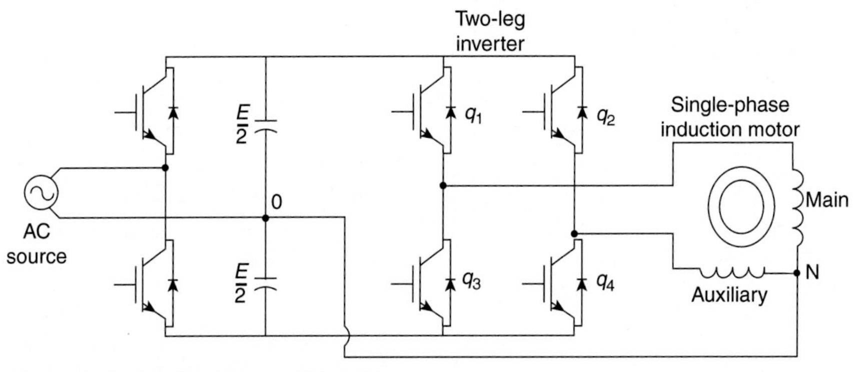

Figure 4.18 Single-phase machine drive.

- No separate modulator is needed, the hysteresis control defines the switch control signals directly.
- There are no PI current controllers. Thus no tuning of the control is required.
- The switching frequency of the transistors is not constant. However, by controlling the width of the tolerance bands the average switching frequency can be kept roughly at its reference value. This also keeps the current and torque ripple small. Thus the torque and current ripple are of the same magnitude as with vector-controlled drives with the same switching frequency.

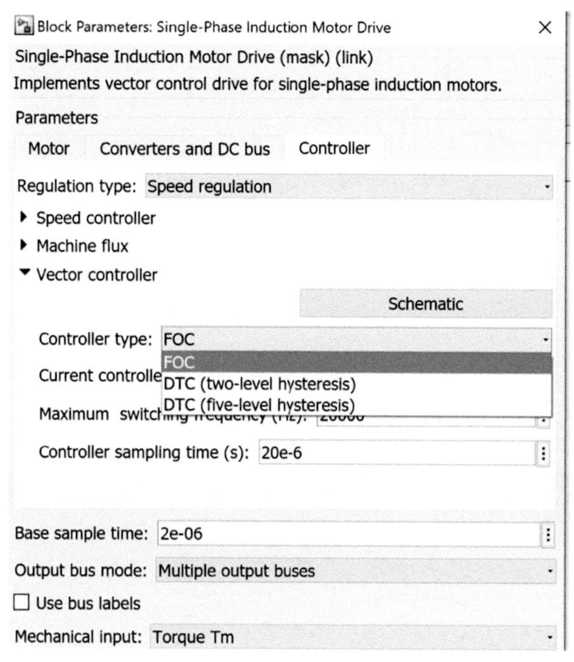

Figure 4.19 Controller setting.

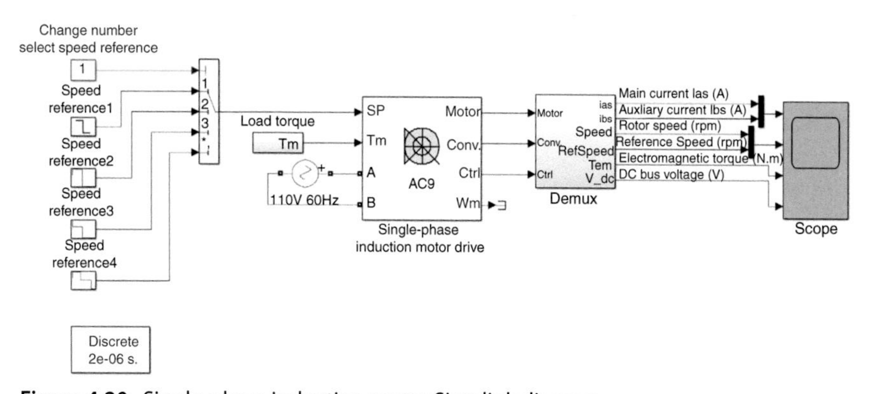

Figure 4.20 Single-phase induction motor Simulink diagram.

- Due to the hysteresis control the switching process is random by nature. Thus there are no peaks in the current spectrum. This further means that the audible noise of the machine is low.
- The intermediate DC circuit's voltage variation is automatically taken into account in the algorithm (in voltage integration). Thus no problems exist due to dc voltage ripple (aliasing) or dc voltage transients.

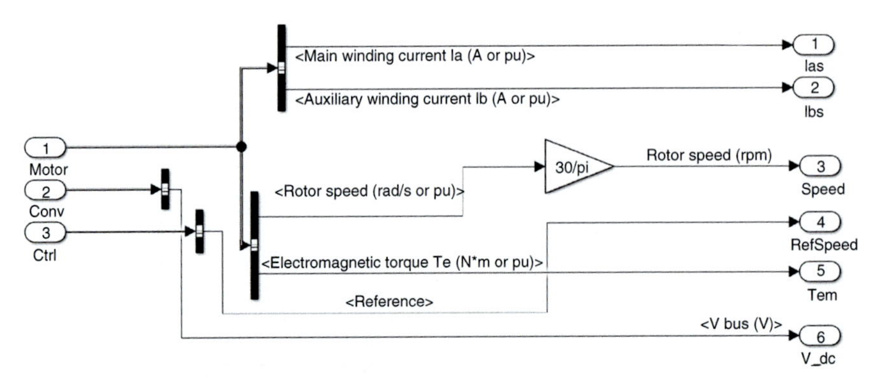

Figure 4.21 Measurements block.

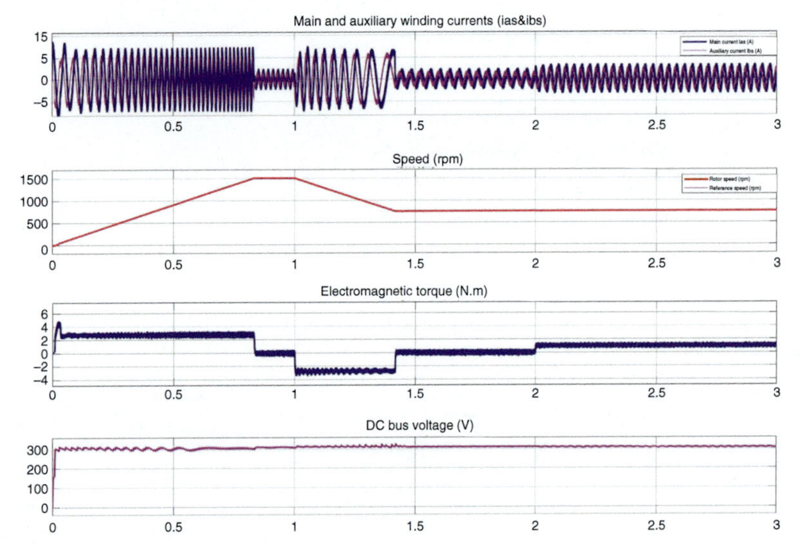

Figure 4.22 Single-phase induction motor results.

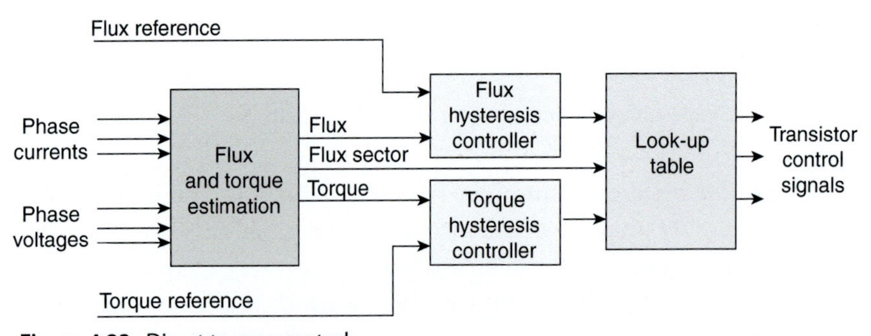

Figure 4.23 Direct torque control.

- Synchronization to rotating machine is straightforward due to the fast control; Just make the torque reference zero and start the inverter. The flux will be identified by the first current pulse.
- Digital control equipment has to be very fast in order to be able to prevent the flux and torque from deviating far from the tolerance bands. Typically the control algorithm has to be performed with 10–30 microseconds or shorter intervals. However, the amount of calculations required is small due to the simplicity of the algorithm.
- The current measuring devices have to be high quality ones without noise because spikes in the measured signals easily cause erroneous control actions. Further complication is that no low-pass filtering can be used to remove noise because filtering causes delays in the resulting actual values that ruins the hysteresis control.
- The stator voltage measurements should have as low offset error as possible in order to keep the flux estimation error down. For this reason, the stator voltages are usually estimated from the measured DC intermediate circuit voltage and the transistor control signals.
- In higher speeds the method is not sensitive to any motor parameters. However, at low speeds the error in stator resistance used in stator flux estimation becomes critical.

These apparent advantages of the DTC are offset by the need for a higher sampling rate (up to 40 kHz as compared with 6–15 kHz for the FOC) leading to higher switching loss in the inverter; a more complex motor model; and inferior torque ripple.

The direct torque method performs very well even without speed sensors. However, the flux estimation is usually based on the integration of the motor phase voltages. Due to the inevitable errors in the voltage measurement and stator resistance estimate, the integrals tend to become erroneous at low speed. Thus, it is not possible to control the motor if the output frequency of the variable frequency drive is zero. However, by careful design of the control system, it is possible to have the minimum frequency in the range 0.5 Hz to 1 Hz that is enough to make possible to start an induction motor with full torque from a standstill situation. A reversal of the rotation direction is possible too if the speed is passing through the zero range rapidly enough to prevent excessive flux estimate deviation.

If continuous operation at low speeds including zero frequency operation is required, a speed or position sensor can be added to the DTC system. With the sensor, high accuracy of the torque and speed control can be

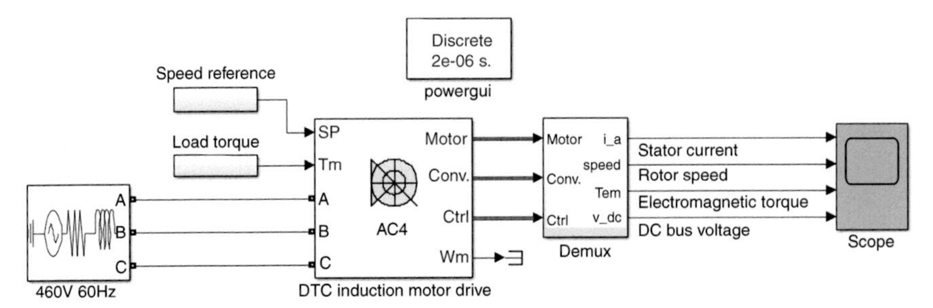

Figure 4.24 Direct torque control Simulink diagram.

maintained in the whole speed range. This circuit uses the AC4 block of Specialized Power Systems library. It models a DTC induction motor drive with a braking chopper for a 200HP AC motor (Fig. 4.24).

The induction motor is fed by a PWM VSI which is built using a Universal Bridge Block. The speed control loop uses a proportional-integral controller to produce the flux and torque references for the DTC block. The DTC block computes the motor torque and flux estimates and compares them to their respective reference. The comparators outputs are then used by an optimal switching table that generates the inverter switching pulses. Motor current, speed, and torque signals are available at the output of the block. The waveforms are given in Figs. 4.25 and 4.26.

4.6 Speed control of induction motor using FOC

AC induction motors offer enviable operational characteristics such as robustness, reliability, and ease of control. They are extensively used in various applications ranging from industrial motion control systems to home appliances. However, the use of induction motors at its highest efficiency is a challenging task because of their complex mathematical model and non-linear characteristic during saturation. These factors make the control of induction motor difficult and call for the use of a high-performance control algorithms such as vector control.

4.6.1 Introduction

Scalar control such as the "V/Hz" strategy has its limitations in terms of performance. The scalar control method for induction motors generates

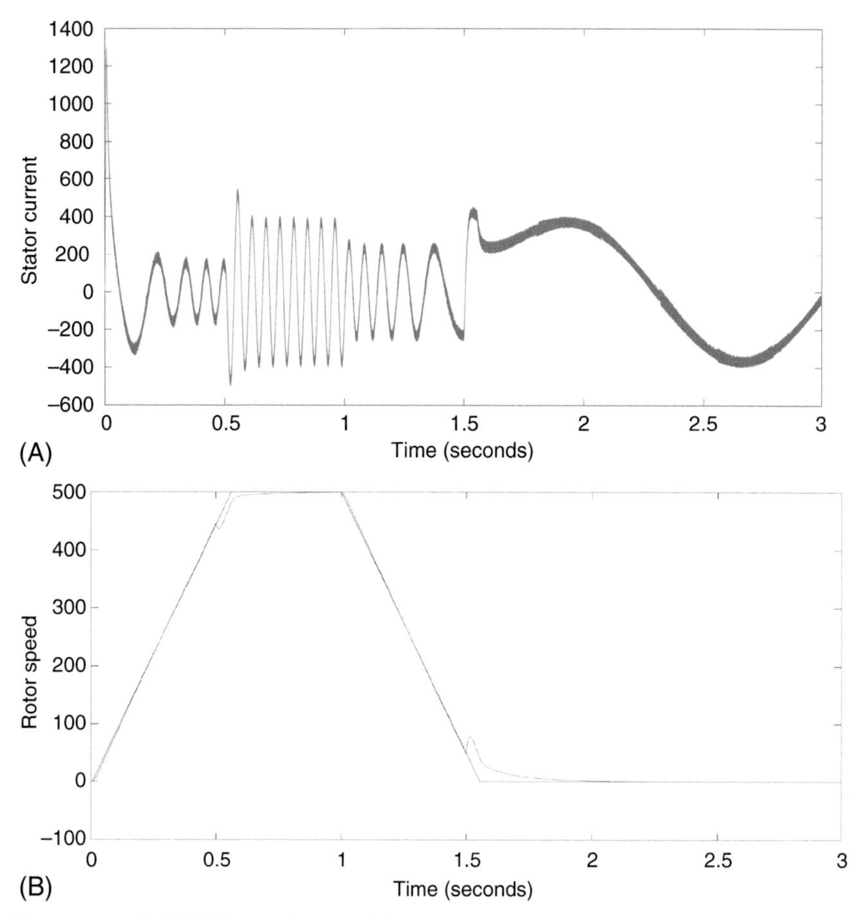

Figure 4.25 (A) DTC-Stator Current, (B) DTC-Rotor Speed.

oscillations on the produced torque. Hence to achieve better dynamic performance, a more superior control scheme is needed for induction motor. With the mathematical processing capabilities offered by the microcontrollers, digital signal processors and FGPA, advanced control strategies can be implemented to decouple the torque generation and the magnetization functions in an AC induction motor. This decoupled torque and magnetization flux is commonly called rotor flux oriented control (FOC).

Field-oriented control describes the way in which the control of torque and speed are directly based on the electromagnetic state of the motor, similar to a DC motor. FOC is the first technology to control the "real"

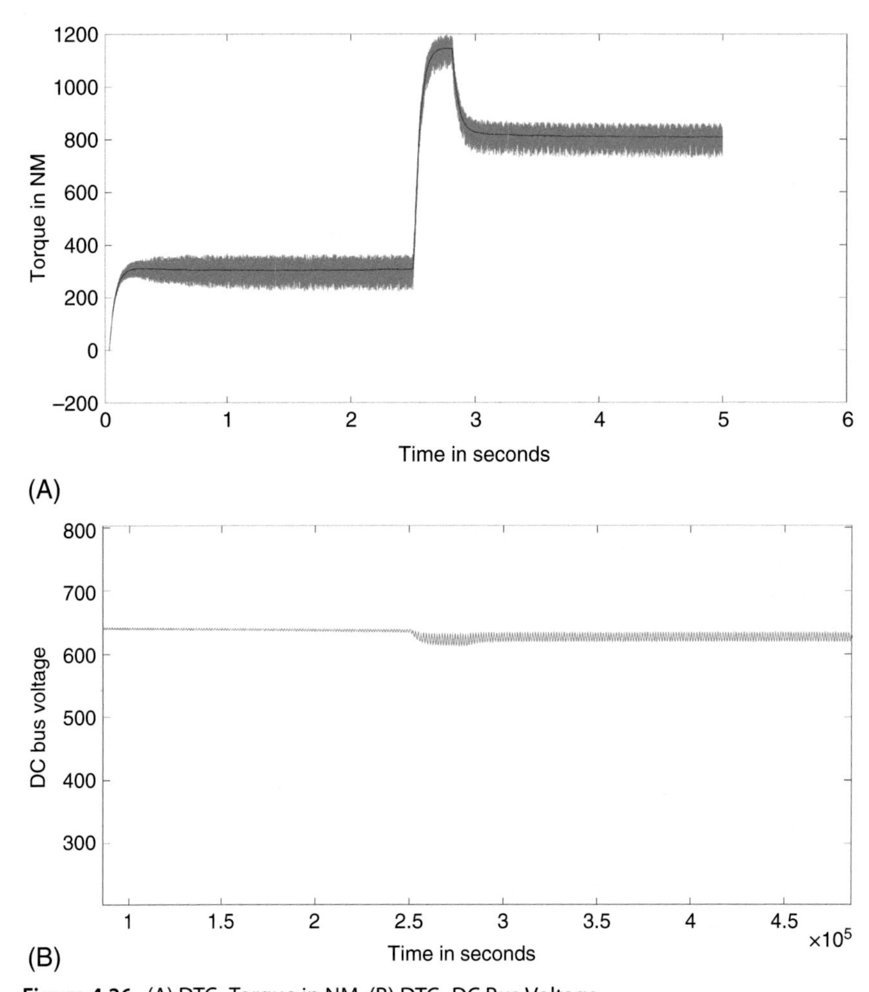

Figure 4.26 (A) DTC- Torque in NM, (B) DTC- DC Bus Voltage.

motor control variables of torque and flux. With decoupling between the stator current components (magnetizing flux and torque), the torque-producing component of the stator flux can be controlled independently. Decoupled control, at low speeds, the magnetization state of motor can be maintained at the appropriate level, and the torque can be controlled to regulate the speed. FOC has been solely developed for high-performance motor applications which can operate smoothly over the wide speed range, can produce full torque at zero speed, and is capable of quick acceleration and deceleration.

4.6.2 Working principle of field-oriented control

The FOC consists of controlling the stator currents represented by a vector. This control is based on projections that transform a three-phase time and speed-dependent system into a two-coordinate (d and q frame) time-invariant system. These transformations and projections lead to a structure similar to that of a DC machine control. FOC machines need two constants as input references: the torque component (aligned with the q coordinate) and the flux component (aligned with d coordinate).

The three-phase voltages, currents, and fluxes of AC-motors can be analyzed in terms of complex space vectors. If we take ia, ib, ic as instantaneous currents in the stator phases, then the stator current vector is defined as follows:

$$\vec{i}_s = i_a + i_b e^{j2\pi/3} + i_c e^{j4\pi/3}$$

where (a, b, c) are the axes of three-phase system. This current space vector represents the three-phase sinusoidal system. It needs to be transformed into a two-time invariant coordinate system. This transformation can be divided into two steps:

(a, b, c) → (α, β) (the Clarke transformation), which gives outputs of two coordinate time-variant system.

(a, β) → (d, q) (the Park transformation), which gives outputs of two coordinate time-invariant system.

4.6.3 The (a, b, c) → (α, β) projection (Clarke transformation)

Three-phase quantities either voltages or currents, varying in time along the axes a, b, and c can be mathematically transformed into two-phase voltages or currents, varying in time along the axes α and β by the following transformation matrix:

$$i_{\alpha\beta o} = \left(\frac{2}{3}\right) * \begin{bmatrix} 1 & -\frac{1}{2} & -\frac{1}{2} \\ 0 & \frac{\sqrt{3}}{2} & -\frac{\sqrt{3}}{2} \\ \frac{1}{2} & \frac{1}{2} & \frac{1}{2} \end{bmatrix}$$

Assuming that the axis a and the axis α are along same direction and β is orthogonal to them, we have the following vector diagram:

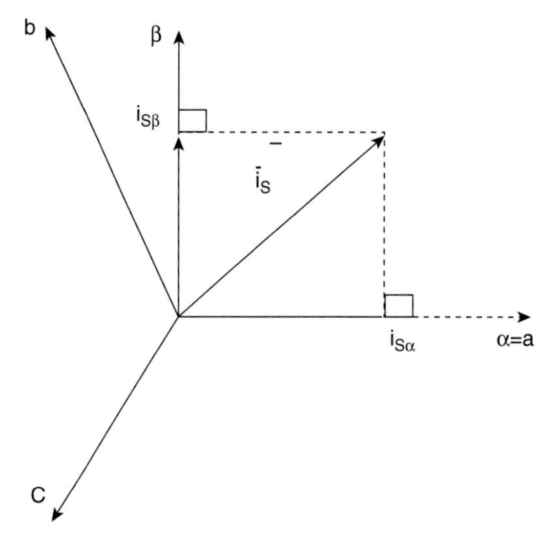

The above projection modifies the three-phase system into the (α, β) two-dimension orthogonal system as stated below:

$$i_{s\alpha} = i_a$$

$$i_{s\beta} = \frac{i_a}{\sqrt{3}} + 2\frac{i_b}{\sqrt{3}}$$

But these two phases (α, β) currents still depend upon time and speed.

4.6.4 The $(\alpha, \beta) \rightarrow$ (dq) projection (Park transformation)

This is the most important transformation in the FOC. In fact, this projection modifies the two-phase fixed orthogonal system (α, β) into d, q rotating reference system. The transformation matrix is given below:

$$i_{dqo} = \left(\frac{2}{3}\right) * \begin{bmatrix} \cos\theta & \cos\left(\theta - \frac{2\pi}{3}\right) & \cos\left(\theta + \frac{2\pi}{3}\right) \\ \sin\theta & \sin\left(\theta - \frac{2\pi}{3}\right) & \sin\left(\theta + \frac{2\pi}{3}\right) \\ \frac{1}{2} & \frac{1}{2} & \frac{1}{2} \end{bmatrix}$$

where θ is the angle between the rotating and fixed coordinate system.

If you consider the d axis aligned with the rotor flux, the figure shows the relationship from the two reference frames for the current vector:

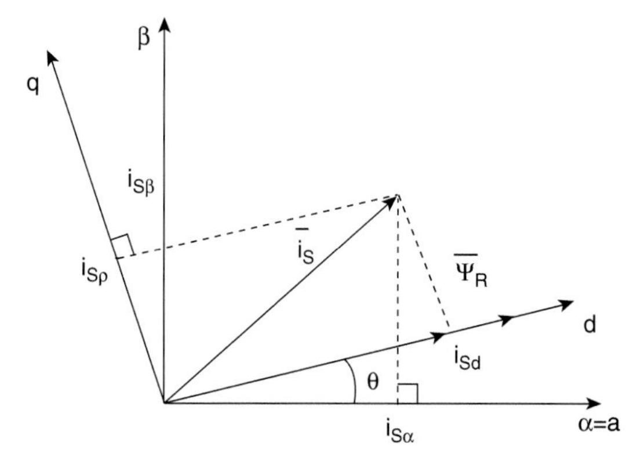

where θ is the rotor flux position. The torque and flux components of the current vector are determined by the following equations:

$$i_{sq} = i_{s\alpha} \sin\theta + i_{s\beta} \cos\theta$$

$$i_{sd} = i_{s\alpha} \cos\theta + i_{s\beta} \sin\theta$$

These components depend on the current vector (α, β) components and on the rotor flux position. If you know the accurate rotor flux position then, by the above equation, the d, q component can be easily calculated. At this instant, the torque can be controlled directly because flux component (i_{sd}) and torque component (i_{sq}) are independent now.

4.6.5 Basic module for FOC

Stator phase currents are measured. These measured currents are fed into the Clarke transformation block. The outputs of this projection are entitled $i_{s\alpha}$ and $i_{s\beta}$. These two components of the current enter into the Park transformation block that provide the current in the d, q reference frame. The i_{sd} and i_{sq} components are contrasted to the references: i_{sdref} (the flux reference) and i_{sqref} (the torque reference). At this instant, the control structure has an advantage: it can be used to control either synchronous or induction machines by simply changing the flux reference and tracking rotor flux position. In case of PMSM the rotor flux is fixed determined by the magnets so there is no need to create one. Therefore, while controlling

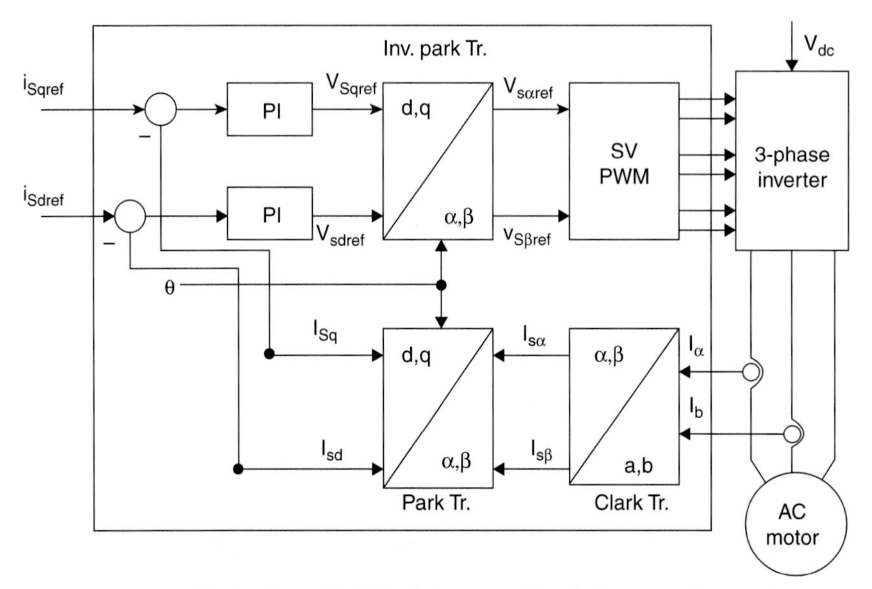

Figure 4.27 Simplified indirect FOC block diagram. *FOC*, field-oriented control.

a PMSM, i_{sdref} should be equal to zero. As induction motors need a rotor flux creation in order to operate, the flux reference must not be equal to zero. This easily eliminates one of the major shortcomings of the "classic" control structures: the portability from asynchronous to synchronous drives. The outputs of the PI controllers are V_{sdref} and V_{sqref}. They are applied to the inverse Park transformation block.

The outputs of this projection are $V_{s\alpha ref}$ and $V_{s\beta ref}$ are fed to the space vector PWM algorithm block. The outputs of this block provide signals that drive the inverter. Here both Park and inverse Park transformations need the rotor flux position. Hence rotor flux position is the essence of FOC (Fig. 4.27). The evaluation of the rotor flux position is different if we consider the synchronous or induction motor.

1. In the case of synchronous motor(s), the rotor speed is equal to the rotor flux speed. Then rotor flux position is directly determined by position sensor or by integration of rotor speed.
2. In the case of asynchronous motor(s), the rotor speed is not equal to the rotor flux speed because of slip; therefore a particular method is used to evaluate rotor flux position (θ). This method utilizes current model, which needs two equations of the induction motor model in d, q rotating reference frame.

4.6.6 Classification of field-oriented control

FOC for the induction motor drive can be broadly classified into two types: indirect FOC and direct FOC schemes. In DFOC strategy rotor flux vector is either measured by means of a flux sensor mounted in the air-gap or by using the voltage equations starting from the electrical machine parameters. But in the case of IFOC rotor flux vector is estimated using the **FOC** equations (current model) requiring a rotor speed measurement. Among both schemes, IFOC is more commonly used because in the closed-loop mode it can easily operate throughout the speed range from zero speed to high-speed field-weakening.

Advantages of field-oriented control

1. Improved torque response.
2. Torque control at low frequencies and low speed.
3. Dynamic speed accuracy.
4. Reduction in size of motor, cost and power consumption.
5. Four quadrant operation.
6. Short-term overload capability.

4.6.7 FOC simulation model using MATLAB

This circuit uses the AC3 block of Specialized Power Systems library. It models a FOC induction motor drive with a braking chopper for a 200HP AC motor. The induction motor is fed by a PWM VSI, which is built using a Universal Bridge Block. The speed control loop uses a PI controller to produce the flux and torque references for the FOC controller. The FOC controller computes the three reference motor line currents corresponding to the flux and torque references and then feeds the motor with these currents using a three-phase current regulator. Motor current, speed, and torque signals are available at the output of the block (Fig. 4.28). The results are presented in Fig. 4.29.

4.7 A VSI fed induction motor drive system using PSIM

VSI fed induction motor drive system using PSIM software is presented in Fig. 4.30. The motor supply currents are measure by ammeters. The IGBT is used as a switch in the VSI. PSIM software also has all the blocks and components required in its library to design induction motor drives. The simulated waveforms are presented in Figs. 4.31 and 4.32.

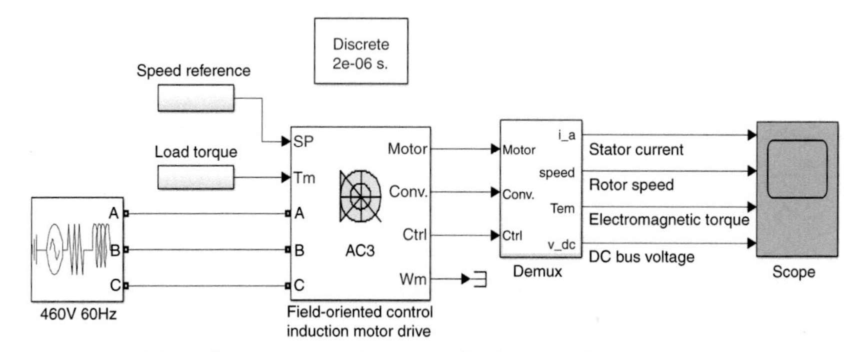

Figure 4.28 FOC induction motor drive. *FOC*, field-oriented control.

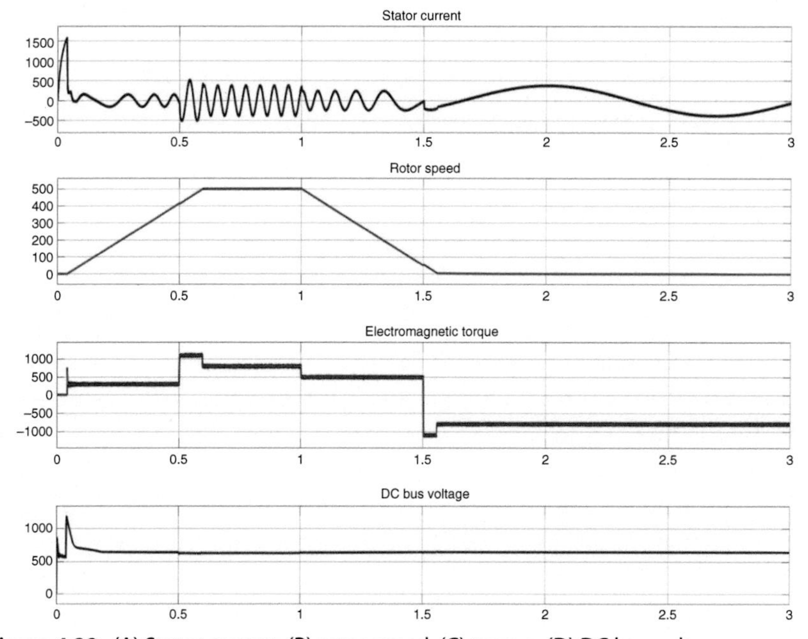

Figure 4.29 (A) Stator current, (B) rotor speed, (C) torque, (D) DC bus voltage.

4.8 Field-oriented control of induction motor drive using PSIM

FOC of induction motor drive using PSIM is shown in Fig. 4.33. The waveforms after simulation of the designed circuit are presented in this section, as shown in Figs. 4.34 and 4.35. The PI controller block is used to generate the pulses. The dq-abc conversion block is also utilized for performing the induction motor drive control.

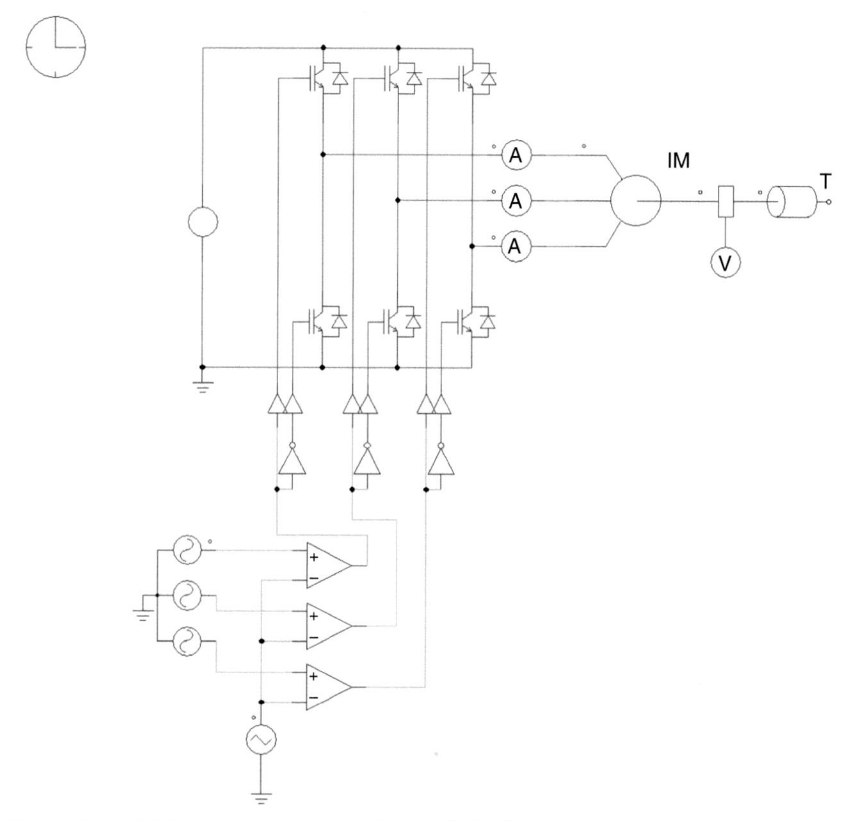

Figure 4.30 VSI fed induction motor drive in PSIM. *VSI*, voltage source inverter.

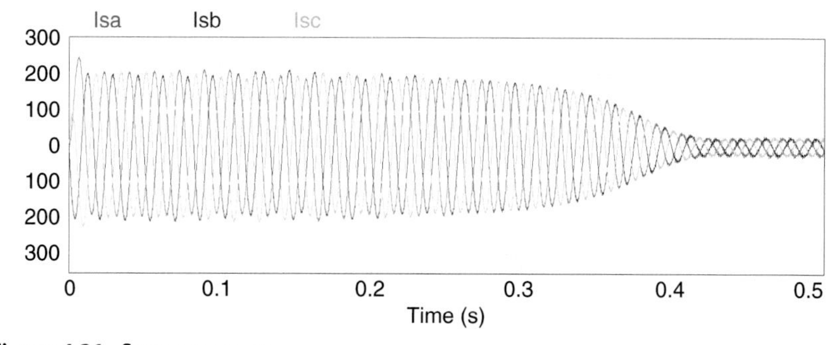

Figure 4.31 Stator currents.

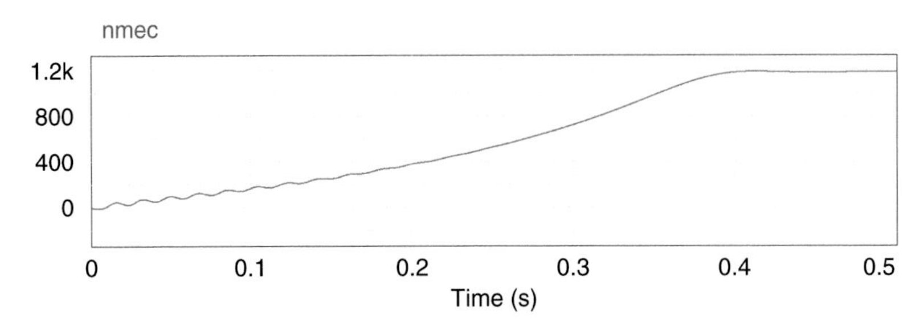

Figure 4.32 Speed waveform.

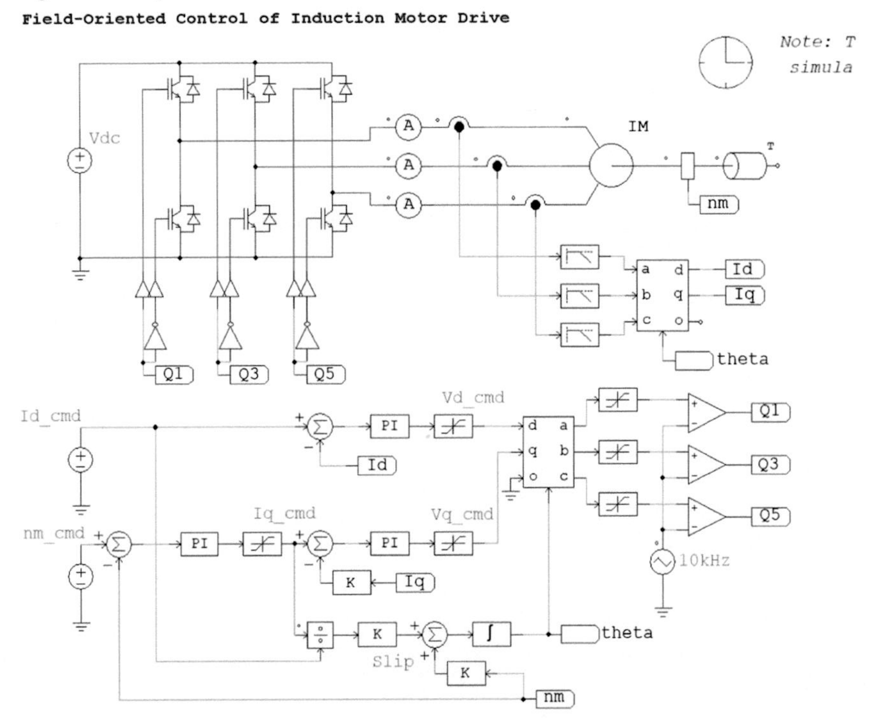

Figure 4.33 FOC control of induction motor drive. *FOC*, field-oriented control.

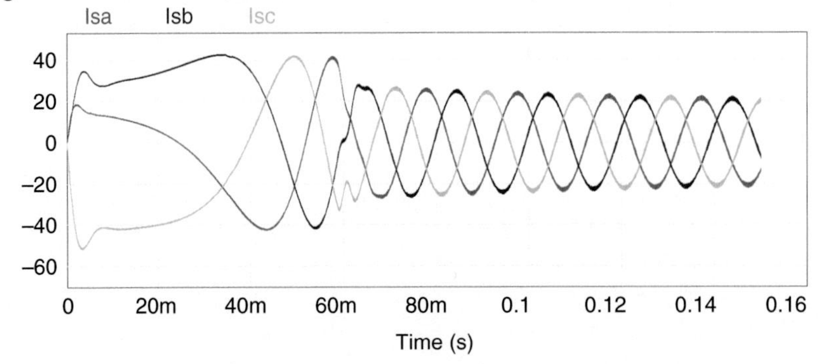

Figure 4.34 Stator current–FOC-based control. *FOC*, field-oriented control.

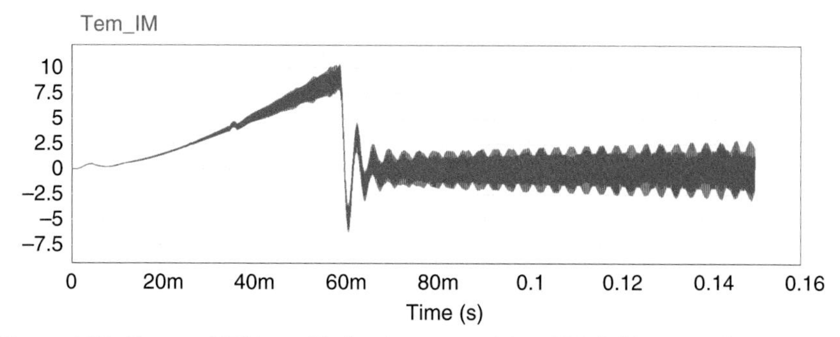

Figure 4.35 Torque–FOC-based induction motor drive. *FOC*, field-oriented control.

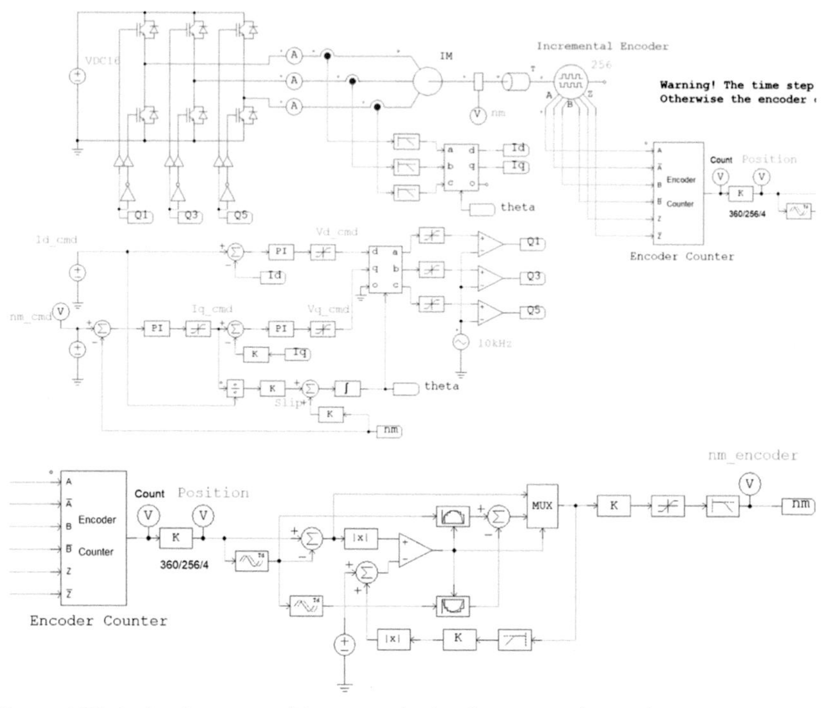

Figure 4.36 Induction motor drive control using incremental encoder.

4.9 Field-oriented control of induction motor drive using the incremental encoder using PSIM

FOC of induction motor drive using the incremental encoder is shown in Fig. 4.36. Incremental encoder and encoder counter blocks are used to control the speed of the induction motor drive. The stator current passing through the motor is presented in Fig. 4.37. The performance characteristics the motor drive are shown in Figs. 4.38–4.41.

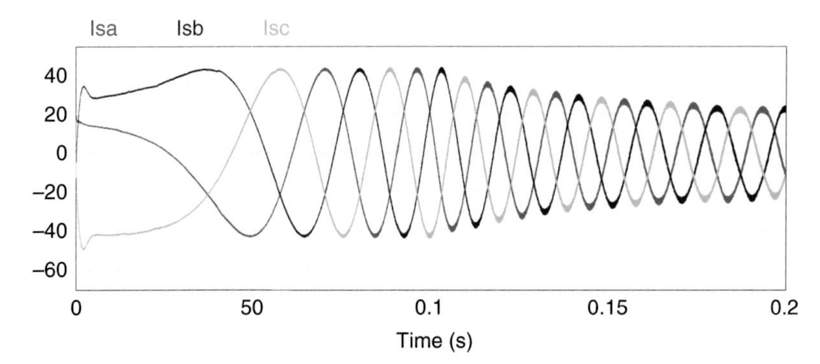

Figure 4.37 Stator currents.

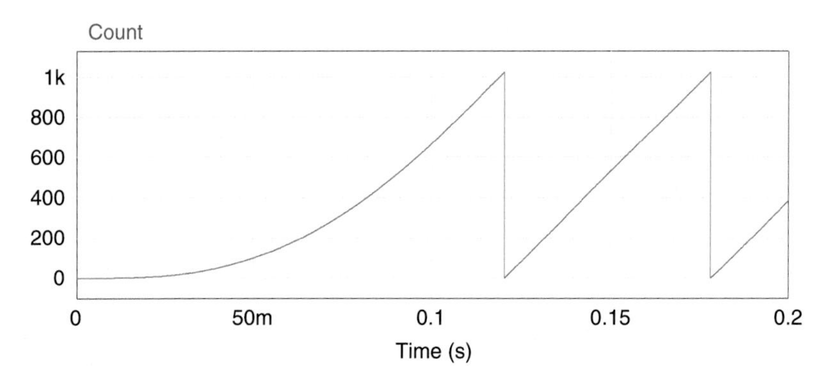

Figure 4.38 Encoder count.

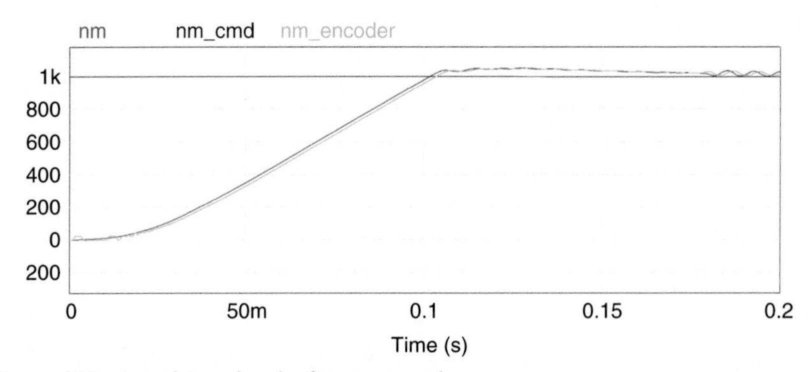

Figure 4.39 Actual speed and reference speed.

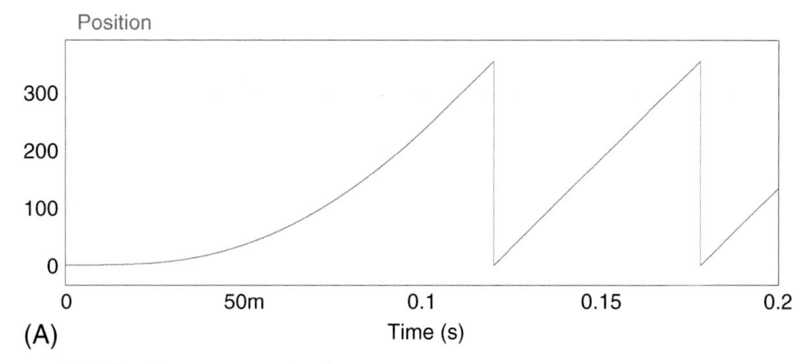

Figure 4.40 Position sensor output.

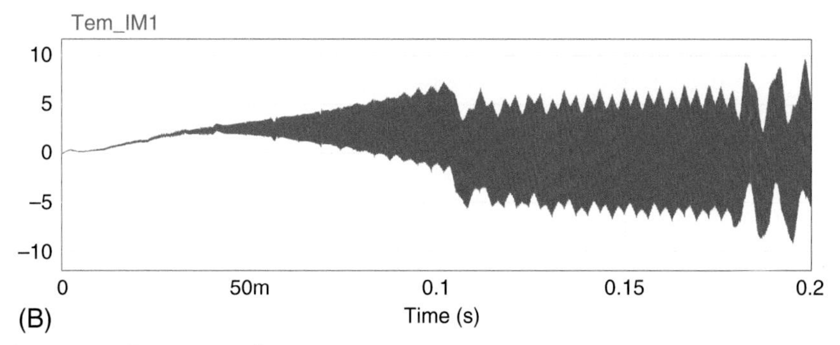

Figure 4.41 Torque waveform.

4.10 Practice questions

1. Design induction motor drive fed from VSI and simulate the complete drive using MATLAB.
2. Design induction motor drive fed from three-phase PWM inverter and simulate the complete drive using MATLAB.
3. Explain about FOC of three-phase induction motor drive with modeling.
4. Explain about FOC of single-phase induction motor drive with modeling of the machine.
5. Explain about direct control of three-phase induction motor drive with modeling.
6. Simulate and plot the waveforms of FOC-based induction motor drive control using Simulink.
7. Simulate and plot the waveforms of DTC-based induction motor drive control.

8. Design induction motor drive fed from VSI and simulate the complete drive using PSIM.
9. Simulate and plot the waveforms of FOC-based induction motor drive control using PSIM.
10. Simulate and plot the waveforms of FOC-based induction motor drive control using incremental encoder in PSIM environment.

CHAPTER 5

Synchronous motor drives and its simulation

Contents

5.1 Introduction to synchronous motor drives

Synchronous motors, as their name implies, revolve at the same speed. The fundamental advantage of synchronous motors is that they work on three AC supplies and provide a DC feed to the rotor when running at synchronous speeds. We can remark that if synchronous motors are only meant to run at rated synchronous speeds, introducing drives to them is pointless. The explanation is straightforward: synchronous motor drives (Fig. 5.1) ensure a smooth and trouble-free start, pull in, and braking procedure. We will go over each one separately.

5.1.1 Starting synchronous motors

The issue with synchronous motors is that they do not start on their own. Before examining the motor's starting method, it is important to understand the type of supply as well as the motor's rotor and stator. The stator of synchronous motors is similar to that of induction motors; the only difference is that the synchronous motors' rotor is supplied with DC power.

Before we learn how synchronous motors are started, it is important to understand why they do not start on their own. When a three-phase supply

Electric Motor Drives and Their Applications with Simulation Practices.
DOI: https://doi.org/10.1016/B978-0-323-91162-7.00008-4

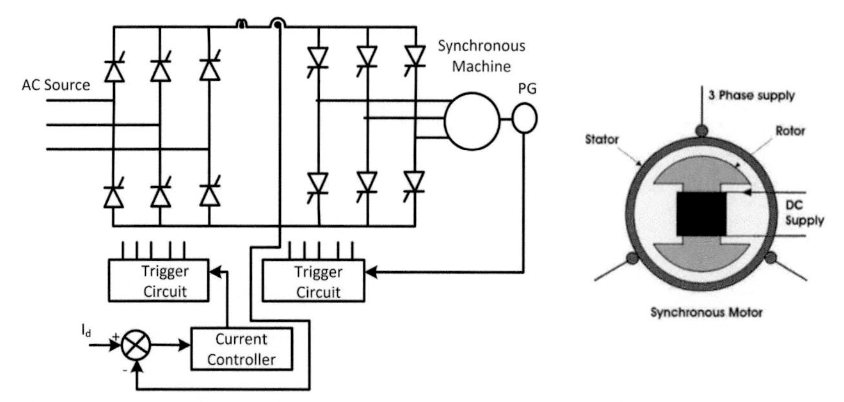

Figure 5.1 Synchronous motor drive. nullhttpqlacwwwgacykssuugsqassgsiggqskeegku iwocqquwmqgeq.

is applied to the stator, a rotating magnetic flux rotates at synchronous speed; when a DC supply is applied to the rotor, the rotor operates as a magnet with two salient poles, producing a revolving magnetic flux at synchronous speed.

The rotor cannot follow the magnetic field, which is rotating at synchronous speed, because it is at a stop. Synchronous motors are not self-starting because the rotor stacks at its position due to the opposite poles moving so quickly that the rotor locks. Finally, we will look at how synchronous motors are started. The synchronous motors are started like standard induction motors at initially, with no DC supply to the rotor. When the motor approaches the rotor, a draw-in occurs, which will be detailed later.

An external motor can also be used to start the synchronous motor drives. The synchronous motor's rotor is rotated by an external motor in this way, and when the rotor speed approaches synchronous speed, the DC-field is turned on, and pull in occurs. The starting torque is quite low with this manner, and it is also not very popular.

5.1.2 Pull in of synchronous motors

The DC field supply is turned on and the pull in process commences when the synchronous motors' rotors approach near synchronous speed. Various disturbances are detected in the motor during the switching on of the DC supply due to the phase angle and torque angle, and there are also multiple slips of poles of air-gap flux. The rotor achieves synchronous speed once the pull in procedure is complete. The DC supply should be turned on at the most suitable angle to finish the pull in as quickly as possible. While the synchronous motor is used as an induction motor, the DC supply should be fed when the induction motor is running at top speed; this is the optimal time because the speed differential is the smallest.

5.1.3 Braking of synchronous motors

Regenerative, dynamic, and plugging type braking are the three forms of braking that we are familiar with. However, only dynamic braking is possible with synchronous motor drives, while theoretically plugging is possible. Because they require a higher speed than synchronous speed, regenerative braking cannot be used on them. The motor is disconnected from the power supply and connected across a three-phase resistor for dynamic braking. At that point, the motor acts as a synchronous generator, dissipating energy through the resistors. Because excessive plugging current might cause significant line disturbance and damage, synchronous motors do not require plugging.

5.1.4 Speed control of synchronous motor

Synchronous motors are those that have a constant rotational speed. They operate at the supply's synchronous speed. They are commonly utilized to improve the power factor by operating at a constant speed under no-load conditions. At a given rating, synchronous motors lose less energy than induction motors.

The speed of a synchronous motor is given by

$$N_s = (120\,f)/p$$

where $f =$ supply frequency and $p =$ number of poles.

As you can see, synchronous speed is determined by the supply frequency and the number of rotor poles. We do not employ that strategy because changing the number of poles is a pain. The frequency of the current provided to the synchronous motor can be changed now that solid-state electronics have been invented. We can change the frequency of the supply to the synchronous motor to alter its speed. To control the speed of synchronous motors, we can use a combination of rectifiers and inverters (Fig. 5.2). They can be put to two different uses.

5.1.5 Inverter fed open-loop synchronous motor drive

The synchronous motor in this method is powered by an open-loop variable frequency inverter (Fig. 5.3). The term "open loop" refers to the absence of supply feedback. The rotor's present position is unknown to the inverter. When precise speed control is not necessary, this method is preferred. The

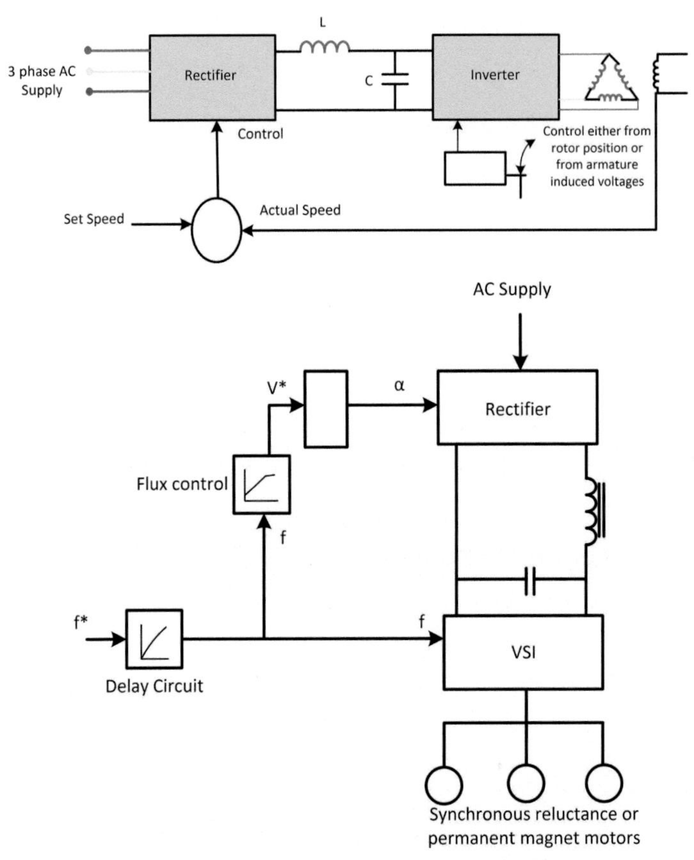

Figure 5.2 Self-control of synchronous motor fed from square wave inverter.

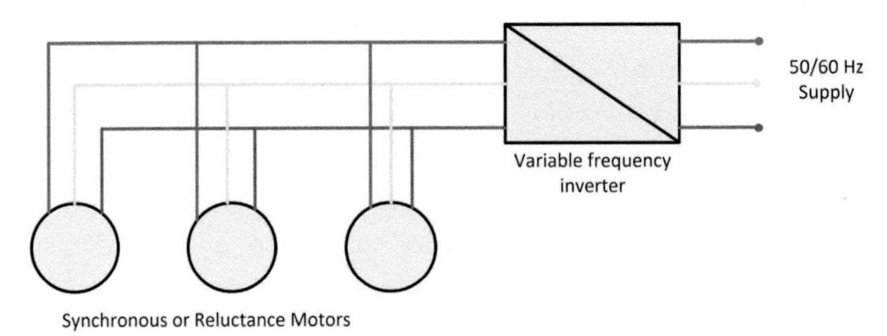

Figure 5.3 Inverter fed open loop synchronous motor drive.

rectifier inverter set receives power from the mains and converts it to the desired frequency. The synchronous speed of the motor can be changed depending on the frequency.

The block diagram of the speed control drive is shown in Fig. 5.4. Rectifiers convert the three-phase power from the mains to DC power.

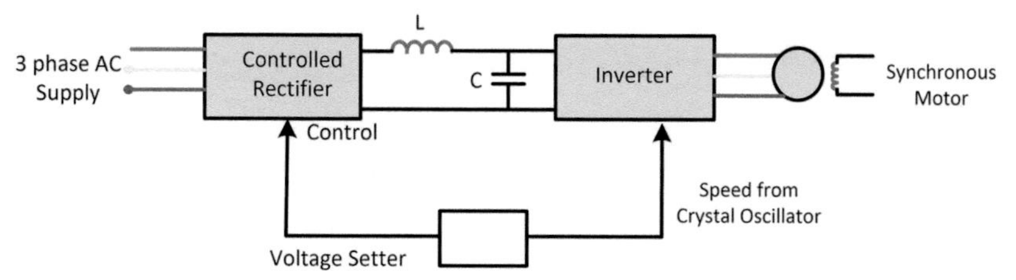

Figure 5.4 Separate control of synchronous motor fed from square wave inverter.

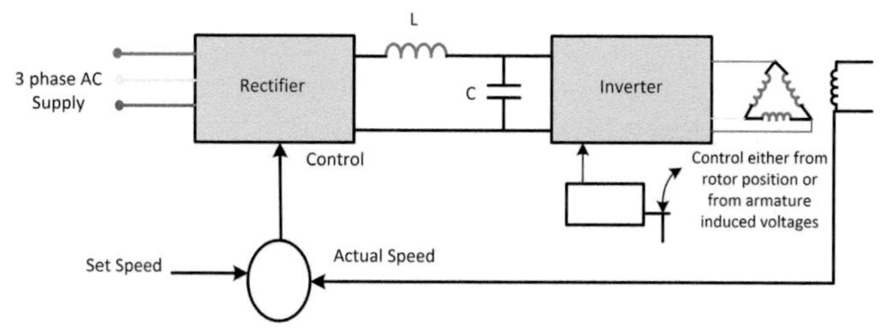

Figure 5.5 Self-control of synchronous motor fed from square wave inverter.

The rippling DC is then smoothed with LC filters. The inverters are fed by DC. Voltage source inverters and current source inverters are two types of inverters that can be used. The frequency of the supply delivered to the motor can be adjusted, and the synchronous motor's speed can be controlled appropriately. When a multitude of motors must run at the same speed, open-loop operation is advantageous. However, there is a drawback to this strategy. This strategy creates hunting or spontaneous oscillation.

5.1.6 Self-synchronous (closed loop) operation

When precise speed control is necessary, we use self-synchronous (closed-loop) operation (Fig. 5.5). The inverter output frequency is determined by the rotor speed in this approach. The rotor's speed is sent back into the differentiator. The rectifier receives the difference between the preset and actual speeds. As a result, the inverter adjusts the motor's speed and frequency. With closed-loop operation, we can have more precise control over the motor speed. When speed is reduced (because to increased load), the stator supply frequency is reduced to keep the rotor in sync with the stator magnetic field. This approach does not have any spontaneous oscillation or hunting.

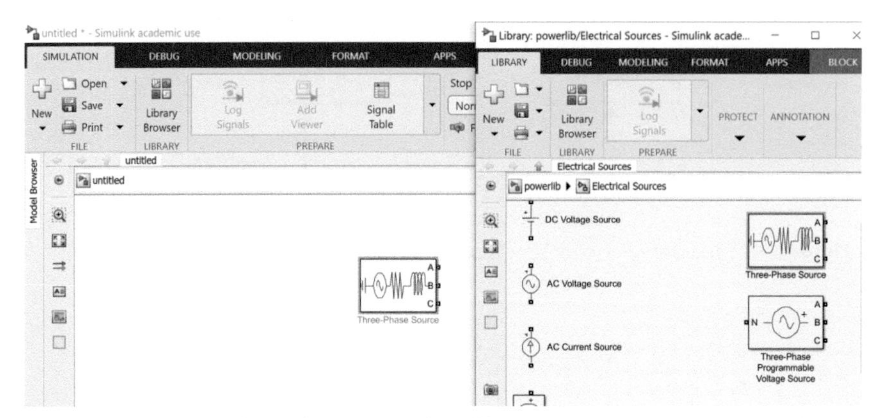

Figure 5.6 Selecting three-phase source from library.

5.1.7 MATLAB simulation of self-control of synchronous motor drive

>>Libraries>>Powerlibrary>>Electrical sources>>Three-phase source

Select three-phase source from library (Fig. 5.6) with the help of the above command. The parameters of three-phase sources are given as per Fig. 5.7.

After selecting three-phase source, we need to choose the synchronous motor drive as per the requirement from the electric drive library with the command as

>>Libraries>>Simscape>>SpecializedPowerSystems>>ElectricalDrives >>ACdrives>>SelfControlled synchronous motor drive.

Once the motor drive is chosen (Fig. 5.8), the parameters play an important role here. Enter the parameters values as given in Fig. 5.9.

We may use the Powergui block to change the initial states and start the simulation from any starting point. So Powergui should be selected in any simulation circuit by using the below command.

>>Libraries>>Simspace>>Powerlib>>Powergui.

In Powergui (Fig. 5.10), we need to select the simulation type as discrete and give sample time as T_s (Fig. 5.11).

Select the subsystem for speed reference as given below with the help of command (Fig. 5.12).

>>Libraries>>Simulink>>commonlyusedblocks>>Subsystems.

Inside the subsystem, we need to include stair generator and output ports. The parameters value of stair case to be selected as given in Fig. 5.13.

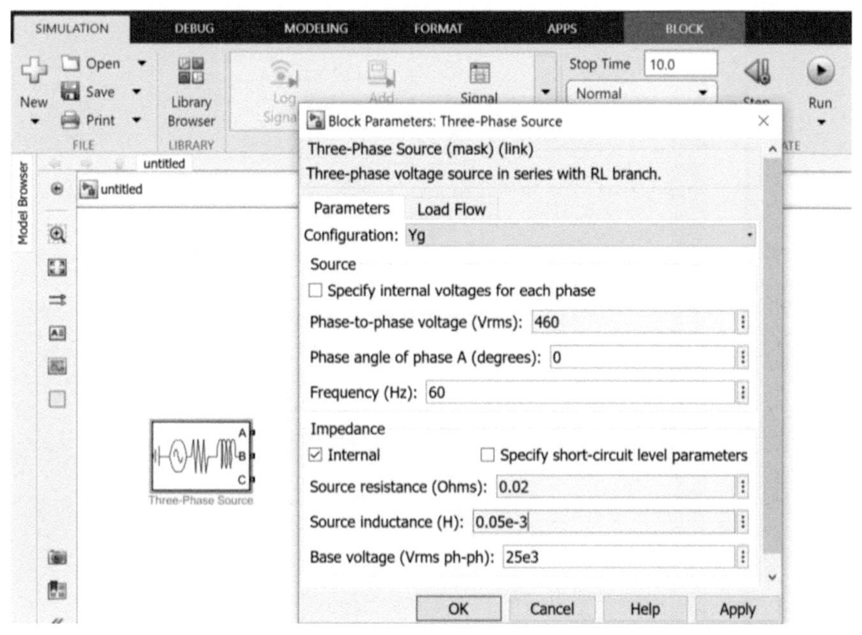

Figure 5.7 Selecting diode from library.

One more subsystem is required here, so select the subsystem from the library (Fig. 5.14) and named it as load torque. In this subsystem, we include demux, bus selector, and all. The parameters of all the blocks are given below.

The parameters of the stair generator for the load torque subsystem are given in Fig. 5.15.

After completing the circuit for the load torque subsystem, the parameters need to be set. In Fig. 5.16, the mod_flux block expression is given.

The gain value is set as 30/pi for the load torque subsystem (Fig. 5.17).

The bus selector block outputs the components from the input bus that you pick. The selected elements can be output independently or in a new virtual bus by the block. In this circuit, we need to select the required signal in the bus. In the subsystem, we have three buses, select the appropriate signals for each bus as the requirement. These are given in Figs. 5.18–5.20.

We have demux in the subsystem, the demux has been drawn as per Fig. 5.21.

In the Demux subsystem, two input buses and three output buses are there. Now will see the signal of what we need to select for those buses in the coming figures (Figs. 5.22–5.25).

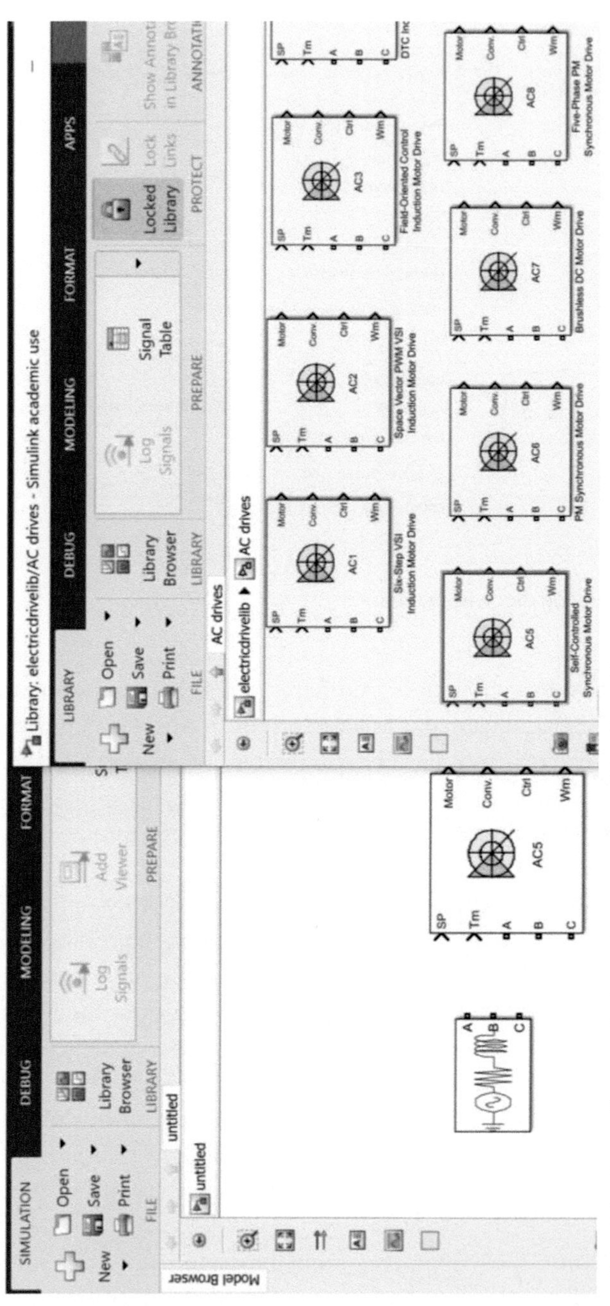

Figure 5.8 Selecting self-controlled synchronous motor drive from library.

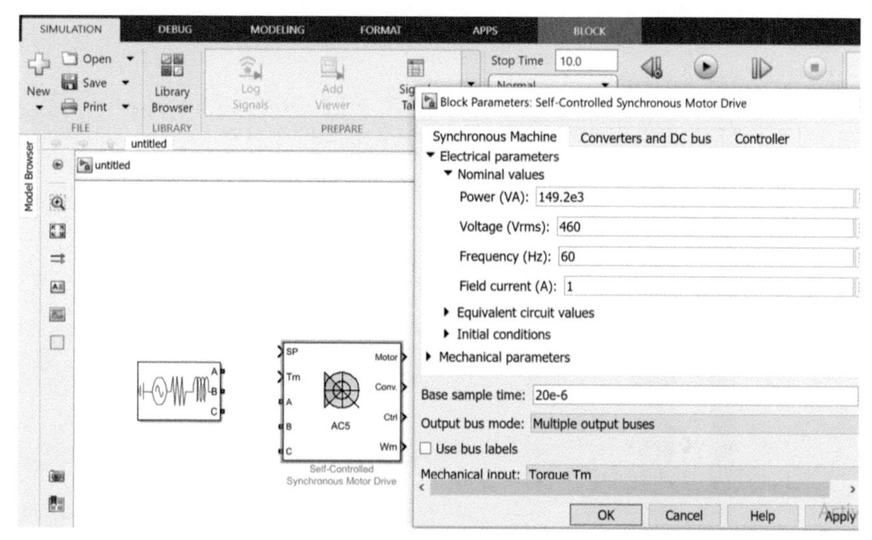

Figure 5.9 Setting parameters value of self-controlled synchronous motor drive.

Select the terminator from library with the help of command (Fig. 5.26). Then connect the terminator to the point W_m in the self-controlled synchronous motor drive.

>>Libraries>>Simulink>>Commonly used blocks >>Terminator

To display all the output, we need scope. So select the scope (Fig. 5.27) by the command >>Libraries>>Simulink>>Commonly used blocks>>Scope.

After including all the elements from the library. Do the connection as per the diagram given below (Fig. 5.28).

After giving run command we will get the simulation results of the circuit (Fig. 5.29).

5.2 Current source inverter fed synchronous motor drives

When V/f and E/f are kept constant and armature resistance is ignored, a synchronous motor draws a stator current that is independent of stator frequency. The motor generates a consistent torque as well. The flux is constant as well. As a result, we can regulate both flux and torque by regulating the stator current of a synchronous motor. Current control is easy and uncomplicated, as discussed in the case of the induction motor. A current source inverter fed synchronous motor drive powers a synchronous motor. A synchronous motor can be controlled separately or autonomously.

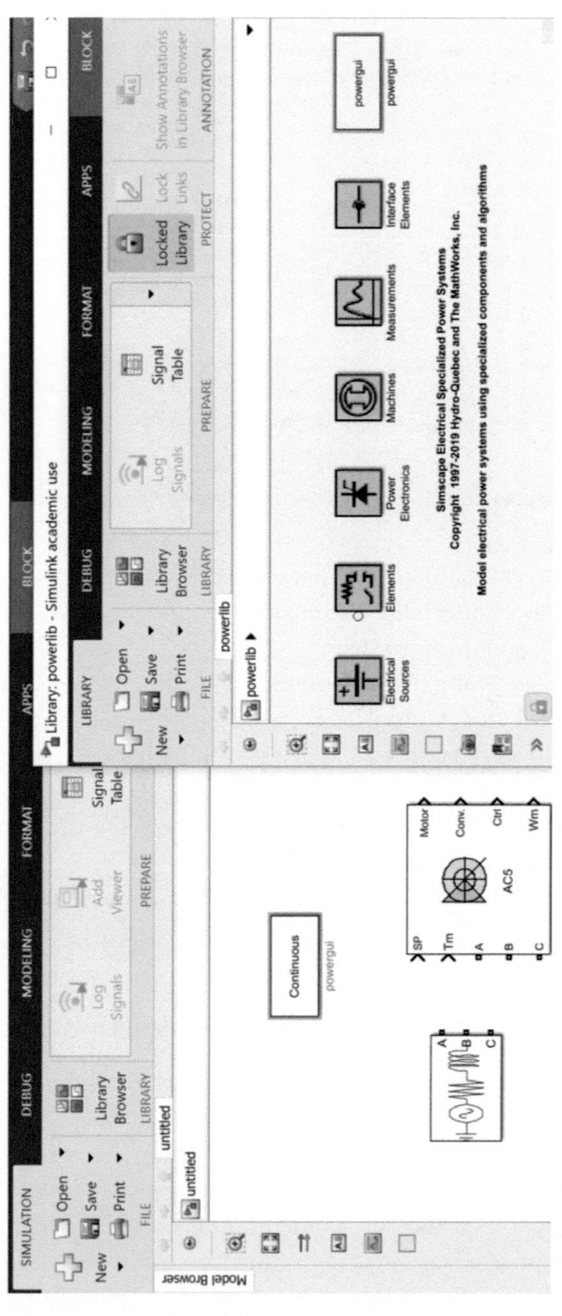

Figure 5.10 Selecting Powergui from library.

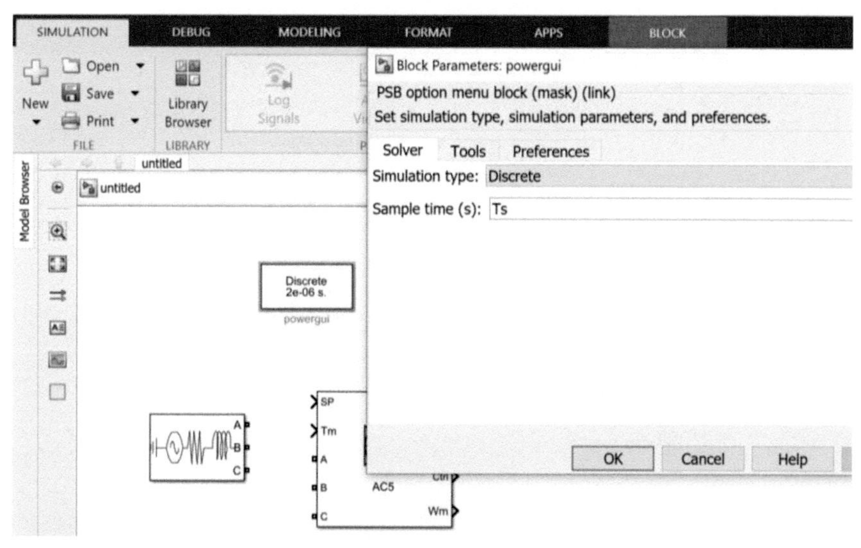

Figure 5.11 Setting parameters of Powergui.

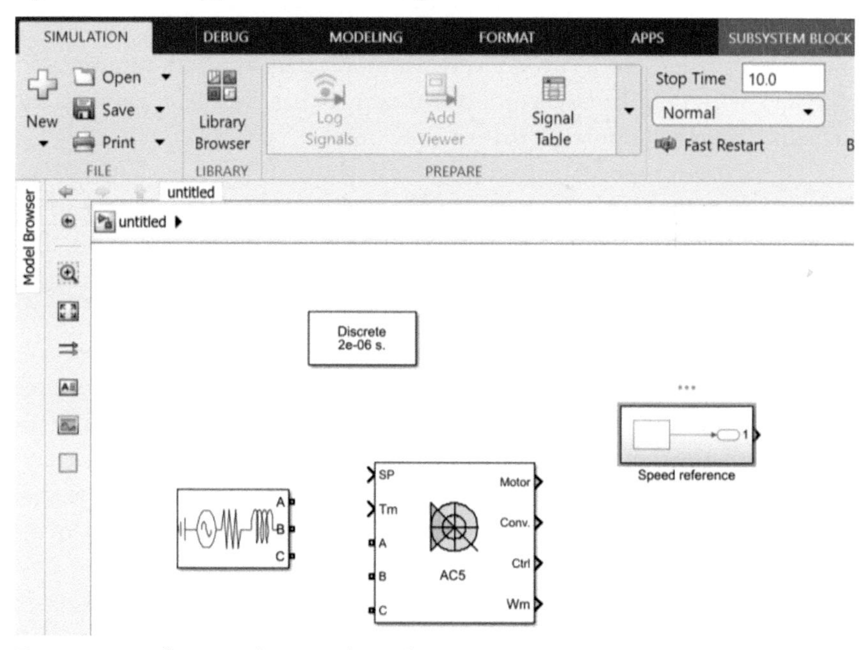

Figure 5.12 Selecting subsystem from library.

Self-control is usually used because of the stable functioning, and it is done via either rotor position sensing or induced voltage sensing. CLM mode is used by the motor. The synchronous motor can be operated at leading power factor when fed from a CSI, allowing the inverter to be commutated using

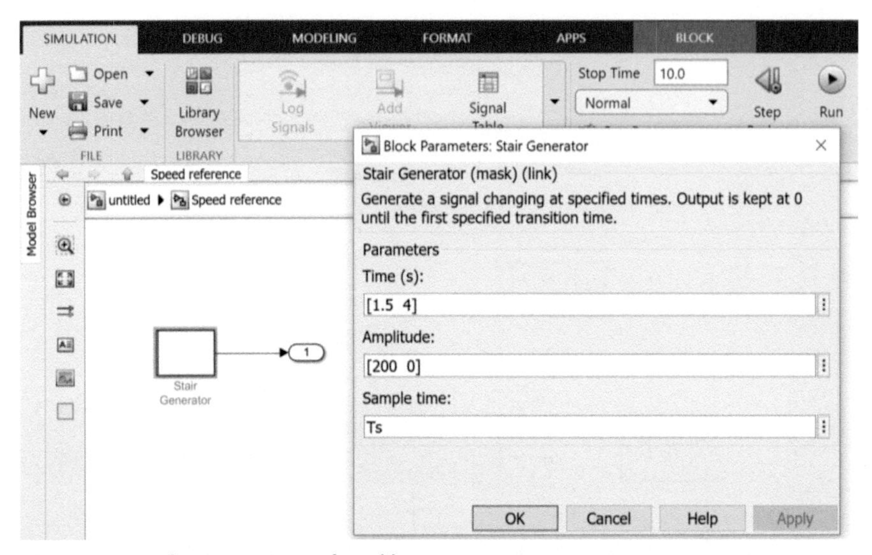

Figure 5.13 Selecting staircase from library.

machine voltages. The term "converter motor" refers to a load-commutated, CSI-fed self-controlled synchronous motor. It has excellent stability and dynamic behavior, similar to that of a DC motor.

The working speed range often starts above 10% of base speed and extends up to base speed due to machine commutation. The lower speed restriction can be increased to zero by utilizing forced commutation. The machine can run at UPF in the speed range of 0%–10% of base speed (beyond which load commutation is possible).

The synchronous motor is provided with variable frequency and variable amplitude currents when fed from a CSI. The DC-link current is permitted to alternatively flow across the motor's phases. If the commutation is instantaneous, the motor currents are quasi-square wave. Square wave currents have a significant impact on motor behavior. Additional losses and heating are caused by harmonics in the stator current. At low speeds, they can create torque pulsations, which are annoying.

The regeneration ability of a CSI-fed synchronous motor drive is built-in (Fig. 5.30). There is no need for an additional converter, and four-quadrant operation is simple.

The machine power factor is leading due to over-excitation. The motor is used less. At retarded angles of firing, the power factor becomes low due to phase control on the line side converter for current control in the DC link. Due to the lack of a commutation circuit, the inverter has a medium

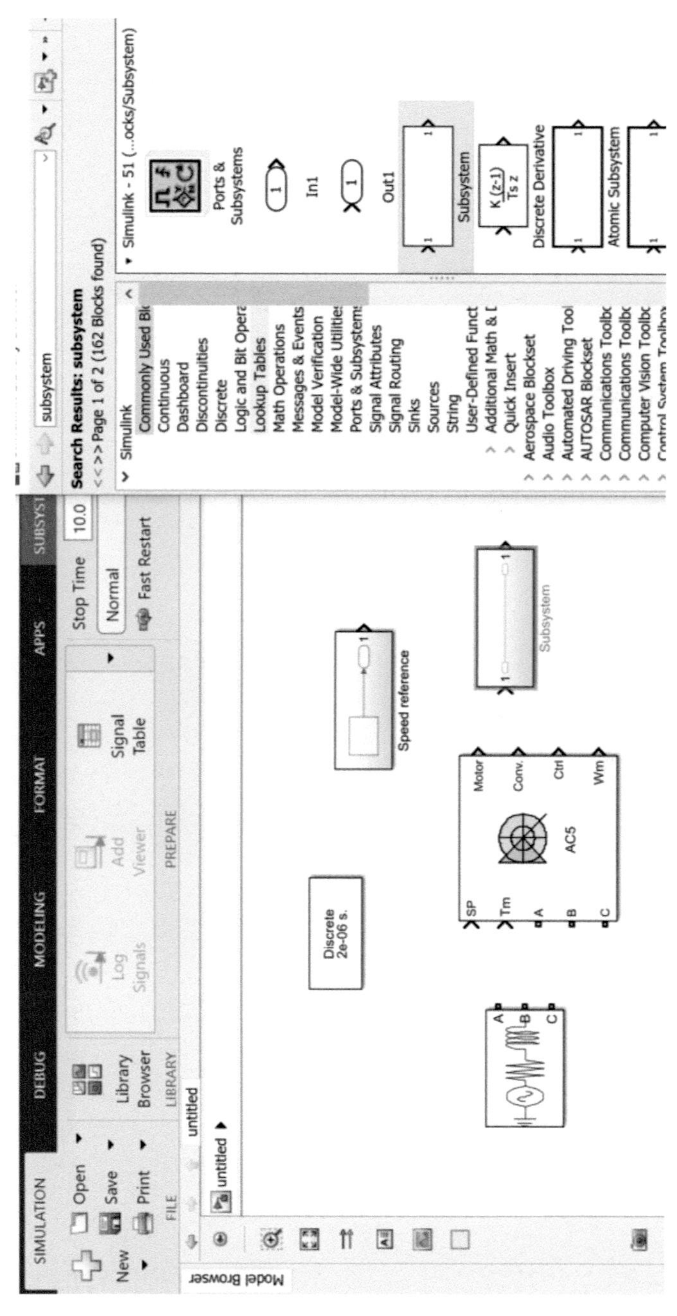

Figure 5.14 Selecting subsystem from library.

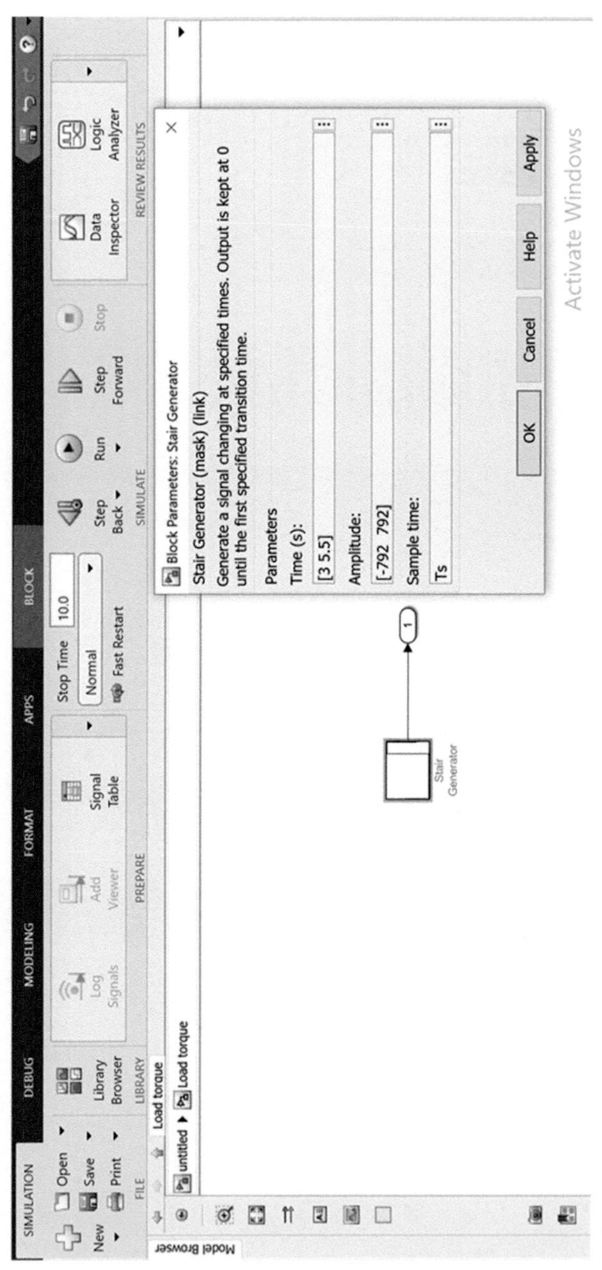

Figure 5.15 Setting parameters of staircase generator.

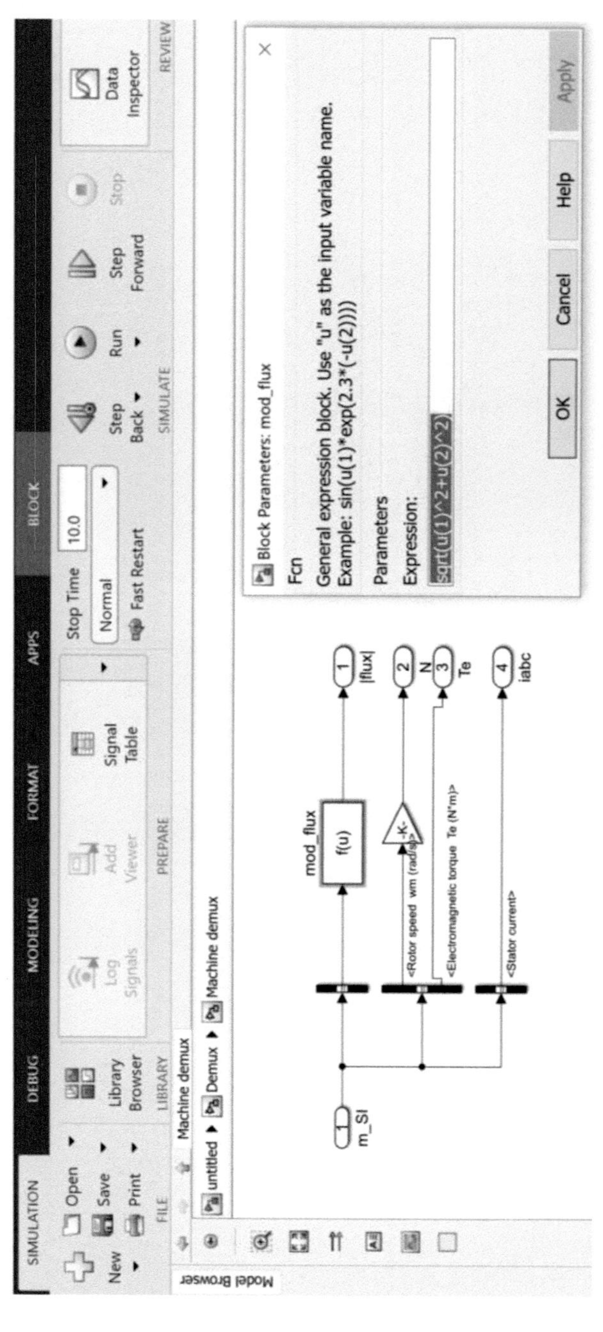

Figure 5.16 Setting expression for mod_flux block.

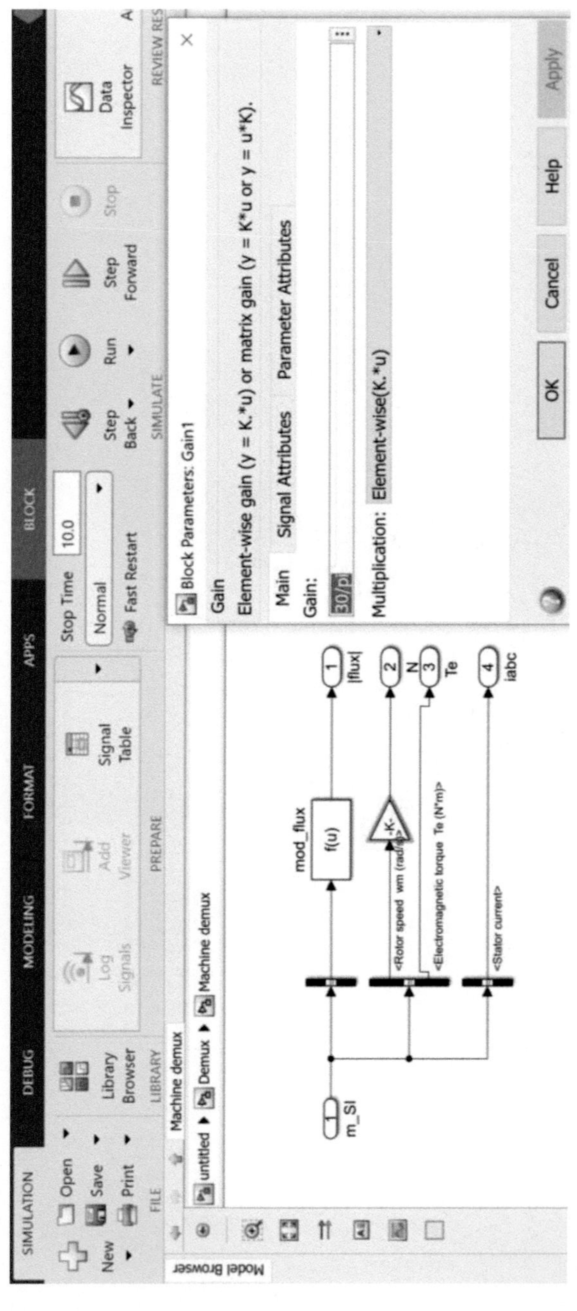

Figure 5.17 Setting value for gain.

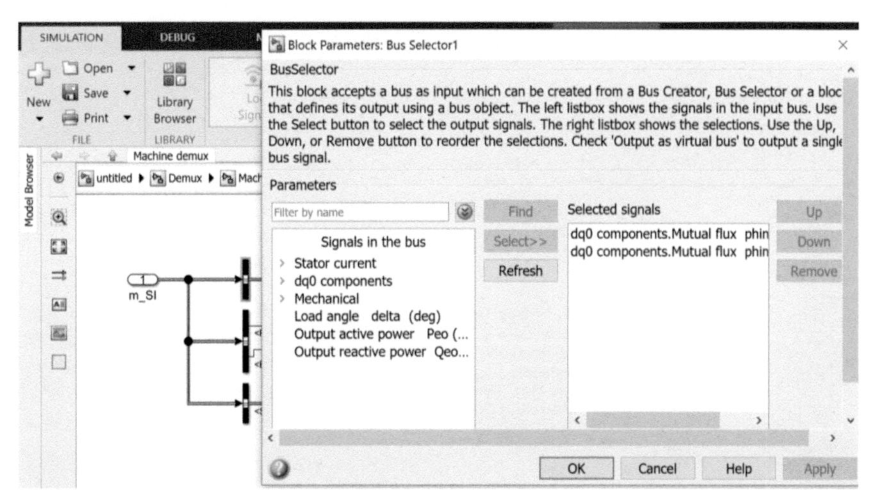

Figure 5.18 Selecting signals for bus1.

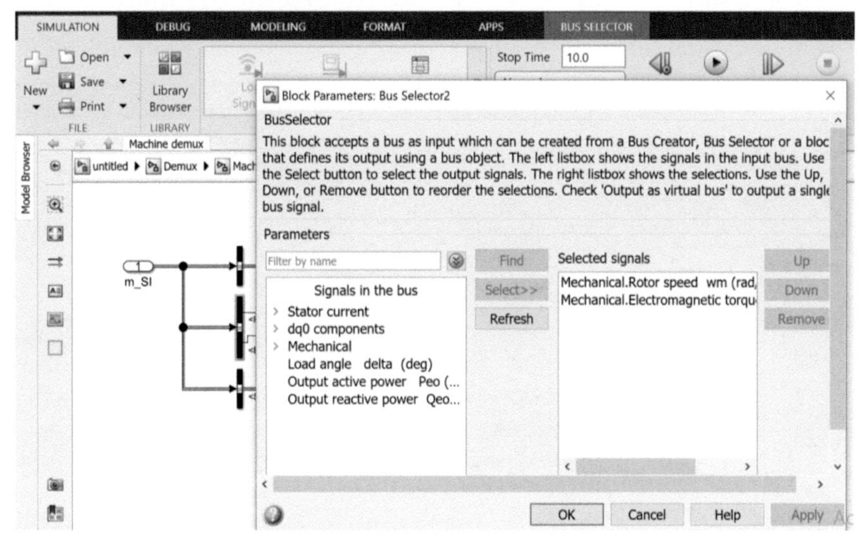

Figure 5.19 Selecting signals for bus2.

cost. The drive is popular as CLM in the medium to high power range and has a reasonably high efficiency. During commutation, voltage spikes occur in the terminal voltage. These are affected by the motor's insulation and are dependent on the subtransient leakage reactance. To prevent voltage spikes, the motor must contain damper windings. This sort of drive is used in gas turbine starting, pumped hydro turbine starting, pump and blower drives, and other applications.

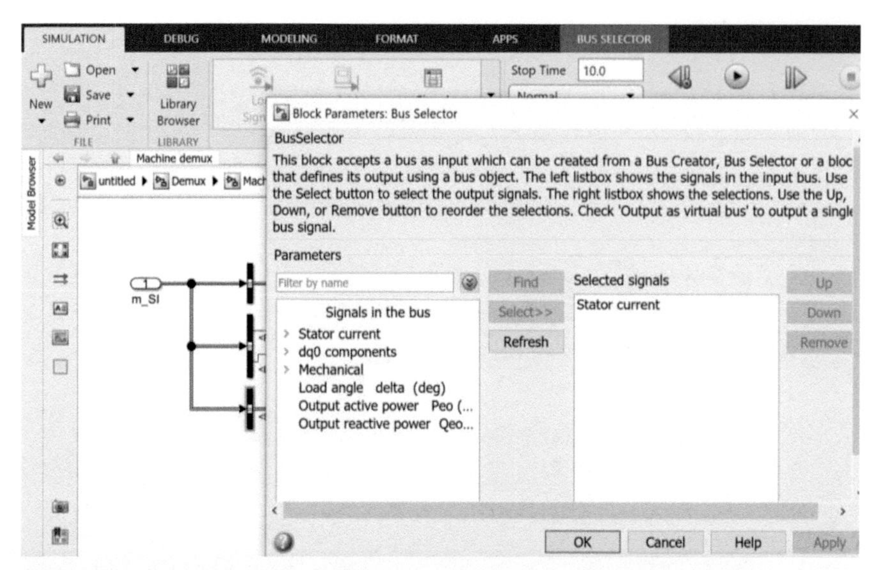

Figure 5.20 Selecting signals for bus3.

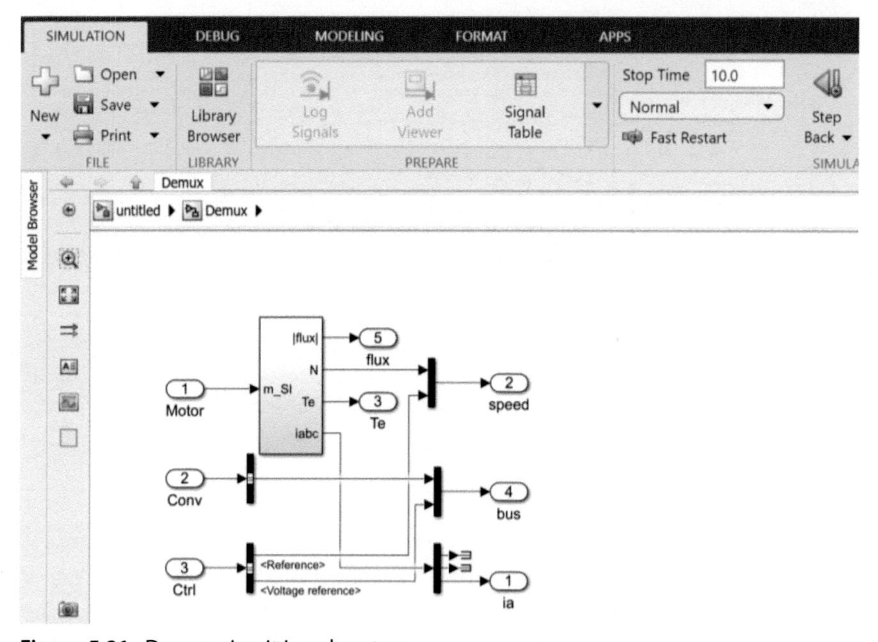

Figure 5.21 Demux circuit in subsystem.

For use with synchronous motors at low speeds for starting, the third harmonic ASC CSI forced commutation for the inverter may be necessary on occasion. The speed range can obviously be extended from zero to base speed. The subject of regeneration, harmonics, and torque pulsations applies

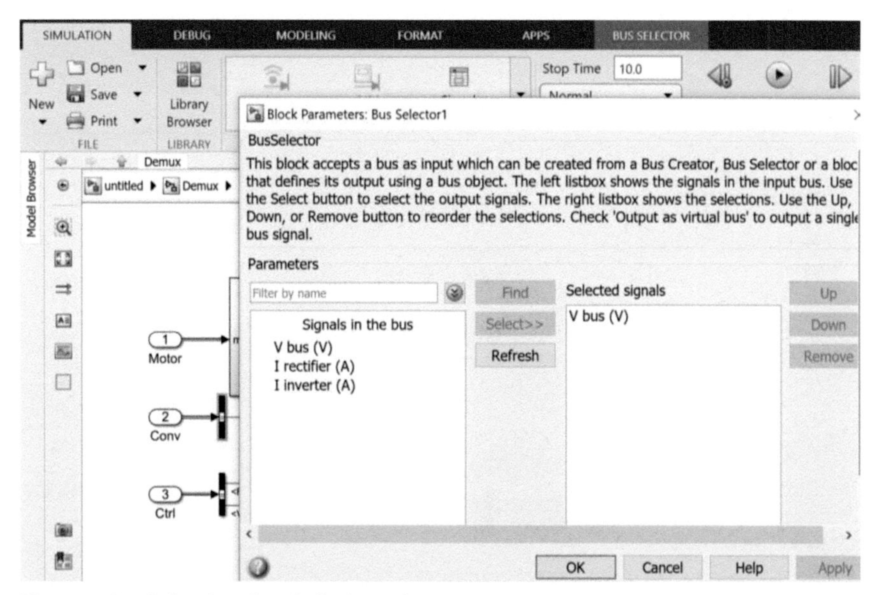

Figure 5.22 Selecting signals for input bus1.

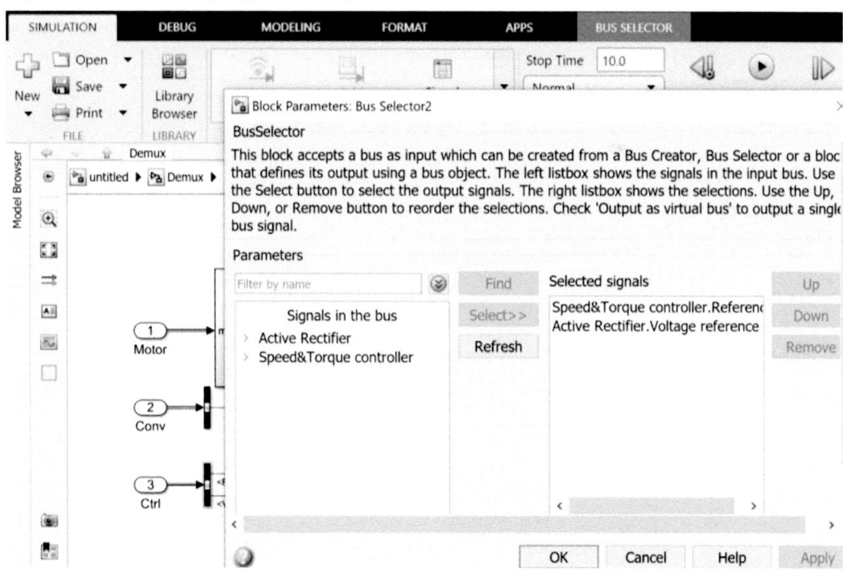

Figure 5.23 Selecting signals for input bus2.

here as well. The power factor of the line is insufficient. The machine, on the other hand, is used at UPF to achieve the previously mentioned benefits. As a result of forced commutation, the inverter's price rises. The drive's efficiency is high, and it is a common choice for low- to medium-power applications in CLM mode. The drive cannot be used in an open-loop configuration.

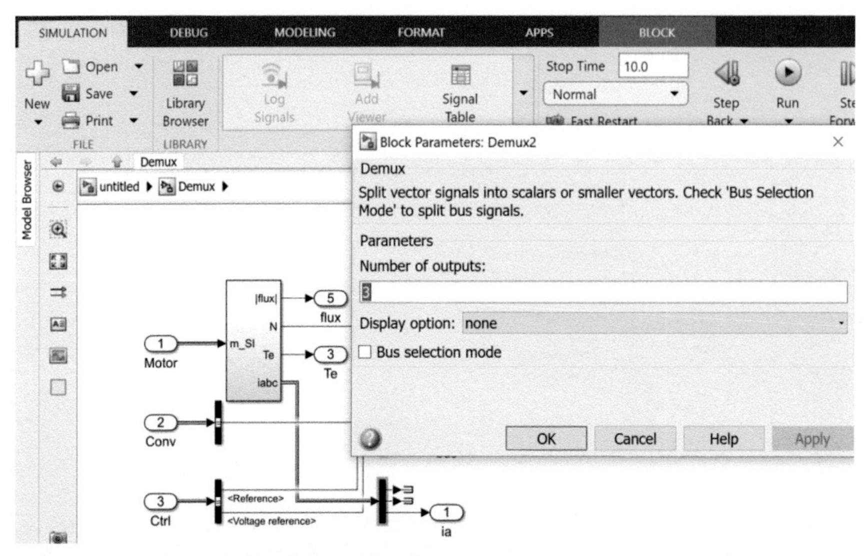

Figure 5.24 Selecting signals for output bus1.

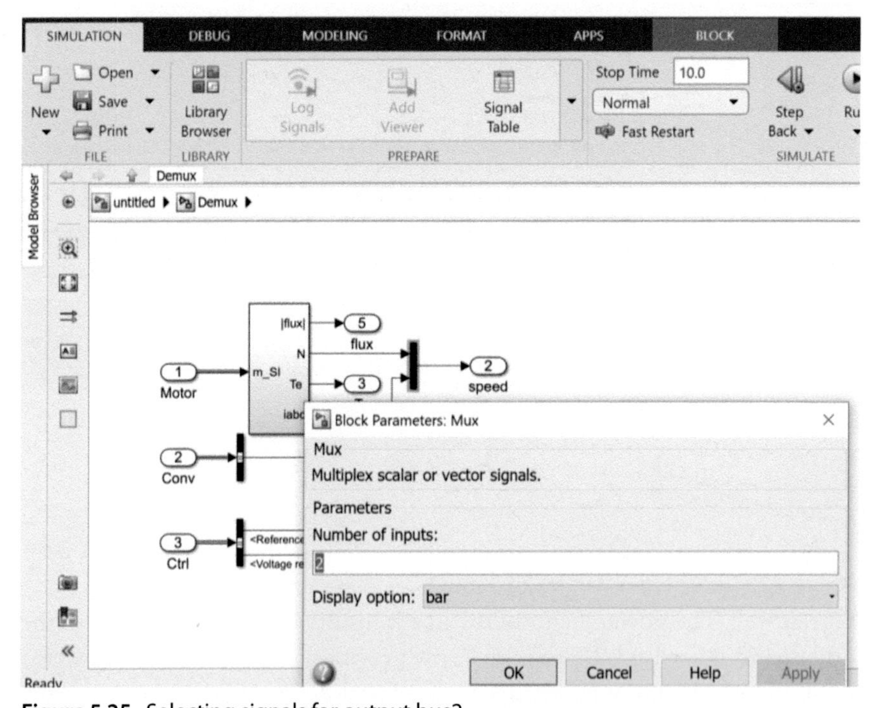

Figure 5.25 Selecting signals for output bus2.

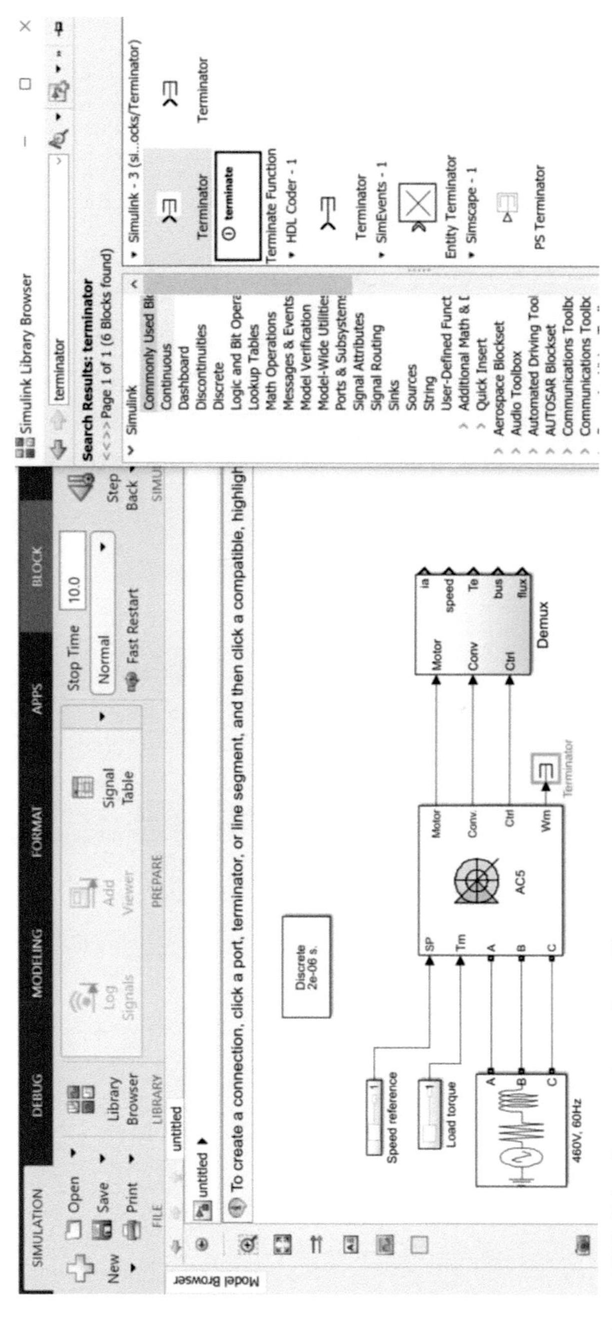

Figure 5.26 Selecting terminator from library.

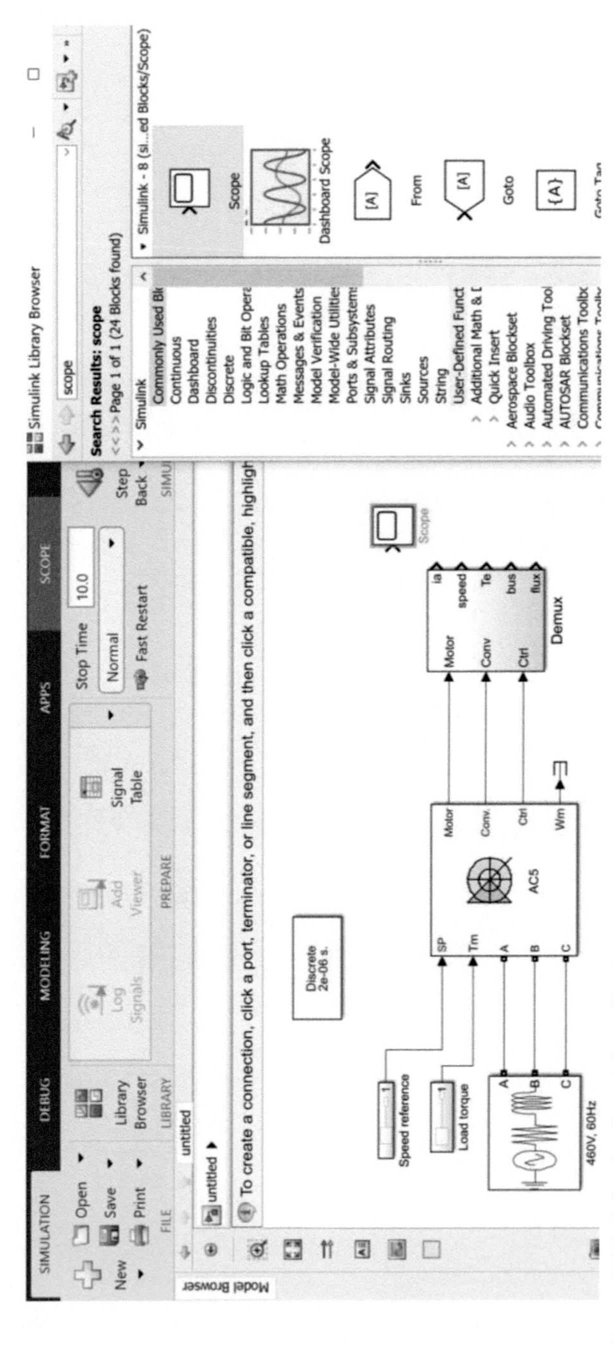

Figure 5.27 Selecting scope from library.

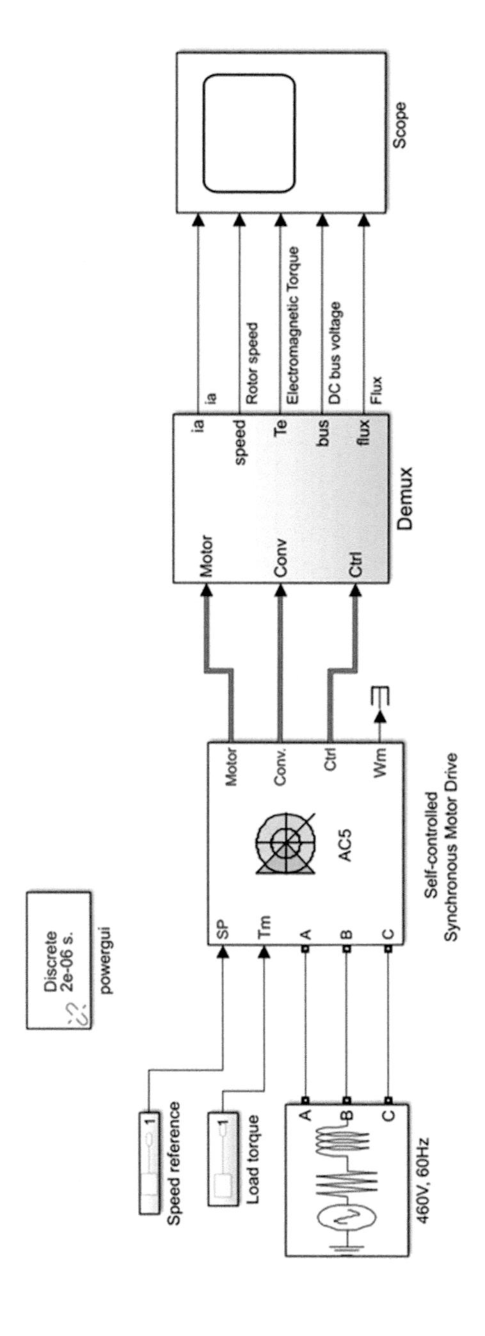

Figure 5.28 MATLAB/Simulink model of self-controlled synchronous motor drive. nullU

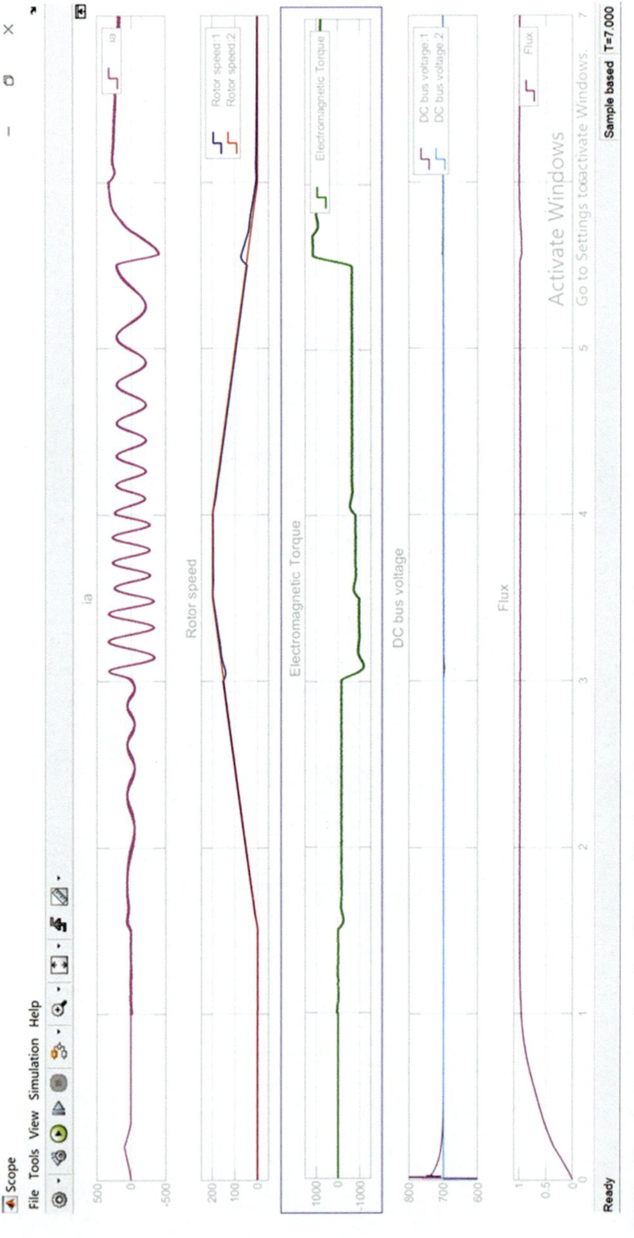

Figure 5.29 Simulation results of self-controlled synchronous motor drive.

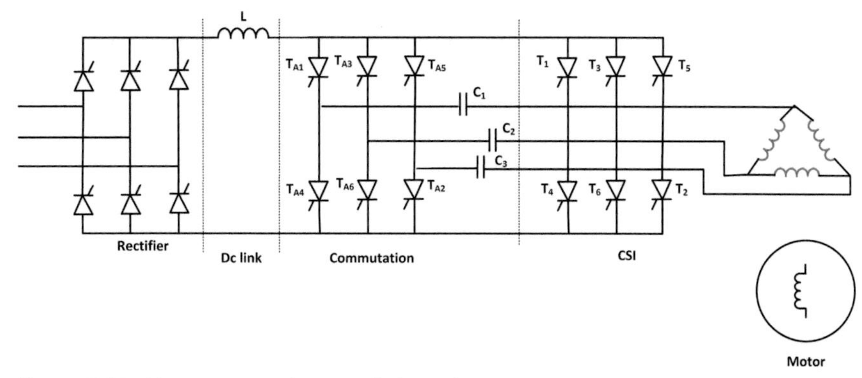

Figure 5.30 Current source inverter fed synchronous motor.

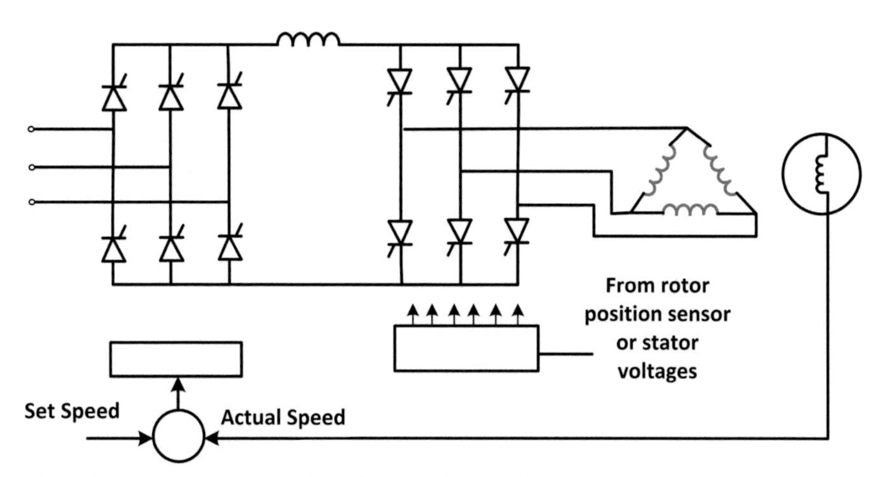

Figure 5.31 Self-control mode with current source inverter.

In CLM mode, stability improves. Voltage spikes are a concern here as well, and commutation is straightforward. The self-control mode with the current source inverter is given in Fig. 5.31.

5.3 Voltage source inverter fed synchronous motor drives

An inverter fed synchronous motor has been very popular as a converter motor in which the synchronous motor is fed from a CSI having load commutation. Understanding the behavior of synchronous motors fed from a voltage source inverter has recently received increasing attention. These drives can also be developed to have self-control, using a rotor position sensor

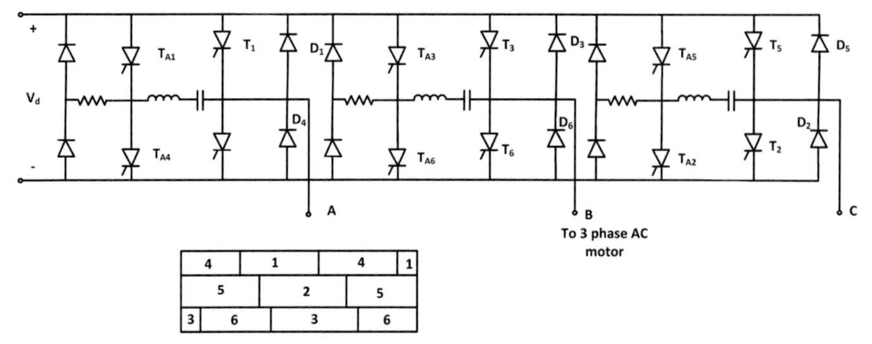

Figure 5.32 Voltage source inverter fed synchronous motor drives.

or phase control methods. It has been suggested in the literature that these drives may cause fewer issues in terms of machine and system design. Forced commutation is required for a typical VS1 with 180° thyristor conduction, and load commutation is not possible.

A typical power circuit of a voltage source inverter is shown in Fig. 5.32. Three combinations are possible, to provide a variable voltage variable frequency supply to a synchronous motor (Fig. 5.33). The voltage control can be obtained external to the inverter using a phase-controlled rectifier. The link voltage is variable. This has the disadvantage that commutation is difficult at very low speeds. As the output voltage is a square wave the inverter is called variable voltage inverter or square wave inverter. The second alternative is to have voltage control in the inverter itself, using principles of PWM or PSM. The inverter is fed from a constant link voltage. A diode rectifier would be sufficient on the line side. This does not have difficulties of commutation at low speeds. Very low speeds up to zero can be obtained. The third alternative is to interpose a DC chopper in between the rectifier and the inverter. The system may appear cumbersome at first sight, but it has advantages. Three simple converters are used to give the desired result. It is possible to reduce the size of link inductance by having a synchronous control of the chopper.

A voltage source inverter feeding a synchronous motor can have either separate control or self-control. In the former, the speed of the motor is determined by external frequency from a crystal oscillator. Open-loop control is possible. The motor has instability problems and hunting, similar to a conventional motor. In latter, the inverter is controlled by means of firing pulses obtained from a rotor position sensor or induced voltage sensor.

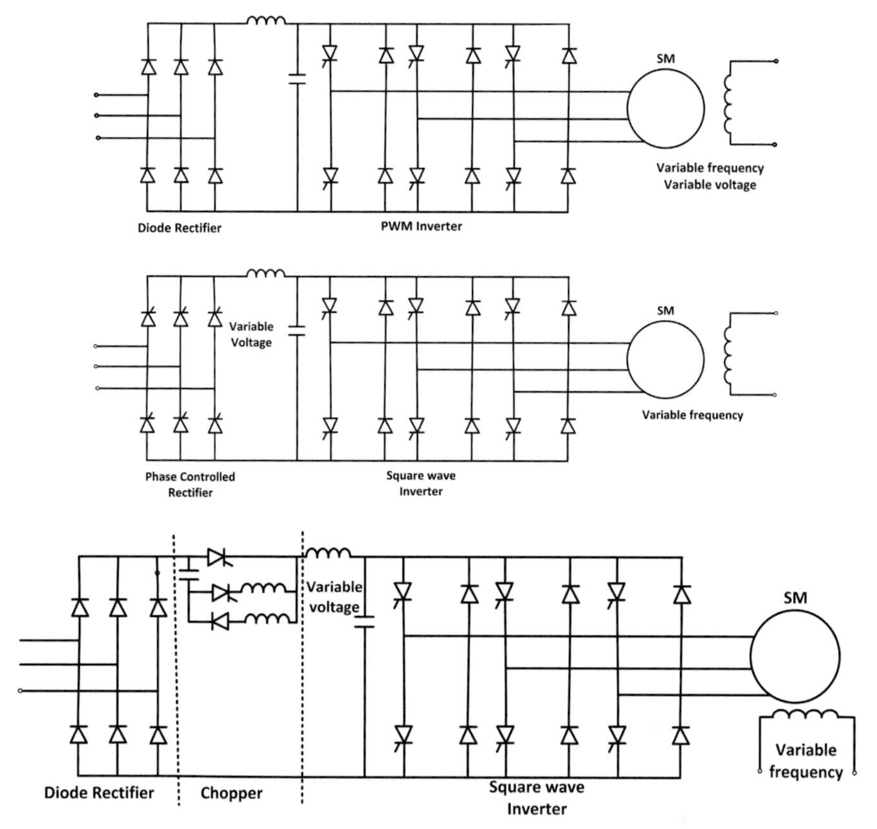

Figure 5.33 Variable voltage, variable frequency supply to feed a synchronous motor.

The motor is in the CLM mode and has better stability characteristic (Fig. 5.34).

The output voltage of the inverter is nonsinusoidal. The behavior of the motor supplied from the inverter is entirely different from the behavior of the motor operating on a conventional sinusoidal supply. Knowledge of the behavior is essential. The steady-state performance enables one to have a proper choice of the thyristors, and also to determine the effects of nonsinusoidal waveforms on torque developed and machine losses.

The stator current drawn by the motor when fed from the square wave inverter has sharp peaks and is rich in harmonic content. These harmonics can cause additional losses and heating of the motor. They also cause pulsating torques that are objectionable at low speeds. Thus the performance

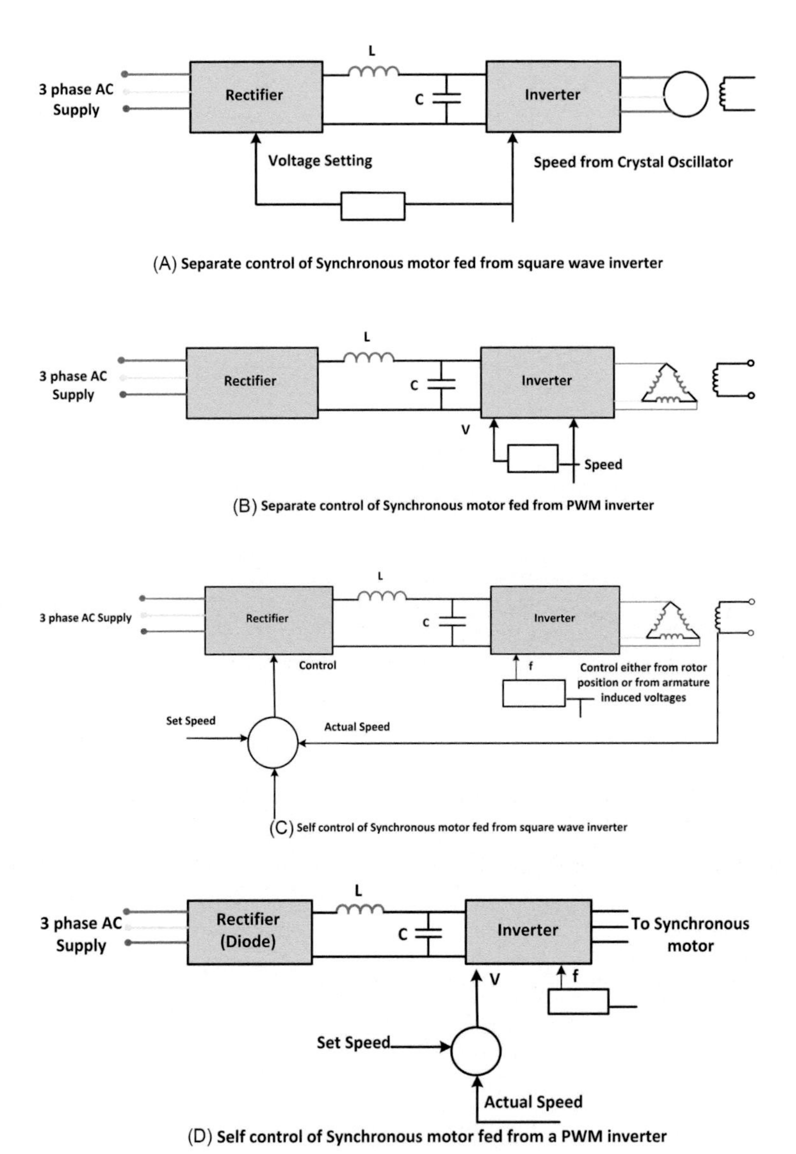

Figure 5.34 Principles of separate and self-control.

with respect to additional heating due to harmonics, and pulsating torques is similar to that of an induction motor.

When a PWM inverter is used, these harmonic effects are reduced. The stator currents are less peaky and have reduced harmonic content. Accordingly, additional losses due to harmonics, consequent motor heating, and torque pulsations are decreased. These effects become minimal.

The discussion on regeneration given for induction motors holds good for these cases also. With the square wave inverter, another phase-controlled rectifier is required on the line side. Dynamic braking can be employed. There are two scenarios that can occur when a PWM inverter is employed. The inverter can be powered by a steady DC supply, in which case regeneration is simple. A diode rectifier can provide the inverter with a DC supply. On the line side, a phase-controlled converter is required in this situation.

A square wave inverter drive must have a phase-controlled converter on the line side. Due to phase control, the line power factor is very poor. A diode rectifier is sufficient in the case of PWM inverter. The line pf. improves to unity. In either case the machine p.f. can be improved by field control. With a view to minimizing the inverter size as well as losses in the inverter and motor, it is advantageous to operate the motor at UPF.

A VSI drive provides reasonably good efficiency. Converter cost is high and multi motor operation is possible. Open-loop (separate) control may pose stability problems at low speeds. CLM mode is very stable. PWM drive has a better dynamic response than a square wave drive. This finds application as a general-purpose industrial drive for low and medium powers.

5.3.1 Simulation of PWM VSI fed synchronous motor drive

To do the simulation of synchronous motor drive, we need to select the components. First, we select the IGBT. To activate these semiconductor devices, we need DC voltage source. We can select the DC voltage sources as previously.

>>Libraries>>Simscape>>SpecializedPowerSystems >>Fundamentalblocks>> PowerElectronics>>IGBT

After selecting the DC voltage source and IGBT (Fig. 5.35), connect it as per given in Fig. 5.36.

Select repeating sequence (Fig. 5.37) with the help of the below command and give the values of parameter as per requirement.

>>Libraries>>Simulink>>Sources>>Repeating Sequence.

Select three sine wave as per the command and give the parameter values are per Fig. 5.38–5.40.

>>Libraries>>Simulink>>Sources>>sine wave

>>Libraries>>Simulink>>Quickinsert>>Logicandbitoperations >>Relational Operator.

Select the relational operator (Fig. 5.41) from the library as per the above command. We require three relational operator in the circuit (Fig. 5.42).

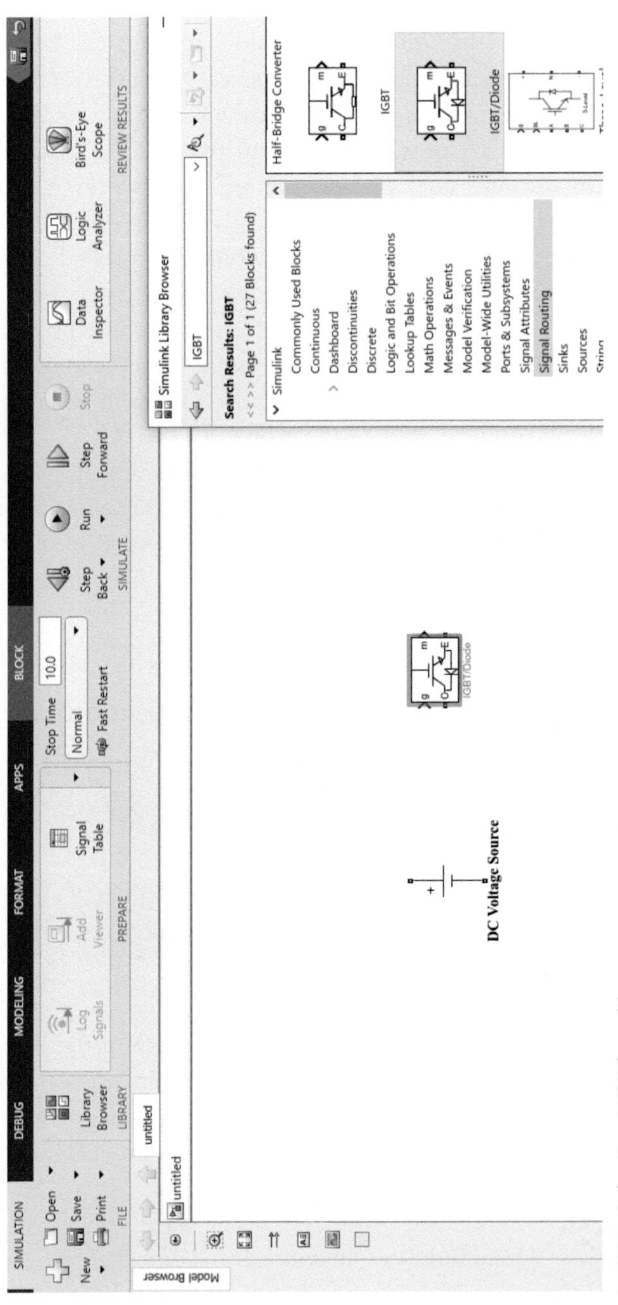

Figure 5.35 Selecting IGBT from library.

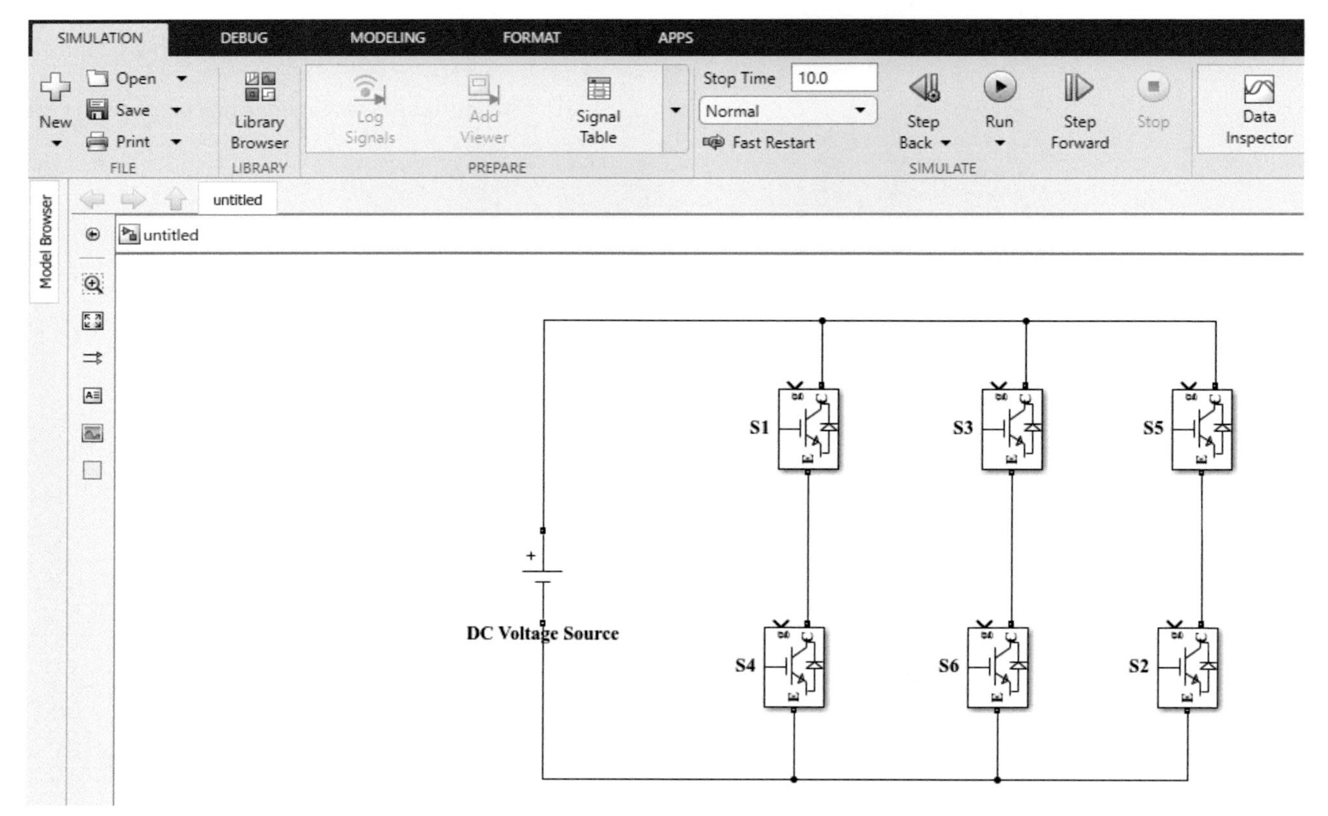

Figure 5.36 Inverter circuit for synchronous motor drive.

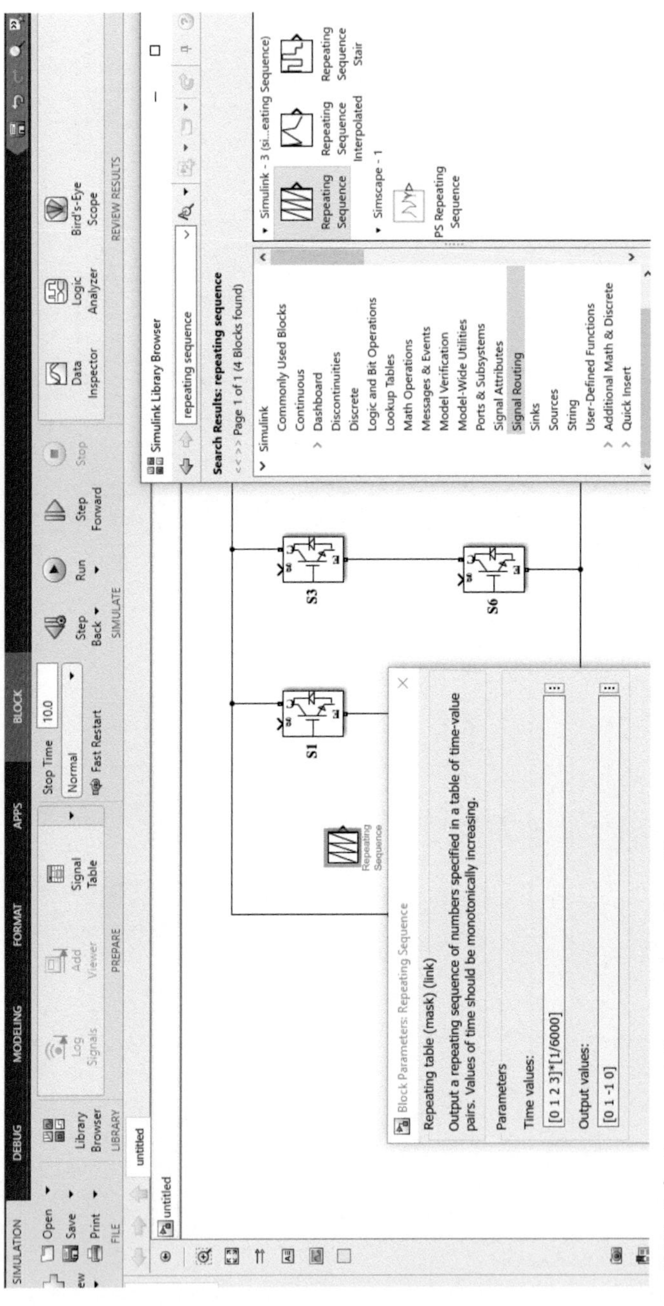

Figure 5.37 Selecting repeating sequence from library.

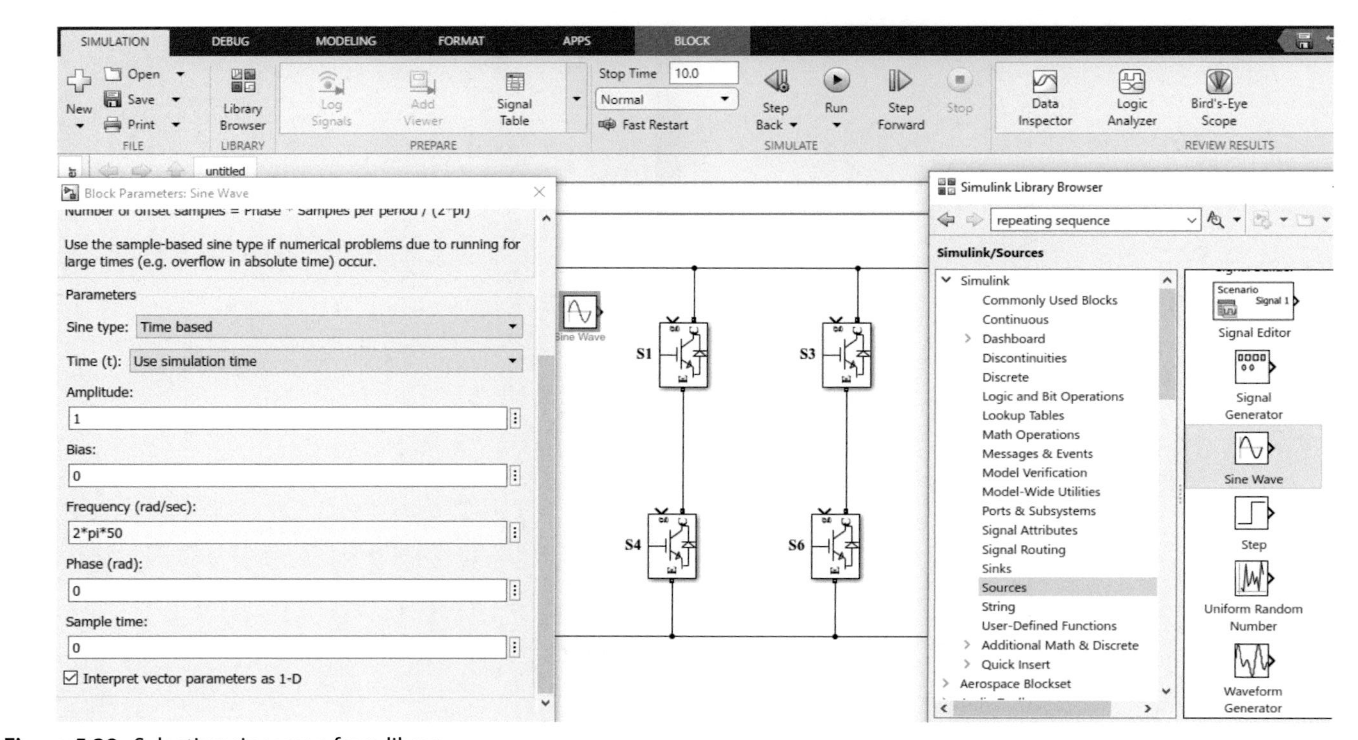

Figure 5.38 Selecting sine wave from library.

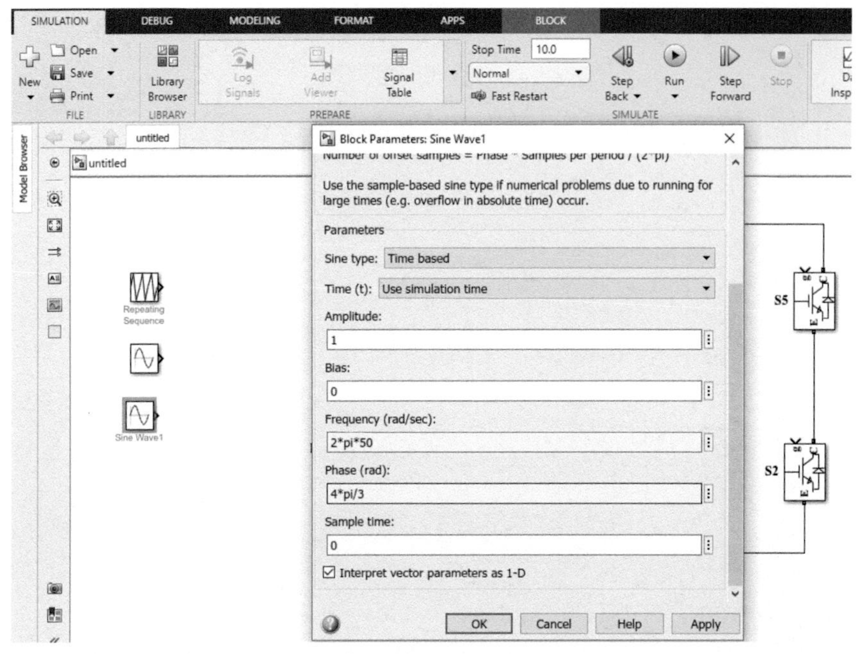

Figure 5.39 Selecting sine wave1 from library.

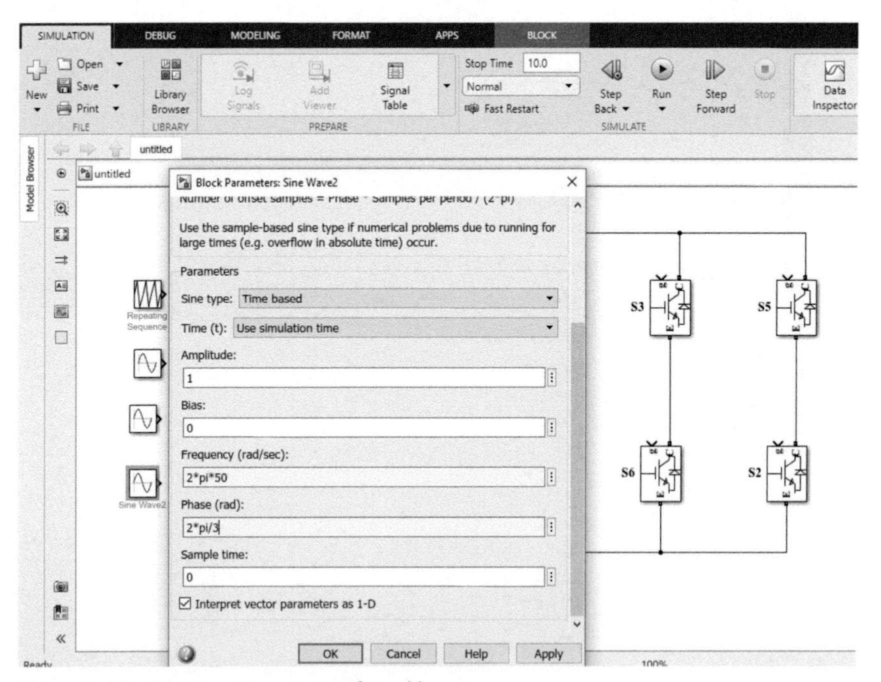

Figure 5.40 Selecting sine wave 2 from library.

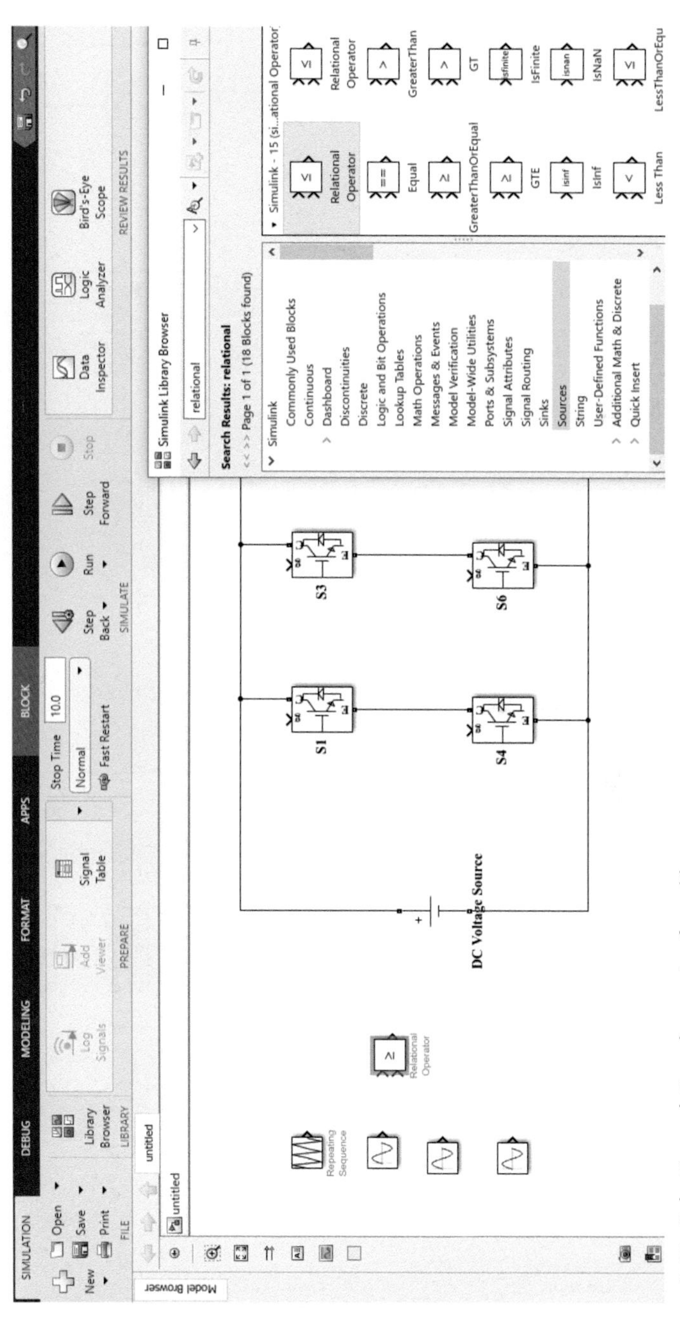

Figure 5.41 Selecting relational operator from library.

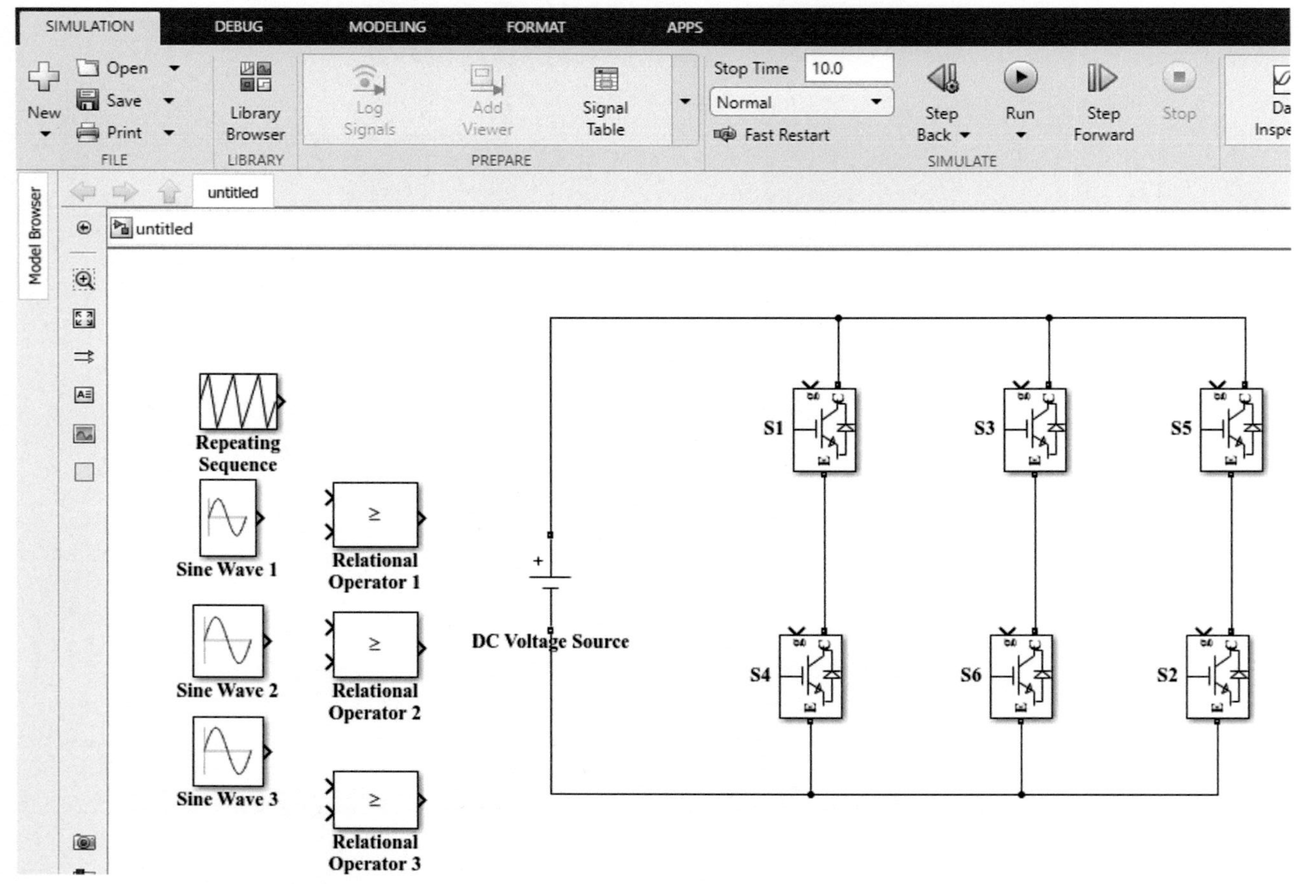

Figure 5.42 Selection of three-relational operator from library.

>>Libraries>>Simulink>>commonlyusedblocks>>Mux.

Select the multiplexer (Fig. 5.43) as per the command and connect the circuit as below. To display all the output, we need scope. So select the scope by the command (Fig. 5.44) >>Libraries>>Simulink>>Commonly used blocks>>Scope.

Select Goto from library. Libraries>>Simulink>>Signal routing >>Goto

Select six Goto operator (Fig. 5.45) and logical operator as shown in Fig. 5.46.

Connect the six Goto along with logical operator as given below (Fig. 5.47).

Select From from library. Libraries>>Simulink>>Signal routing >>From

In Fig. 5.49, it is explained the detail connection of the circuit with From (Fig. 5.48), Goto, and logical operators.

To find the voltage and current across the IGBT, three-phase voltage–current measurement is needed. So select V–I measurement from library (Fig. 5.50).

Connecting the V–I measurement across the IGBTs as per Fig. 5.51.

Select Demux (Fig. 5.52) as per previous.

Libraries>>Simulink>>Simscape>>Electrical>>Specialized Power Systems>>Electrical Machines>>Permanent Magnet synchronous machine.

Select Electrical machines from library like Fig. 5.53. Connect the bus selector to the permanent magnet (PM) synchronous machine and select three parameters as per Fig. 5.54.

>>Libraries>>Simulink>>Sources>>Step.

Select step from the library as per Fig. 5.55.

>>Libraries>>Simspace>>Powerlib>>Powergui

In Powergui (Fig. 5.56), we need to select the simulation type as Discrete and give sample time as 10e-6.

Connect the components as per given in Fig. 5.57.

After simulation, you can see the graph (Figs. 5.58–5.61) in scope by clicking it.

5.4 Cycloconverter fed synchronous motor drives

A cycloconverter is a device that converts fixed AC frequency to variable frequency. A self-controlled synchronous motor fed via a three-phase cycloconverter is shown in the block diagram below in Fig. 5.62.

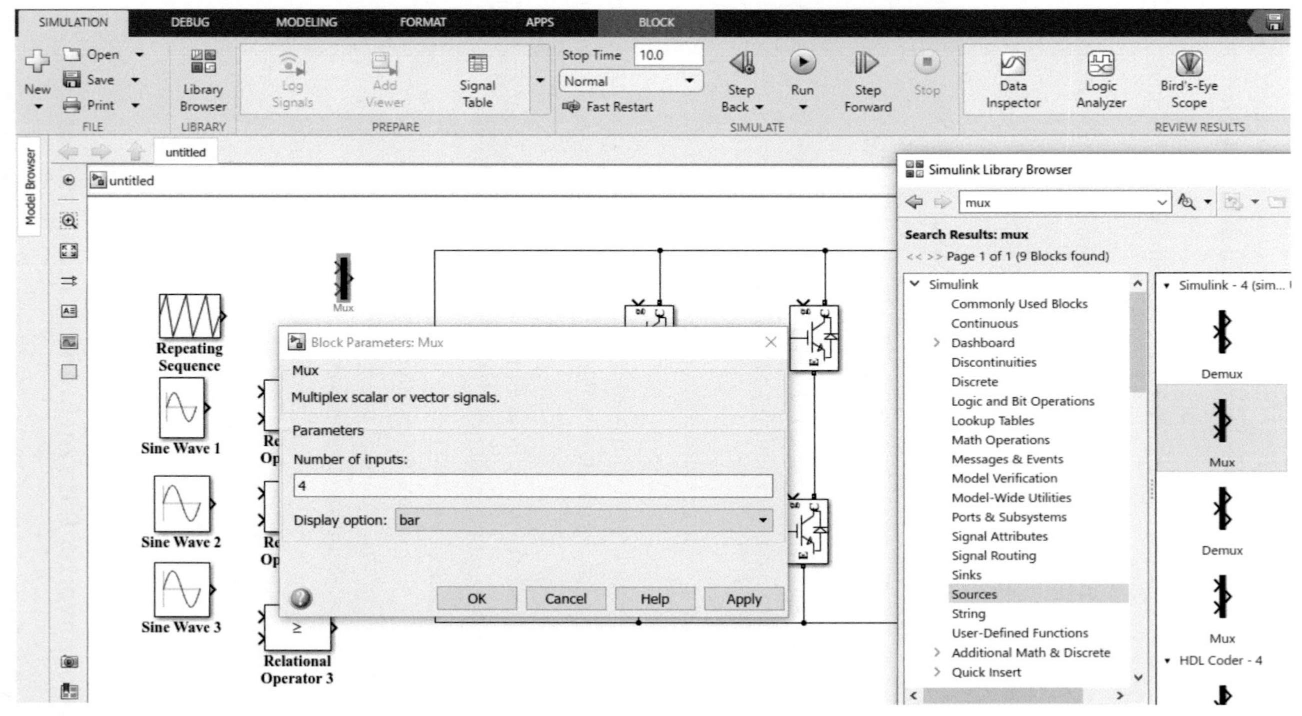

Figure 5.43 Selecting mux from library.

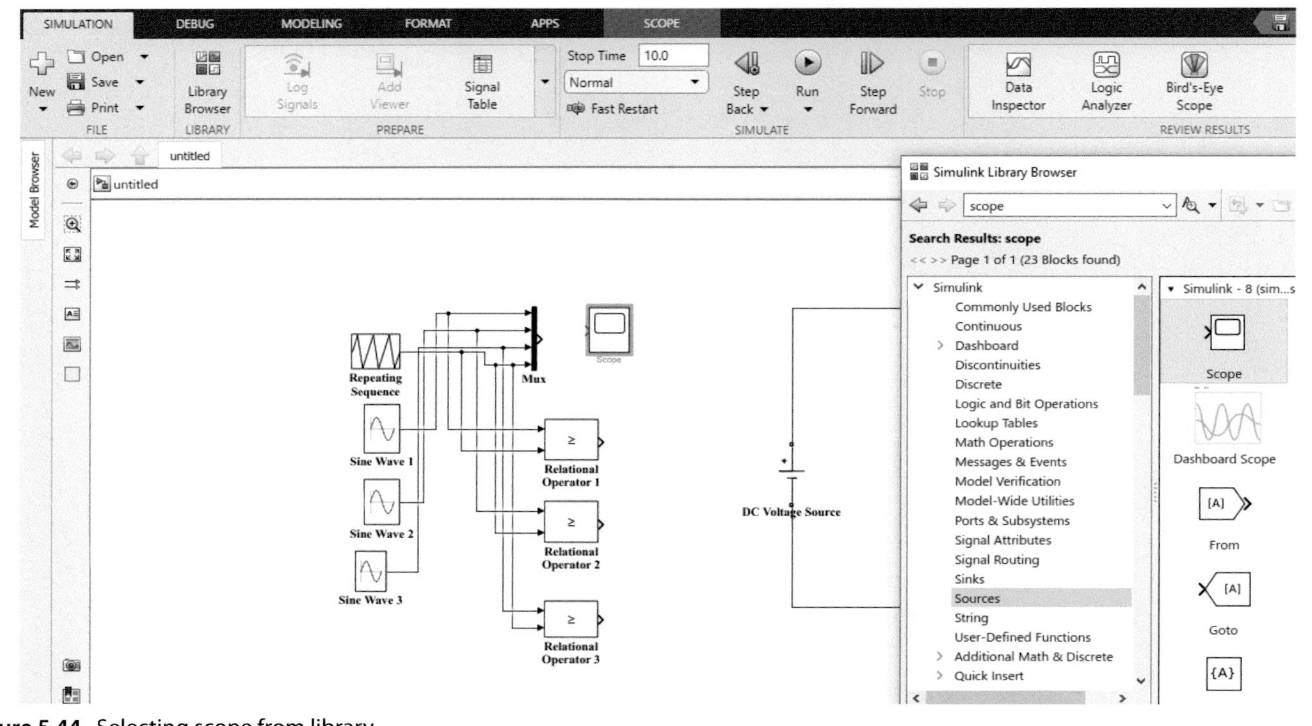

Figure 5.44 Selecting scope from library.

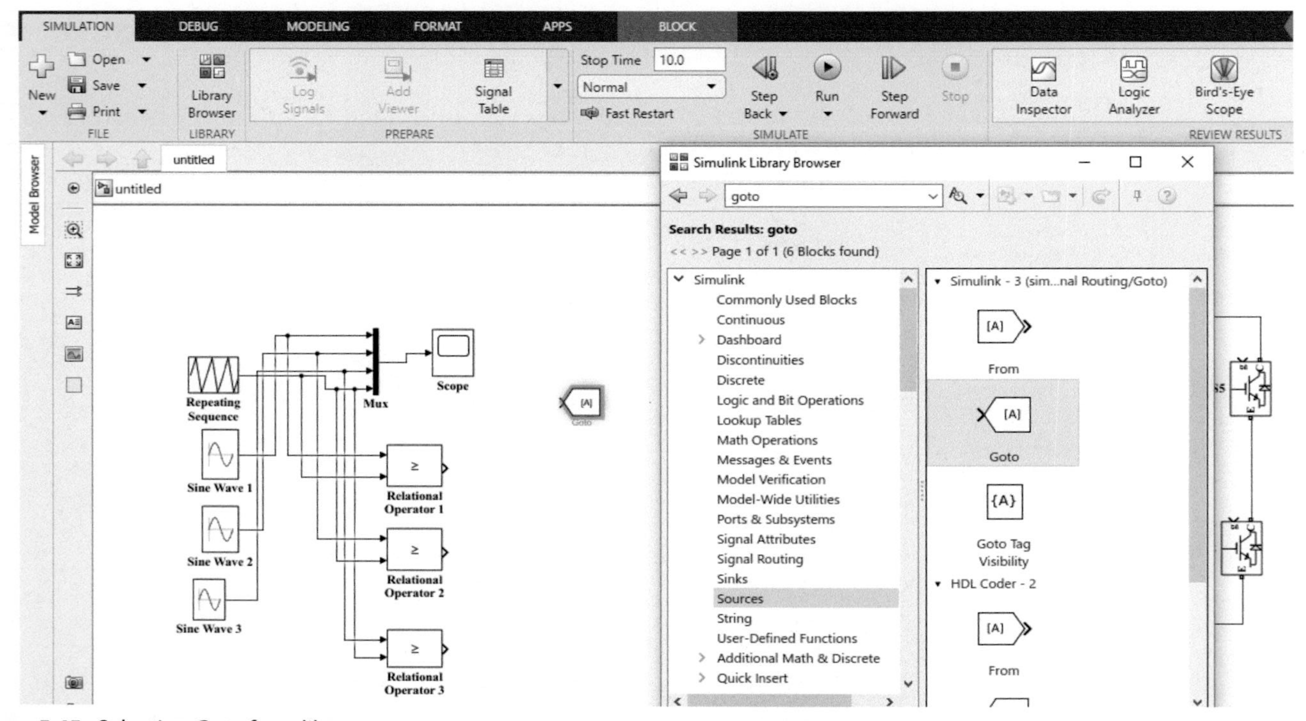

Figure 5.45 Selecting Goto from library.

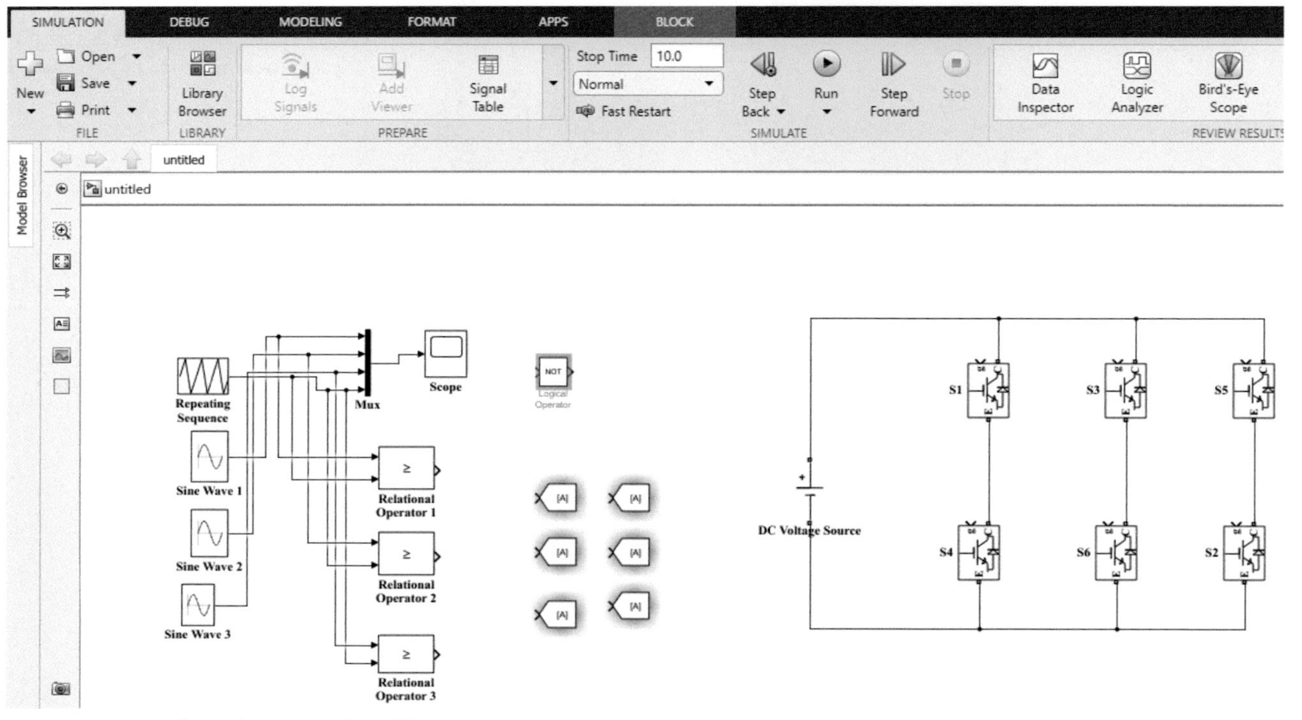

Figure 5.46 Selecting logical operator from library.

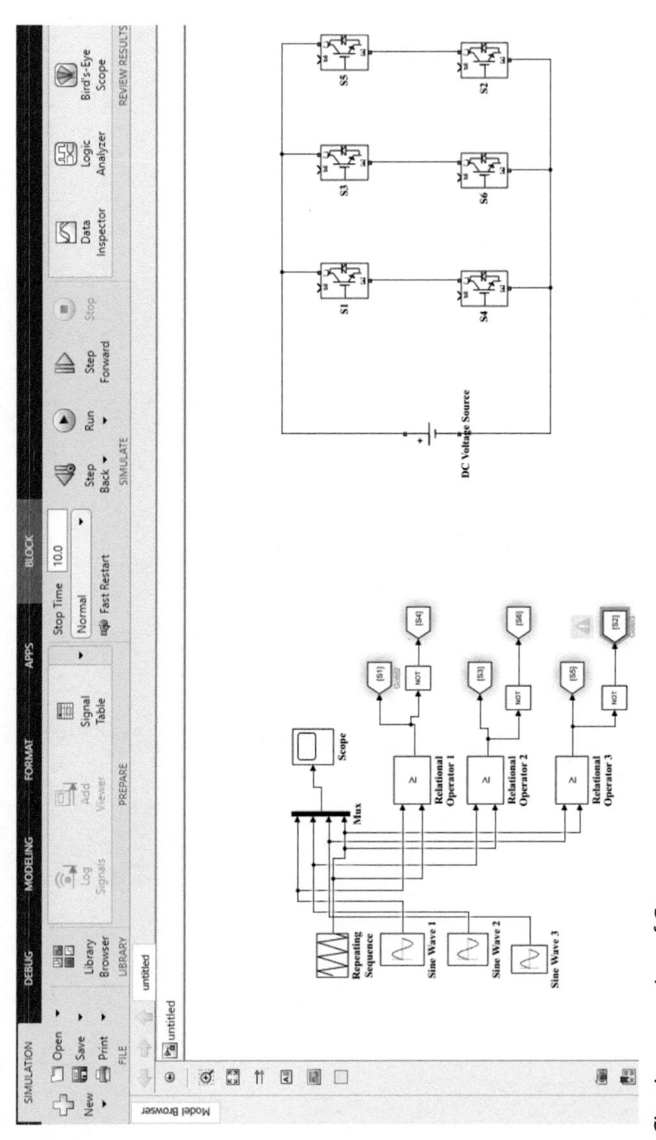

Figure 5.47 Circuit connection of Goto.

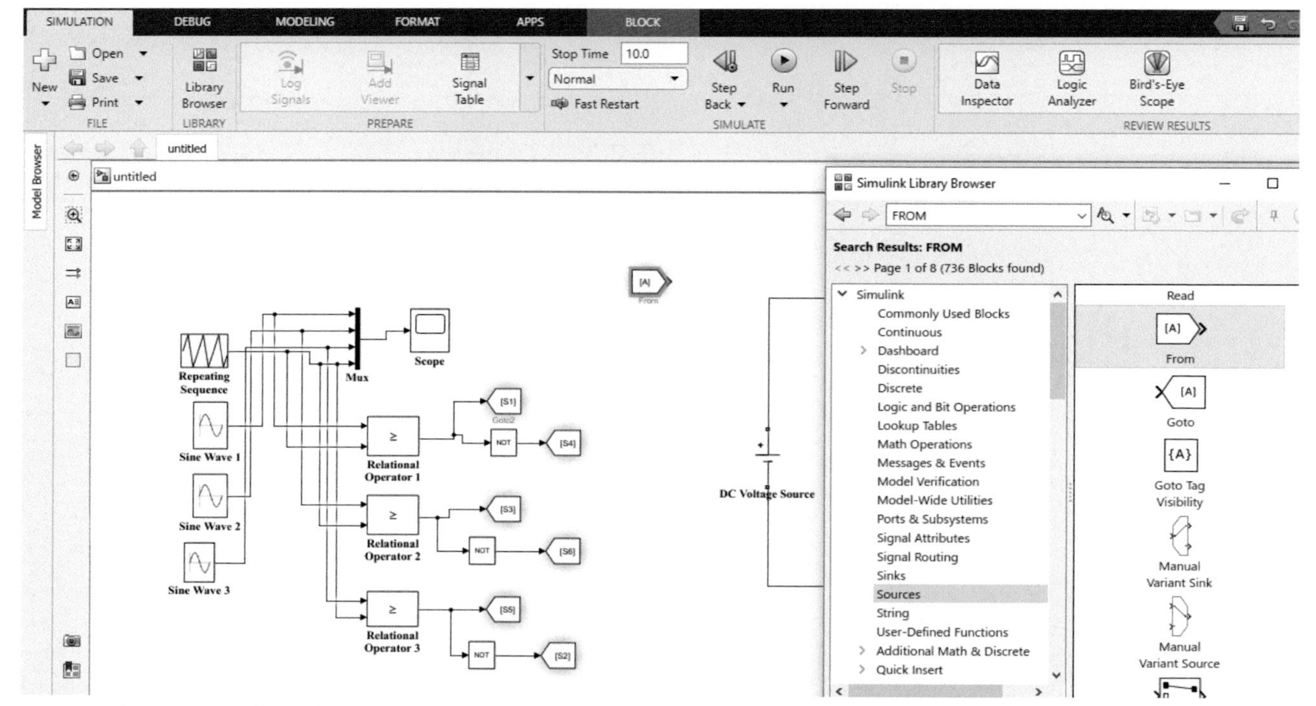

Figure 5.48 Selecting From from library.

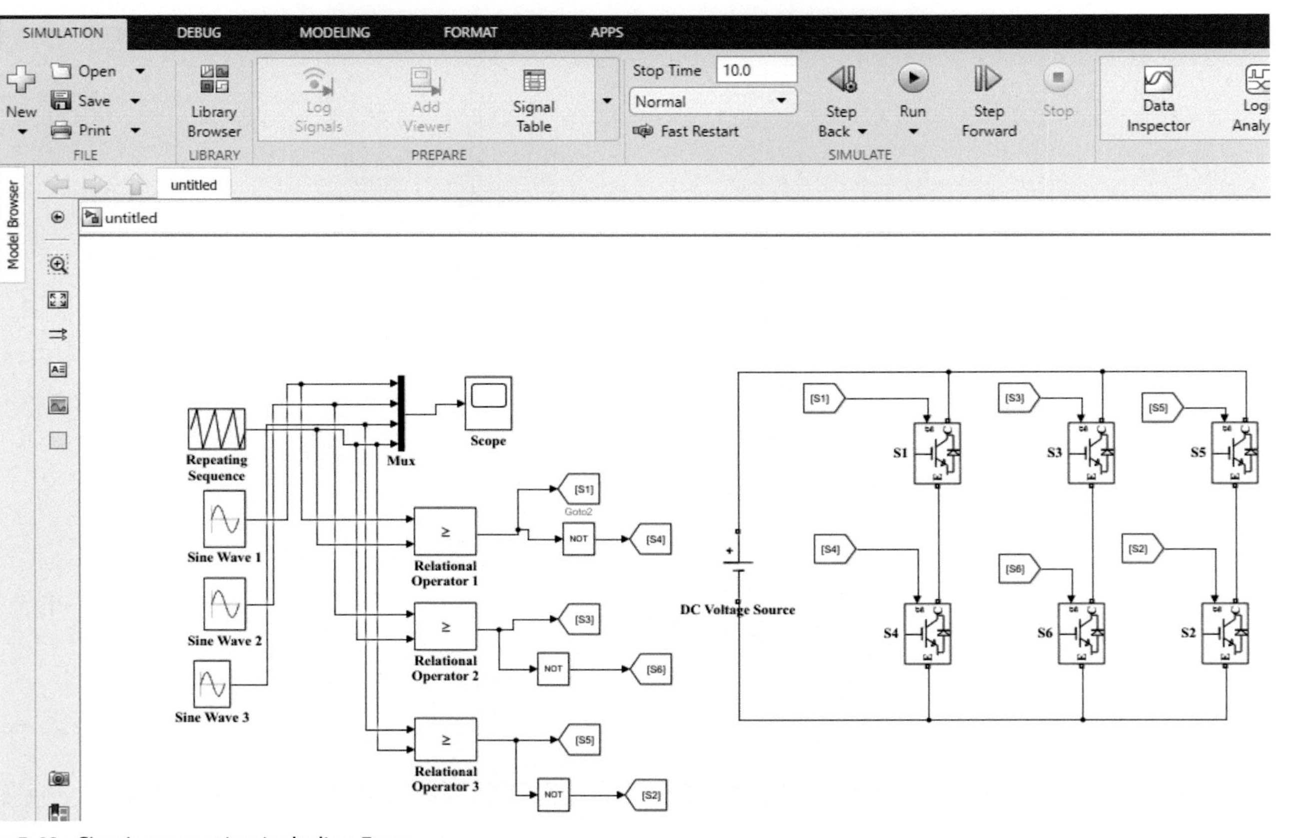

Figure 5.49 Circuit connection including From.

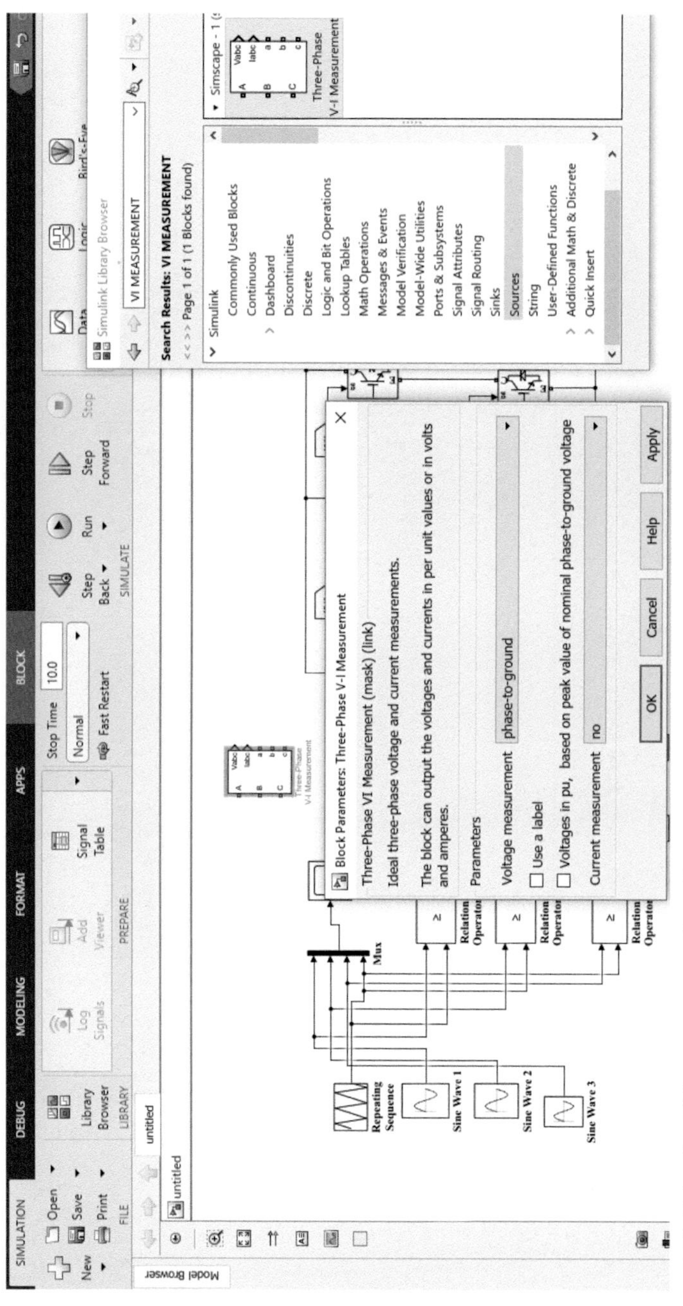

Figure 5.50 Selecting V–I measurement from library.

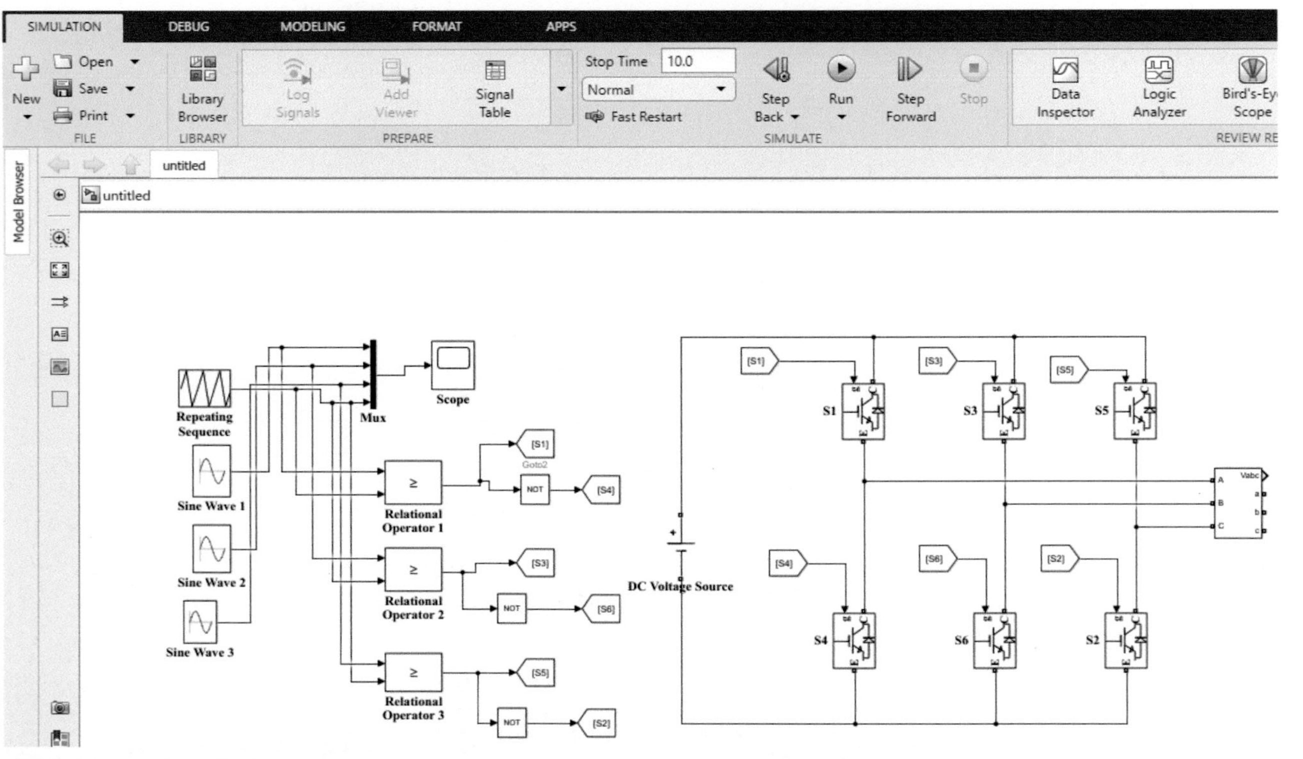

Figure 5.51 Connection of V–I measurement.

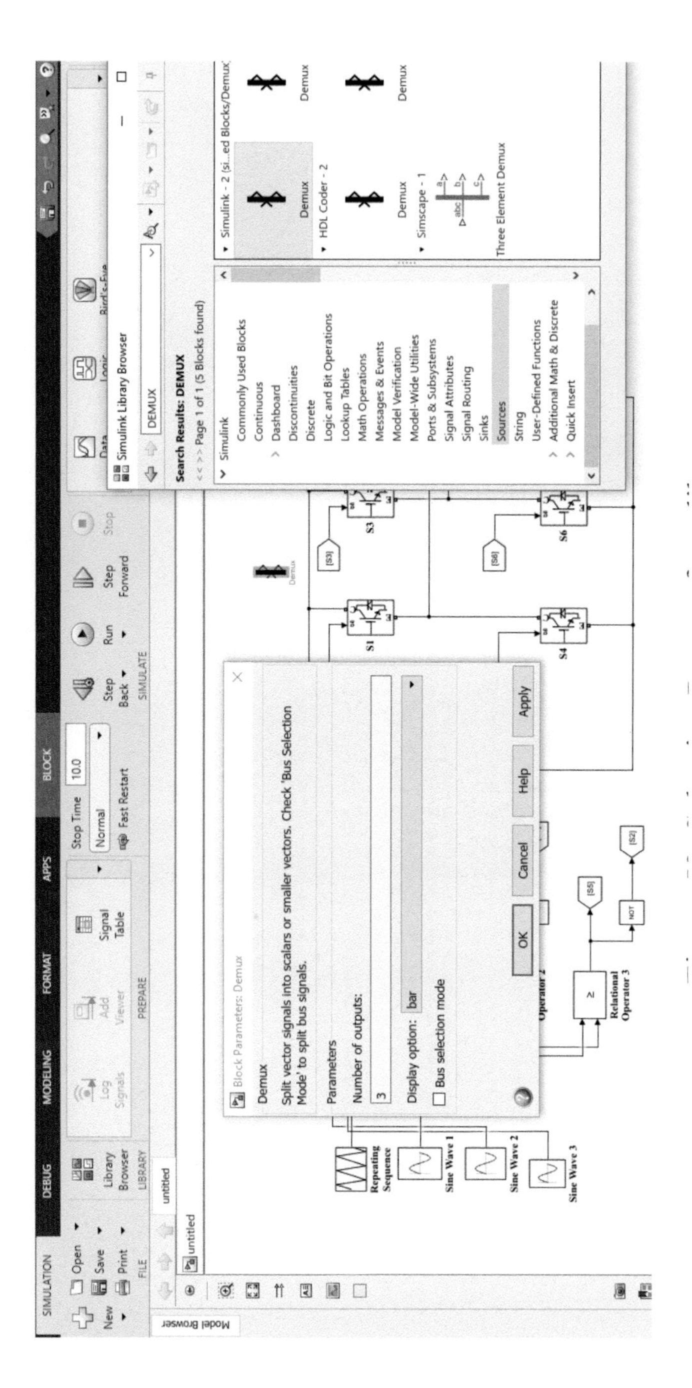

Figure 5.52 Selecting Demux from library.

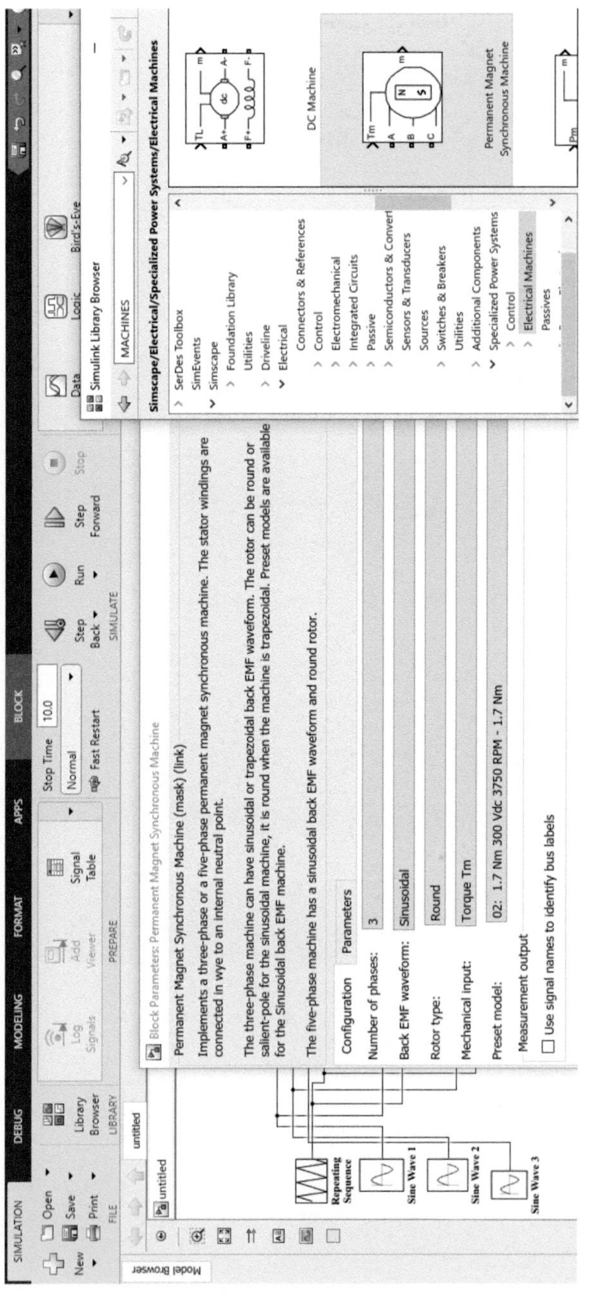

Figure 5.53 Selecting electrical machines from library.

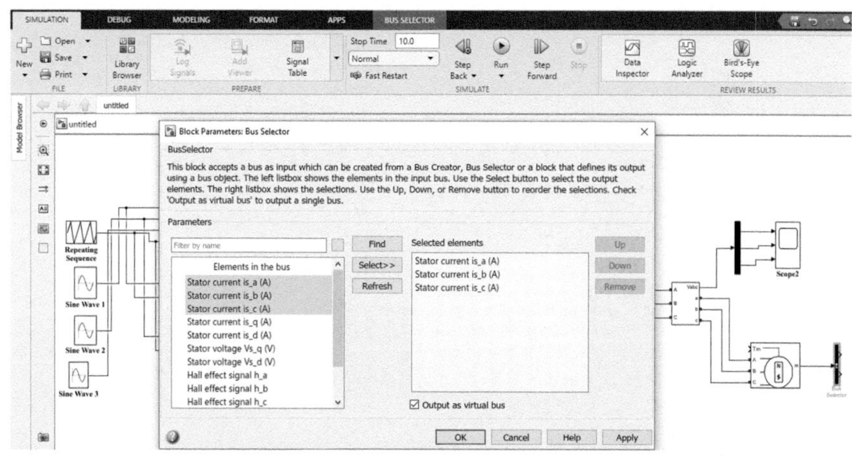

Figure 5.54 Selecting parameters for bus selector.

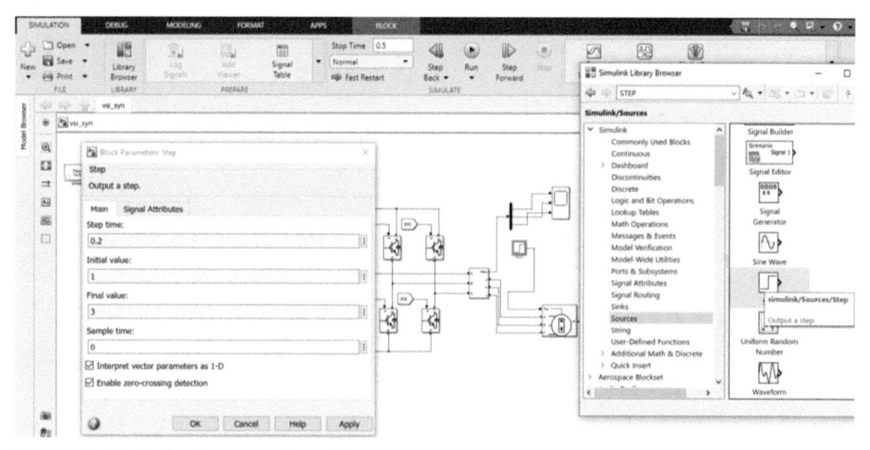

Figure 5.55 Selecting step from library.

The three-phase supply is sent to the cycloconverter, which then feeds the output to the synchronous motor. The rotor position sensor is an essential component of a self-controlled synchronous motor.

A DC link converter is a two-stage converter that produces a variable voltage and variable frequency supply. A cycloconverter, which is a single-stage conversion device, may provide variable voltage, variable frequency supply. Fig. 5.63 depicts the power circuit of a three-phase synchronous motor fed by a cycloconverter. When compared to a DC link converter, this has a numerous differences.

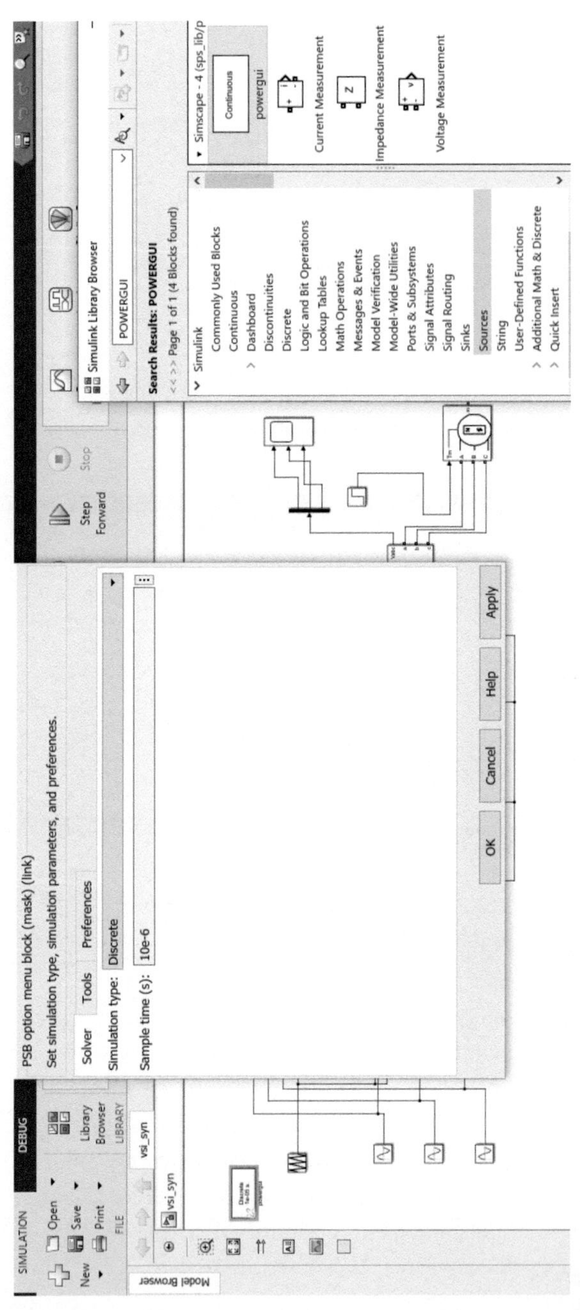

Figure 5.56 Selecting Powergui from library.

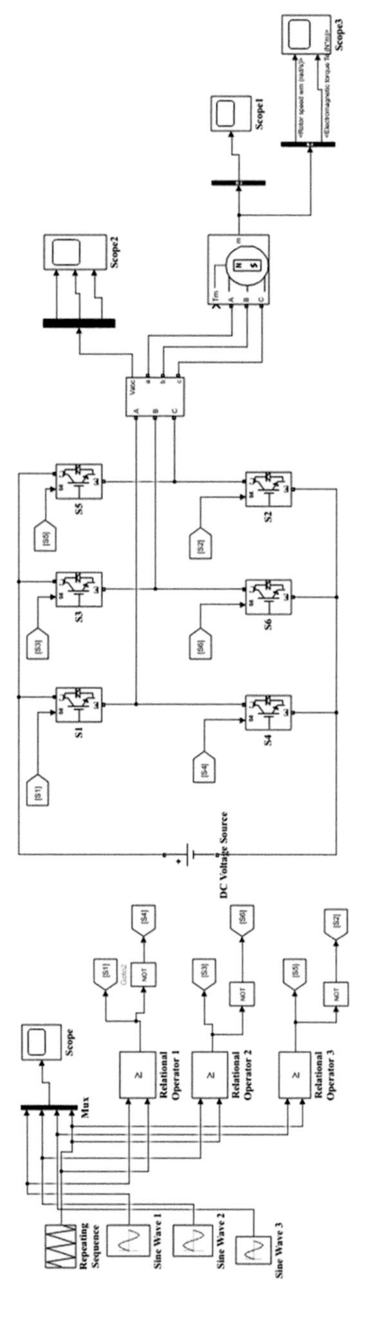

Figure 5.57 MATLAB/Simulink model of voltage source inverter fed synchronous motor drive.

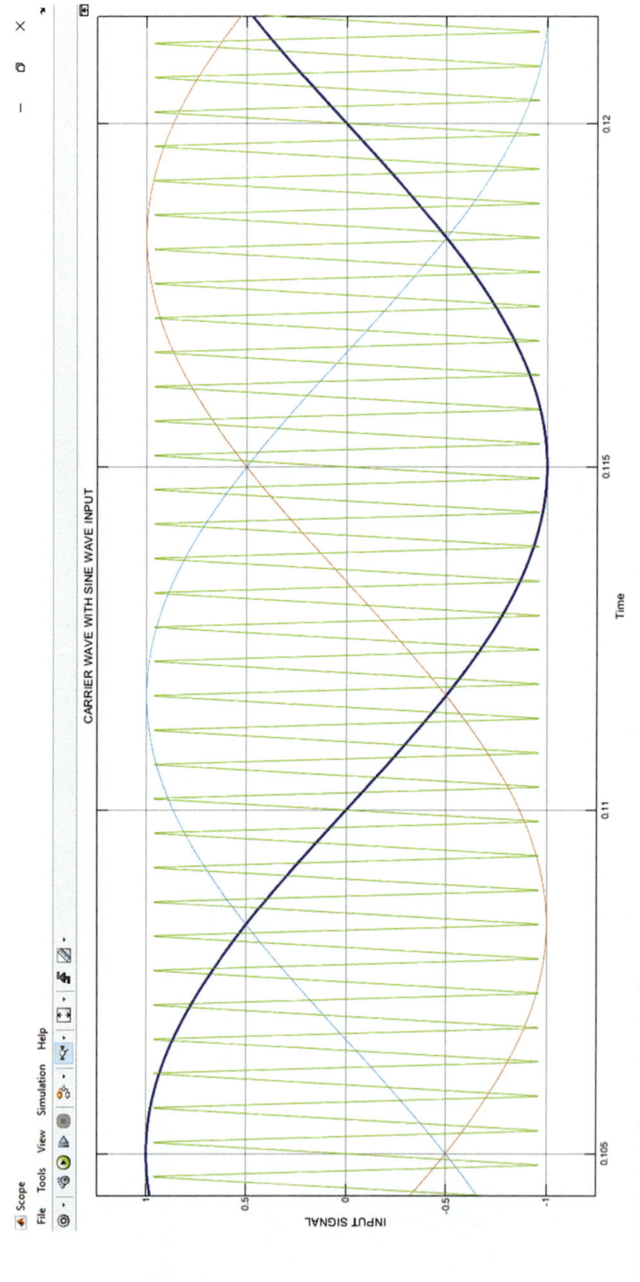

Figure 5.58 Simulation results from mux Scope.

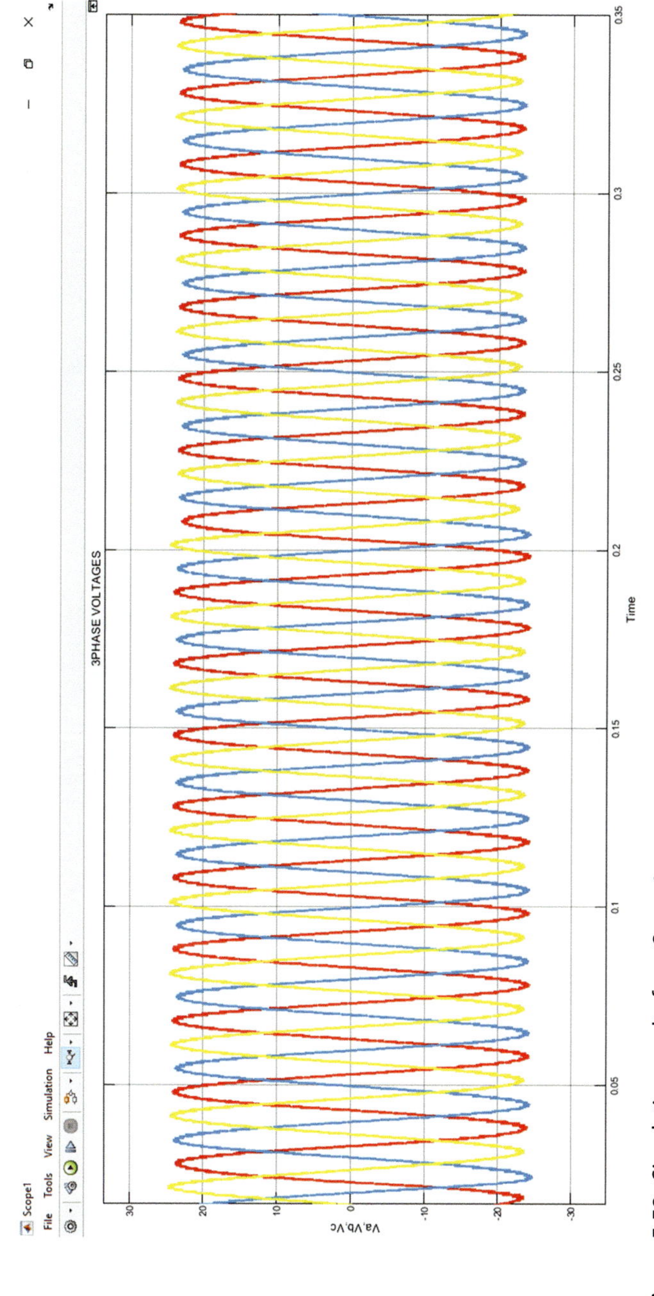

Figure 5.59 Simulation results from Scope1.

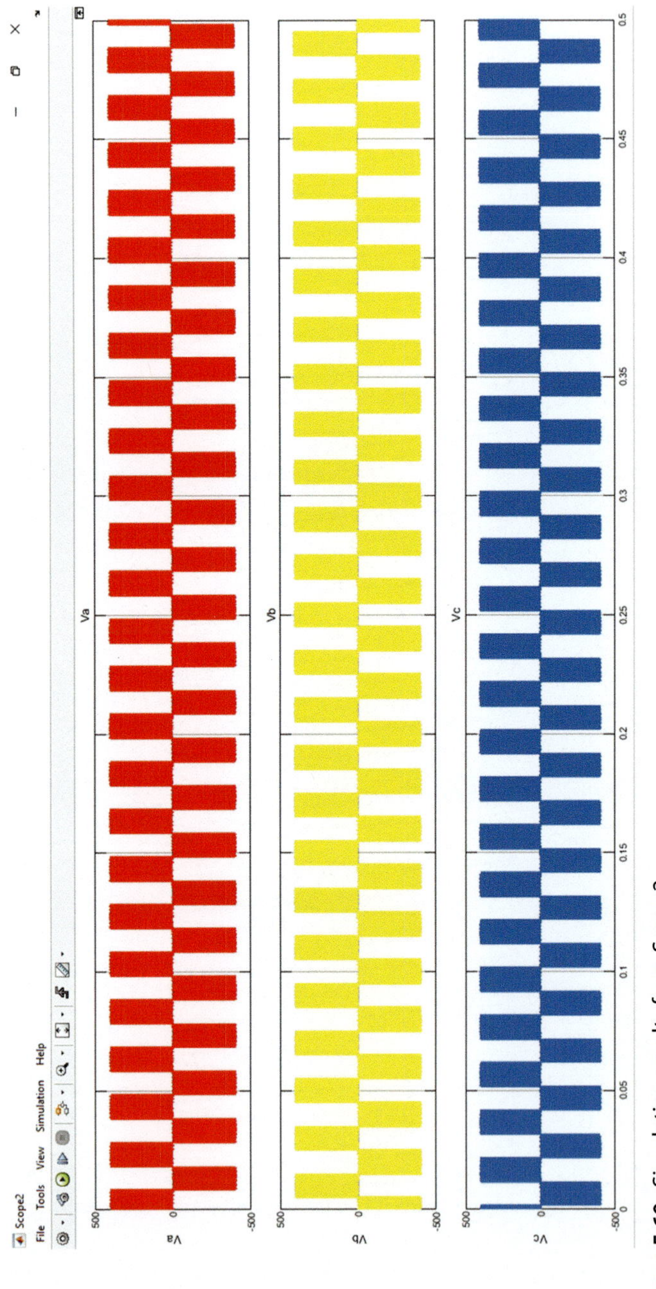

Figure 5.60 Simulation results from Scope2.

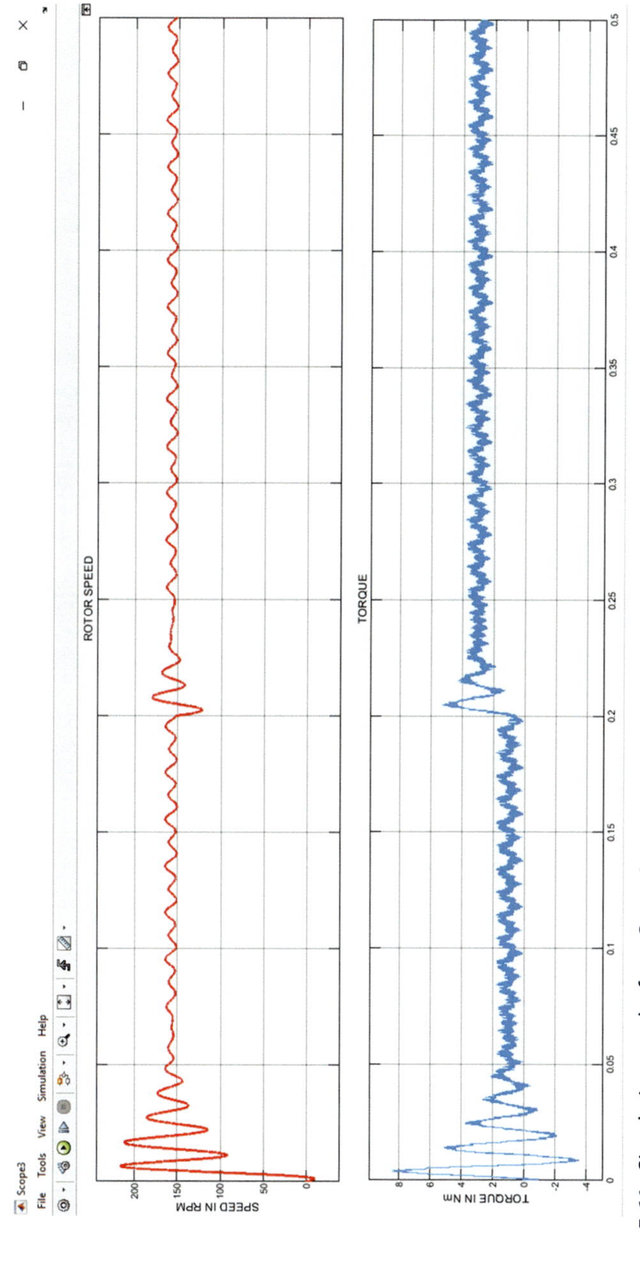

Figure 5.61 Simulation results from Scope3.

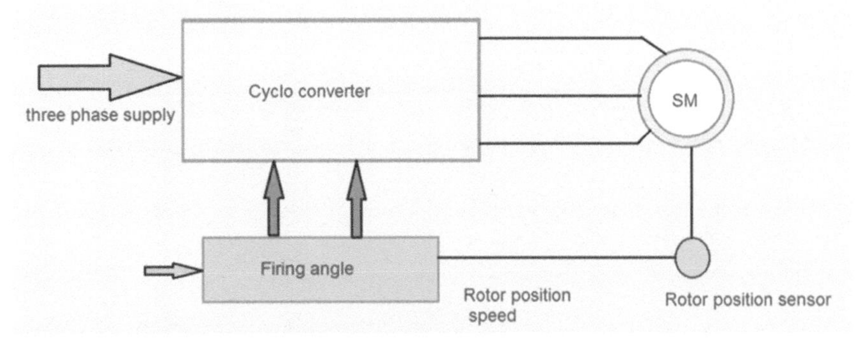

Figure 5.62 Cycloconverter fed synchronous motor.

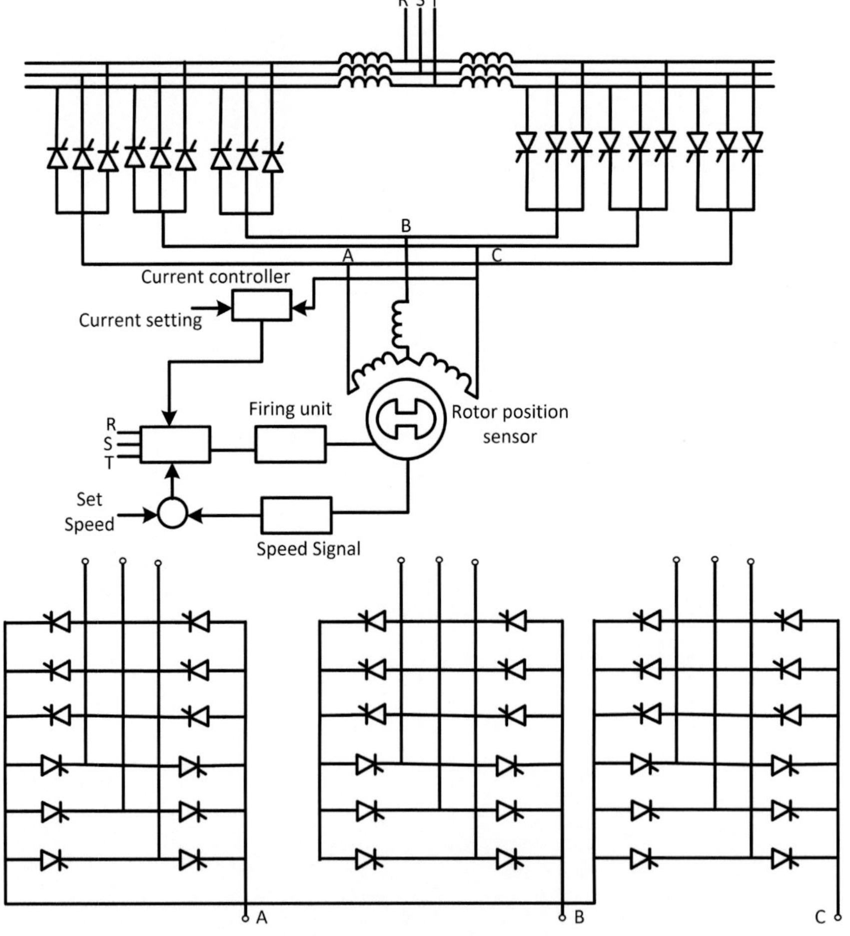

Figure 5.63 Cycloconverter feeding a three phase synchronous motor.

The line voltages are made use of to commutate the thyristors of a cycloconverter. The output frequency can be varied from 0 to 1/3 of the input frequency. The range of speed control is therefore limited, extending from 0 to 1/3 base speed. Cycloconverters are inherently capable of power transfer in both directions. Four-quadrant operation is simple.

A cycloconverter in the above speed range gives a high-quality sinusoidal output voltage. The resulting currents are also nearly sinusoidal. The harmonic content of the current is small. Consequent effects of harmonic current, such as losses, heating, and torque pulsations are minimal. The line power factor is somewhat better because the machine power factor can be made unity.

A synchronous motor fed from cycloconverter requires a large number of thyristors and its control circuitry is complex. Converter grade thyristors are sufficient but the cost of the converter is high. The efficiency is good and the drive has a good dynamic behavior. The operation in CLM mode is popular.

Synchronous motors fed by cycloconverter drives are popular in big, low-speed reversing mills that require quick acceleration and deceleration. Large gearless drives, such as those for reversing mills, mining hoists, and so on, are typical applications.

If the load is capable of delivering the required reactive power for the inverter, a cycloconverter can also be commutated utilizing the load voltages. The required reactive power can be provided by an overexcited synchronous motor. As a result, a load commutation cycloconverter feeding such a motor is possible. The speed control range is medium to base speed. Load commutation is not achievable at very low speeds. If line commutation is utilized at low speeds, the speed range can be expanded to zero. Four-quadrant operation is simple. The problems associated with harmonics are minimal due to high quality of the output. The line power factor depends on the angle of firing and is poor. The cost of the converter is high with complex control. Its efficiency is good and the drive has a fast response. It finds application in high power pump and blower type drives.

5.5 Load commutated synchronous motor drives

Fig. 5.64 depicts a load commutated inverter fed synchronous motor drive. The inverter is a current source inverter that uses T1–T6 thyristors. The voltages induced in the synchronous motor's armature are used to commutate the inverter thyristor. To get a variable Dc voltage V_{ds} from a set source

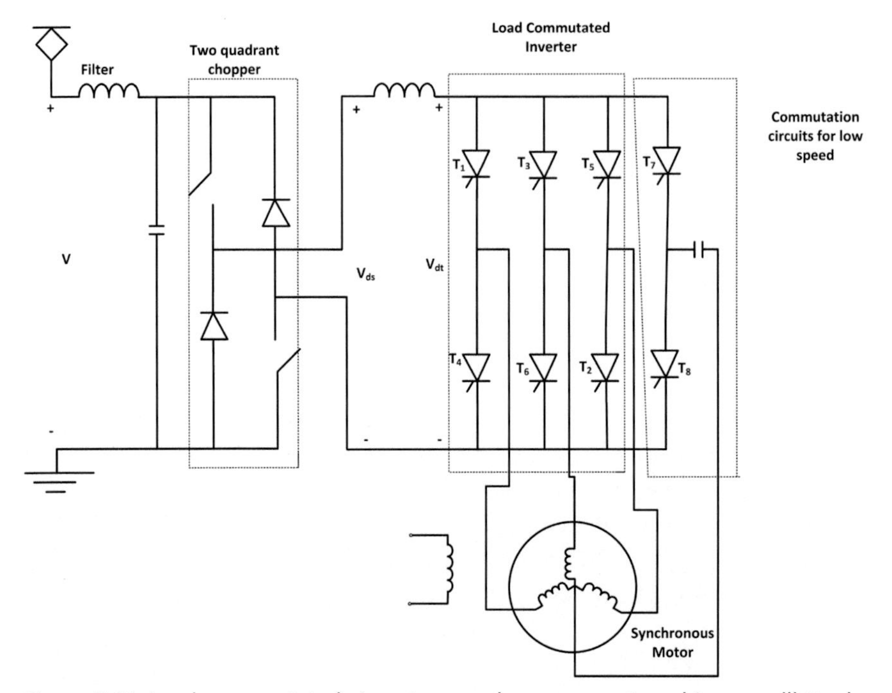

Figure 5.64 Load commutated inverter synchronous motor drives. nullhttpqla cwwwgacykssuugsqassgsiggqskeegkuiwocqquwmqgeqswkegagakcqyqcomugsasyme.

voltage V, a chopper is employed. V_{ds} is adjusted in relation to V_{dl} such that the DC link, and hence the motor, receives the necessary current.

During motoring, power is sent from the DC mains to the motor via the chopper, DC link, and inverter. The voltage V_{dl} reverses when the inverter firing angle is altered from near to 180°–0°. When the chopper function is altered to make V_{ds} negative but less than V_{dl} in magnitude, power flows from the load to the DC mains via the machine, inverter, and chopper, resulting in regenerative braking. The setup for dynamic braking is not shown here, but it can be added in the same way that it is previously shown.

At low speeds, such as stationary, armature-induced voltages are insufficient to commutate inverter thyristors. At low speeds, thyristors T_7 and T_8, as well as capacitor C, are employed to commutate inverter thyristors. Around 10% of the base speed gate pulses are removed from T_7 and T_8, and load commutation is used instead.

The inverter is essentially a current source inverter due to the existence of L_d. As a result, each traction motor has its own inverter. Four of these inverters will be required if there are four traction motors.

Furthermore, the inverters can be connected in series but not parallel due to current source characteristics. When four traction motors are employed, one option is to connect all four inverters in series and feed them from a single chopper. Adhesion will be harmed as a result of this series connection. Alternatively, two inverters can be connected in series, with each pair receiving its own chopper.

In comparison to a PWM VSI induction motor drive, this load commutated inverter fed synchronous motor drive has the following features:

- The converter efficiency is lower due to the additional power stage (i.e., chopper), but the motor efficiency is higher.
- The drive has a delayed dynamic response due to the huge inductance L_d, resulting in poor adhesion.
- More volume and weight.
- Each motor should be equipped with its own inverter, which can be linked in series but not parallel. When large traction motors are used, the drive becomes costly, and series connection has a negative effect on adhesion.
- Due to the lack of a shoot-through fault, the inverter is more reliable.
- The acceleration is not smooth due to torque pulsations caused by harmonics. This has a negative impact on adhesion.

5.6 Line commutated cycloconverter-fed synchronous motor drives

It is a frequency converter with just one stage. The frequency of a synchronous motor can be derived using a cycloconverter. It is a phase-controlled, line-commutated system.

- Waveform distortion happens when the output frequency is high. So, in order to maintain the limited output frequency, they allowed a frequency that was about one-third of the supply frequency, which is acceptable for low-speed and low-frequency operations.
- A cycloconverter allows power to flow in both directions, allowing for the development of a four-quadrant drive with regeneration capability. By segmenting the output voltage waveform, the harmonic content is reduced to a minimum.

  Merits:

- The current waveforms are smoothed by machine inductance.
- Harmonics have a minor impact.
- Torque pulsations are kept to a minimum.
- Increased efficiency.

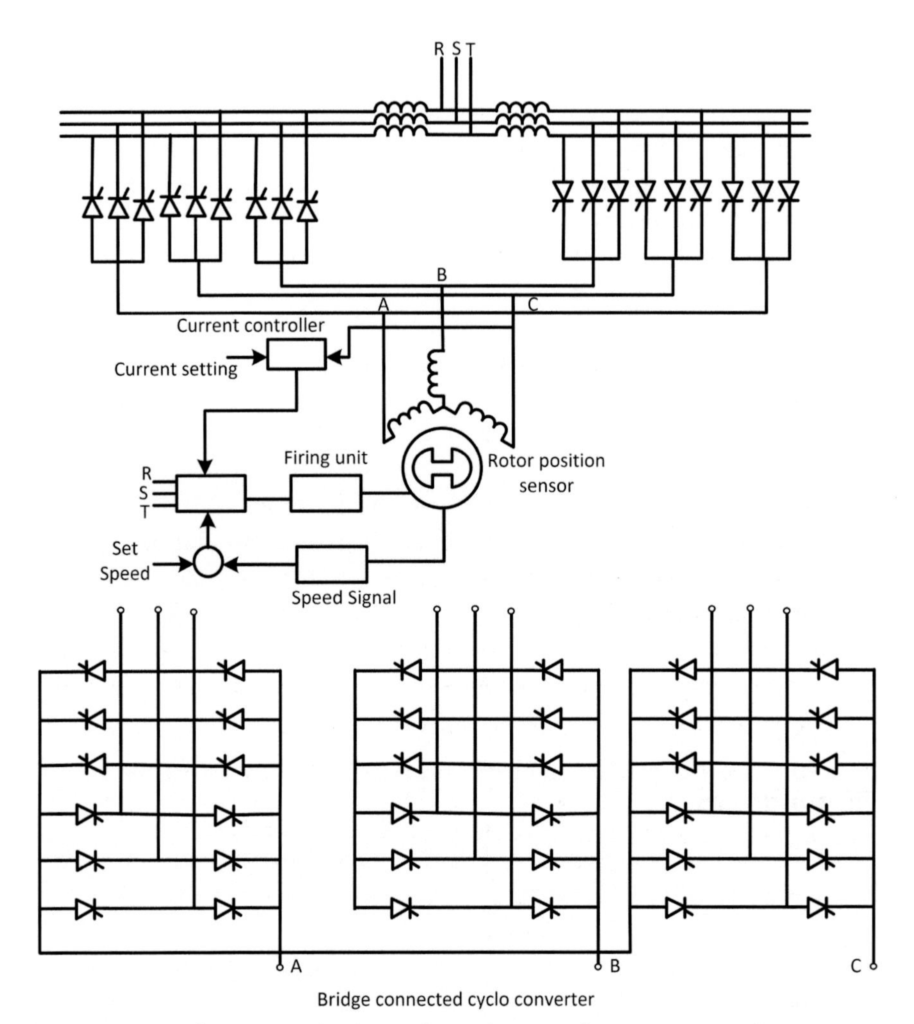

Figure 5.65 Cycloconverter feeding a three-phase synchronous motor—line commutation.

- By decreasing line power factor, machine power factor will be improved.
  Demerits:
- Only at low speeds is the drive accepted.
- A rotor position sensor is required.

A DC link converter is a two-stage converter that generates a variable voltage, variable frequency supply. A cycloconverter, which is a single-stage conversion device, may provide variable voltage, variable frequency supply. Fig. 5.65 depicts the power circuit of a three-phase synchronous motor fed by a cycloconverter. When compared to a DC link converter, this has a numerous differences.

The line voltages are made use of to commutate the thyristors of a cycloconverter. The output frequency can be varied from 0 to 1/3 of the input frequency. The range of speed control is therefore limited, extending from 0 to 1/3 base speed. Cycloconverters are inherently capable of power transfer in both directions. Four-quadrant operation is simple.

A cycloconverter in the above speed range gives a high-quality sinusoidal output voltage. The resulting currents are also nearly sinusoidal. The harmonic content of the current is small. Consequent effects of harmonic current, such as losses, heating and torque pulsations are minimal. The line power factor is somewhat better because the machine power factor can be made unity.

A synchronous motor fed from cycloconverter requires a large number of thyristors and its control circuitry is complex. Converter grade thyristors are sufficient but the cost of the converter is high. The efficiency is good and the drive has a good dynamic behavior. The operation in CLM mode is popular.

A synchronous motor fed from cycloconverter drive is attractive for low-speed operation and is frequently employed in large, low-speed reversing mills requiring rapid acceleration and deceleration. Typical applications are large gearless drives, for example, drives for reversing mills, mine hoists, etc.

A cycloconverter can also be commutated using the load voltages if the load is capable of providing the necessary reactive power for the inverter. An overexcited synchronous motor can provide the necessary reactive power. Hence a cycloconverter feeding such a motor can be load commutated. The range of speed control is from medium to base speed. At very low speeds load commutation is not possible. The speed range can be extended to zero if line commutation is used at low speeds. Four quadrant operation is simple. The problems associated with harmonics are minimal due to the high quality of the output. The line power factor depends on the angle of firing and is poor. The cost of the converter is high with complex control. Its efficiency is good and the drive has a fast response. It finds application in high power pump and blower type drives.

5.7 Case studies

1. An imitation model of the electric drive of an oil-pumping unit based on synchronous motor STD-4000 was developed in the MATLAB Simulink software complex (Fig. 5.66).

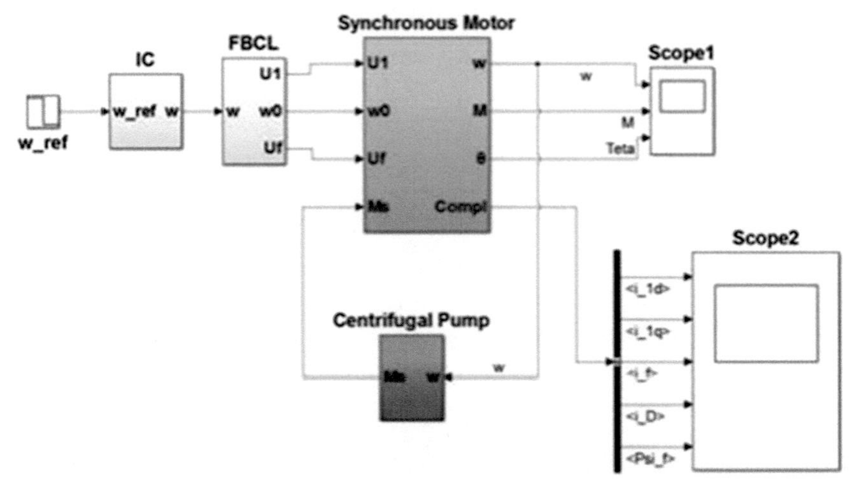

Figure 5.66 Simulation model of the synchronous electric drive. 5

2. Combined rotor position estimation and temperature monitoring in sensorless synchronous motor drives.

The current vector control technique for PMSM and SynchRel motor drives is given in Fig. 5.67.

5.8 Numerical solutions with simulation

Construct the synchronous motor drive during speed regulation using MATLAB\SIMULINK.

The AC6 block from the specialized power systems library is used in this circuit. For a 3HP motor, it simulates a PM synchronous motor drive with a braking chopper (Fig. 5.68).

A PWM voltage source inverter, built with a Universal Bridge Block, powers the PM synchronous motor. The flux and torque references for the vector control block are produced by a PI regulator in the speed control loop. The vector control block calculates the three reference motor line currents for the flux and torque references and then feeds these currents to the motor via a three-phase current regulator.

The block's output provides current, speed, and torque indications from the motor.

5.9 Summary

In this chapter, synchronous motor drives simulations using MATLAB/Simulink are provided in step by step manner. The Simulink diagrams

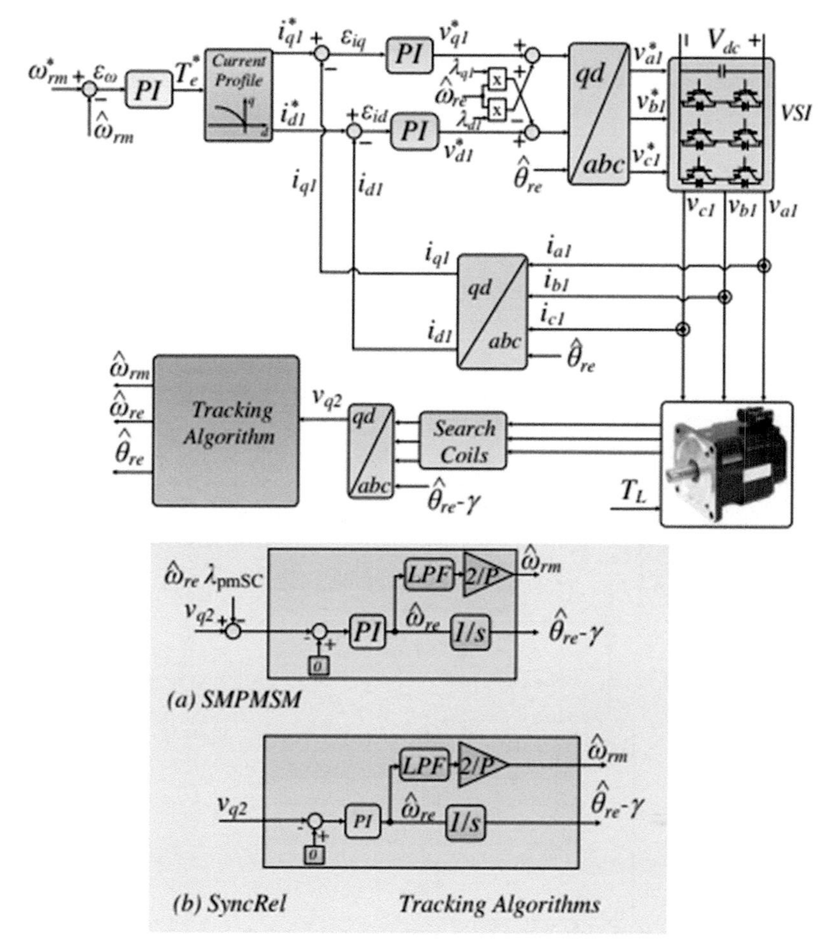

Figure 5.67 Current vector control technique for PMSM and SynchRel motor drives. nullS

and results are clearly given here. It is easy for everyone to simulate the synchronous motor drive circuits using MATLAB software.

Multiple Choice Questions

1. Which motor is suitable to drive the rotary compressor?
   a) DC Shunt Motor
   b) Synchronous Motor
   c) DC Series Motor
   d) Universal Motor

   Answer: Option (b)

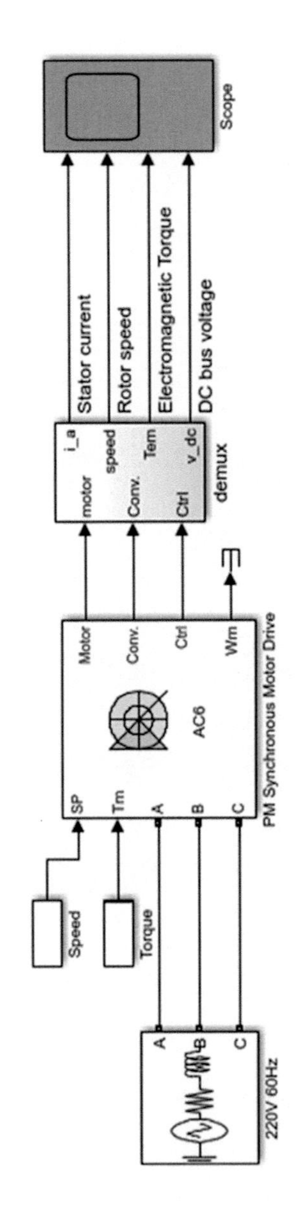

Figure 5.68 MATLAB/Simulink model of synchronous motor drive during speed regulation.

2. Which is more advantageous in the synchronous motor in addition to its constant speed when compared with three phase induction motor?
   a) High power factor
   b) Better efficiency
   c) None of the above
   d) Both a) & b)
   Answer: Option (d)
3. At what speed the rotary Kilns of the synchronous motor runs
   a) Ultra Low speed
   b) Medium speed
   c) Ultra High speed
   d) Low speed
   Answer: Option (a)
4. A synchronous motor is more economical, when the load is above______
   a) 20kW
   b) 50kW
   c) 2kW
   d) 100kW
   Answer: Option (d)
5. Which motor has a least range of speed control.
   a) DC Motor
   b) Induction motor
   c) Synchronous Motor
   d) DC Series motor
   Answer: Option (c)
6. When, the negative phase sequence exists in synchronous motor_________
   a) Underloaded
   b) Supplied with unbalanced load
   c) Over loaded
   d) None of the above
   Answer: Option (b)
7. In the synchronous motor, increases in _________, reduces motor size.
   a) Speed
   b) Horse power rating
   c) Flux density
   d) All of the above
   Answer: Option (c)

8. The synchronous motor's direction of rotation can be reversed by changing
   a) Current to field winding
   b) Supply phase sequence
   c) Polarity of rotor poles
   d) None of the above

   Answer: Option (b)
9. At which load angle the synchronous motor develops maximum power?
   a) 60
   b) 45
   c) 90
   d) 120

   Answer: Option (c)
10. When synchronous motor can be used as synchronous capacitor?
    a) Over excited
    b) Under excited
    c) Over loaded
    d) Under loaded

    Answer: Option (a)

References

https://www.eeeguide.com/load-commutated-inverter-fed-synchronous-motor-drive/.

https://www.electrical4u.com/synchronous-motor-drives/.

Mathworks.com.

Ershov M., Sidorenko M. "The research of frequency-controlled synchronous drive transient processes" IEEE 2018 X International Conference on Electrical Power Drive Systems (ICEPDS), Novocherkassk.

Scelba, G., Tornello, L.D., Scarcella, G., Testa, A., Foti, S., Pulvirenti, M., 2018. Combined rotor position estimation and temperature monitoring in sensorless synchronous motor drives. In: IEEE 9th International Symposium on Sensorless Control for Electrical Drives (SLED). Helsinki.

CHAPTER 6

BLDC-based drives control and simulation

Contents

6.1 Introduction to BLDC

6.1.1 Definition

A brushless DC motor consists of a rotor in the form of a permanent magnet and stator in the form of polyphase armature windings. It differs from the conventional DC motor in such that it does not contain brushes and the commutation is done using electrically, using an electronic drive to feed the stator windings.

Basically, a BLDC motor can be constructed in two ways—by placing the rotor outside the core and the windings in the core and another by placing the windings outside the core. In the former arrangement, the rotor magnets act as an insulator and reduce the rate of heat dissipation from the motor and operate at a low current. It is typically used in fans. In the latter arrangement, the motor dissipates more heat, thus causing an increase in its torque. It is used in hard disk drives (Fig. 6.1).

6.1.2 Fundamentals of operation

The BLDC motor's electronic commutator sequentially energizes the stator coils generating a rotating electric field that 'drags' the rotor around with it. N "electrical revolutions" equates to one mechanical revolution, where

Electric Motor Drives and Their Applications with Simulation Practices.
DOI: https://doi.org/10.1016/B978-0-323-91162-7.00003-5

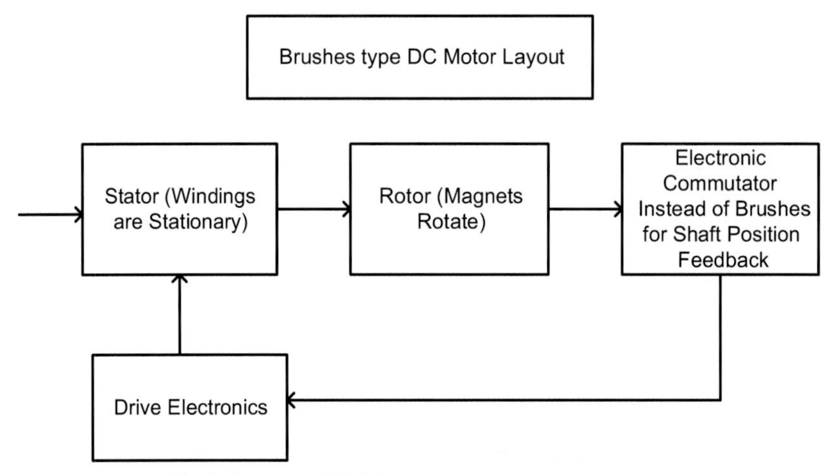

Figure 6.1 Basic block diagram of BLDC motor.

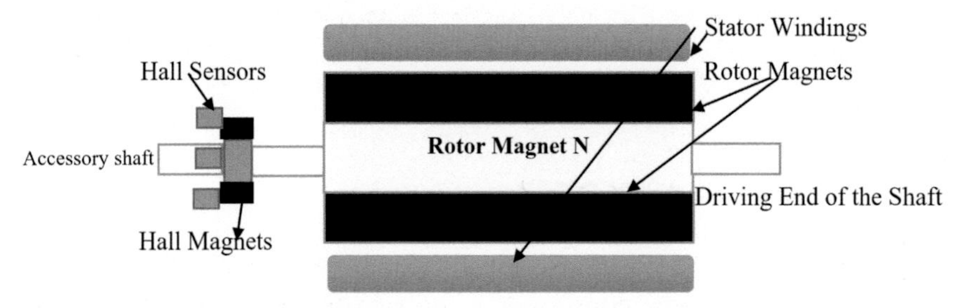

Figure 6.2 Hall sensors with BLDC.

N is the number of magnet pairs. For a three-phase motor, three Hall effect sensors are embedded in the stator to indicate the relative positions of stator and rotor to the controller so that it can energize the windings in the correct sequence and at the correct time. The Hall sensors are usually mounted on the nondriving end of the unit (Fig. 6.2).

Hall sensors are embedded in the stator of a BLDC motor to determine the winding energizing sequence. When the rotor magnetic poles pass the Hall sensors, a high (for one pole) or low (for the opposite pole) signal is generated. As discussed in detail below, the exact sequence of commutation can be determined by combining the signals from the three sensors.

All electric motors generate a voltage potential due to the movement of the windings through the associated magnetic field. This potential is known as an electromotive force (EMF) and, according to Lenz's law, it gives rise to a current in the windings with a magnetic field that opposes the original change in magnetic flux. In simpler terms, this means the EMF tends to resist the rotation of the motor and is therefore referred to as "back" EMF. For a given motor of fixed magnetic flux and number of windings, the EMF is proportional to the angular velocity of the rotor.

But the back EMF, while adding some "drag" to the motor, can be used for an advantage. By monitoring the back EMF, a microcontroller can determine the relative positions of stator and rotor without the need for Hall effect sensors. This simplifies motor construction, reducing its cost as well as eliminating the additional wiring and connections to the motor that would otherwise be needed to support the sensors. This improves reliability when dirt and humidity are present.

However, a stationary motor generates no back EMF, making it impossible for the microcontroller to determine the position of the motor parts at start-up. The solution is to start the motor in an open-loop configuration until sufficient EMF is generated for the microcontroller to take over motor supervision. These so-called "sensorless" BLDC motors are gaining in popularity.

6.1.3 Four Pole 2 phase motor operation

The brushless DC motor is driven by an electronic drive that switches the supply voltage between the stator windings as the rotor turns. The rotor position is monitored by the transducer (optical or magnetic) which supplies information to the electronic controller and based on this position, the stator winding to be energized is determined. This electronic drive consists of transistors (2 for each phase) which are operated via a microprocessor.

The magnetic field generated by the permanent magnets interacts with the field induced by the current in the stator windings, creating a mechanical torque. The electronic switching circuit or the drive switches the supply current to the stator so as to maintain a constant angle of 0 to 90 degrees between the interacting fields. Hall sensors are mostly mounted on the stator or on the rotor. When the rotor passes through the hall sensor, based on the North or South Pole, it generates a high or low signal. Based on the combination of these signals, the winding to be energized is defined. In order to keep the motor running, the magnetic field produced by the windings should shift position, as the rotor moves to catch up with the stator field.

In a 4 pole, 2-phase brushless DC motor, a single hall sensor is used, which is embedded on the stator (Fig. 6.3). As the rotor rotates, the hall sensor senses the position and develops a high or low signal, depending on the pole of the magnet (North or South). The hall sensor is connected via a

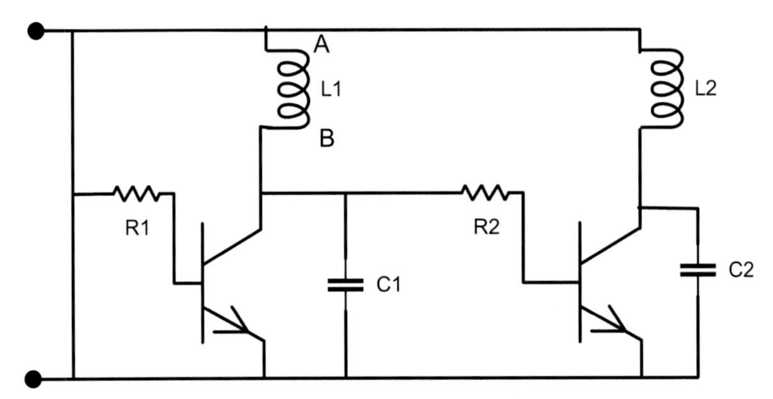

Figure 6.3 Four Pole, 2 phase brushless DC motor.

resistor to the transistors. When a high voltage signal occurs at the output of the sensor, the transistor connected to coil A starts conducting, providing the path for the current to flow and thus energizing coil A. The capacitor starts charging to the full supply voltage. When the hall sensor detects a change in polarity of the rotor, it develops a low voltage signal at its output and since transistor 1 does not get any supply, it is in cutoff condition. The voltage developed around the capacitor is Vcc, which is the supply voltage to the 2nd transistor, and coil B is now energized, as current passes through it.

BLDC motors have fixed permanent magnets, which rotate and a fixed armature, eliminating the problems of connecting current to the moving armature. And possibly more poles on the rotor than the stator or reluctance motors. The latter may be without permanent magnets, just poles that are induced on the rotor then pulled into an arrangement by timed stator windings. An electronic controller replaces the brush/commutator assembly of the brushed DC motor, which continually switches the phase to the windings to keep the motor turning. The controller performs comparative timed power distribution by using a solid-state circuit instead of the brush/commutator system.

6.1.4 Construction of BLDC motor

The main design difference between a brushed and brushless motors is the replacement of mechanical commutator with an electric switch circuit. Keeping that in mind, a BLDC motor is a type of synchronous motor in the sense that the magnetic field generated by the stator and the rotor revolve at the same frequency. Brushless motors are available in three configurations: single-phase, two-phase and three-phase. Out of these, the three-phase

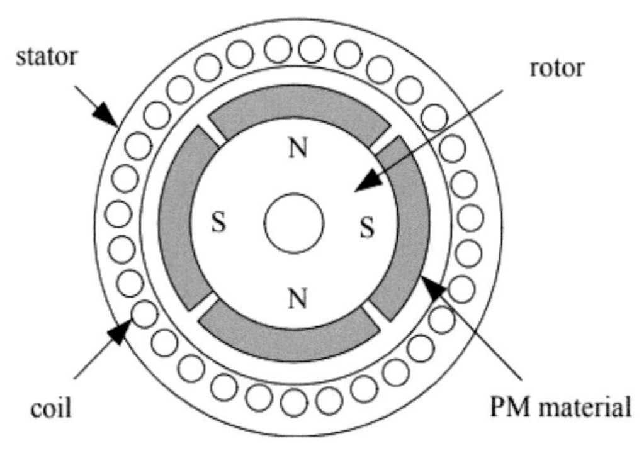

Figure 6.4 BLDC cross-sectional view.

Figure 6.5 BLDC winding.

BLDC is the most common one. The following image shows the cross-section of a BLDC Motor (Fig. 6.4).

The structure of the stator of a BLDC motor is similar to that of an induction motor. It is made up of stacked steel laminations with axially cut slots for winding. The winding in BLDC is slightly different than that of the traditional induction motor (Fig. 6.5).

Generally, most BLDC motors consists of three stator windings that are connected in star or "Y" fashion (without a neutral point). Additionally, based on the coil interconnections, the stator windings are further divided into trapezoidal and sinusoidal motors (Fig. 6.6). In a trapezoidal motor, both the drive current and the back EMF are in the shape of a trapezoid

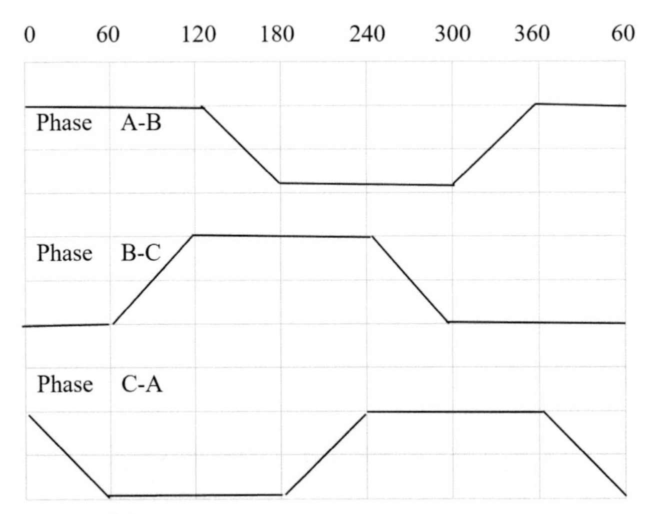

Figure 6.6 Trapezoidal motors.

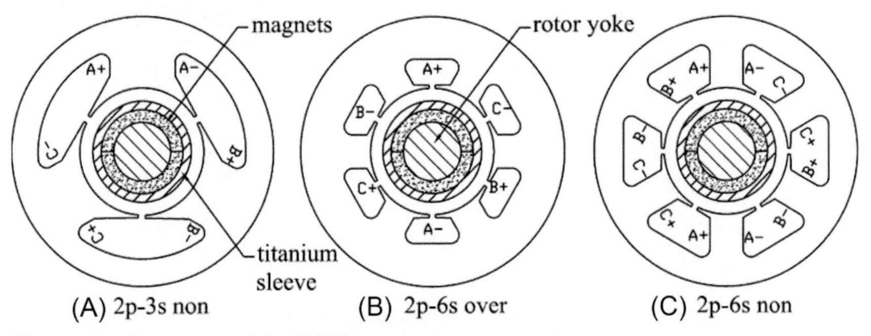

Figure 6.7 Rotor part of the BLDC motor.

(sinusoidal shape in case of sinusoidal motors). Usually, 48 V (or less) rated motors are used in automotive and robotics (hybrid cars and robotic arms).

The rotor part of the BLDC motor is made up of permanent magnets, as shown in Fig. 6.7. (usually, rare earth alloy magnets like Neodymium (Nd), Samarium Cobalt (SmCo), and alloy of Neodymium, Ferrite and Boron (NdFeB)). Based on the application, the number of poles can vary between two and eight with North (N) and South (S) poles placed alternately. The following image shows three different arrangements of the poles. In the first case, the magnets are placed on the outer periphery of the rotor

The second configuration is called magnetic-embedded rotor, where rectangular permanent magnets are embedded into the core of the rotor. In the third case, the magnets are inserted into the iron core of the rotor. Since there are no brushes in a BLDC motor, the commutation is controlled

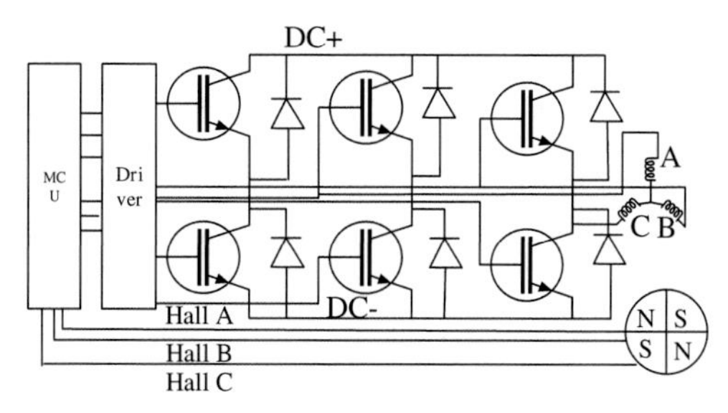

Figure 6.8 BDLC using an 8-bit microcontroller.

electronically. In order to rotate the motor, the windings of the stator must be energized in a sequence and the position of the rotor (i.e., the North and South poles of the rotor) must be known to precisely energize a particular set of stator windings (Fig. 6.8).

A position sensor, which is usually a Hall Sensor (that works on the principle of Hall effect), is generally used to detect the position of the rotor and transform it into an electrical signal. Most BLDC motors use three Hall sensors that are embedded into the stator to sense the rotor's position. The output of the Hall sensor will be either HIGH or LOW depending on whether the North or South pole of the rotor passes near it. By combining the results from the three sensors, the exact sequence of energizing can be determined.

Advantages of brushless DC motors:

- Better speed versus torque characteristics.
- High dynamic response.
- High efficiency.
- Long operating life due to a lack of electrical and friction losses.
- Noiseless operation.
- Higher speed ranges.

Applications:

The cost of the brushless DC motor has declined since its presentation, because of progressions in materials and design. This decrease in cost, coupled with the numerous focal points it has over the brush DC motor, makes the brushless DC motor a popular component in numerous distinctive applications. Applications that use the BLDC Motor include, yet are not constrained to:

- Consumer electronics.
- Transport.
- Heating and ventilation.

- Industrial engineering.
- Model engineering.

6.1.5 Principle of working

The principles for the working of BLDC motors are the same as for a brushed DC motor, that is, the internal shaft position feedback. In the case of a brushed DC motor, feedback is implemented using a mechanical commutator and brushes. Within BLDC motor, it is achieved using multiple feedback sensors. In BLDC motors, we mostly use a Hall effect sensor, whenever rotor magnetic poles pass near the hall sensor, they generate a HIGH or low-level signal, which can be used to determine the position of the shaft. If the direction of the magnetic field is reversed, the voltage developed will reverse too.

BLDC engines are kind of self-synchronous engines, the magnetic field produced by the stator and magnetic field created by the rotor pivot at a same frequency. Subsequently BLDC motor do not encounter the slip that in normal induction motor. The stator looks a considerable measure like that of an induction motor, however, the windings are appropriated in various conduct BLDC engines come in single-stage, three-stage design, comparing to it compose the stator has a similar number of windings, out of these 3-phase motor are the most mainstream and generally utilized in light of the fact that they produce high torque and for the most part utilized as a part of high-power applications. The brushless engine has surface-mounted magnet on the rotor not at all like the DC brushed motor; the stator of the motor is made out of stationary electromagnets. The major advantage of brushless engine is that the rotor conveys just the lasting magnets, it needs no power by any means. No connection should be finished with rotor; accordingly, no brush commutator combine should be made, because it requires no brushes so it is called brushless motor.

One of the upsides of BLDC motor is that it does not require carbon forgets about which wear quick subsequently it can perform silent and start free task. In addition, brushless motors are more equipped as far as power consumptions. There is contrast in the hypothesis of task of brushed DC motors with perpetual magnets like how the commutator is made, and how the coil changes polarity during rotation. In any case, brushless motors have no commutator no brushes. In this manner, there is have to know where each time the rotor is; however, there are a few approaches to discover where the rotor is. In some cases, utilizing rotating encoders, alongside their controllers and know precisely the point the rotor is, others utilize Hall effect sensors. The hall sensor is put in appropriate position.

Normally three lobby sensors are utilized which are set at 120Ú electrical separated from each other. It can detect if before it is the North or South Pole correspondingly the hall sensor will give the high or low flag to the controller of the motor. The controller will then turn on or off the fitting loops required keeping in mind the end goal to deliver pivoting motion and give torque. Any BLDC engine has two essential parts, the rotor as appeared in Figs. 2.1 and 2.2 the turning part having perpetual magnet and the stator the stationary part having armature coils. There are two fundamental BLDC engine outlines inward rotor and external rotor plan.

6.1.6 Controlling a BLDC motor

The control unit is implemented by microelectronic has several high-tech choices. This may be implemented using a micro-controller, a dedicated micro-controller, a hard-wired microelectronic unit, a PLC, or similar another unit.

The analog controller is still using, but they cannot process feedback messages and control accordingly. With this type of control circuits, it is possible to implement high-performance control algorithms, such as vector control, field-oriented control, high-speed control all of which are related to electromagnetic state of the motor. Furthermore, outer loop control for various dynamics requirements such as sliding motor controls, adaptive control, predictive control, etc., are also implemented conventionally.

Besides all these, we find high-performance power integrated circuit (PIC), application-specific integrated circuits, etc., that can greatly simplify the construction of the control and the power electronic unit both. For example, today we have complete pulse width modulation (PWM) regulator in a single IC that can replace the entire control unit in some systems. Compound driver IC can provide the complete solution of driving all six power switches in a three-phase converter. There are numerous similar integrated circuits with more and more adding day by day. At the end of the day, system assembly will possibly involve only a piece of control software with all hardware coming to the right shape and form.

PWM waves can be used to control the speed of the motor. Here the average voltage is given or the average current flowing through the motor will change depending on the ON and OFF time of the pulses controlling the speed of the motor, that is, the duty cycle of the wave controls its speed. On changing the duty cycle (ON time), we can change the speed. By interchanging output ports, it will effectively change the direction of the motor.

Speed control of the BLDC motor is essential for making the motor work at the desired rate. The speed of a brushless DC motor can be controlled by controlling the input DC voltage. The higher the voltage, the more is the speed. When motor works in normal mode or runs below rated speed, the input voltage of armature is changed through the PWM model. When a motor is operated above rated speed, the flux is weakened by means of advancing the exiting current.

The speed control can be closed-loop or open-loop speed control.

Open-loop speed control—It involves simply controlling the DC voltage applied to motor terminals by chopping the DC voltage. However, this results in some form of current limiting.

Closed-loop speed control—It involves controlling the input supply voltage through the speed feedback from the motor. Thus, the supply voltage is controlled depending on the error signal.

The closed-loop speed control consists of three basic components.

1. A PWM circuit to generate the required PWM pulses. It can be either a microcontroller or a timer IC.
2. A sensing device to sense the actual motor speed. It can be a hall effect sensor, an infrared sensor, or an optical encoder.
3. A motor drive to control the motor operation.

This technique of changing the supply voltage based on the error signal can be either through the PID controlling technique or using fuzzy logic. While BLDC motors are mechanically relatively simple, they do require sophisticated control electronics and regulated power supplies. The designer is faced with the challenge of dealing with a three-phase high-power system that demands precise control to run efficiently.

Fig. 6.8 shows a typical arrangement for driving a BLDC motor with Hall effect sensors. (The control of a sensorless BLDC motor using back EMF measurement will be covered in a future article.) This system shows the three coils of the motor arranged in a "Y" formation, a Microchip PIC18F2431 microcontroller, an insulated-gate bipolar transistor (IGBT) driver, and a three-phase inverter comprising six IGBTs, MOSFETs can also be used for the high-power switching). The output from the microcontroller (mirrored by the IGBT driver) comprises pulse width modulated (PWM) signals that determine the average voltage and average current to the coils (and hence motor speed and torque). The motor uses three Hall effect sensors (A, B, and C) to indicate rotor position. The rotor itself uses two pairs of permanent magnets to generate the magnetic flux.

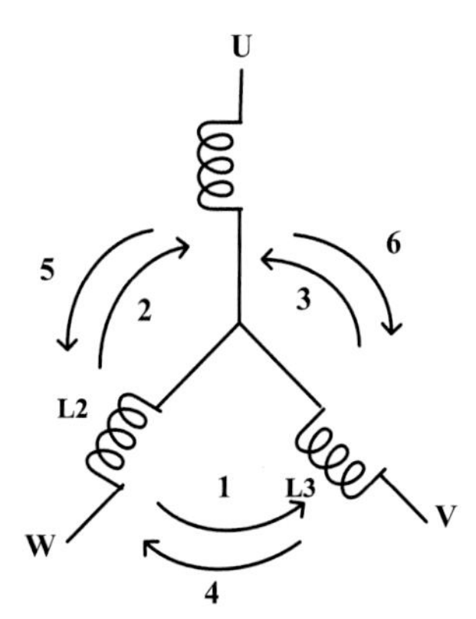

Figure 6.9 Coil-energizing sequence for one electrical revolution.

The system employs a six-step commutation sequence for each electrical revolution. Because the motor has two pairs of magnets, two electrical revolutions are required to spin the motor once. Fig. 6.4 shows the current flow in an identical arrangement of coils to the motor in Fig. 6.3 (this time labeled U, V, and W) for each of the six steps, and Fig. 6.9 shows the subsequent Hall effect sensor outputs and coil voltages.

A pair of Hall effect sensors determine when the microcontroller energizes a coil. In this example, sensors H1 and H2 determine the switching of coil U. When H2 detects an N magnet pole, coil U is positively energized; when H1 detects a N magnet pole, coil U is switched open; when H2 detects a S magnet pole coil U is switched negative, and finally, when H1 detects a S magnet pole, coil U is again switched open. Similarly, sensors H2 and H3 determine the energizing of coil V, with H1 and H3 looking after coil W (Fig. 6.10).

At each step, two phases are on with one-phase feeding current to the motor, and the other providing a current return path. The other phase is open. The microcontroller controls which two of the switches in the three-phase inverter must be closed to positively or negatively energize the two active coils. For example, switching Q1 positively energizes coil A and switching Q2 negatively energizes coil B to provide the return path. Coil C remains open.

Designers can experiment with 8-bit microcontroller-based development kits to try out control regimes before committing on the design of a full-size motor. For example, Atmel has produced an inexpensive starter kit, the ATAVRMC323, for BLDC motor control based on the ATxmega128A1 8-bit microcontroller.

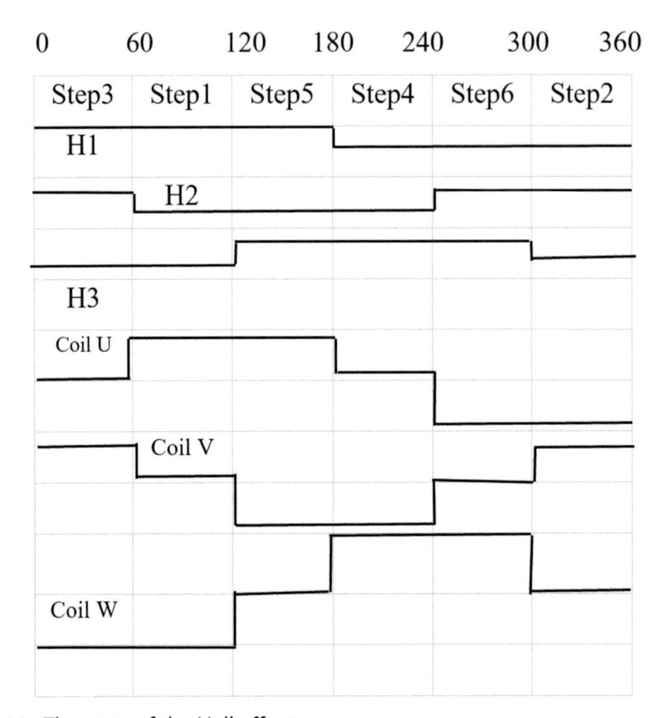

Figure 6.10 The state of the Hall effect sensors.

Table 6.1 Hall sequence for counter clockwise rotation of magnetic field.

H_A	H_B	Hc	halleffect A	halleffectB	halleffect X
1	0	1	+	−	NX
1	0	0	+	X	−
1	1	0	NX	+	−
0	1	0	−	+	NX
0	1	1	−	NX	+
0	0	1	NX	−	+

The most generally utilized sensors are Hall effect sensors. In the commutation system, one that depends on the situation of engine distinguished utilizing criticism sensors: two of three electrical windings are empowered at once, Fig. 6.11 demonstrates the back EMF, phase current, and Hall motion for one entire electrical revolution.

From the above figure, Tables 6.1 and 6.2 are derived and show phase excitation corresponding to hall signals. If the phases are excited in given sequence rotating magnetic field will be produced in clockwise and counterclockwise direction, respectively.

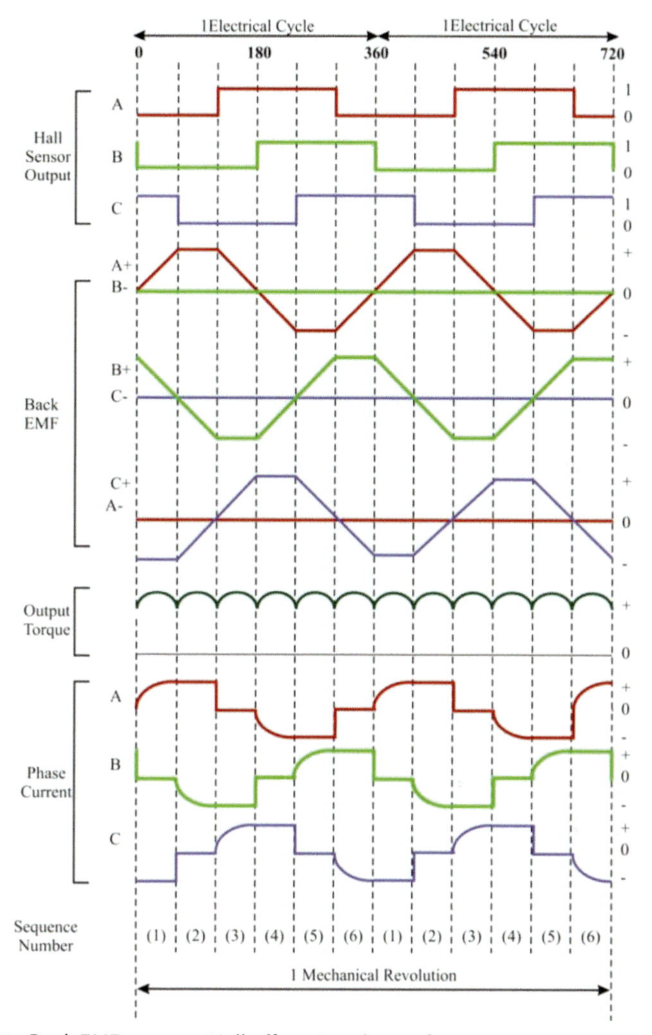

Figure 6.11 Back EMF, current, Hall effect signal waveforms.

Table 6.2 Hall sequence for clockwise rotation of magnetic field.

H_A	H_B	Hc	**Hall effect** A	**Hall effect** B	**Hall effect**
0	0	1	NX	−	+
0	1	1	−	NX	+
0	1	0	−	+	NX
1	0	1	+	−	NX
1	0	0	+	X	−
1	1	0	NX	+	−

Fig. 6.12 demonstrates the substitution relating to hall signals. There are add up to six possible results of Hall signals, relating to each flag there is a settled excitation of stages. This figure additionally gives data about star associated three-phase system, which two stages are energized relating to given Hall flag. Change of excitation of stage current is called commutation. For BLDC engine compensation happens at whatever point there is the change in Hall signal and its comparing stage excitation. This figure demonstrates the counter

6.1.7 Driving a BLDC motor

While an 8-bit microcontroller allied to a three-phase inverter is a good start, it is not enough for a complete BLDC motor control system. To complete the job requires a regulated power supply to drive the IGBT or MOSFETs (the "IGBT Driver" shown in Fig. 6.13). Fortunately, the job is made easier because several major semiconductor vendors have specially designed integrated driver chips for the job. These devices typically comprise a step-down ("buck") converter (to power the microcontroller and other system power requirements), gate driver control, and fault handling, plus some timing and control logic.

This predriver supports up to 2.3 A sink and 1.7 A source peak current capability, and requires a single power supply with an input voltage of 8 to 60 V. The device uses automatic handshaking when high-side or low-side IGBTs or MOSFETs are switching to prevent current shoot through.

ON semiconductor offers a similar chip, the LB11696V. In this case, a motor driver circuit with the desired output power (voltage and current) can be implemented by adding discrete transistors in the output circuits. The chip also provides a full complement of protection circuits, making it suitable for applications that must exhibit high reliability. This device is designed for large BLDC motors such as those used in air conditioners and on-demand water heaters.

6.2 BLDC position control

Brushless DC motor are synchronous motors powered by DC electricity via an inverter. Brushless DC motors complete commutating by switching power supply with an inverter, instead of with the help of carbon brush. Because of the absence of mechanical structures like carbon brush and slip ring, BLDC motors can prevent problems like friction that are caused by carbon brush. Both brushless DC motors and steppers are DC synchronous

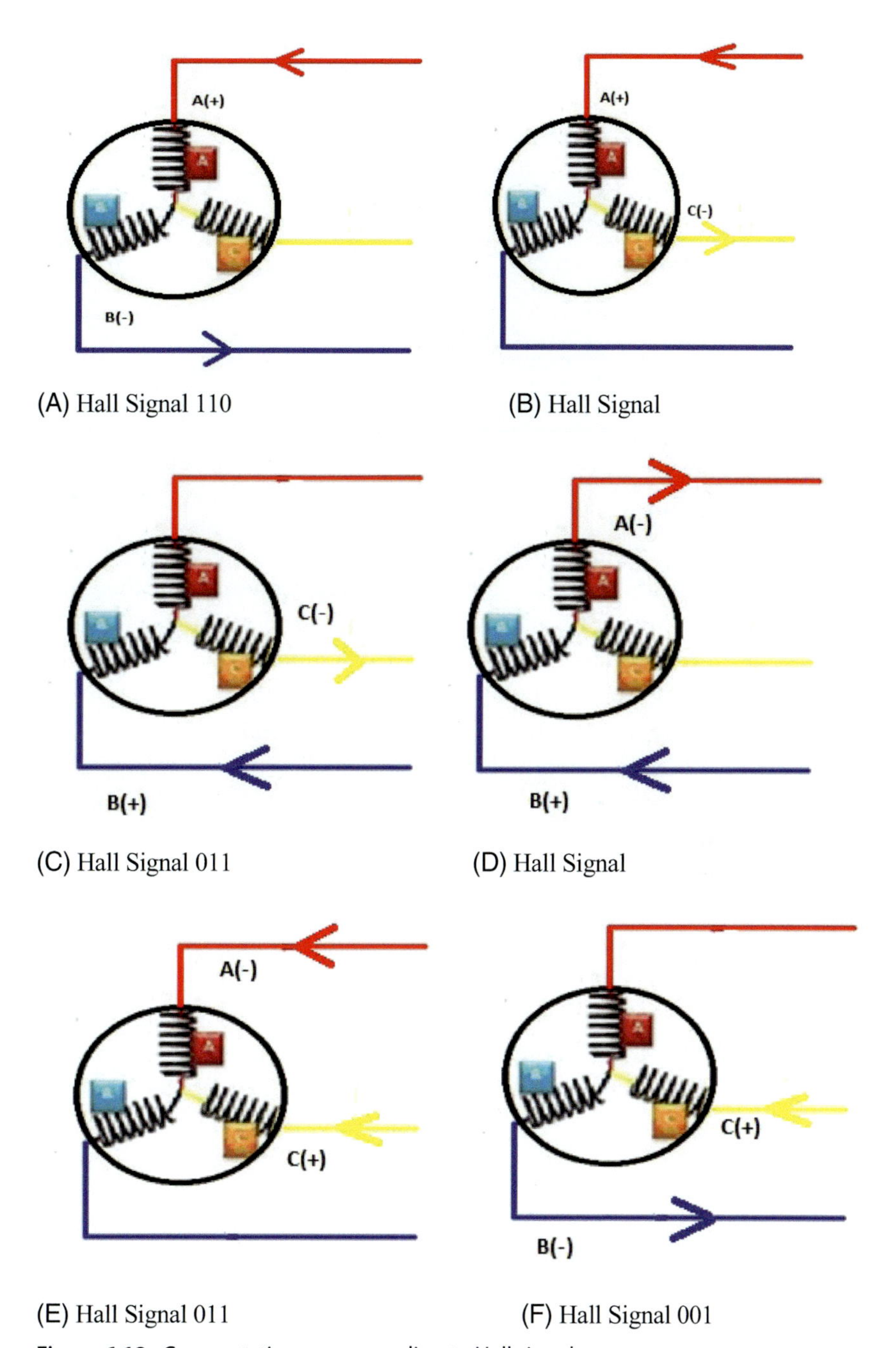

(A) Hall Signal 110 (B) Hall Signal

(C) Hall Signal 011 (D) Hall Signal

(E) Hall Signal 011 (F) Hall Signal 001

Figure 6.12 Commutation corresponding to Hall signals.

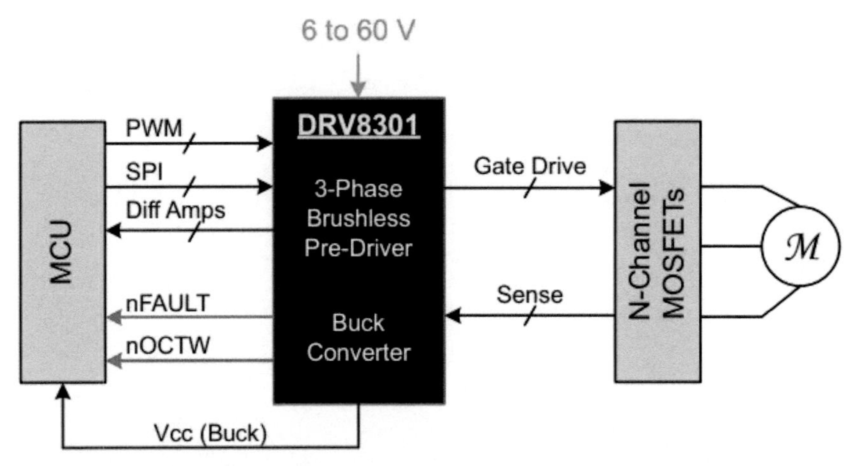

Figure 6.13 DRV8301 motor driver.

motors, they generally share the same mechanical structures, but they have many different features. Brushless DC motors perform well in acceleration and they are less likely to generate remarkable amount of noise or heat. Steppers are made for position control, but they sometimes step out, what's more, they generate a lot of heat and noise while running. Brushless DC motors are usually used when high speed and small size are required, while Steppers are used for positioner. If we can use brushless DC motors for position control, in one way, we can solve the stepping out problem of steppers and managed to achieve higher accuracy and performance control of position, in the other way, circumstances that need the small size and low heat and position control, which may have bothered us a lot, may have a convenient economic solution by using BLDC motors. Although BLDC motors are not made for position control, but the mechanical structure similarity with steppers make it possible to realize some kind of position control theoretically. By now, there are many ways (with sensor and without sensor) to detect the position of the brushless DC motors' rotor's position, which make the idea of using BLDC MOTORS for position control possible in practice. Generally, ways with sensors detect the rotor position by placing a sensor in the motor, and those without sensors usually calculate the rotor position by detect the voltage and current of the motor. This simulation presents our efforts in applying sensorless rotor positioning technology in driving the brushless DC motors and in using the method of sensorless positioning to realize position control (Fig. 6.15). The simulation results are given in Fig. 6.16A and B.

(A)

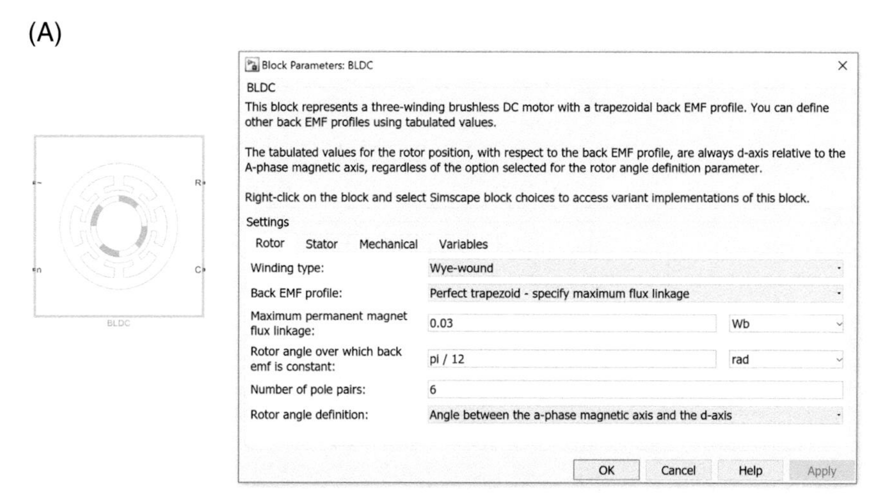

Figure 6.14A MATLAB/Simulink library—BLDC motor.

The following are the important Blocks used to design a BLDC motor Control (Fig. 6.14A–E). The MATLAB Library: Simscape/Electrical/Electromechanical/Permanent Magnet

6.3 BLDC hysteresis current control

This control method keeps the motor current in a range around the desired value, as shown in Fig. 6.17, by controlling switching ON and OFF states. When the current tends to exceed the upper band, the corresponding phase connects to the DC voltage source's negative terminal to let the current decrease or discharge. In a similar manner, when the current tends to exceed the lower band, the switch is closed to let the current pass to increase its amplitude or charge. In BLDC motors, since the back-EMF and the current drawn from the motor are, respectively, proportional to speed and voltage difference between the DC-link and back-EMF; thus, the current and switching frequency decrease with increase in speed; and at low speeds they increase. However, it is possible to decrease the switching frequency in low speeds by increasing the hysteresis band, but this will increase the torque ripple.

The BLDC motor drive strategy with the hysteresis current control method with and without reference current shaping is shown in Fig. 6.18. The reference currents are generated by using processed speed error in PI controller with regard to the explanations given in the PWM approaches

(B)

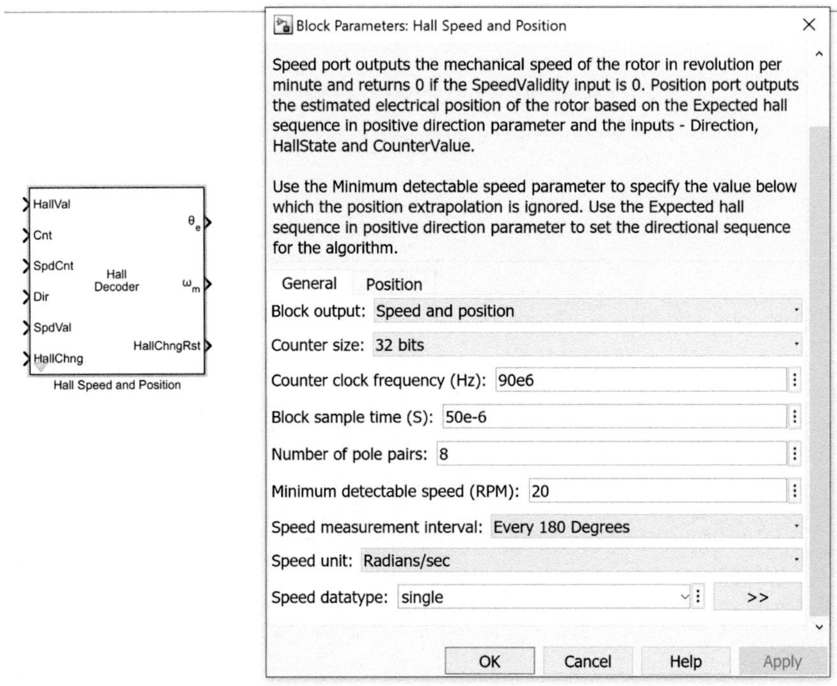

Figure 6.14B MATLAB/Simulink library—Hall sensor.

(C)

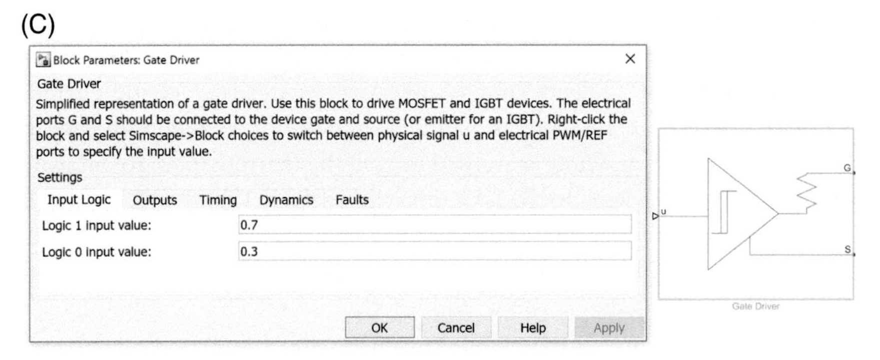

Figure 6.14C MATLAB/Simulink library—gate driver circuit.

section. Then real values of stator currents are measured and compared with the reference currents and the error current is obtained. Finally, by using the rotor position data sensed to improve the commutation and proper switching pattern, all the three hysteresis controllers independently send the required

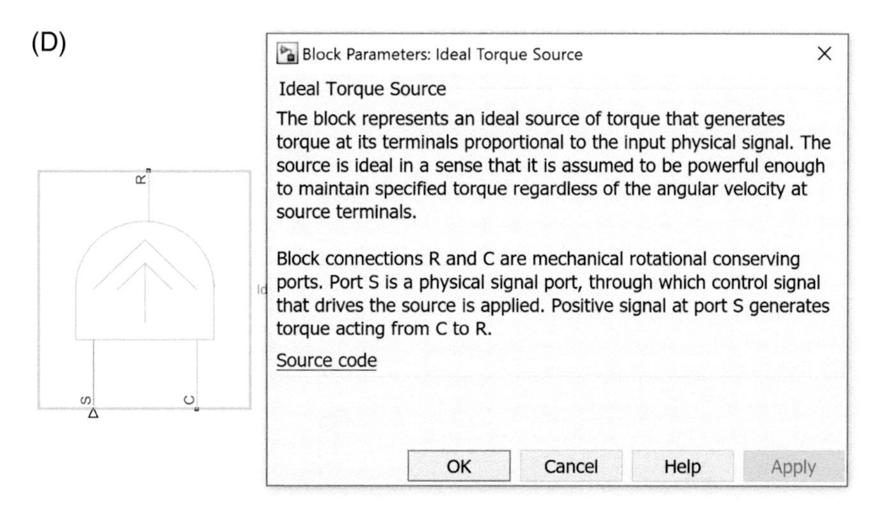

Figure 6.14D MATLAB/Simulink library—torque source.

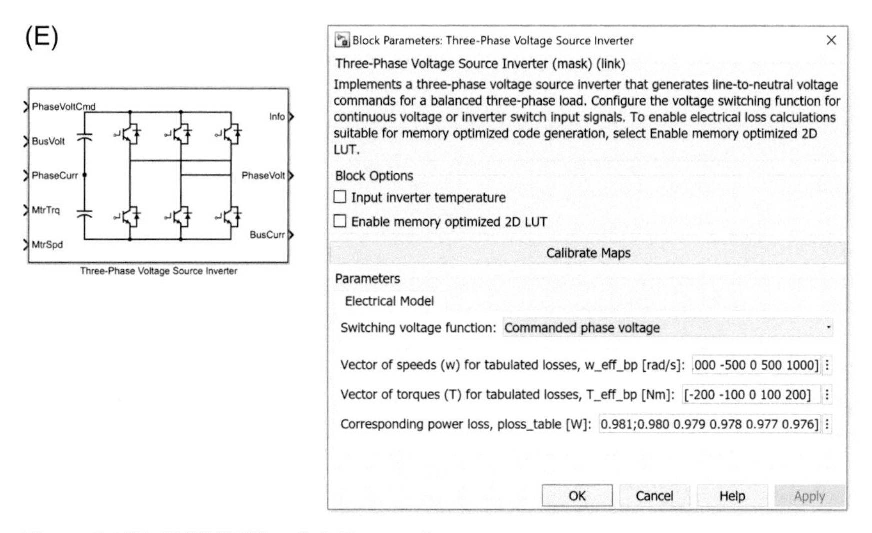

Figure 6.14E MATLAB/Simulink library—inverter setup.

command to the power switches in order to adjust each of the phase currents. It is certainly possible that no position data are sensed for the BLDC motor to operate in three-phase conduction mode. In this case, the error data, alone, are sufficient for the controller.

A DC voltage source feeds the BLDC through a controlled three-phase inverter. A ramp of the current request is provided to the motor

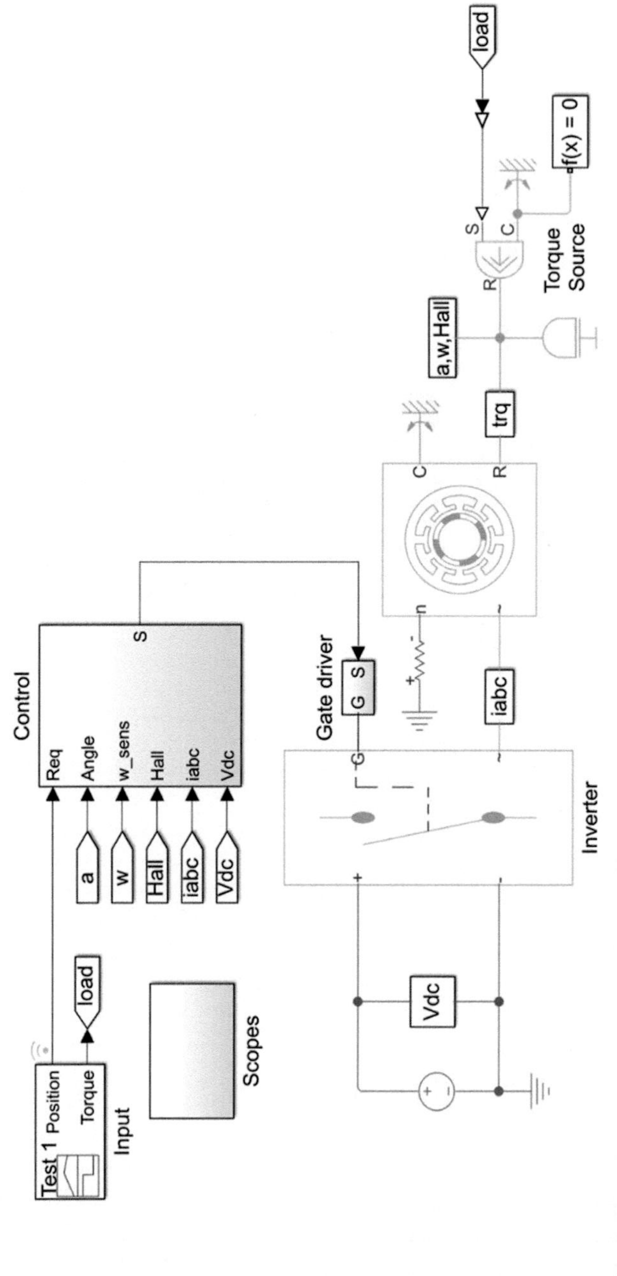

Figure 6.15 BLDC position control using MATLAB.

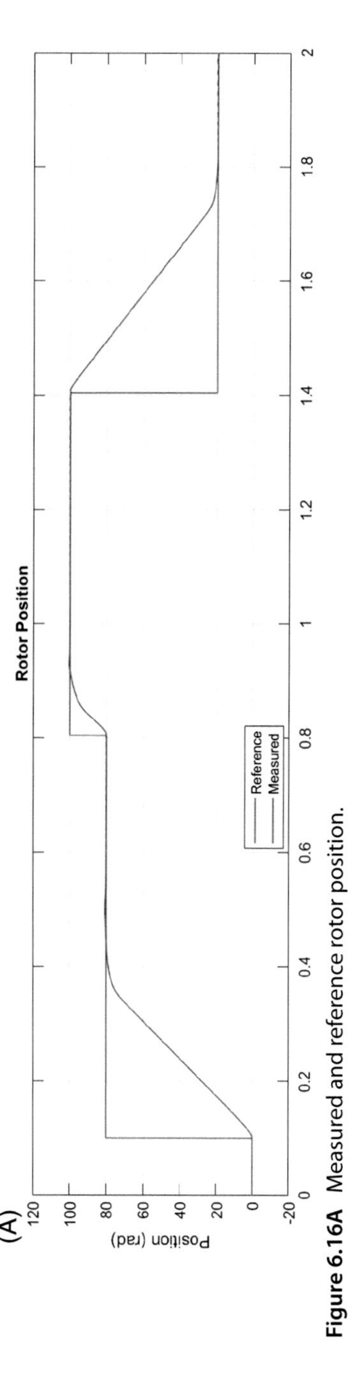

Figure 6.16A Measured and reference rotor position.

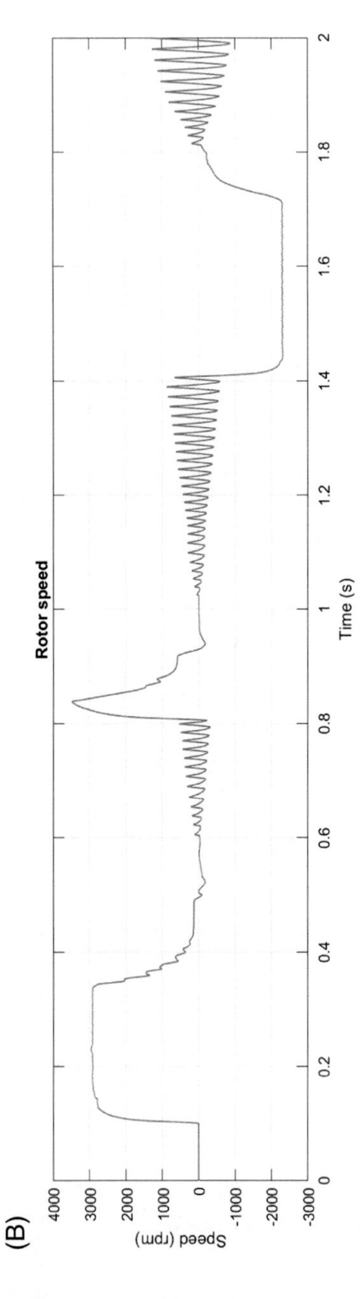

Figure 6.16B Rotor speed.

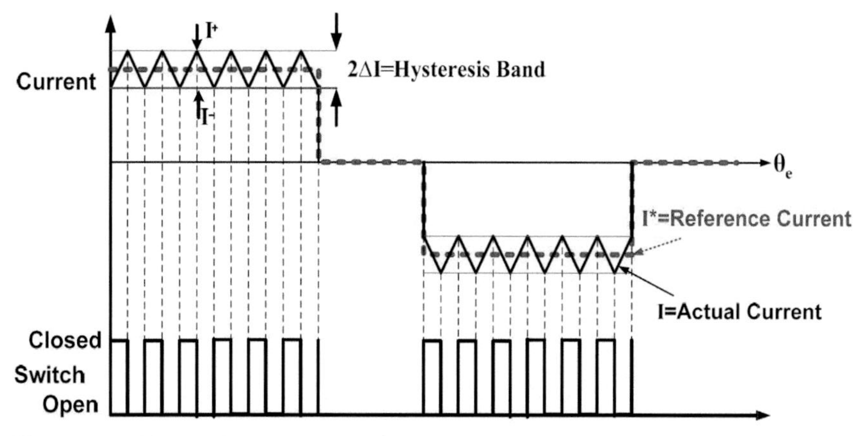

Figure 6.17 Hysteresis current control range.

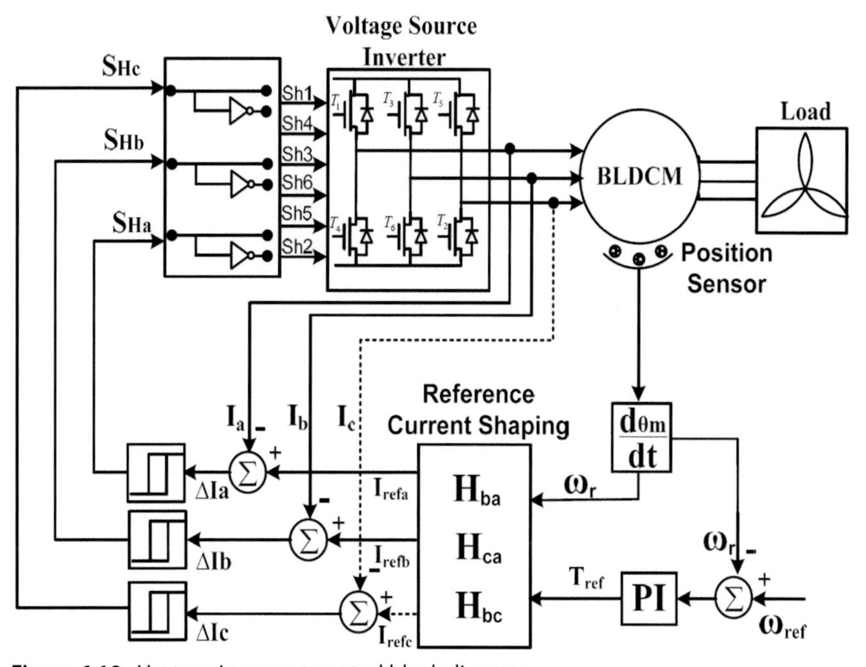

Figure 6.18 Hysteresis current control block diagram.

controller. The load torque is quadratically dependent on the rotor speed. The control subsystem implements the hysteresis-based current control strategy. The simulation diagram in MATLAB and results are presented in Figs. 6.19–6.21.

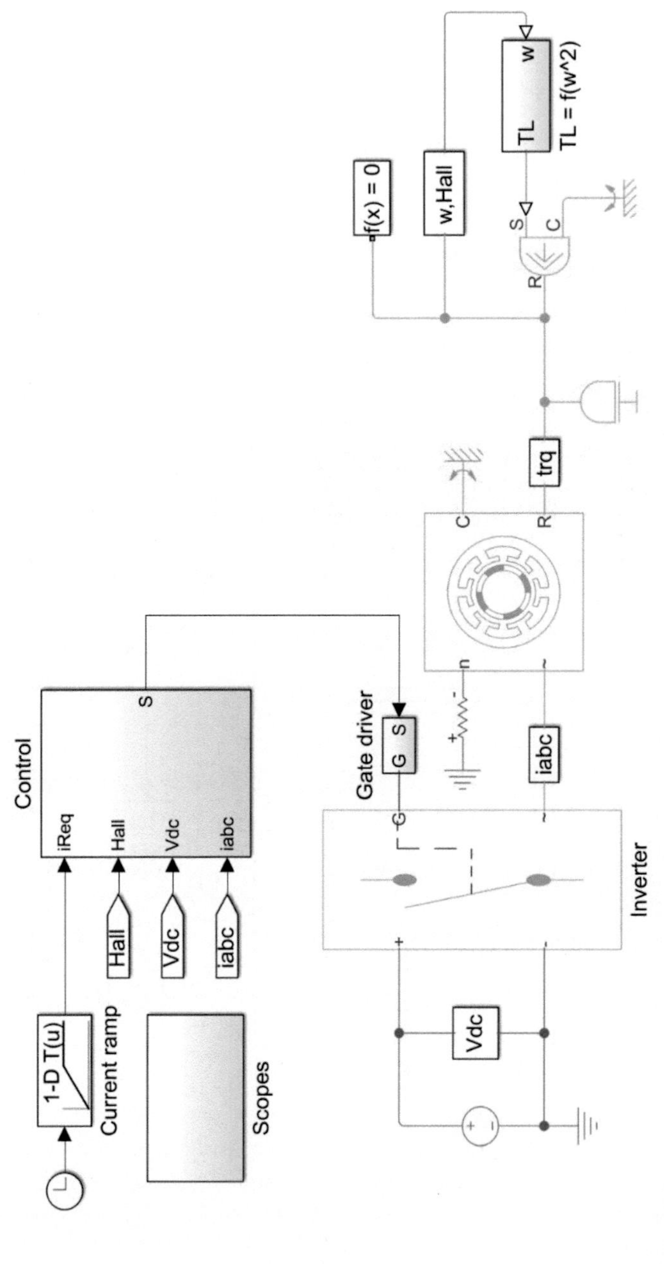

Figure 6.19 Hysteresis current control simulation diagram.

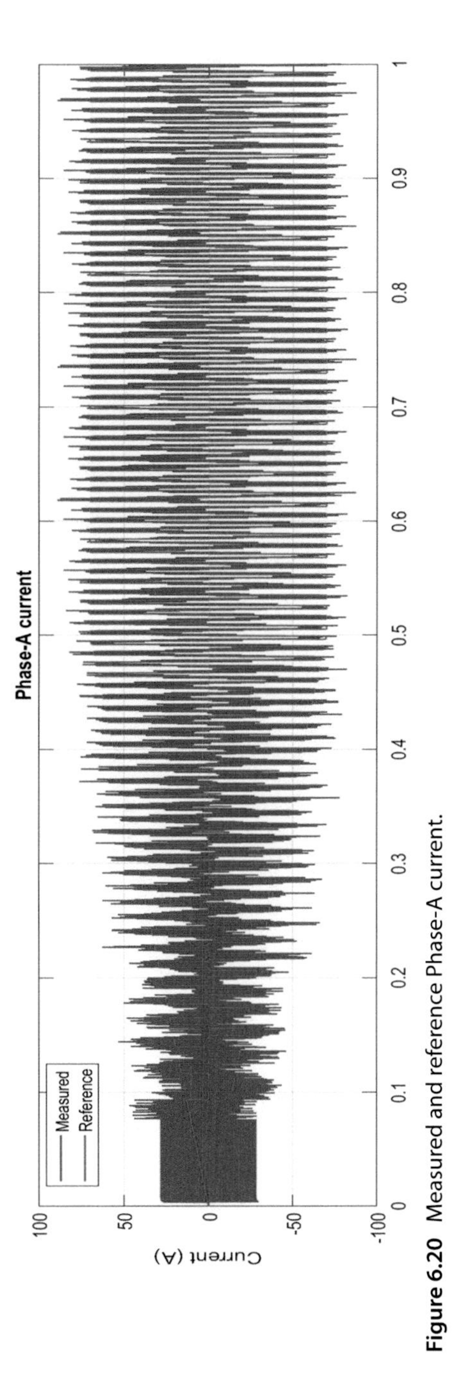

Figure 6.20 Measured and reference Phase-A current.

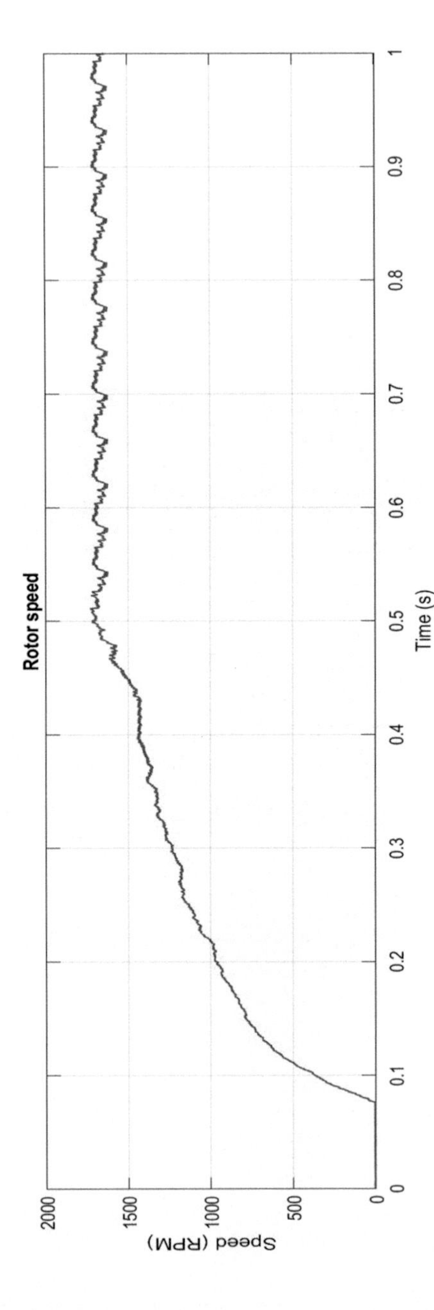

Figure 6.21 Rotor speed.

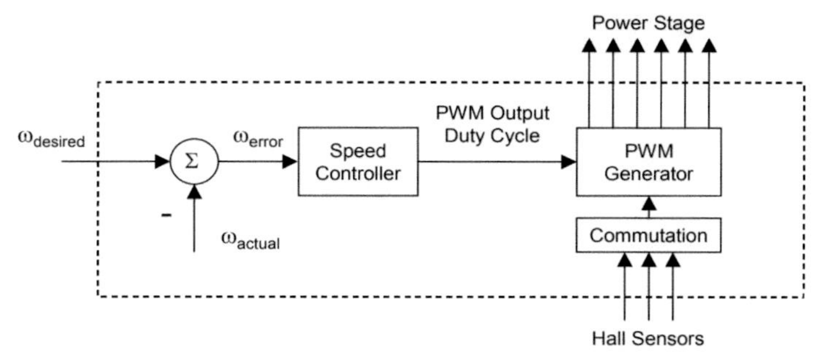

Figure 6.22 Speed control block diagram.

6.4 BLDC speed control

Commutation ensures the proper rotor rotation of the BLDC motor, while the motor speed only depends on the amplitude of the applied voltage. The amplitude of the applied voltage is adjusted using the PWM technique. The required speed is controlled by a speed controller, which is implemented as a conventional proportional-integral (PI) controller (Fig. 6.22). The difference between the actual and required speeds is input to the PI controller which then, based on this difference, controls the duty cycle of the PWM pulses which correspond to the voltage amplitude required to maintain the desired speed.

The speed controller calculates the PI algorithm given in the equation below:

$$u(t) = K_c \left[e(t) + \frac{1}{T} \int_0^t e\,(\tau) d\tau \right]$$

After transforming the equation into a discrete-time domain using an integral approximation with the Backward Euler method, we get the following equations for the numerical PI controller calculation:

$$u(k) = u_p(k) + u_I(k)$$

$$u_p(k) = K_c \,.\, e(k)$$

$$u_1(k) = u_1(k-1) + K_c \frac{T}{T_1} . e(k)$$

e(k) = Input error in step k.
w(k) = Desired value in step k.
m(k) = Measured value in step k.
u(k) = Controller output in step k.

up(k) = Proportional output portion in step k.
uI (k) = Integral output portion in step k.
uI (k-1) = Integral output portion in step k-1.
TI = Integral time constant.
T = Sampling time.
Kc = Controller gain.

An ideal torque source provides the load. The control subsystem uses a PI-based cascade control structure with an outer speed control loop and an inner DC-link voltage control loop. The DC-link voltage is adjusted through a DC–DC buck converter (Fig. 6.23). The BLDC is fed by a controlled three-phase inverter. The gate signals for the inverter are obtained from hall signals. The results are shown in Fig. 6.24A and B.

6.5 Introduction to BLDC in PSIM software

The image of the three-phase brushless DC machine is shown in Fig. 6.25. Two types of parameter inputs are provided: one based on machine model parameters (Tables 6.3 and 6.4), and the other based on manufacturer datasheet information.

The node assignments of the image are Nodes a, b, and c are the stator winding terminals for Phase A, B, and C, respectively. The stator windings are Y connected, and Node n is the neutral point. The shaft node is the connecting terminal for the mechanical shaft. They are all power nodes and should be connected to the power circuit.

Node sa, sb, and sc are the outputs of the built-in 6-pulse hall effect position sensors for Phase A, B, and C, respectively. The sensor output is a bipolar commutation pulse (1, 0, and −1). The sensor output nodes are all control nodes and should be connected to the control circuit.

The equations of the three-phase brushless DC machine are:

$$V_a = R.i_a + (L - M) \cdot \frac{di_a}{dt} + E_a$$

$$V_b = R.i_b + (L - M) \cdot \frac{di_b}{dt} + E_b$$

$$V_c = R.i_c + (L - M) \cdot \frac{di_c}{dt} + E_c$$

where va, vb, and vc are the phase voltages, ia, ib, and ic are the phase currents, R, L, and M are the stator phase resistance, self-inductance, and mutual inductance, and Ea, Eb, and Ec are the back emf of Phases A, B, and

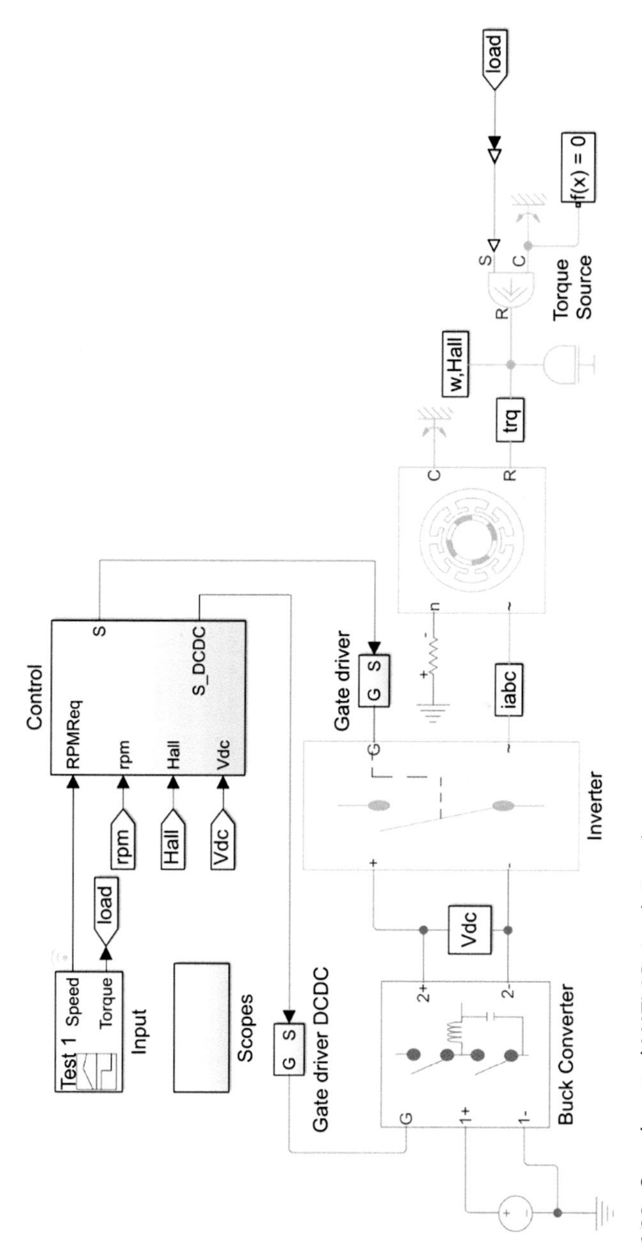

Figure 6.23 Speed control MATLAB simulation diagram.

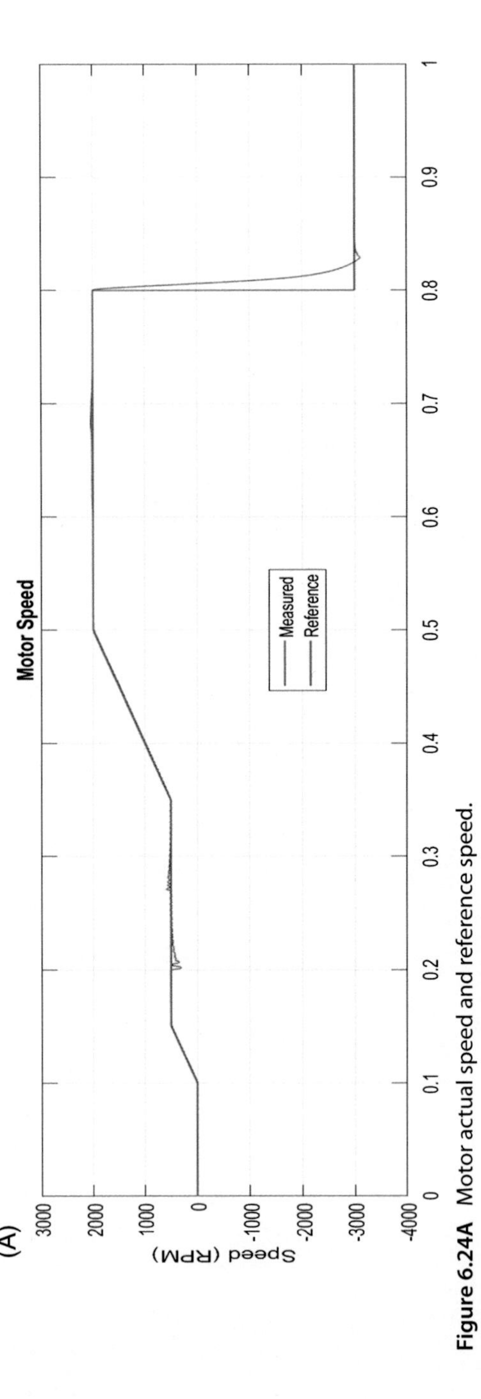

Figure 6.24A Motor actual speed and reference speed.

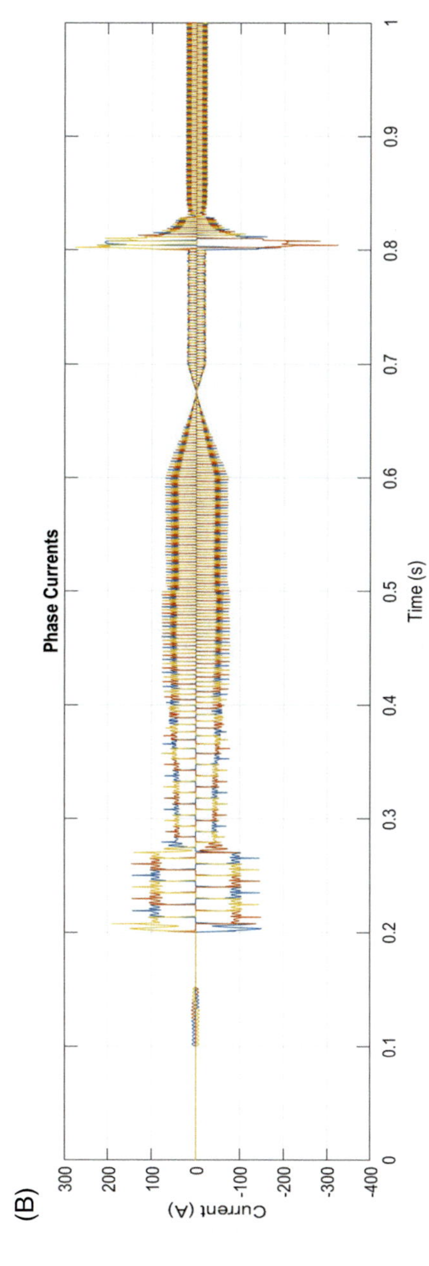

Figure 6.24B Three-phase current waveforms.

Table 6.3 Machine model Parameters 1.

Parameters	Description
R (stator resistance)	Stator phase resistance R, in Ohm
L (stator self ind.)	Stator phase self-inductance L, in H
M (stator mutual ind.)	Stator mutual inductance M, in H
	The mutual inductance M is a negative value. Depending on the winding structure, the ratio between M and the stator self-inductance L is normally between -1/3 and -1/2. If M is unknown, a reasonable value of M equal to -0.4*L can be used as the default value.
Vpk / krpm	Peak line-to-line back emf constant, in V/krpm (mechanical speed)
Vrms / krpm	RMS line-to-line back emf constant, in V/krpm (mechanical speed).
	The values of Vpk/krpm and Vrms/krpm should be available from the machine data sheet. If these values are not available, they can be obtained through experiments by operating the machine as a generator at 1000 rpm and measuring the peak and rms values of the line-to-line voltage
No. of Poles *P*	Number of poles P
Moment of Inertia	Moment of inertia J of the machine, in kg*m^2
Shaft Time Constant	Shaft time constant τ_{shaft}
theta_0 (deg.)	Initial rotor angle θr, in electrical deg.
	The initial rotor angle is the rotor angle at t=0. The zero rotor angle position is defined as the position where Phase A back emf crosses zero (from negative to positive) under a positive rotation speed.
theta_advance (deg.)	Position sensor advance angle $\theta_{advance}$, in electrical deg.
	The advance angle is defined as such that, for a brushless DC machine with a 120° trapezoidal back emf waveform, if the advance angle is 0, the leading edge of the Phase A hall effect sensor signal will align with the intersection of the rising ramp and the flat-top of the back emf trapezoidal waveform
Conduction Pulse Width	Position sensor conduction pulse width, in electrical deg.
	Positive conduction pulse can turn on the upper switch and negative pulse can turn on the lower switch in a full bridge inverter. The conduction pulse width is 120 electrical deg. for 120o conduction mode
Torque Flag	Output flag for internal developed torque *Tem*
Master/Slave Flag	Master/slave flag of the machine (1: master; 0: slave)

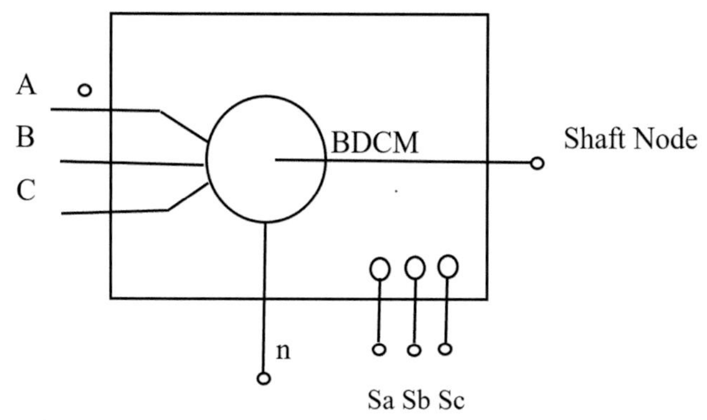

Figure 6.25 Three-phase brushless DC machine.

Table 6.4 Machine model Parameters 2.

Parameters	Description
Resistance (phase–phase)	Phase-to-phase (or line-to-line) resistance, in Ohm
Inductance (phase–phase)	Phase-to-phase (or line-to-line) inductance, in H
Speed constant	Speed constant *Kv*, defined as the ratio between the speed and the applied voltage, in rpm/V
Torque constant	Torque constant *Kt*, defined as the ratio between the generated torque and the applied current, in N*m/A
No. of poles P	Number of poles *P*
Moment of inertia	Moment of inertia *J* of the machine, in kg*m^2
No load speed	The motor speed at no load with the nominal voltage applied, in rpm
No load current	The current under no load operation, in A
Torque flag	Output flag for internal developed torque *Tem*
Master/slave flag	Master/slave flag of the machine (1: master; 0: slave)

C, respectively. The back emf voltages are a function of the rotor mechanical speed ωm and the rotor electrical angle θr, that is:

$$E_a = k_{e\_a} \cdot \omega_m$$

$$E_b = k_{e\_b} \cdot \omega_m$$

$$E_c = k_{e\_c} \cdot \omega_m$$

The coefficients ke_a, ke_b, and ke_c are dependent on the rotor angle θr. In this model, an ideal trapezoidal waveform profile is assumed, as shown

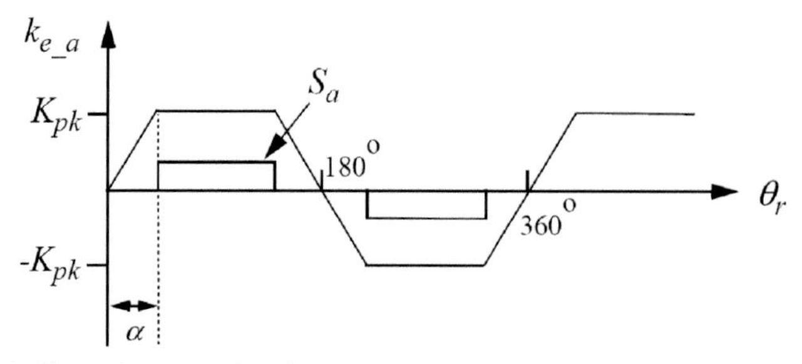

Figure 6.26 Phase A sensor signal.

below for Phase A. Also shown is the Phase A hall effect sensor signal *Sa* (Fig. 6.26). where Kpk is the peak trapezoidal value, in V/(rad/s), which is defined as:

$$K_{pk} = \frac{V_{pk}/Krpm}{2} \cdot \frac{1}{1000.2\pi/60}$$

Given the values of Vpk/krpm and Vrms/krpm, the angle α is determined automatically in the program.

The developed torque of the machine is:

The developed torque of the machine is:

$$T_{em} = (E_a i_a + E_b i_b + E_c i_c)/\omega_m$$

The mechanical equations are:

$$J \cdot \frac{d\omega_m}{dt} = T_{em} - B.\omega_m - T_{load}$$

$$\frac{d\theta_r}{dt} = \frac{P}{2}\,\omega_m$$

where B is the friction coefficient, T_{load} is the load torque, and P is the number of poles. The coefficient B is calculated from the moment of inertia J and the shaft time constant τ_{shaft} as below:

$$B = \frac{J}{\tau_{shaft}}$$

The shaft time constant τ_{shaft}, therefore, reflects the effect of the friction and windage of the machine. Note that when the shaft time constant is set to 0, the friction term is $B^{*}\omega m$ is ignored.

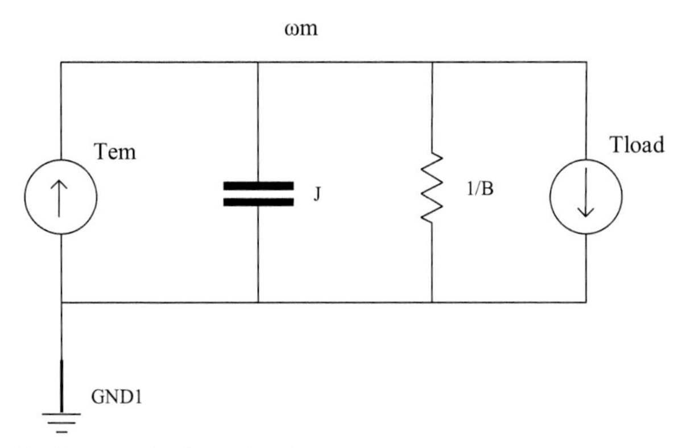

Figure 6.27 Machine equivalent circuit.

To better understand the definition of the shaft time constant, we can represent the mechanical equation with the following equivalent circuit (Fig. 6.27).

This circuit shows that the shaft time constant τ_{shaft} is equal to the RC time constant of the resistor 1/B and the capacitor *J*. Therefore, the shaft time constant can be measured by the following test:

Connect the machine to an external mechanical source. With the stator side in open circuit, drive the machine to a certain speed. Remove the mechanical source. The shaft time constant will be equal to the time that it takes the machine to decelerate to 36.8% of its initial speed.

A hall effect position sensor consists of a set of hall switches and a set of trigger magnets.

The hall switch is a semiconductor switch (e.g., MOSFET or BJT) that opens or closes when the magnetic field is higher or lower than a certain threshold value. It is based on the hall effect, which generates an emf proportional to the flux-density when the switch is carrying a current supplied by an external source. It is common to detect the emf using a signal conditioning circuit integrated with the hall switch or mounted very closely to it. This provides a TTL-compatible pulse with sharp edges and high noise immunity for connection to the controller via a screened cable. For a three-phase brushless DC motor, three hall switches are spaced 120 electrical deg. apart and are mounted on the stator frame.

The set of trigger magnets can be a separate set of magnets, or it can use the rotor magnets of the brushless motor. If the trigger magnets are separate, they should have the matched pole spacing (with respect to the rotor magnets), and should be mounted on the shaft in close proximity to

Table 6.5 Maxon motor datasheet.

Values at nominal voltage	Nominal voltage (V)32
	No load speed (rpm) 38700
	No load current (mA)327
Characteristics	Terminal resistance phase to phase (Ohm)0.363
	Terminal inductance phase to phase (mH)0.049
	Torque constant (mNm/A)7.85
	Speed constant (rpm/V) 1220
	Rotor inertia (gcm 2) 4.2
Other specifications	Number of poles pairs1
	Number of phases 3

Table 6.6 Motor parameters in PSIM.

Resistance (phase–phase)	0.363
Inductance (phase–phase)	0.049 m
Speed constant	1220
Torque constant	7.85 m
No. of poles P	2
Moment of inertia	4.2e-7
No load speed	38700
No load current	327m
Torque flag	1
Master/slave flag	1

the hall switches. If the trigger magnets use the rotor magnets of the machine, the hall switches must be mounted close enough to the rotor magnets, where they can be energized by the leakage flux at the appropriate rotor positions.

Defining brushless DC motor parameters from manufacturer datasheet
This example illustrates how to define brushless DC motor parameters from manufacturer datasheet. Below is the information provided on the datasheet of the brushless DC motor *Maxon EC-22-16730* (32V, 50W) from Maxon Motor (Table 6.5).

Using the element based on manufacturer datasheet information, and after converting all the quantities to the SI units, the motor parameters in PSIM are defined in Table 6.6.

6.6 Brushless DC motor drive with 6-pulse operation using PSIM

Open-loop operation of BLDC motor fed from 6 pulse inverter simulation diagram is shown in Fig. 6.28. Three-phase currents given to the BLDC

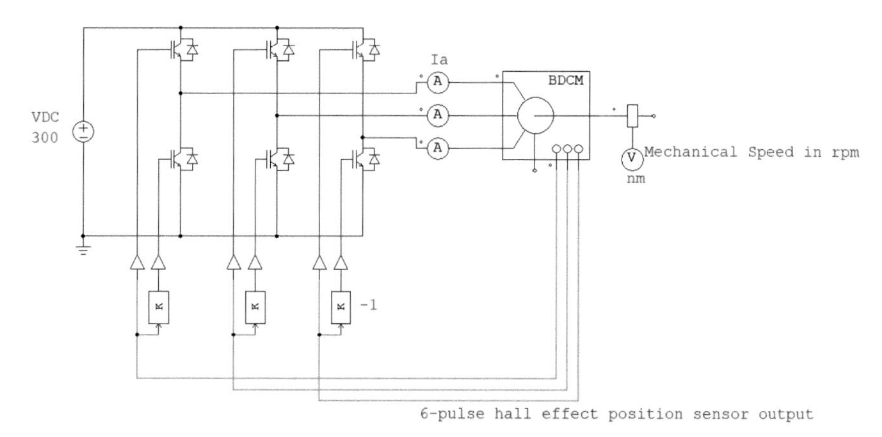

Figure 6.28 Simulation of BLDC in open loop.

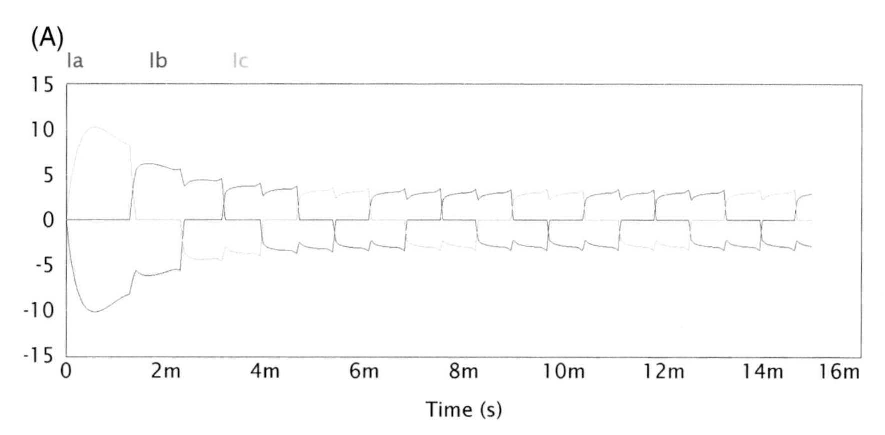

Figure 6.29A Three-phase current waveform.

motor is measured and presented in Fig. 6.29A. The output from the BLDC is presented in Fig. 6.29B. The internal torque flag is presented in Fig. 6.29C.

6.7 Brushless DC motor drive with speed feedback (6-pulse operation) using PSIM

Open-loop operation of BLDC motor fed from 6 pulse inverter with speed feedback simulation diagram is shown in Fig. 6.30. Three-phase currents given to the BLDC motor are measured and presented in Fig. 6.31A. The output from the BLDC is presented in Fig. 6.31B. The internal torque flag is presented in Fig. 6.31C.

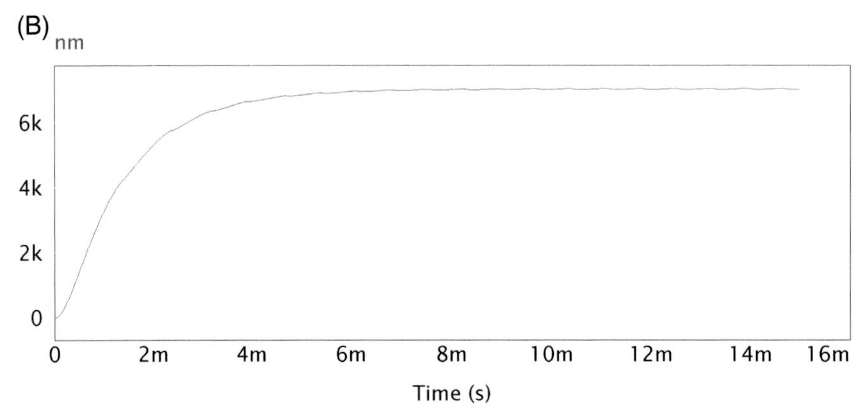

Figure 6.29B Mechanical speed measurement.

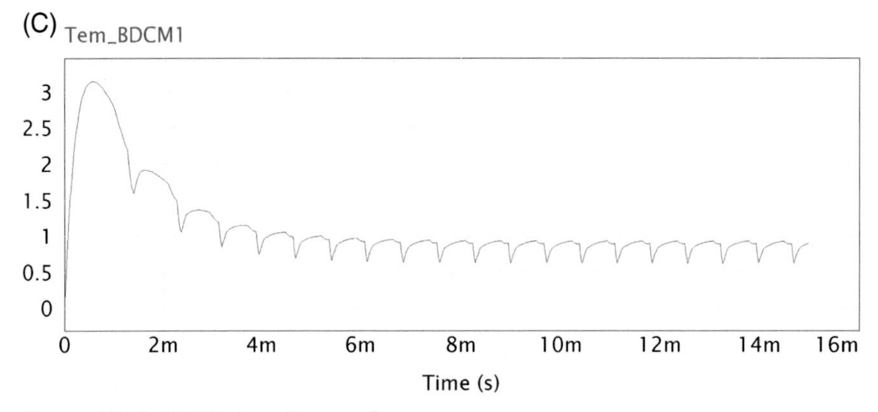

Figure 6.29C BLDC internal torque flag.

6.8 Brushless DC motor drive using the Hall effect sensor using PSIM

The PSIM simulation of BLDC motor with speed feedback using Hall effect sensor is presented in this section (Fig. 6.32). The simulation model is presented in the figure. Three-phase currents given to the BLDC motor and DC input current are measured and presented in Figure. The output from the BLDC is presented in Fig. 6.33A–E.

Advantages of using a Hall effect sensor in a BLDC motor controller

- Hall effect sensor is a very simple device comprising of magnets, hence, very cost-effective for motor control systems.
- For the same reason, these sensors are easy to implement in advanced motor control systems for EVs and other automotive solutions.

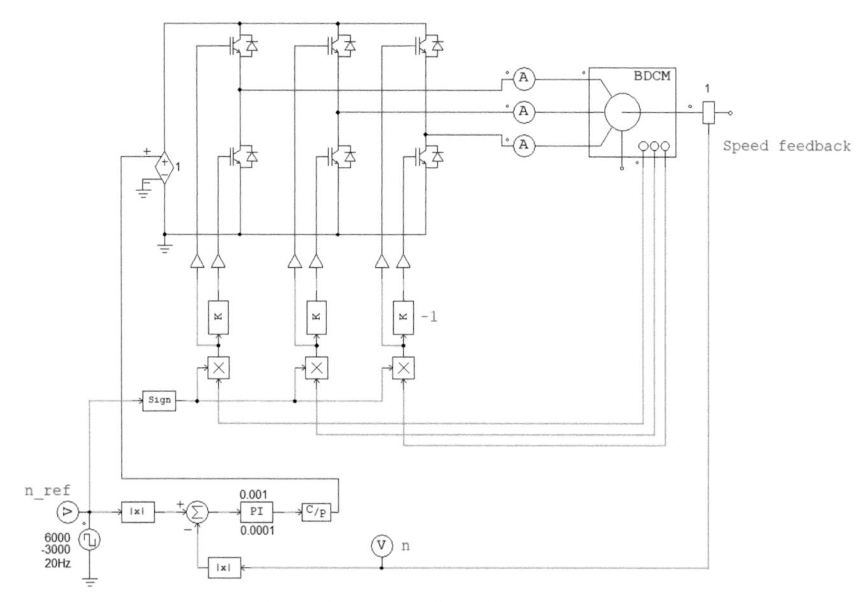

Figure 6.30 Simulation of BLDC with speed feedback.

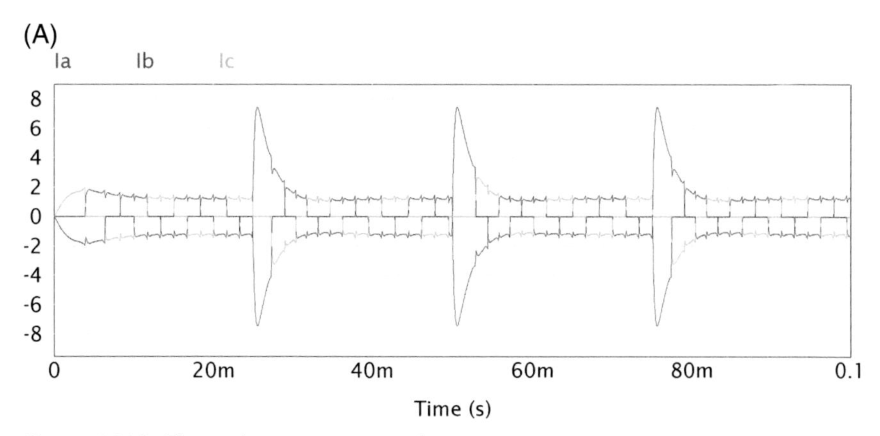

Figure 6.31A Three-phase current waveform.

- Most BLDC motors come equipped with these sensors.
- Hall Effect Sensors are mostly immune to environmental conditions like humidity, temperature, dust, and vibration.

6.9 Summary

BLDC motors offer a number of advantages over conventional motors. The removal of brushes from a motor eliminates a mechanical part that otherwise

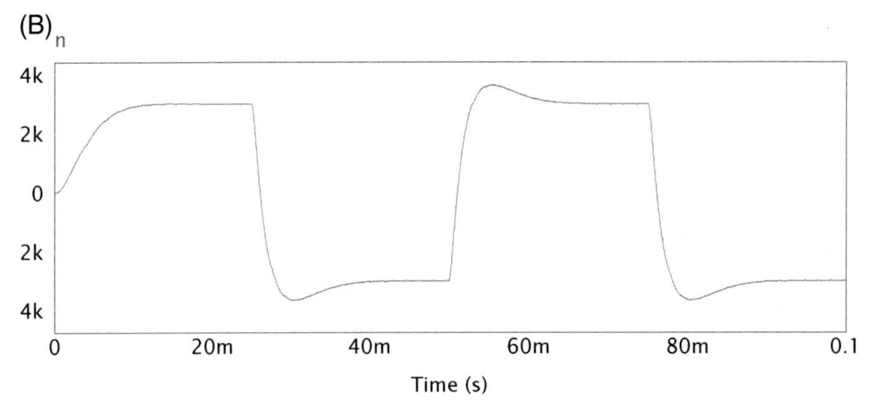

Figure 6.31B Mechanical speed measurement.

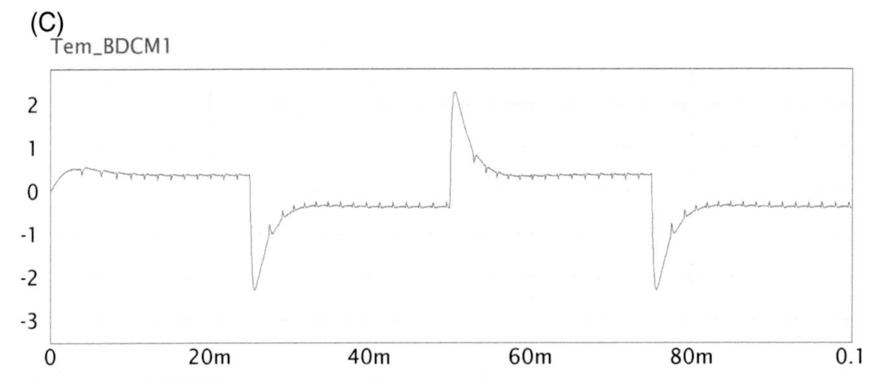

Figure 6.31C BLDC internal torque flag.

reduces efficiency, wears out, or can fail catastrophically. In addition, the development of powerful rare earth magnets has allowed the production of BLDC motors that can produce the same power as brush-type motors while fitting into a smaller space.

One perceived disadvantage is that BLDC motors, unlike the brush type, require an electronic system to supervise the energizing sequence of the coils and provide other control functions. Without the electronics, the motors cannot operate. However, the proliferation of inexpensive, robust electronic devices specially designed for motor control means that designing a circuit is relatively simple and inexpensive. In fact, a BLDC motor can be set up to run in a basic configuration without even using a microcontroller by employing a modest three-phase sine- or square-wave generator.

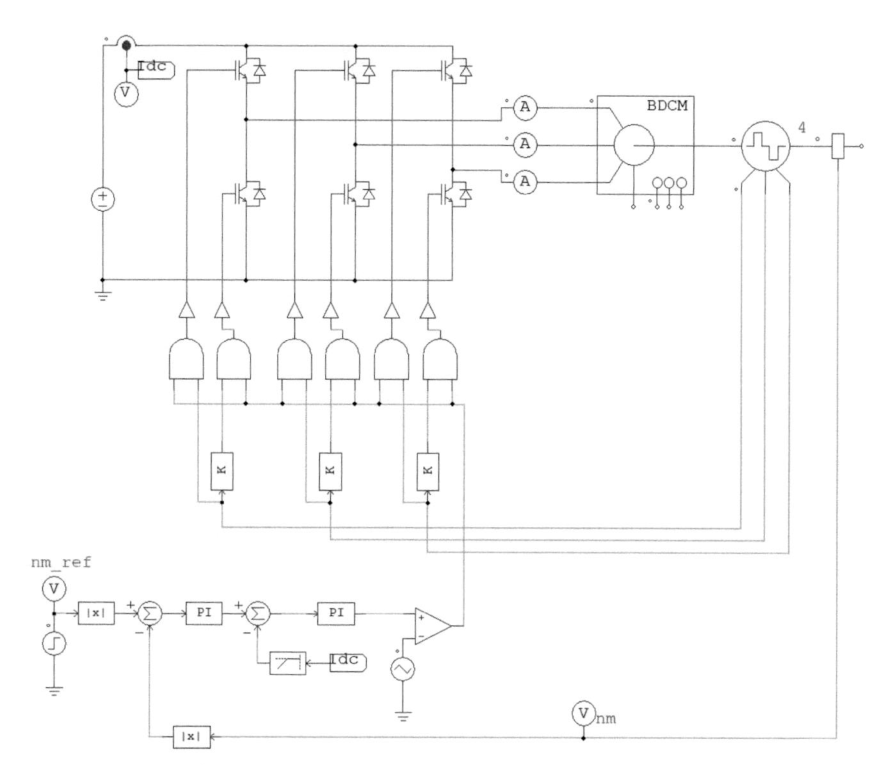

Figure 6.32 Simulation BLDC with Hall sensors.

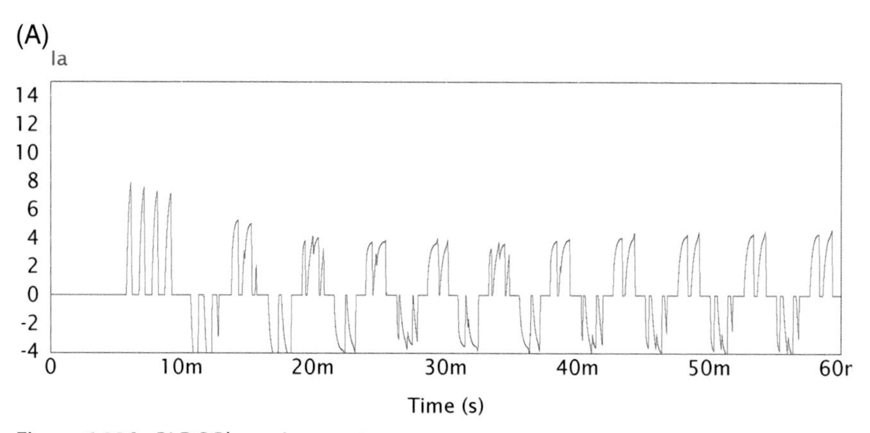

Figure 6.33A BLDC Phase-A current.

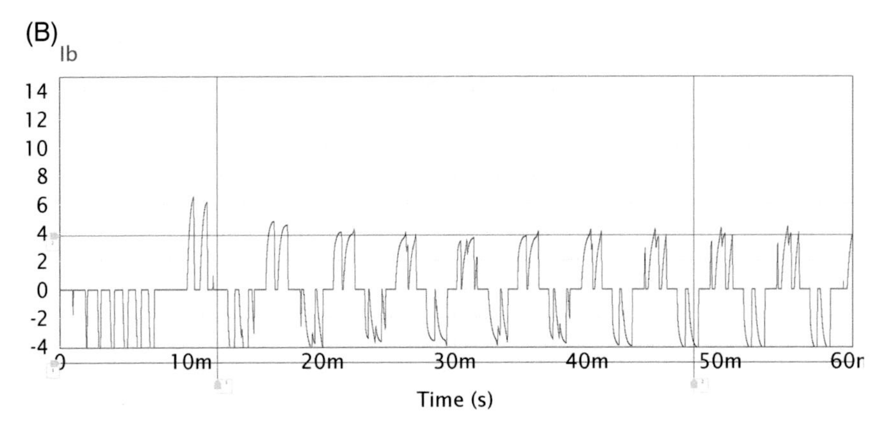

Figure 6.33B BLDC Phase-B current.

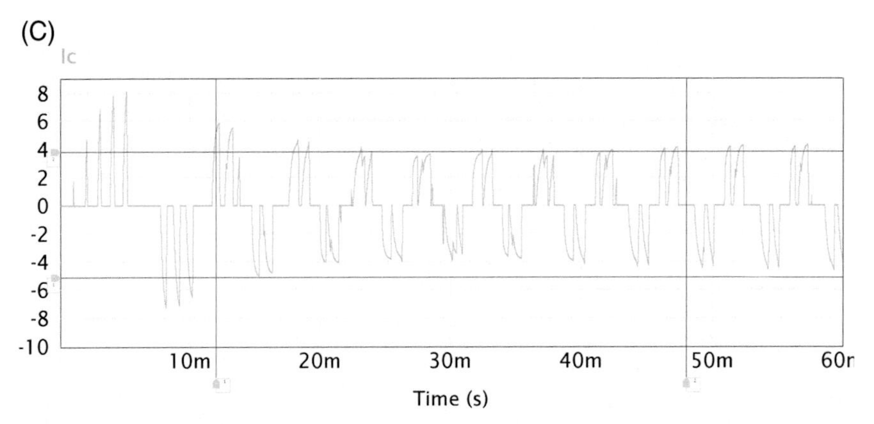

Figure 6.33C BLDC Phase-C current.

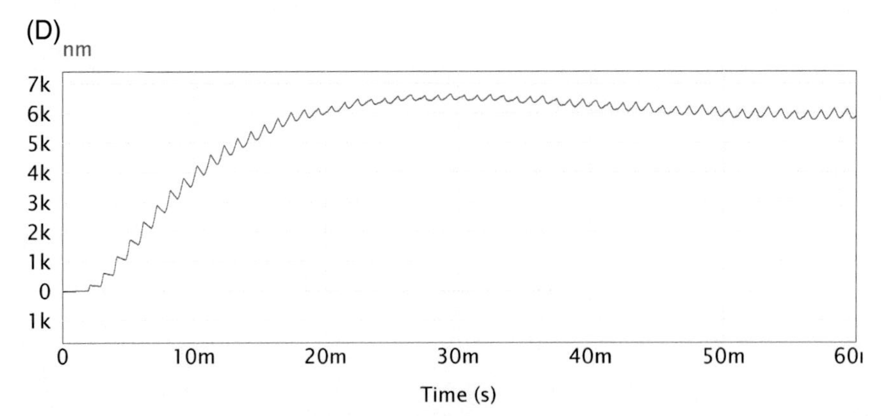

Figure 6.33D The actual speed from the motor.

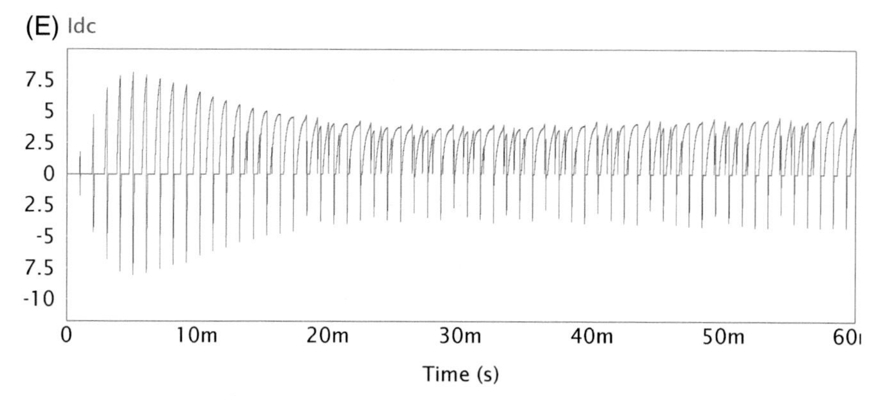

Figure 6.33E DC current waveform.

Review Questions

1. List the applications of BLDC motor.
2. Give the control methods of BLDC motor in detail.
3. What are the advantages and disadvantages of BLDC motor?
4. What are the challenges need to be faced while using Hall effect sensor in BLDC.
5. Construct the BLDC Motor with its speed control and simulate it using MATLAB/Simulink
6. Construct the BLDC Motor with its hysteresis current control and simulate it using MATLAB/Simulink
7. Construct the BLDC Motor with its position control and simulate it using MATLAB/Simulink
8. Do the simulation of Brushless DC Motor Drive in open-loop using PSIM software environment and obtain the results.
9. Do the simulation of Brushless DC Motor Drive with speed feedback in PSIM software environment and obtain the results.
10. Do the simulation of Brushless DC Motor Drive Using the Hall Effect Sensor in PSIM software environment and obtain the results.

CHAPTER 7

PMSM drives control and simulation using MATLAB

Contents

7.1 Introduction to PMSM

A permanent magnet synchronous motor (PMSM) is an electric motor with a permanent magnet inductor. The rotor is the primary distinction between a PMSM and an induction motor. According to studies, the PMSM has a 2% higher efficiency than a highly efficient (IE3) induction electric motor when the stator is of the same design and the same variable frequency drive is utilized for control. In this scenario, permanent magnet synchronous electric motors outperform other electric motors in terms of power/volume, torque/inertia, and other factors [2].

The PMSM is an AC synchronous motor with a sinusoidal back EMF waveform and field excitation provided by permanent magnets. It has the same rotor and stator as an induction motor, but the rotor is a permanent magnet that generates a magnetic field. As a result, no field winding on the rotor is required. A 3-phase brushless permanent sine wave motor is another name for it. The schematic of a PMSM is illustrated below (Fig. 7.1).

The PMSM is a hybrid of a brushless DC motor and an induction motor. It has a permanent magnet rotor and stator windings, just like a brushless DC motor. The stator structure, which uses windings to produce a sinusoidal flux density in the machine's air gap, is similar to that of an induction motor.

Electric Motor Drives and Their Applications with Simulation Practices.
DOI: https://doi.org/10.1016/B978-0-323-91162-7.00007-2

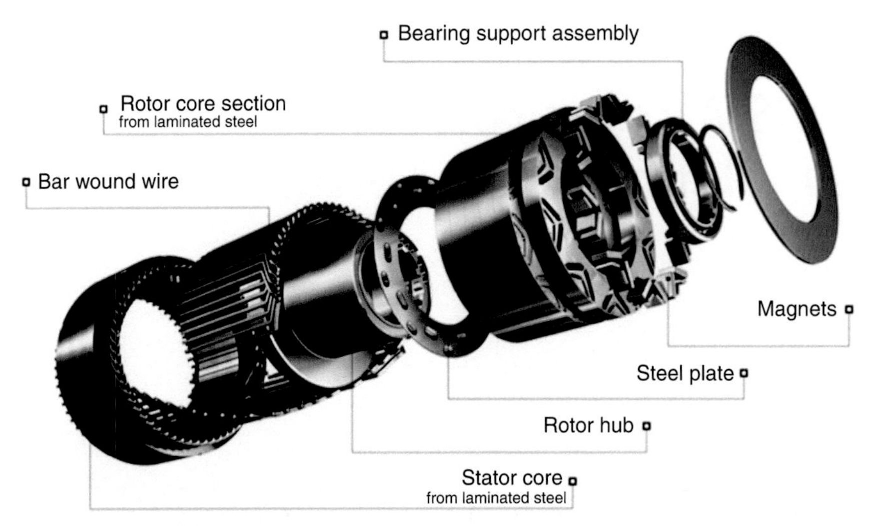

Figure 7.1 Interior permanent magnet synchronous motor.

Because there is no stator power allocated to magnetic field creation, it has a higher power density than induction motors with similar ratings.

Synchronous motors, which transform electrical energy into mechanical energy, are the most extensively used steady-state 3-phase AC motors in an electrical system. This motor rotates at a constant synchronous speed that is synchronized with the supply frequency, and the period of rotation is equal to the integral number of AC cycles. That is, the motor's rotational speed is equal to the spinning magnetic field's rotational speed. This motor is primarily used to increase the power factor in power systems. There are two types of synchronous motors: nonexcited and DC excited, which function based on the magnetic power of the motor. Nonexcited synchronous motors include reluctance motors, hysteresis motors, and permanent magnet motors.

The interaction of the rotating magnetic field of the stator and the constant magnetic field of the rotor provides the basis for synchronous motor functioning. The rotating magnetic field of a synchronous motor's stator is analogous to that of a three-phase induction motor.

According to Ampere's Law, the magnetic field of the rotor interacts with the synchronous alternating current of the stator windings to produce torque, which causes the rotor to rotate (more).

A steady magnetic field is created by permanent magnets on the rotor of the PMSM. The rotor poles interlock with the revolving magnetic field of the stator when the rotor rotates at a synchronous speed with the stator field. When the PMSM is connected directly to the three-phase current network, it is unable to start (current frequency in the power grid 50Hz).

The PMSM can generate torque at zero speed using permanent magnets, but it requires a digitally controlled inverter to operate. PMSMs are commonly utilized in motor drives with great performance and efficiency. Smooth rotation across the whole speed range of the motor, full torque control at zero speed, and quick acceleration and deceleration are all characteristics of high-performance motor control.

A control system, such as a variable frequency drive or a servo drive, is required for a PMSM. There are many different control strategies used in control systems. The task that is placed in front of the electric drive determines which control method is best. The table below shows the major methods for regulating a PMSM.

Popular methods to control PMSMs.

Control				**Advantages**	**Disadvantages**
Sinusoidal	Scalar			Simple control scheme	Control is not optimal, not suitable for tasks where the variable load, loss of control is possible
	Vector	Field ori-ented control	With position sensor	Smooth and precise setting of the rotor position and motor rotation speed, large control range	Requires rotor position sensor and powerful microcontroller inside the control system
			Without position sensor	No rotor position sensor required. Smooth and precise setting of the rotor position and motor rotation speed, large control range, but less than with position sensor	Sensorless field oriented control over *full speed range* is possible only for PMSM with salient pole rotor, a powerful control system is required
		Direct torque control		Simple control circuit, good dynamic performance, wide control range, no rotor position sensor required	High torque and current ripple
Trapezoidal	Open loop			Simple control scheme	Control is not optimal, not suitable for tasks where the variable load, loss of control is possible
	Closed loop	With position sensor (Hall sensors)		Simple control scheme	Hall sensors required. There are torque ripples. It is intended for control of PMSM with trapezoidal back EMF, when controlling PMSM with sinusoidal back EMF, the average torque is lower by 5%
		Without sensor		More powerful control system required	Not suitable for low speed operation. There are torque ripples. It is intended for control of PMSM with trapezoidal back EMF, when controlling PMSM with sinusoidal back EMF, the average torque is lower by 5%

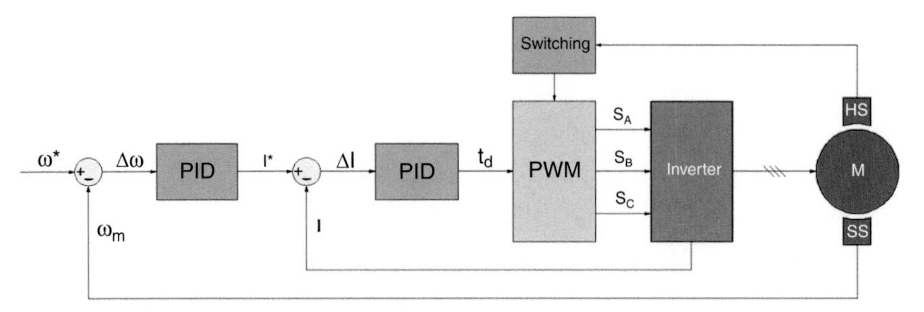

Figure 7.2 Trapezoidal control by Hall sensors.

Trapezoidal control with Hall sensors is commonly used to perform simple jobs (for example, computer fans). Field-oriented control (FOC) is typically chosen to handle challenges that require maximum performance from the electric drive [2].

7.1.1 Trapezoidal control

Trapezoidal control is one of the most basic methods of controlling a PMSM. The PMSM with trapezoidal back EMF is controlled using trapezoidal control. This method also allows you to operate the PMSM with a sinusoidal back EMF, but the average torque of the electric drive will be reduced by 5%, and the torque ripples will be 14% of their maximum value. A trapezoidal control is available, both with and without feedback from the rotor position.

The open-loop control (without feedback) is inefficient and may cause the PMSM to be released from synchronism, resulting in controllability loss.

There are two types of closed-loop control: h trapezoidal control with a position sensor (typically Hall sensors); and h trapezoidal control without a position sensor (sensorless trapezoidal control).

Three Hall sensors incorporated into an electric motor are widely employed as rotor position sensors for three-phase trapezoidal control, allowing an angle to be determined with an accuracy of 30 degrees (Fig. 7.2). The stator current vector only occupies six positions for one electric period with this regulation, resulting in ripple torque at the output.

7.1.2 Field-oriented control

FOC allows you to manage the speed and torque of a brushless motor in a smooth, accurate, and independent manner (Fig. 7.3). The position of the brushless motor's rotor must be known in order for the FOC method to work.

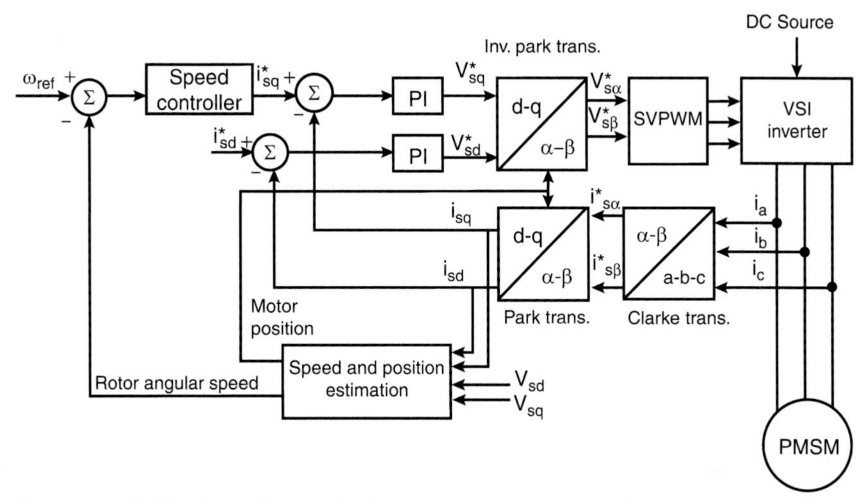

Figure 7.3 Field-oriented control of permanent magnet synchronous motor using a rotor position sensor.

The rotor position can be determined in two ways: by using a position sensor or by using a real-time control system to calculate the angle based on the information provided.

7.1.3 PMSM field-oriented control by position sensor

Angle sensors include the following sorts of sensors:

- inductive (resolver, inductosyn, etc.),
- optical (optical sensors), and
- magnetic (magnetoresistive sensors).

7.1.4 Field-oriented control of PMSM without a position sensor

Because of the rapid growth of microprocessors, sensorless vector control systems for brushless AC motors have been developed since the 1970s. The initial sensorless angle estimation methods relied on the electric motor's ability to create back EMF during rotation. Because the motor back EMF provides information on the rotor's position, you may calculate the rotor's position by computing the value of the back EMF in the stationary coordinate system. However, the back EMF is absent when the rotor is not rotating, and the back EMF has a low amplitude at low speeds, making it difficult to distinguish from noise; thus, this method is not suited for measuring the position of the motor rotor at low speeds.

There are two main sensorless start PMSM techniques:

- Begin with the scalar technique, focusing on a specific property of voltage dependency on frequency. The capabilities of the control system and the parameters of the electric drive as a whole are severely limited by scalar control.
- High-frequency signal injection method—only works with the salient pole PMSM.

Sensorless field-oriented PMSM control in the complete speed range is currently only achievable for motors with salient pole rotors (Fig. 7.4).

7.2 Vector control of PMSM

Because of a high-order (fifth-order) system effect, scalar control is relatively straightforward to implement, but the inherent coupling effect (i.e., both torque and flux are functions of voltage or current and frequency) causes slow response and the system is easily prone to instability. To clarify, if the torque is raised by increasing the slip (i.e., the frequency), the flux will tend to decrease. It is worth noting that the flux variation is always low. The slow flux control loop compensates for the decrease in flux by feeding in additional voltage. The torque sensitivity with slip is reduced and the response time is lengthened by temporarily dipping the flux. For current-fed drives, this expansion is likewise applicable.

Vector or field-oriented control can be used to solve forgetting difficulties. The introduction of vector control in the early 1970s, as well as the proof that an induction motor can be controlled as if it were a separately excited DC motor, ushered in a new era in AC drive high-performance control. Vector control is also known as decoupling, orthogonal, or trans vector control because of its machine-like performance. Both synchronous and induction motor drives can benefit from vector control. Vector control and the accompanying feedback signal processing, especially for modern sensorless vector control, are unquestionably difficult, necessitating the employment of a powerful microcomputer or DSP. Vector control appears to be on the verge of displacing scalar control as the industry-standard control for AC drives.

This section goes through some of the vector control strategies utilized in the PMSM (Fig. 7.5).

7.2.1 Field-oriented control

The vector control technique known as FOC is commonly employed. The FOC can be built with or without a position sensor (direct) (indirect). The

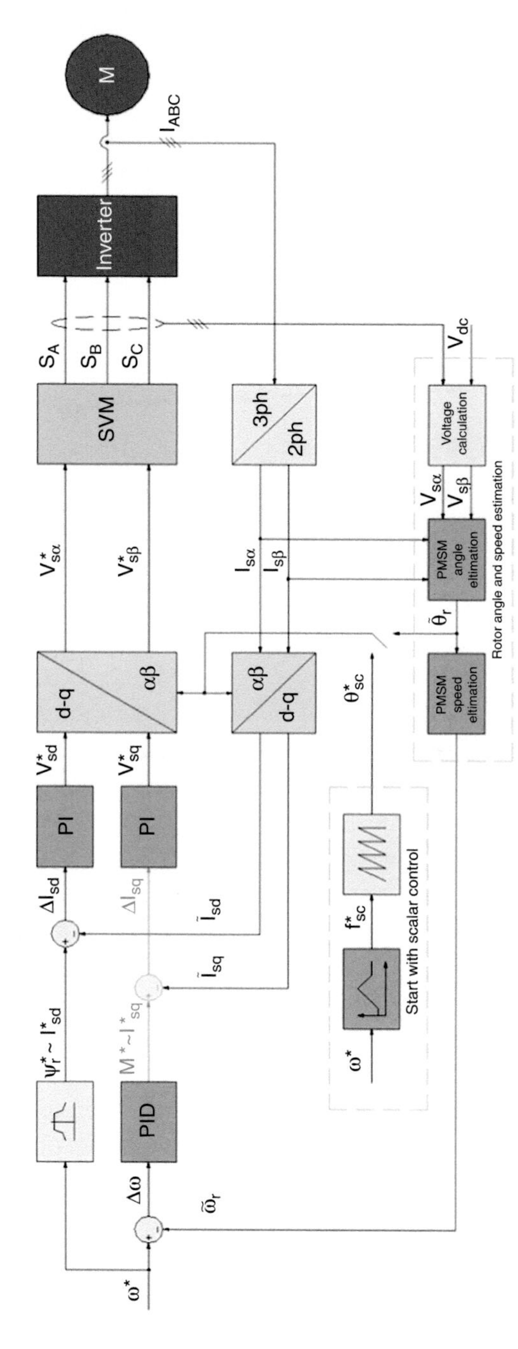

Figure 7.4 Field-oriented control of permanent magnet synchronous motor without rotor position sensor with scalar start.

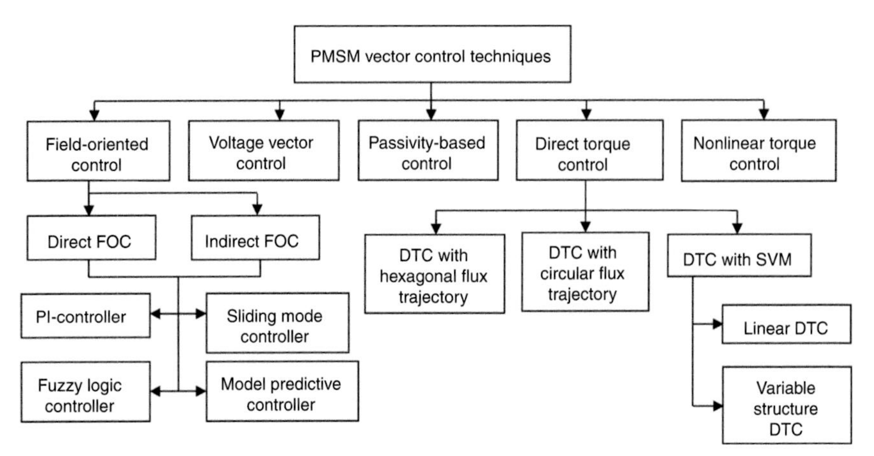

Figure 7.5 Vector control techniques.

mathematical model of the system is simpler in the first model, and the control is faster and more exact.

The currents of the stator phases of the motor are measured in FOC and Clarke transformed into a two-phase $\alpha\beta$ system. The value of the rotor angle from the position sensor (or one estimated indirectly in the absence of a sensor) is used to transform currents from the static reference frame $\alpha\beta$ to the rotational dq one associated with the rotor flux linkage via park transformation. A separate management of the torque and the motor excitation flux is carried out in accordance with the reference speed signals and the d component of the stator current, taking into consideration the feedback signals. This is the foundation of FOC.

7.2.2 Direct torque control

Direct torque control (DTC) circuits are less complicated than FOC circuits since they do not require the usage of a PWM inverter, a speed sensor, or reference frame transformers. Because information regarding the exact position of the rotor is not necessary in such systems, all computations are performed in the stator reference frame (except the start-up case of the synchronous motor). However, their steady-state functioning, especially at low speeds, is characterized by high stator current, flux linkage, and torque ripples, which severely limits their usage for high-precision drives. DTC with space vector modulation (SVM) is designed to solve the drawbacks of DTC. The pulse width modulation technique is used by DTC with SVM to generate the voltage. When compared to DTC, there are various circuit changes that considerably improve the control system's efficiency.

7.2.3 Voltage vector control

Voltage vector control does not require information about the rotor's position; however, reference frame modifications are used, same as in the FOC circuit. Only the flux-forming component of the stator current is modulated at the same time. Both methods provide for a quick reaction to control signals. VVC's circuit construction and computational techniques are simpler than those of the standard FOC, making it less susceptible to changes in motor parameters.

7.2.4 Passivity-based control

The PBC system structure is not simpler than the FOC system structure in terms of complexity. Position feedback, reference frame modifications, and PWM are all used. Although the PBC has reasonable speed dynamics, the torque peaks in transient processes are large. The level of ripples is also noticeably higher than when FOC is used. As a result, the accuracy of the presented method (as well as the DTC and VVC approaches) cannot be compared to that of the FOC.

7.2.5 Mathematical model of PMSM

The wound rotor synchronous motor has a mathematical model that is comparable to this one. There is no need to include the rotor voltage calculations because there is no external source linked to the rotor side and the change in the rotor flux with respect to time is insignificant. The PMSM model is derived using the rotor reference frame. In terms of phase variables, the electrical dynamic equation can be expressed as:

$$va = Raia + p\lambda a$$

$$vb = Rbib + p\lambda b$$

$$vc = Rcic + p\lambda c$$

While the flux linkage equations are:

$$\lambda a = Laaia + Labib + Lacic + \lambda ma$$

$$\lambda b = Labia + Lbbib + Lbcic + \lambda mb$$

$$\lambda b = Lacia + Lbcib + Lccic + \lambda mc$$

Considering the symmetry of mutual inductances such as $Lab = Lba$, self inductances $Laa = Lbb = Lcc$ and flux linkage $\lambda ma = \lambda mb = \lambda mc = \lambda m$. For this model, input power pi can be represented as:

$$pi = vaia + vbib + vcic$$

Applying the transformations equations to voltages, flux linkages, and currents from the equation, we get a set of simple transformed equations as:

$vq = (Rs + Lqp)iq + wrLdid + wr\lambda m$

$vd = (Rs + Ldp)id - wrLqiq$

Ld and Lq are called d- and q-axis synchronous inductances, respectively. wr is motor electrical speed (= 2wm mechanical shaft speed of 4 pole motor). Each inductance is made up of self-inductance (which includes leakage inductance) and contributions from other two-phase currents. Instantaneous power (Pi) can be derived from the above power equation via transformation as:

$Pi = 3/2(vqiq + vdid)$

The produced torque Te which is power divided by mechanical speed can be represented as:

$Te = (3/2)(P/2)(\lambda miq + (Ld - Lq)idiq)$

The produced torque is constituted of two separate mechanisms, as shown by the preceding equation. The first part relates to "the mutual reaction torque" between iq and the permanent magnet, while the second term corresponds to "the reluctance torque" caused by d-axis and q-axis reluctance differences (or inductance). Because $Lq > Ld$, id must be negative in order to produce additive reluctance torque. The motor dynamics equation is as follows: $(Te - Tl) = P\ 2\ (Jpwr + Bwr)$

7.2.6 Vector control

For the control of the inverter-fed AC motor, several scalar1 control techniques have given a good steady-state but poor dynamic response. Until now, control systems have relied on the size and frequency of stator phase currents, but not on their phases. The phase and magnitude of the air-gap flux links deviated from their prescribed values as a result of this. Separately excited DC drives are simpler and have better dynamic responses because they manage flux independently, which leads to independent torque control when kept constant. This is made possible by controlling the field and armature currents separately, which controls the field flux and torque separately. The magnitude and phase of the controlled variables are referred to as vector controls. The control quantities are represented using matrices and vectors (voltages, currents, flux, etc.). The mathematical equations that describe the motor dynamics are taken into account in this procedure. This method necessitates more calculations than a traditional control technique. A computation unit contained in a specialized digital signal processor can be used to solve it (DSP) [4].

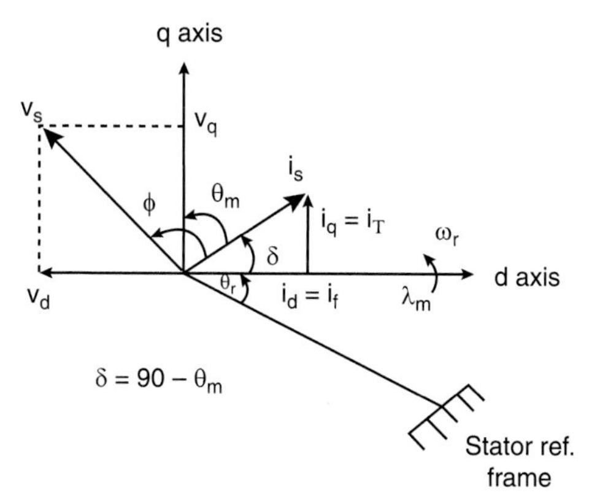

Figure 7.6 Phasor diagram of PMSM. *PMSM*, permanent magnet synchronous motor.

A vector diagram of the PMSM is shown in Fig. 7.6. The reference phase is considered to be phase a. The rotor's instantaneous position (and so rotor flux) is at an angle of θr to phase a. To make it similar to a DC machine, vector control requires that the quadrature axis current iq be in quadrature with the rotor flux. As a result, since id lags iq by 90 degrees in the reference, id must be along with the rotor flux. The d-axis stator flux contributes to the rotor flux when id is in the same direction as the rotor flow, resulting in an increase in the net air-gap flux. If id is negative, however, the stator d-axis flux is in opposition to the rotor flux, resulting in a reduction in air gap flux. The PMSMs are built in such a way that the rotor magnet can produce the needed air-gap flux on its own up to the stipulated speed. As a result, in the constant torque mode of operation, id is usually zero. Consider the following three-phase currents:

ia $=$ is sin(wrt $+ \delta$)

ib $=$ is sin(wrt $+ \delta - 2\pi$ 3)

ic $=$ is sin(wrt $+ \delta + 2\pi$ 3) Where θr $=$ wrt , From phasor diagram we get iq id $=$ is sin δ cos δ iq $=$ Torque-producing component of stator current $=$ iT id $=$ Flux-producing component of stator current $=$ if The electric torque equation becomes:

Te $=$ KiT (19) where, K $=$ (3/2)(P/2)λm

As a result, the electric torque is only determined by the current along the quadrature axis, and a constant torque can be obtained by ensuring that iq remains constant. Up to rated speed, a continuous air gap flux is necessary. Vector control is only possible when the instantaneous rotor flow can be precisely determined. As a result, it is naturally easier in the PMSM than in the induction motor since the rotor flux position in the PMSM is only defined by the rotor position. As a result, vector control can be used to control the torque (iq) and flux (id) producing currents independently.

The reference signal to the PID controller is given as a step signal. The values of the step signal is been assigned as per Fig. 7.7.

PID controller is used in this simulation, the controller parameters are set as per Fig. 7.8.

The saturation limits are considered as 15 and −15 for upper limit and lower limit, respectively (Fig. 7.9).

The dq-abc subsystem consists of multiplexer and functions. The circuit is connected as per Fig. 7.10.

The function $v_a = v_q \cos\theta - v_d \sin\theta$ has been set in f(u) Fcn1 as per Fig. 7.11.

The function $v_b = v_q \cos\left(\theta - \frac{2\pi}{3}\right) - v_d \sin(\theta - \frac{2\pi}{3})$ has been set in f(u) Fcn2 as per Fig. 7.12.

The function $v_c = v_q \cos\left(\theta + \frac{2\pi}{3}\right) - v_d \sin(\theta + \frac{2\pi}{3})$ has been set in f(u) Fcn3 as per Fig. 7.13.

The subsystem consists of relays. These relays are to be connected as per Fig. 7.14.

The relay parameters are given as per Fig. 7.15.

Fig. 7.16 gives the inverter subsystem which has universal bridge circuit.

The universal bridge parameters are given as per Fig. 7.17.

The step values are set as 10 and −10 for initial and final values, respectively. The step time is fixed as 3 (Fig. 7.18).

The heart of the simulation is permanent magnet synchronous machine. After selecting the PMSM need to configure it (Fig. 7.19). Here the rotor type is selected as salient pole type, back emf waveform as Sinusoidal.

The parameters of PMSM values are set as per Fig. 7.20.

Gain value is set as 4 (Fig. 7.21).

After completing all the connections of the selected components, we will get a simulation model of PMSM vector control is like above in Fig. 7.22.

7.3 Modeling and simulation of single-phase PMSM

An ideal torque source provides the load. An H-Bridge feeds the SPPMSM.

In Fig. 7.23, the subsystem of commutation is given. Here Hall signal is used to compute the gate signals.

Gate signals required for four-quadrant chopper are generated using the above circuit (Fig. 7.24).

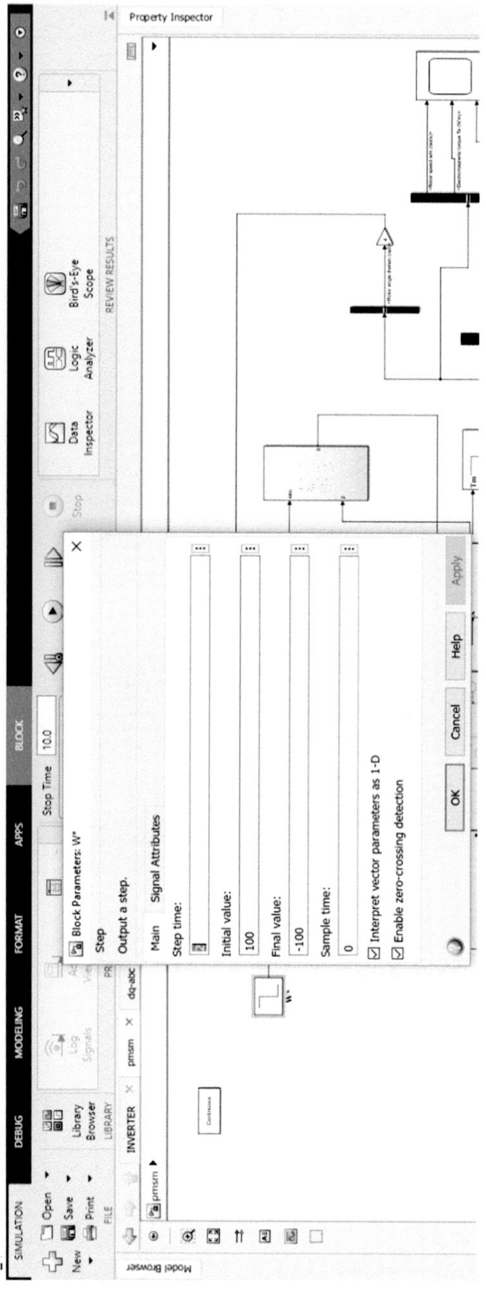

Figure 7.7 Block parameter—step.

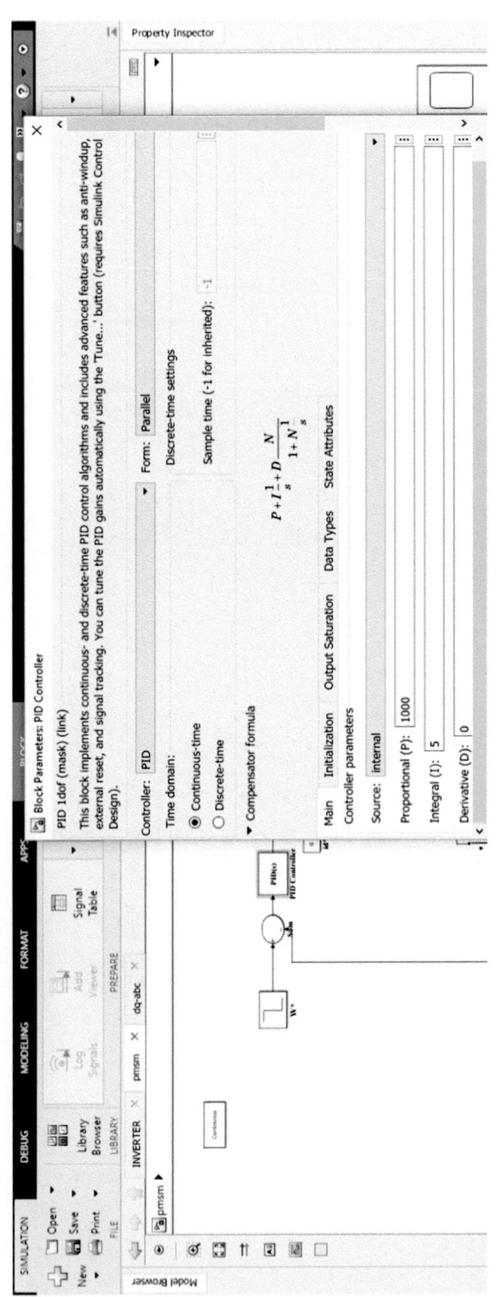

Figure 7.8 Block parameter—PID controller.

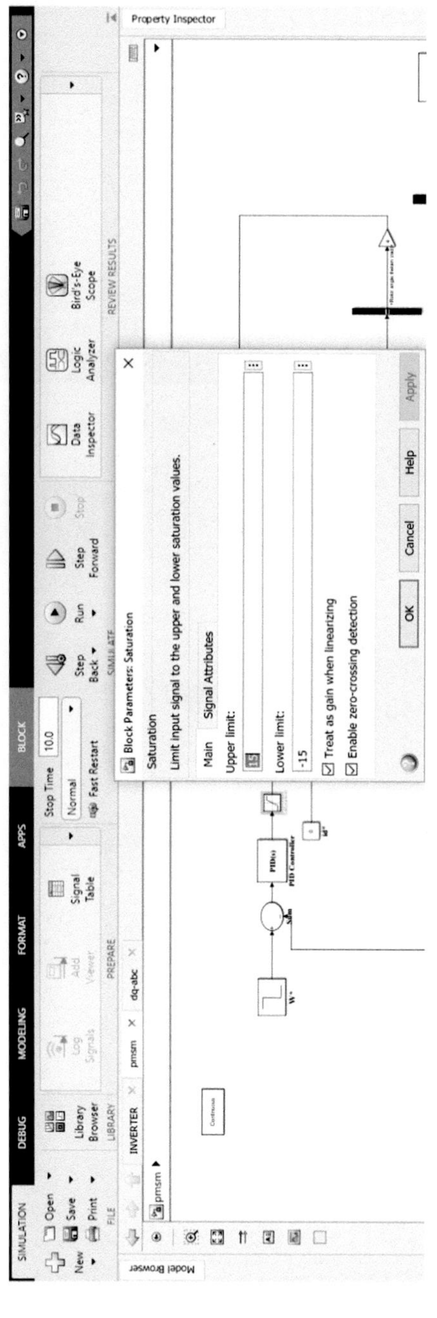

Figure 7.9 Block parameter—saturation.

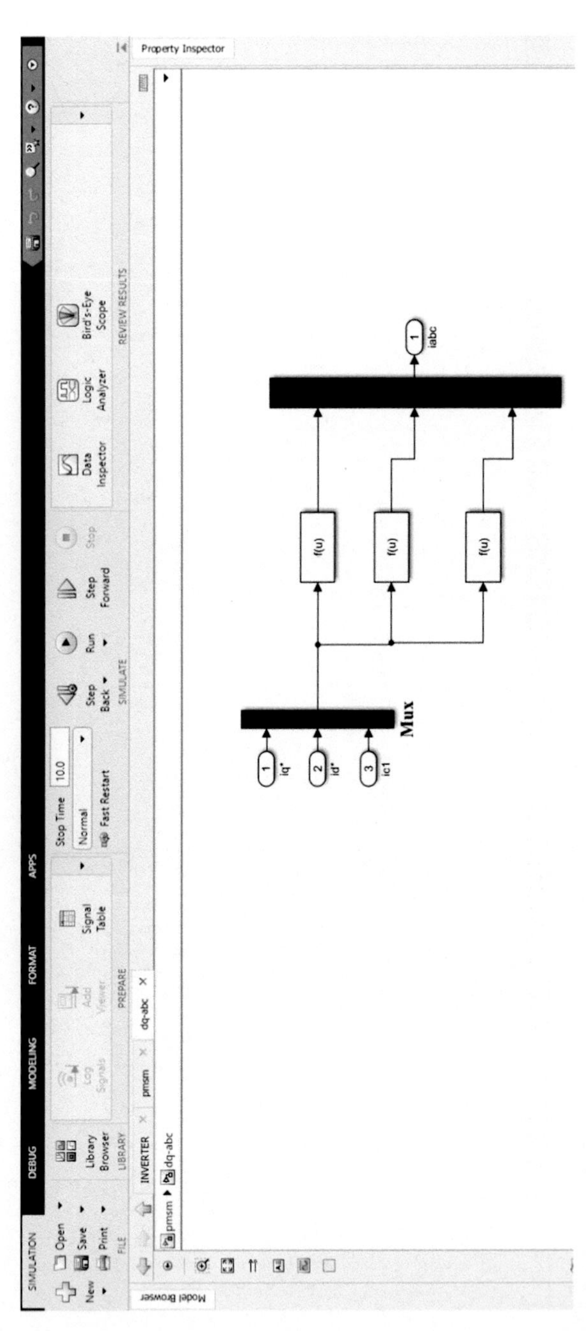

Figure 7.10 Subsystem of dq-abc.

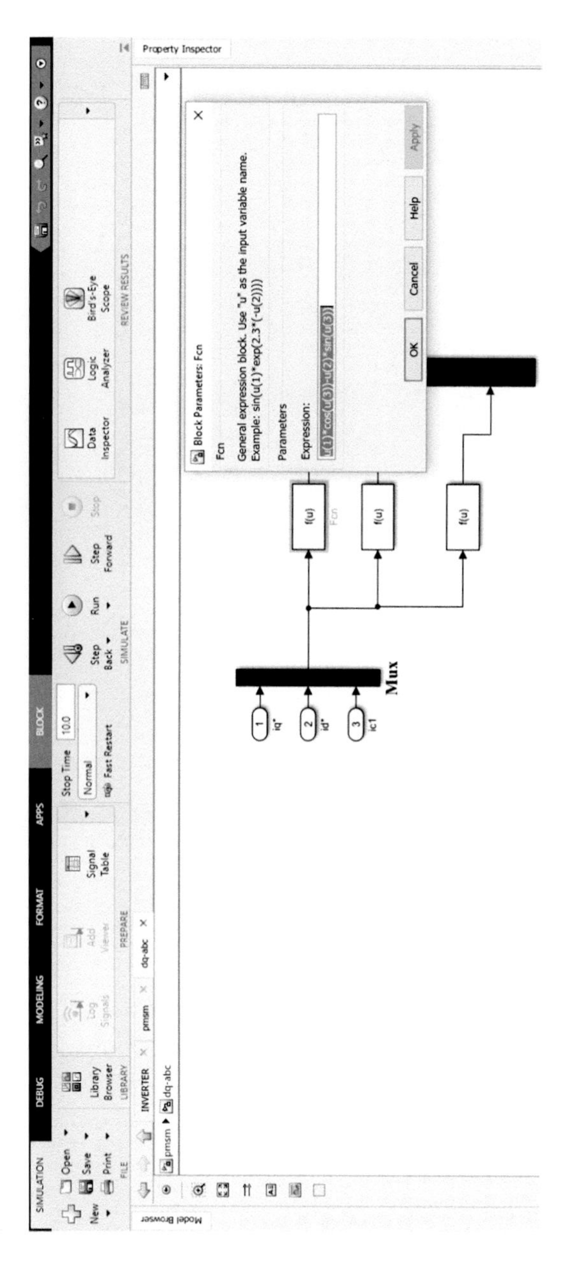

Figure 7.11 Block parameter—Fcn1.

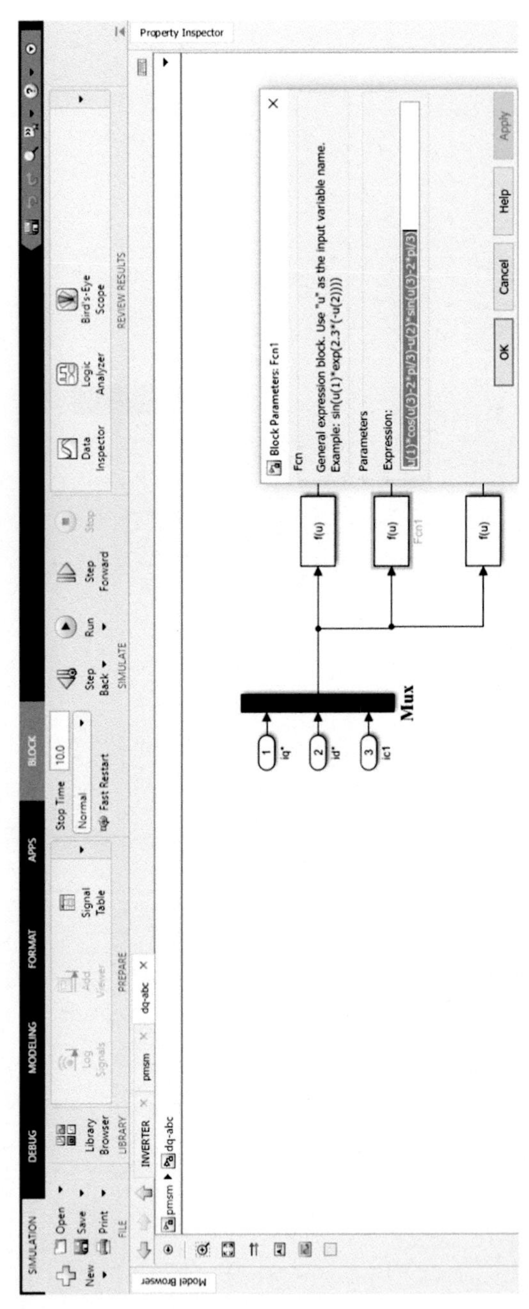

Figure 7.12 Block parameter—Fcn2.

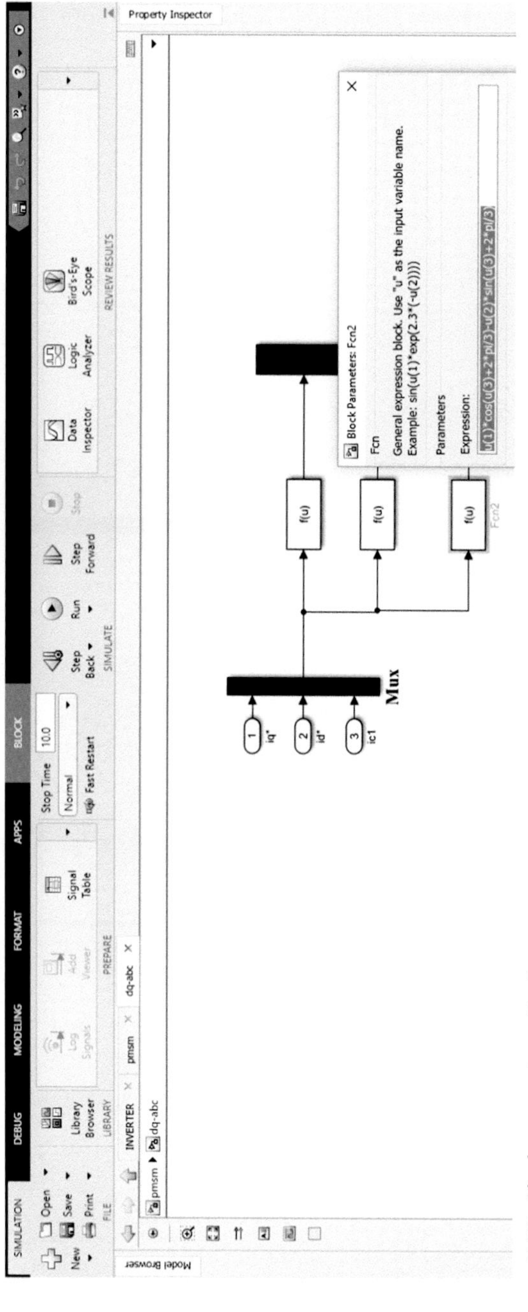

Figure 7.13 Block parameter—Fcn3.

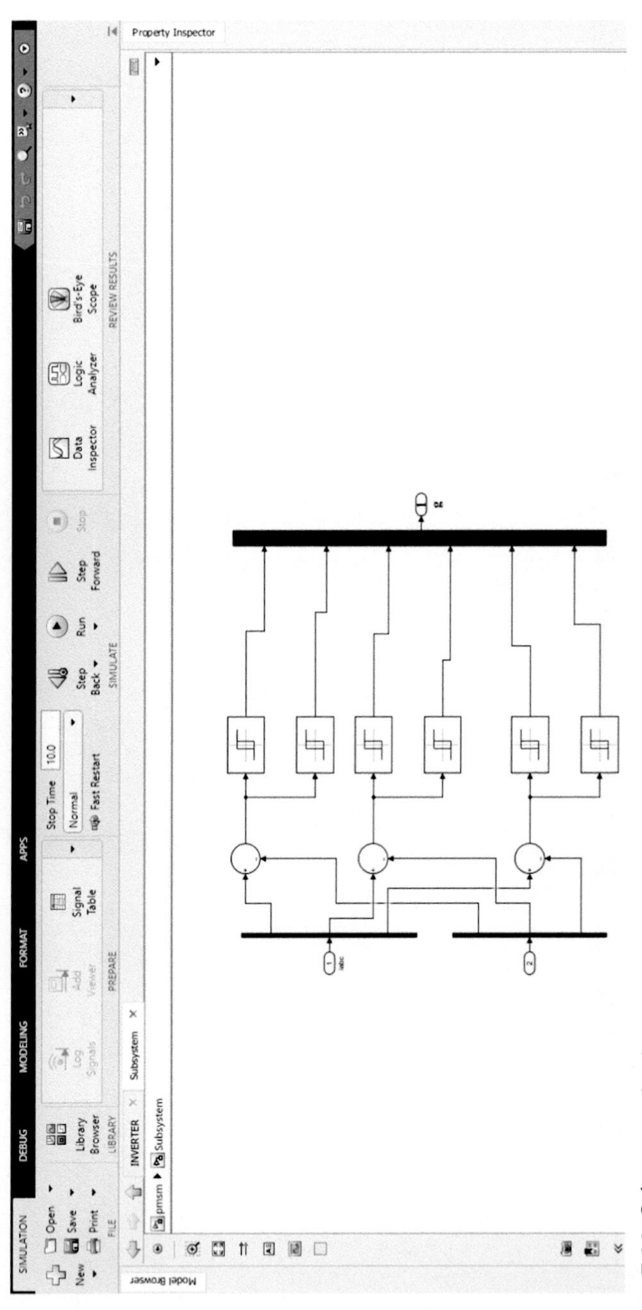

Figure 7.14 Subsystem circuit.

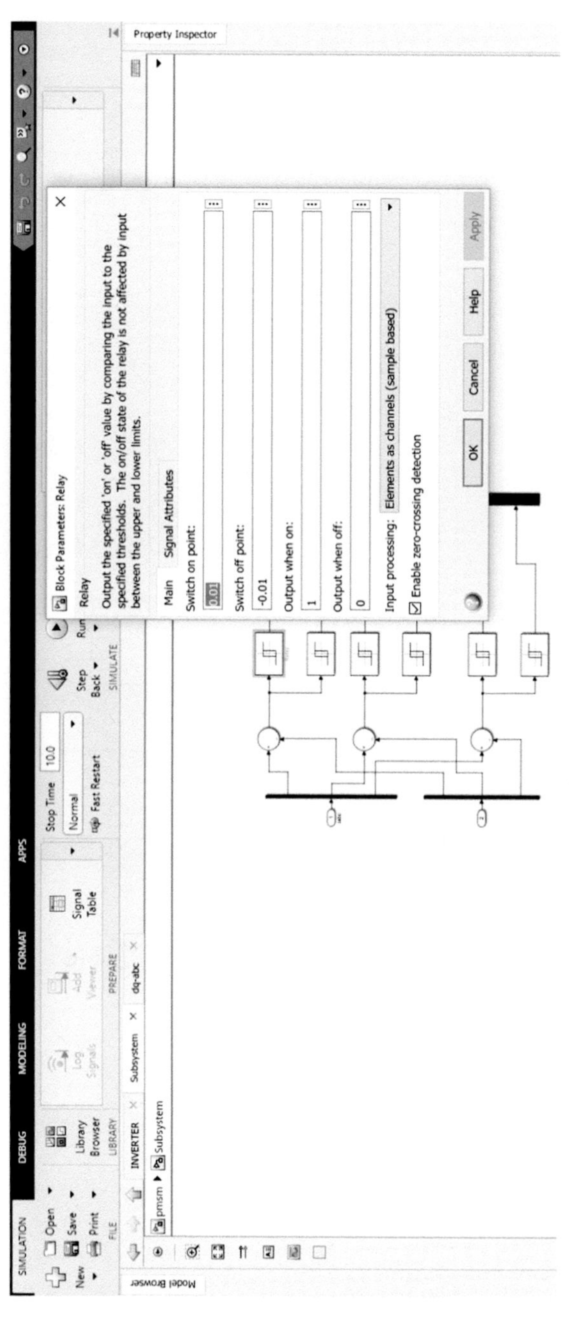

Figure 7.15 Block parameter—relay.

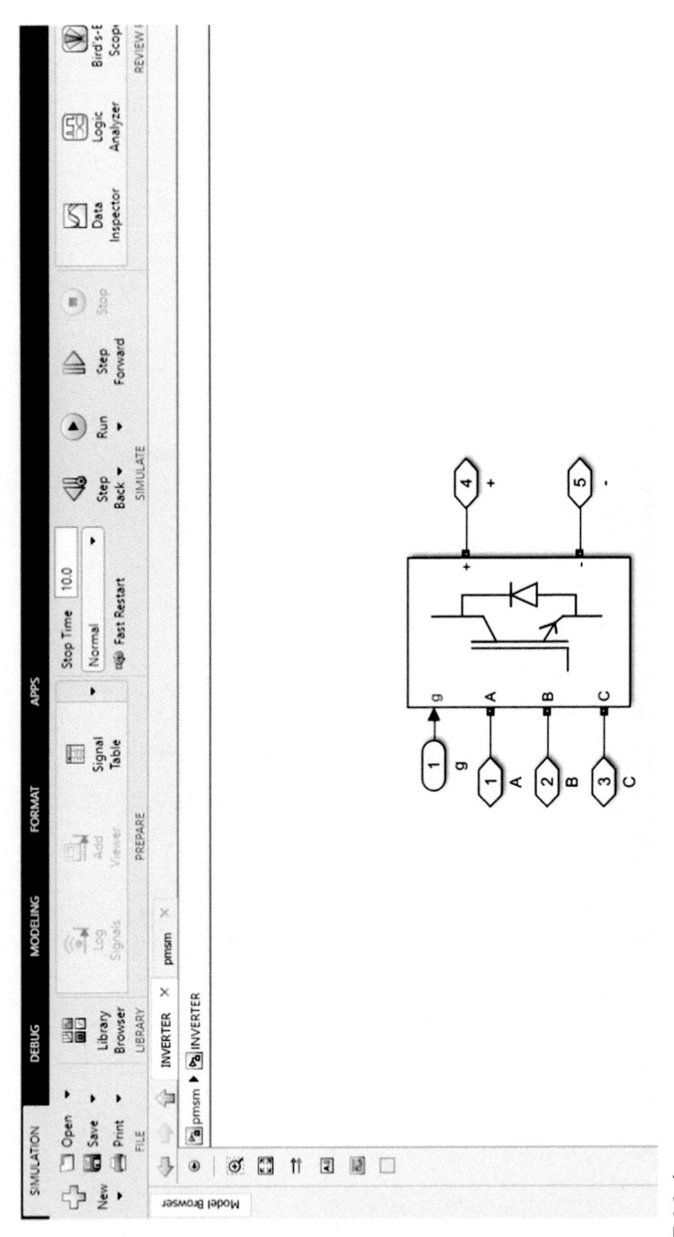

Figure 7.16 Inverter.

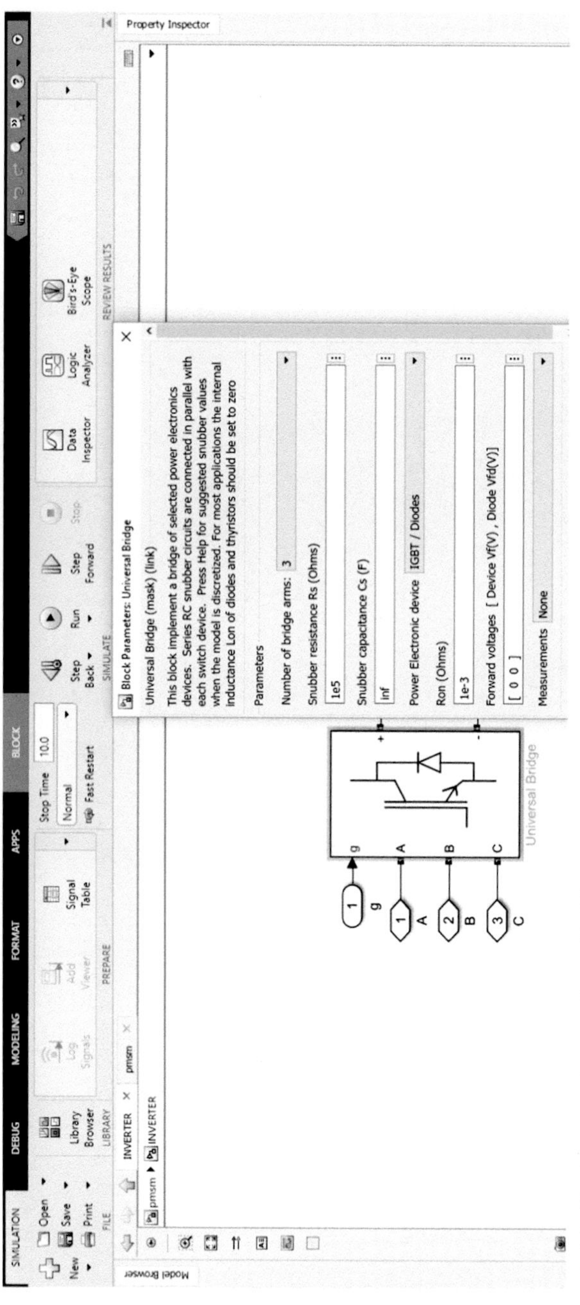

Figure 7.17 Block parameter—universal bridge.

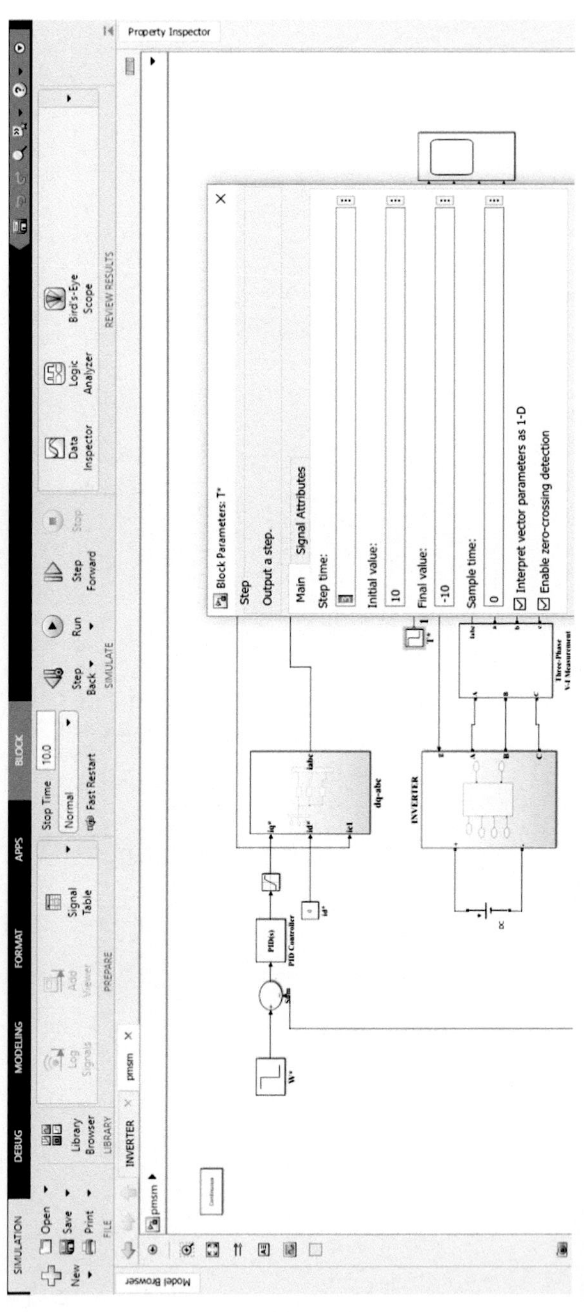

Figure 7.18 Block parameter—step T.

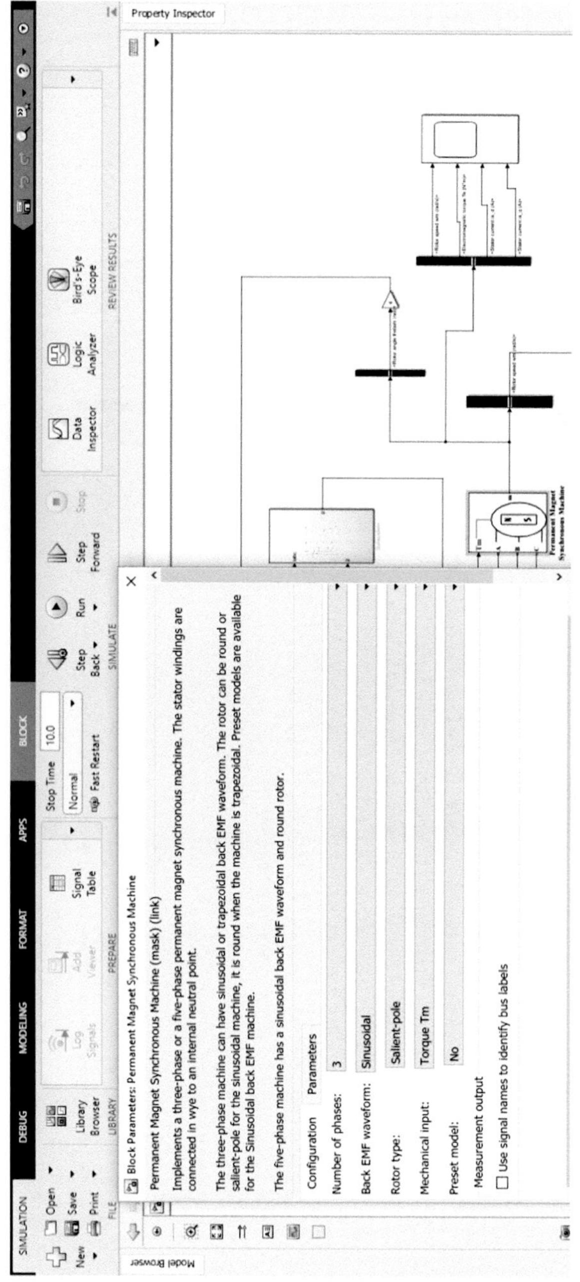

Figure 7.19 Block parameter—PMSM—configuration. *PMSM*, permanent magnet synchronous motor.

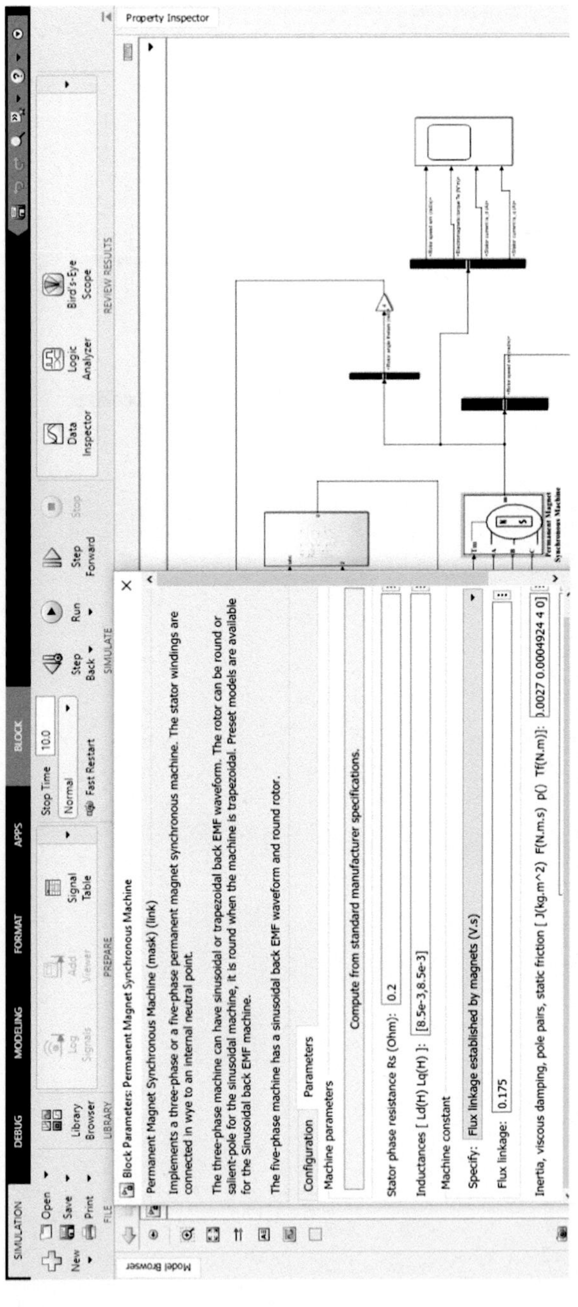

Figure 7.20 Block parameter—PMSM. *PMSM*, permanent magnet synchronous motor.

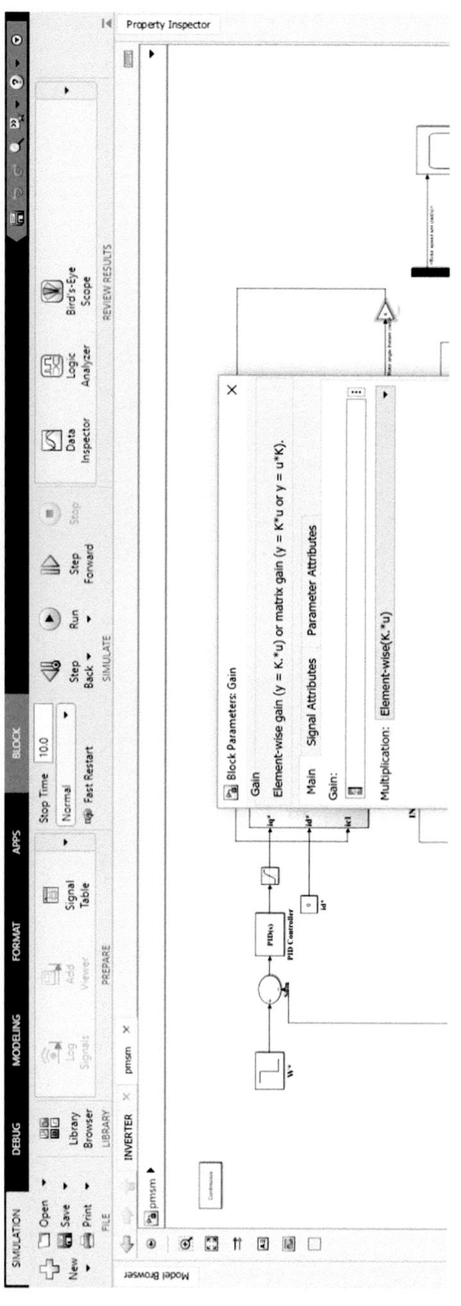

Figure 7.21 Block parameter—Gain.

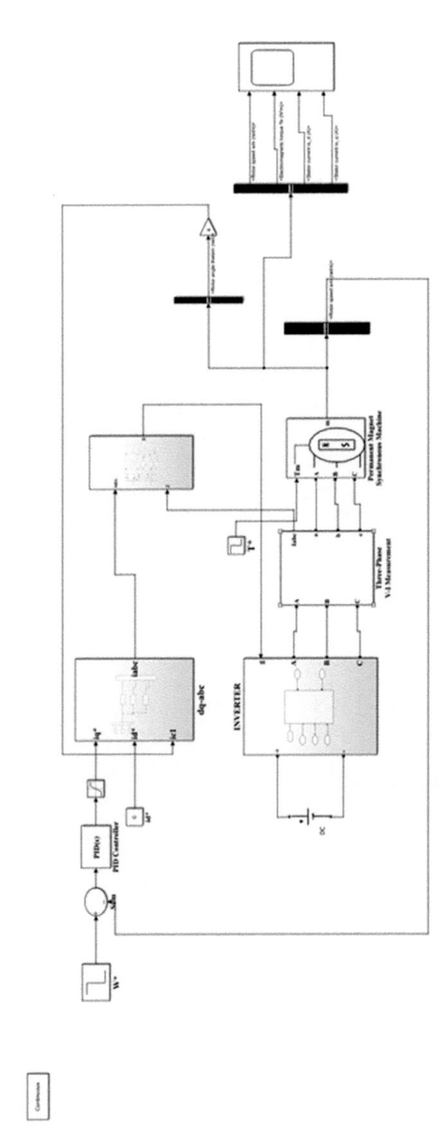

Figure 7.22 MATLAB/Simulink model of vector control PMSM. *PMSM*, permanent magnet synchronous motor.

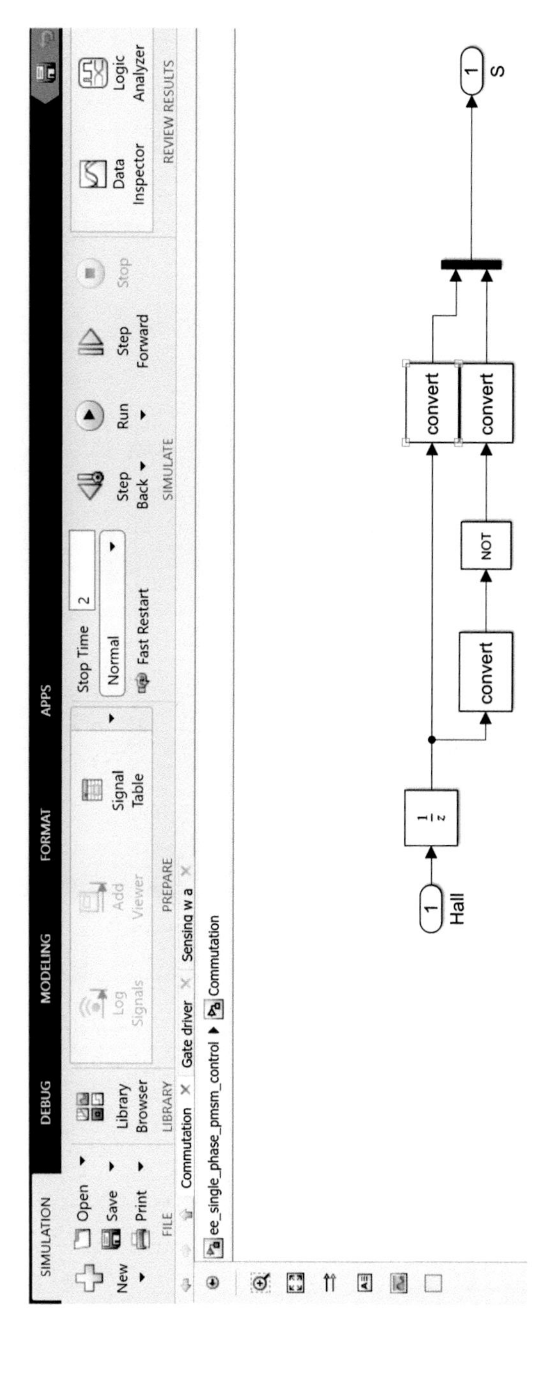

Figure 7.23 Subsystem of commutation.

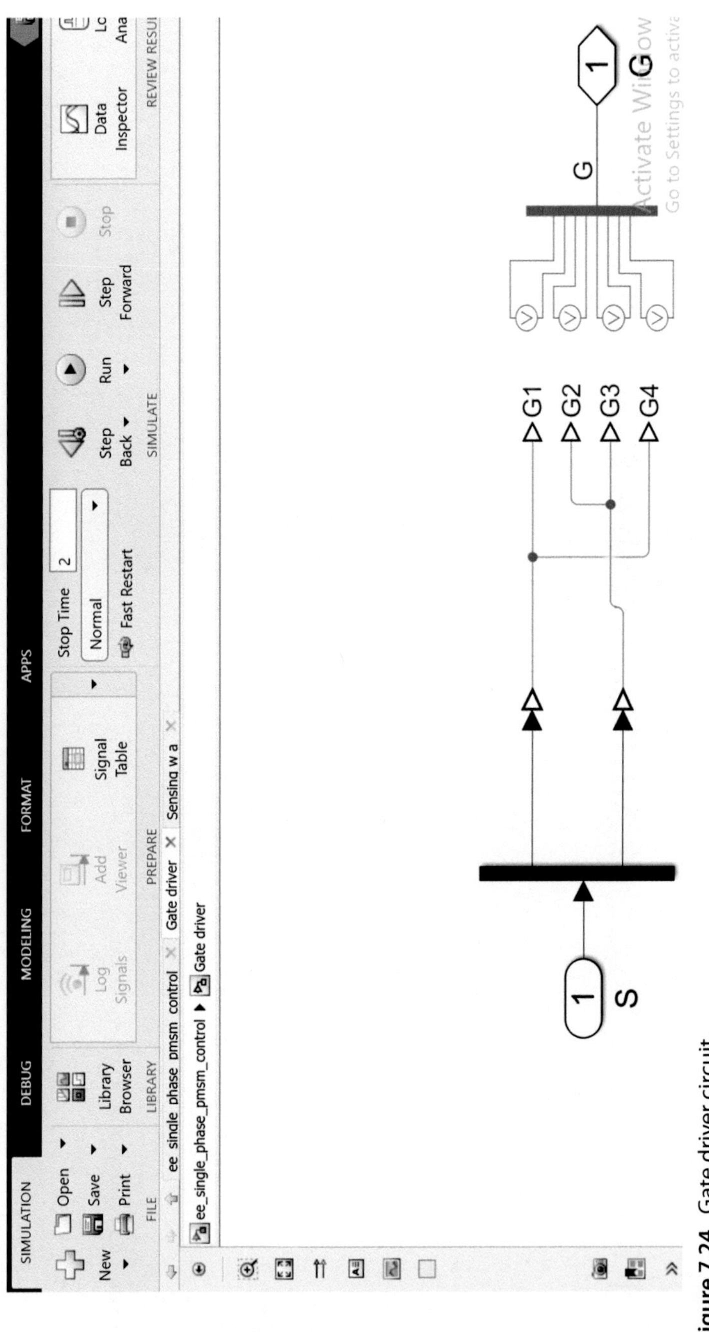

Figure 7.24 Gate driver circuit.

The parameters of the four-quadrant chopper are taken as per Fig. 7.25.

Fig. 7.26 Shows the circuit which senses the current.

In Fig. 7.27, motion sensing circuit is given. In Fig. 7.28, torque-sensing circuit is given.

Inertia value is been set as 29×10^{-7} kgm^2 which is mentioned in Fig. 7.29.

The solver configuration parameters are need to be selected as per Fig. 7.30.

Once the selected components are connected and configured we will get the simulation circuit as per Fig. 7.31. After giving run command the simulated results are visible in the scope as like Figs. 7.32 and 7.33.

7.4 Modeling and simulation of three-phase PMSM

The solve configurations are set as per Fig. 7.34.

Select the rate transition and configure it like above in Fig. 7.35.

The subsystem of PMSM controller is given in the above Fig. 7.36.

The parameters of PMSM FOC need to be set as per Fig. 7.37.

Here three-phase inverter circuit is used (Fig. 7.38).

The circuit which is used to sense the current is given in above Fig. 7.39.

The permanent magnet synchronous machine parameters need to be like Fig. 7.40.

The encoder circuit is given in Fig. 7.41.

At last, will get the simulation diagram of three-phase PMSM (Fig. 7.42). After pressing the run button the simulated results will be like Fig. 7.43.

7.5 PMSM motor control with speed feedback using PSIM

7.5.1 Principle of vector control

The basic principle behind vector control is to use coordinate transformation theory to turn two AC components orthogonal in time phase into two DC components orthogonal in space, effectively dividing the stator current of an AC motor into excitation and torque components. Two distinct DC values lower the degree of coupling of the control system's order, allowing for motor flux and torque regulation. A control technique based on rotor flux orientation (id=0) is employed in vector control to make the stator current vector positioned on the q-axis without the d-axis component, that is, the stator current is used entirely to generate torque.

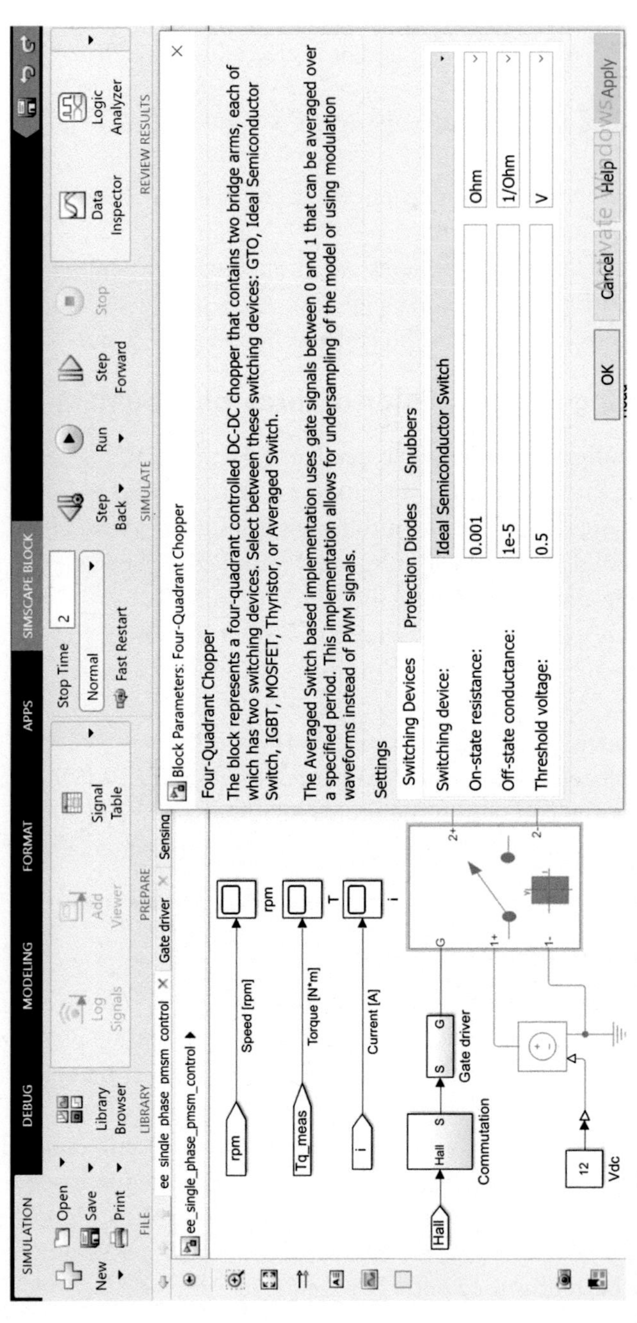

Figure 7.25 Block parameters—four-quadrant chopper.

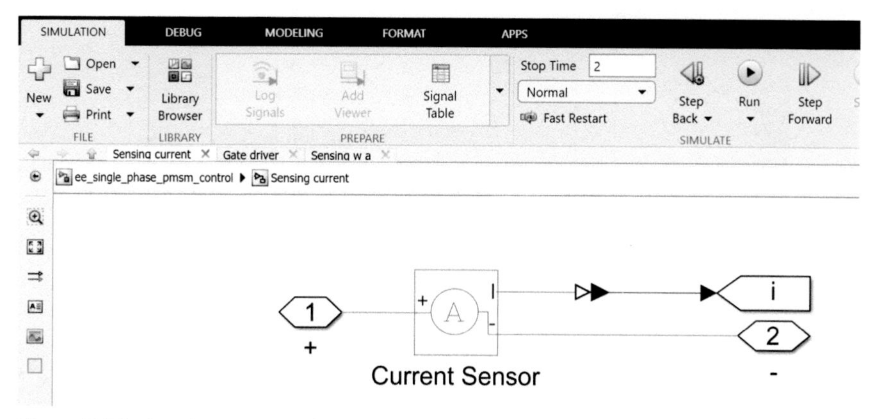

Figure 7.26 Sensing current circuit.

The rise of the PM motor drive business has necessitated the development of simulation tools that can handle motor drive simulations. By decreasing cost and time, simulations have aided the development of innovative technologies, including motor drives. Simulation tools may perform dynamic simulations of motor drives in a visual environment, which can help with the development of novel systems. PSIM is a simulation tool for power electronics and motor control applications. PSIM is a strong simulation environment for power converter analysis, control loop design, and motor drive system studies, thanks to its quick simulation, friendly user interface, and waveform processing. Through Simcoupler blocks, PSIM provides links to third-party applications like as JMAG and MATLAB. Also, through the dynamic link library (DLL) block, we can use C code in either the power circuit or the control circuit. The external DLL blocks enable users to create their own C code, compile it as a DLL using Microsoft C/C++ or Borland C++, then link it with PSIM.

Fig. 7.44 depicts the simulation method.

When the PSIM is opened, will get the above screen as a welcome screen (Fig. 7.45). We can open the new schematic to start.

When new schematic is like the above in Fig. 7.46, there we can start our simulation diagram (Figs. 7.47 and 7.48).

7.6 PMSM motor control with speed feedback using the absolute encoder using PSIM

The absolute encoder determines the exact location of the rotor with an accuracy that is proportional to the encoder's number of bits. It can revolve

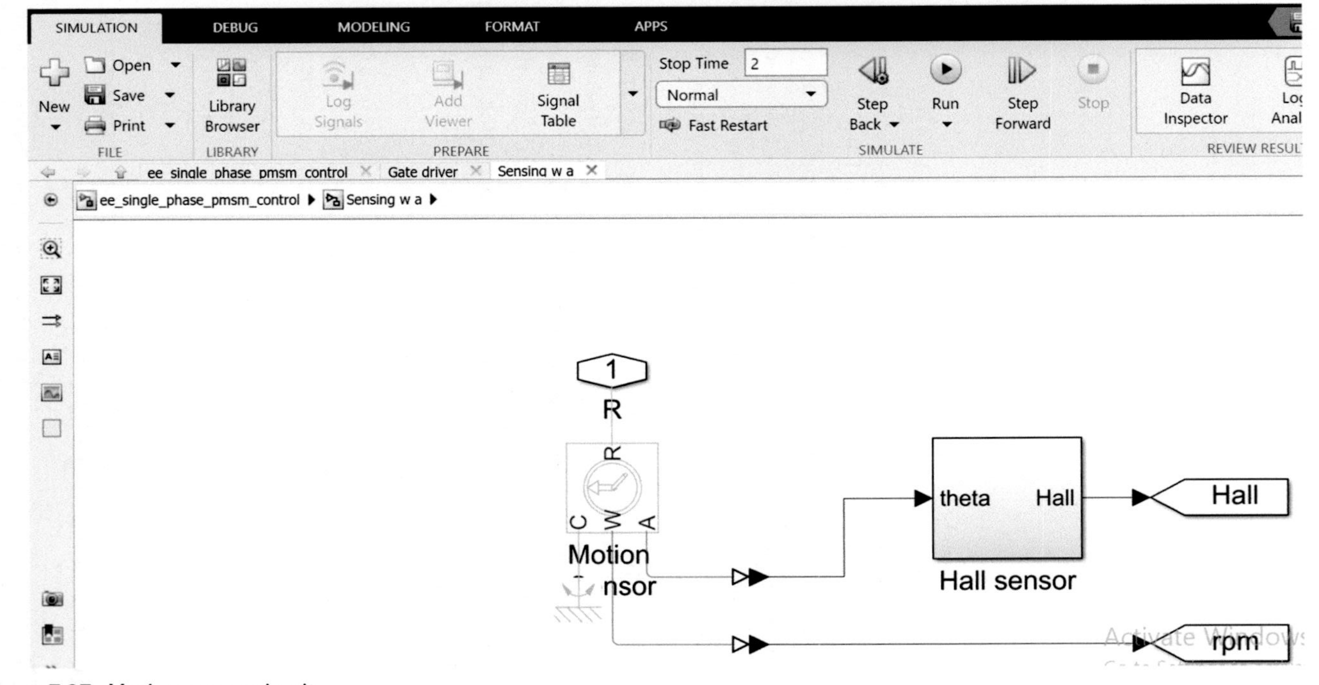

Figure 7.27 Motion sensor circuit.

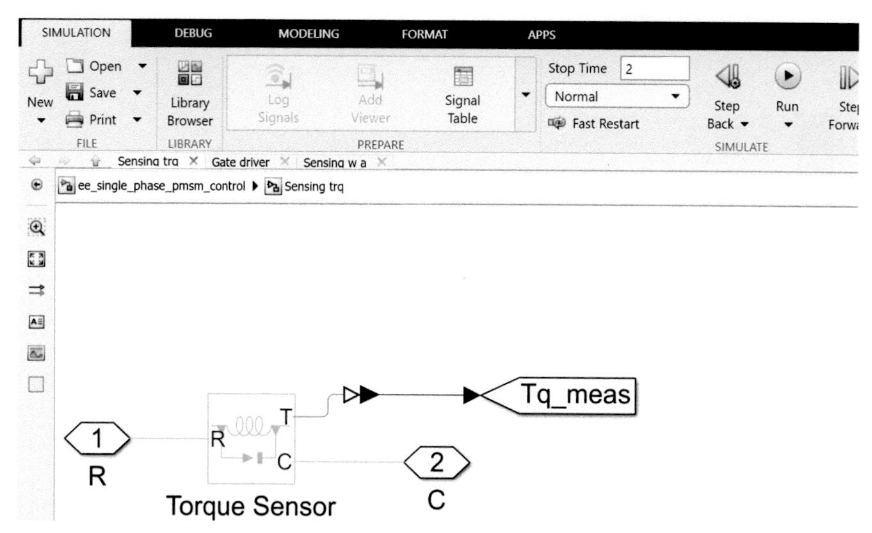

Figure 7.28 Torque sensor circuit.

forever, and the position can be measured or obtained even if the motor stops [5].

Fig. 7.49 shows an absolute encoder that captures the exact location of the rotor with an accuracy proportional to the encoder's number of bits. It can revolve forever, and the position can be measured or obtained even if the motor stops. It generates a "full word" output, with each location represented by a different coding pattern. This code is formed from individual photo detectors' independent tracks on the encoder disc (one for each "bit" of resolution). Depending on the code disc pattern for that particular place, these detectors produce HI (light) or LO (dark) output.

Absolute encoders are utilized in applications such as flood gate control, telescopes, cranes, valves, and other devices that are dormant for long periods of time or move slowly. They are also advised for systems that need to keep track of their position in the event of a power loss (Fig. 7.50).

7.7 PMSM motor control with speed feedback using a resolver using PSIM

Rotating transformers are another name for position revolvers. The primary winding is situated on the rotor, similar to how a transformer works. The induced voltage at the transformer's two secondary windings is altered by

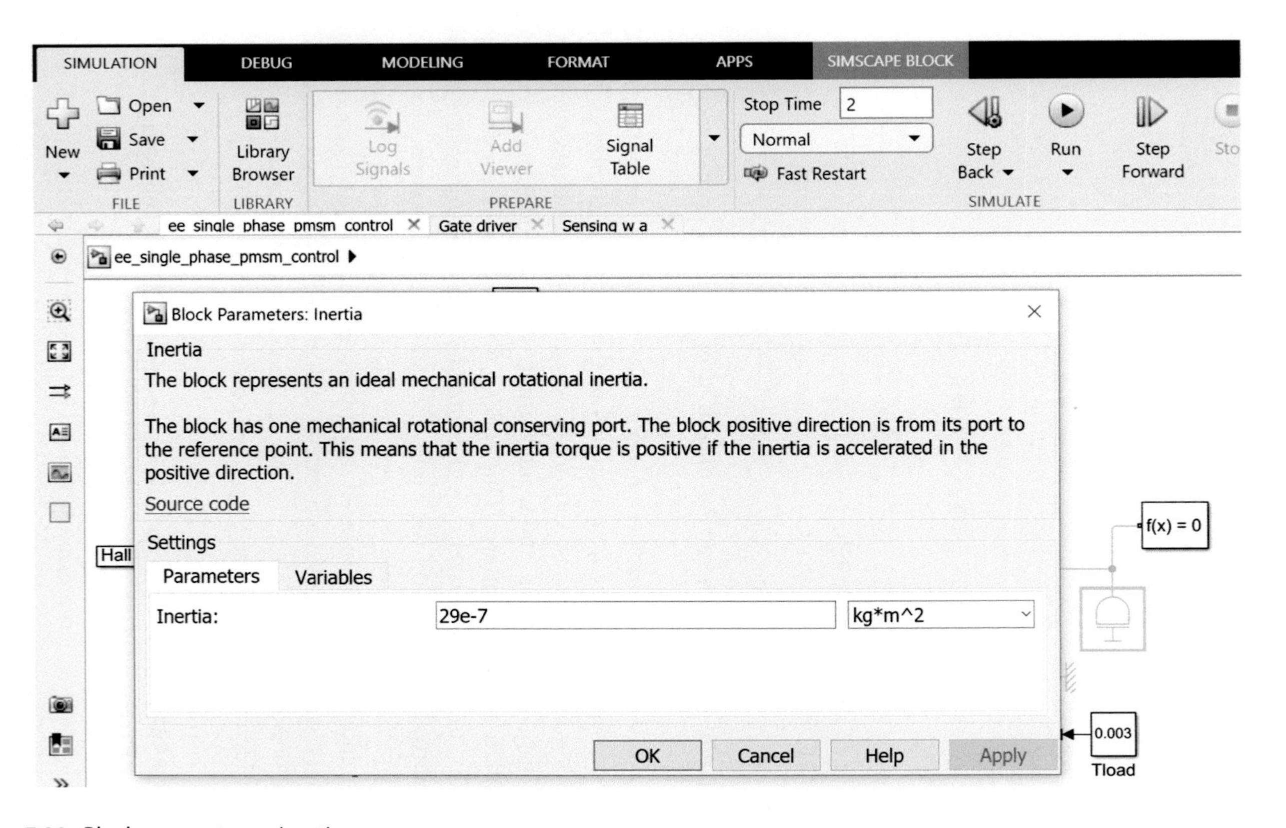

Figure 7.29 Block parameters—inertia.

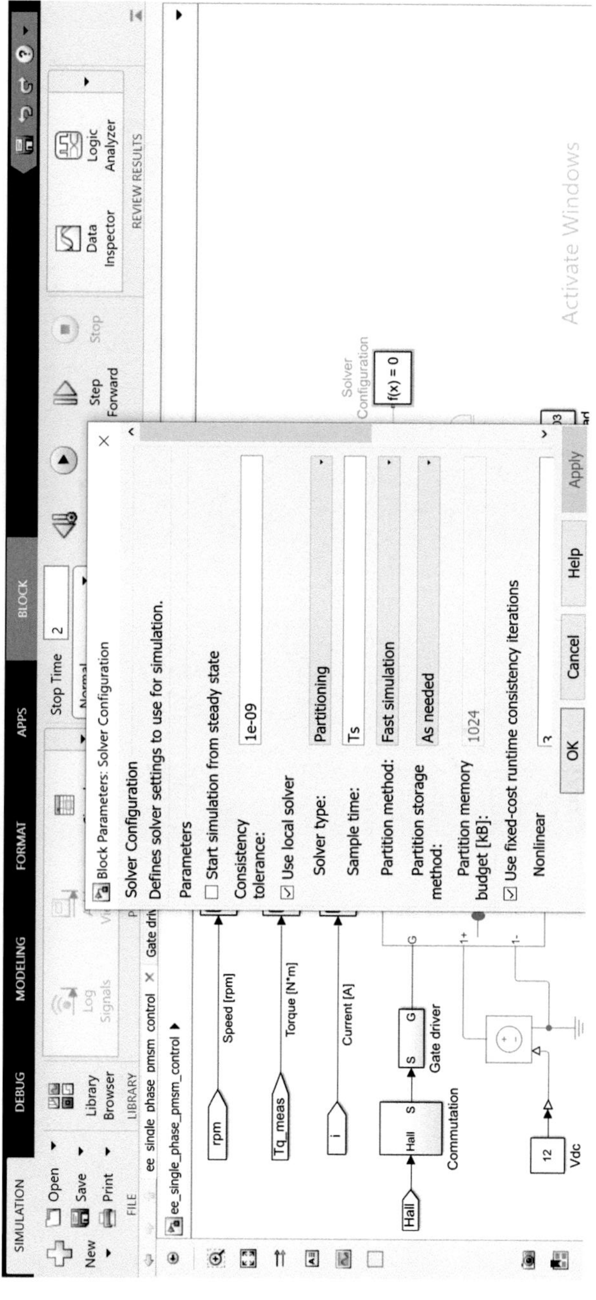

Figure 7.30 Block parameters—solver configuration.

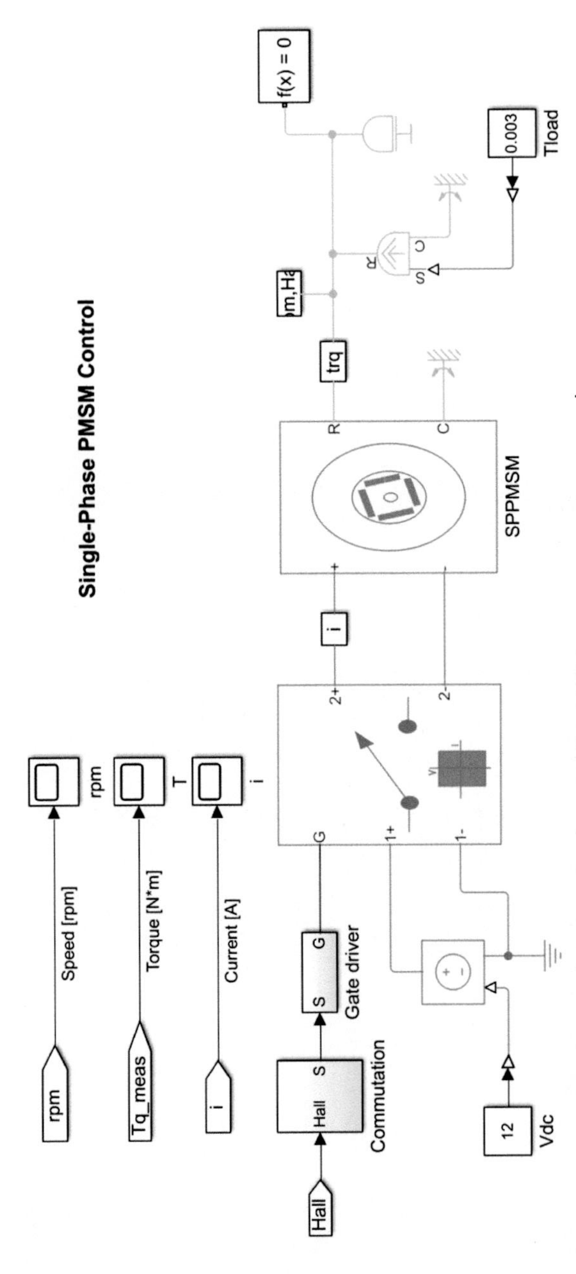

Figure 7.31 MATLAB/Simulink model of single phase PMSM. *PMSM*, permanent magnet synchronous motor.

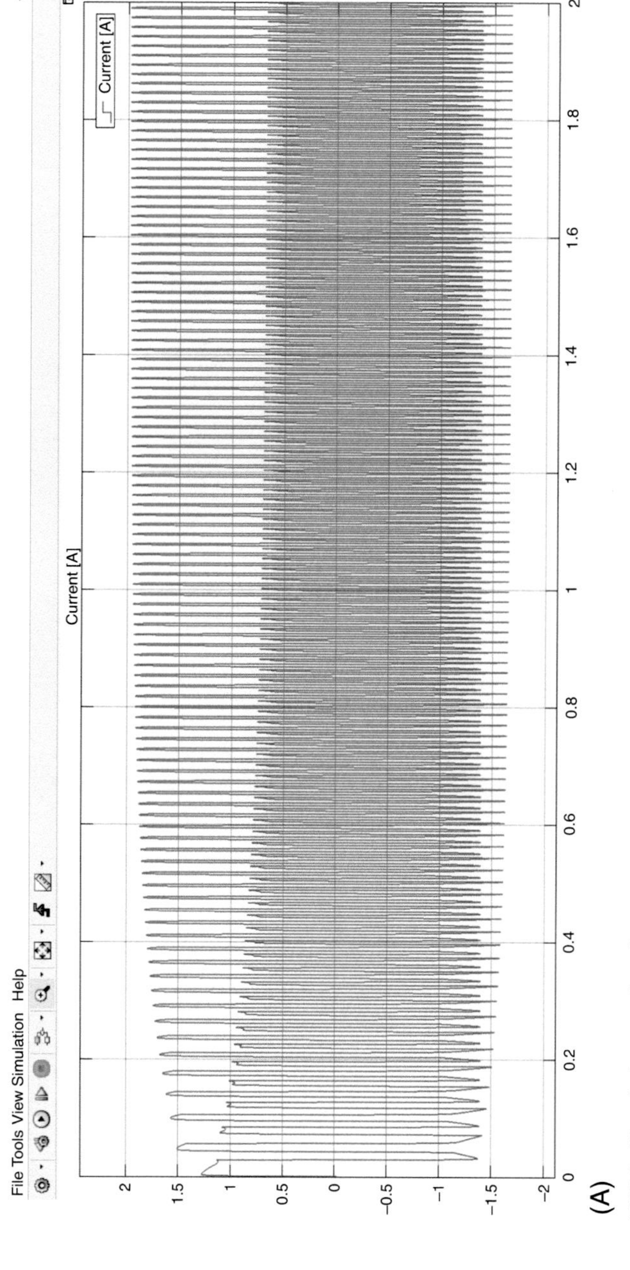

(A)

Figure 7.32A Simulation results of current.

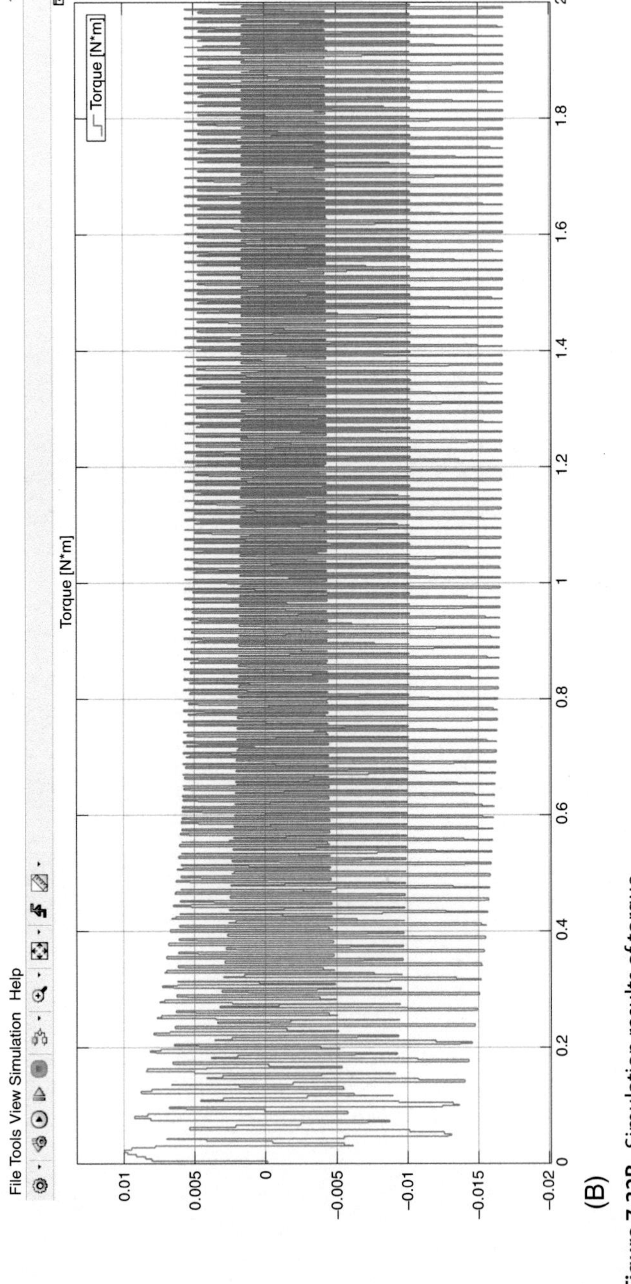

(B)

Figure 7.32B Simulation results of torque.

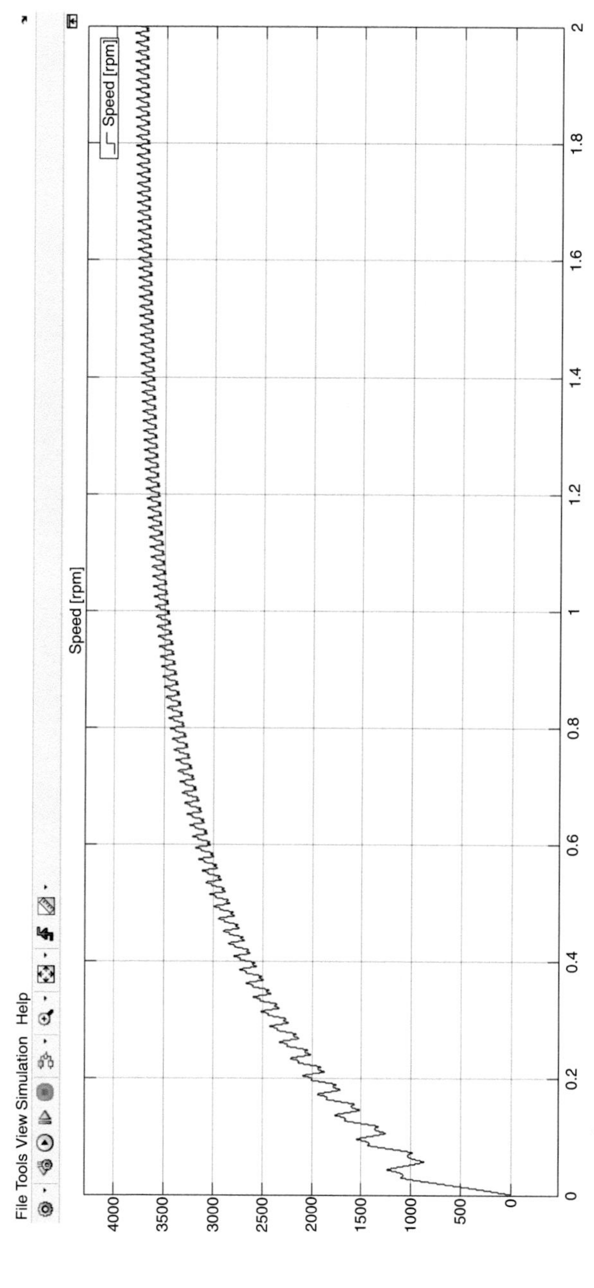

Figure 7.33 Simulation results of speed.

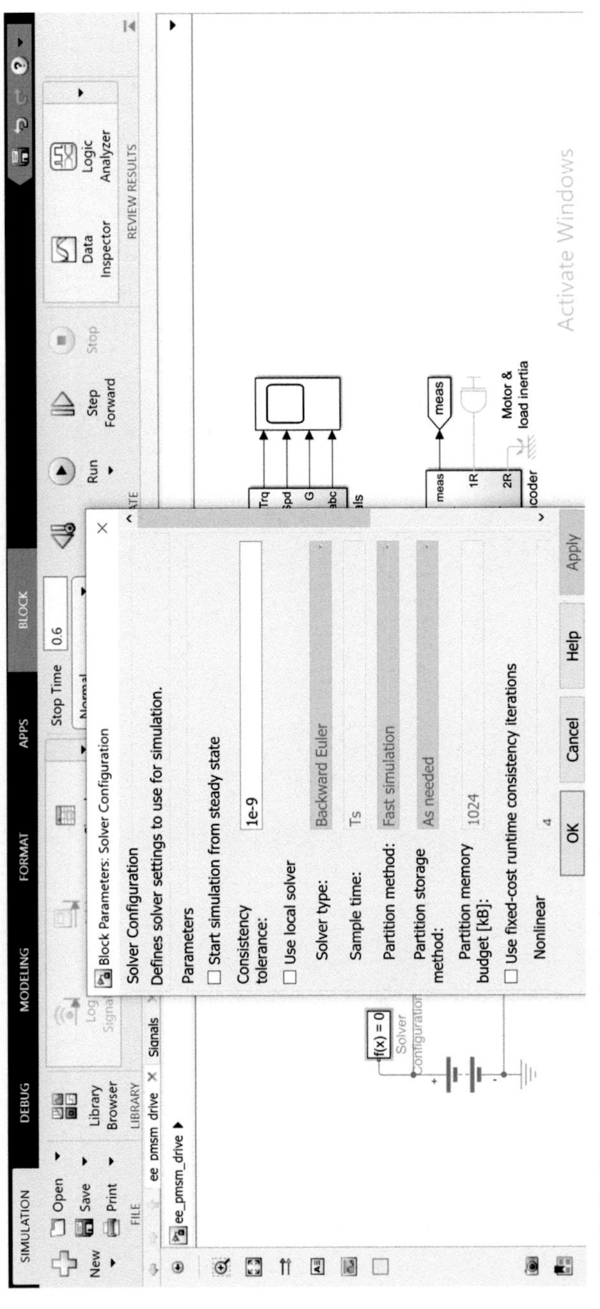

Figure 7.34 Block parameters—solve configuration.

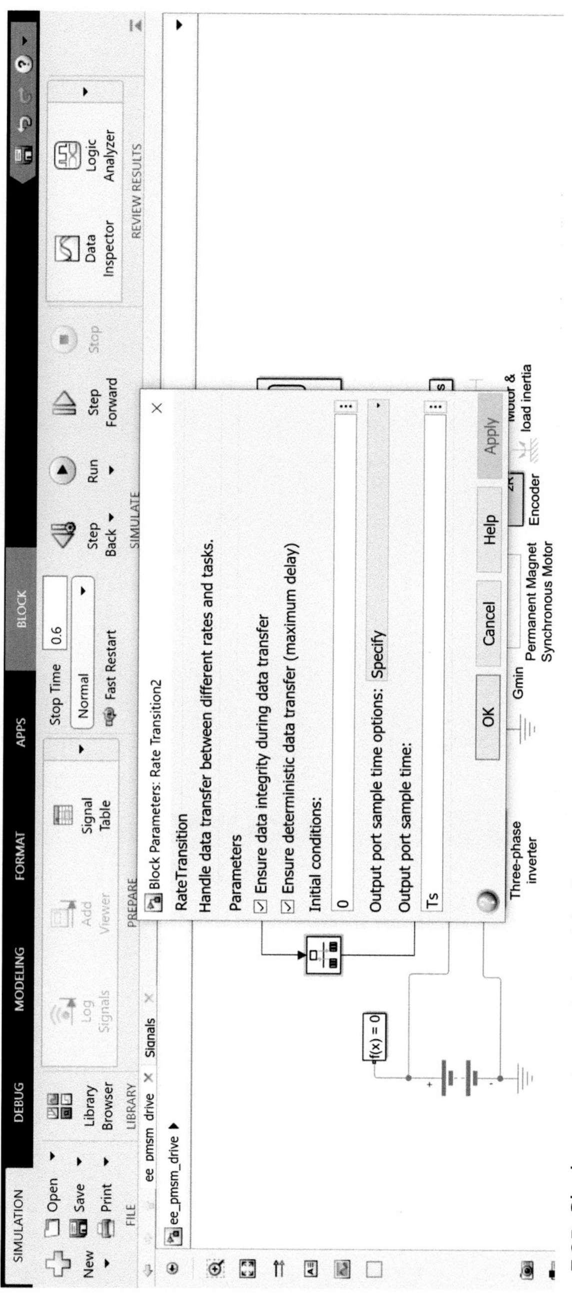

Figure 7.35 Block parameter—rate transition2.

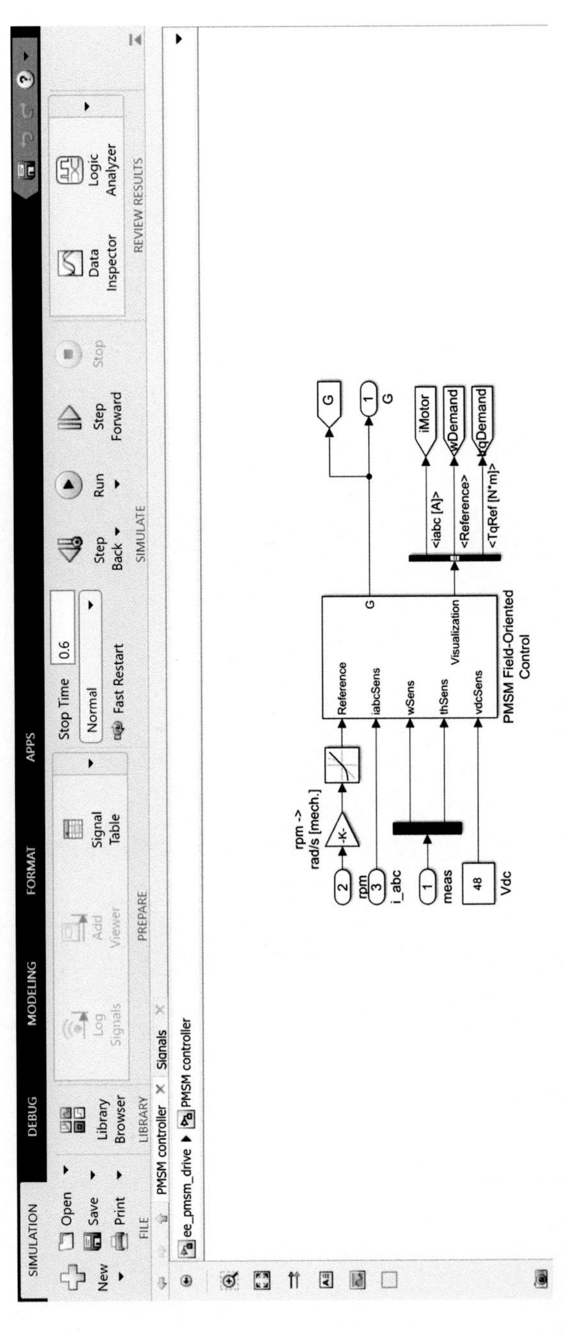

Figure 7.36 PMSM controller. *PMSM*, permanent magnet synchronous motor.

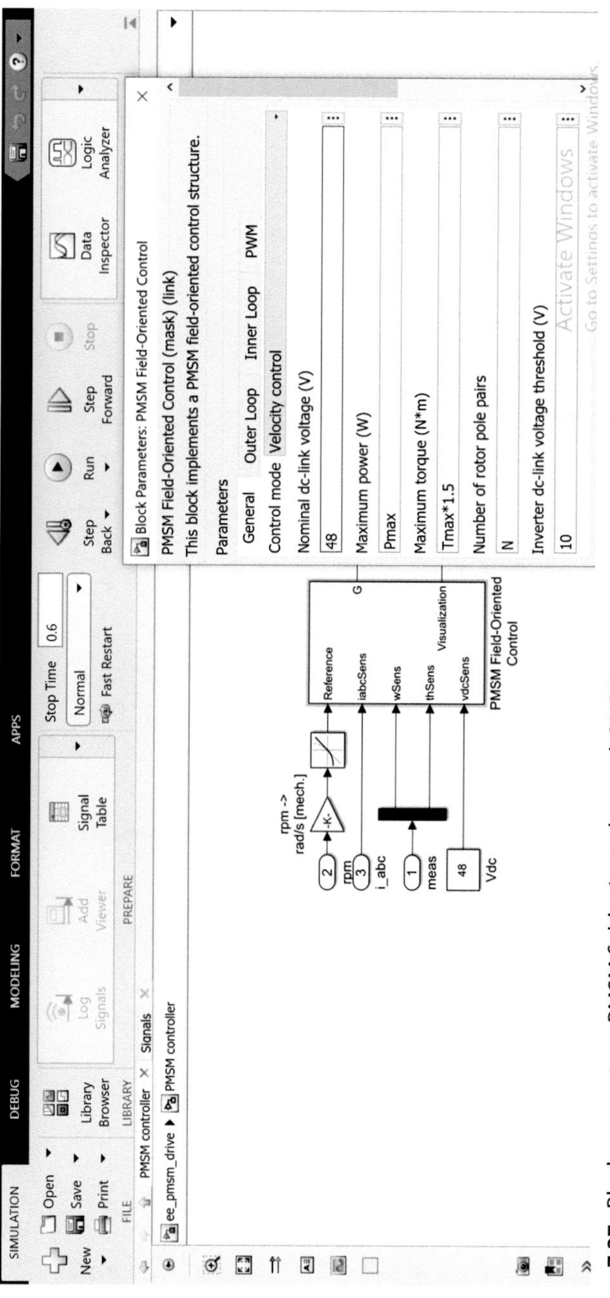

Figure 7.37 Block parameter—PMSM field oriented control. *PMSM*, permanent magnet synchronous motor.

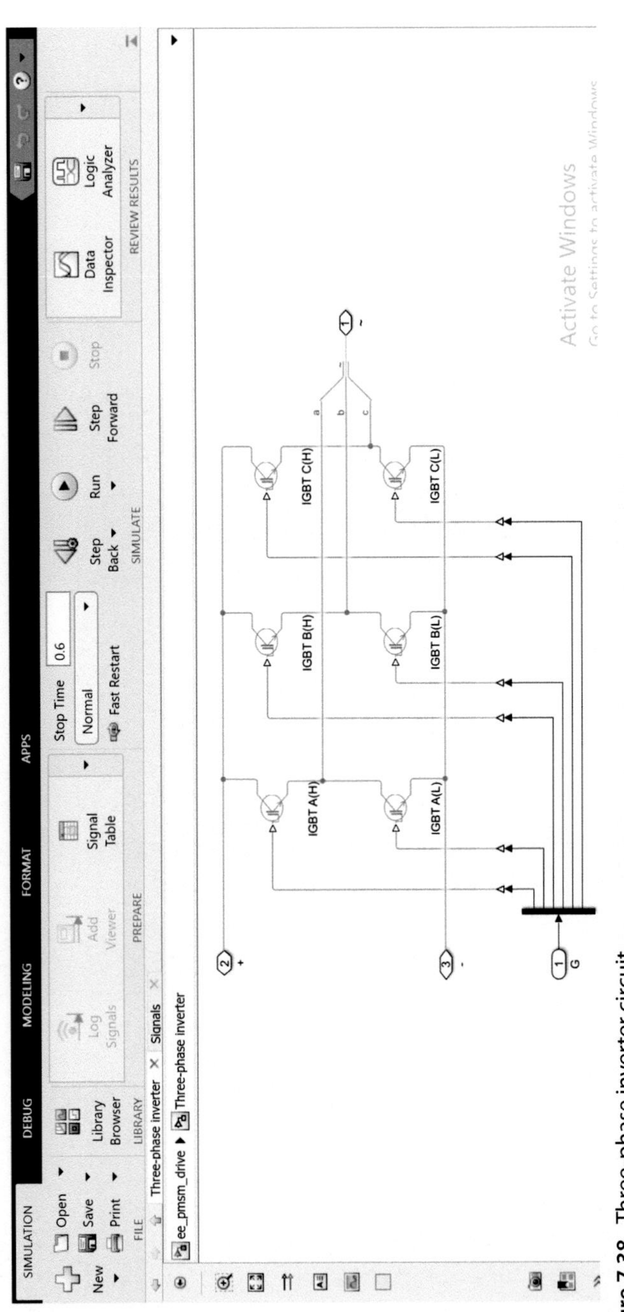

Figure 7.38 Three-phase inverter circuit.

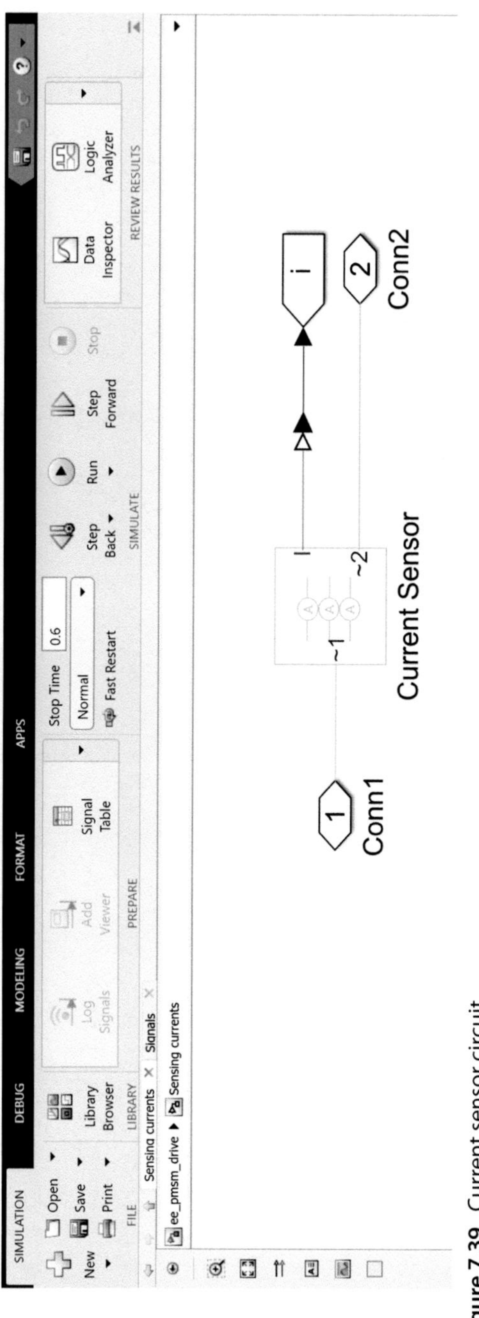

Figure 7.39 Current sensor circuit.

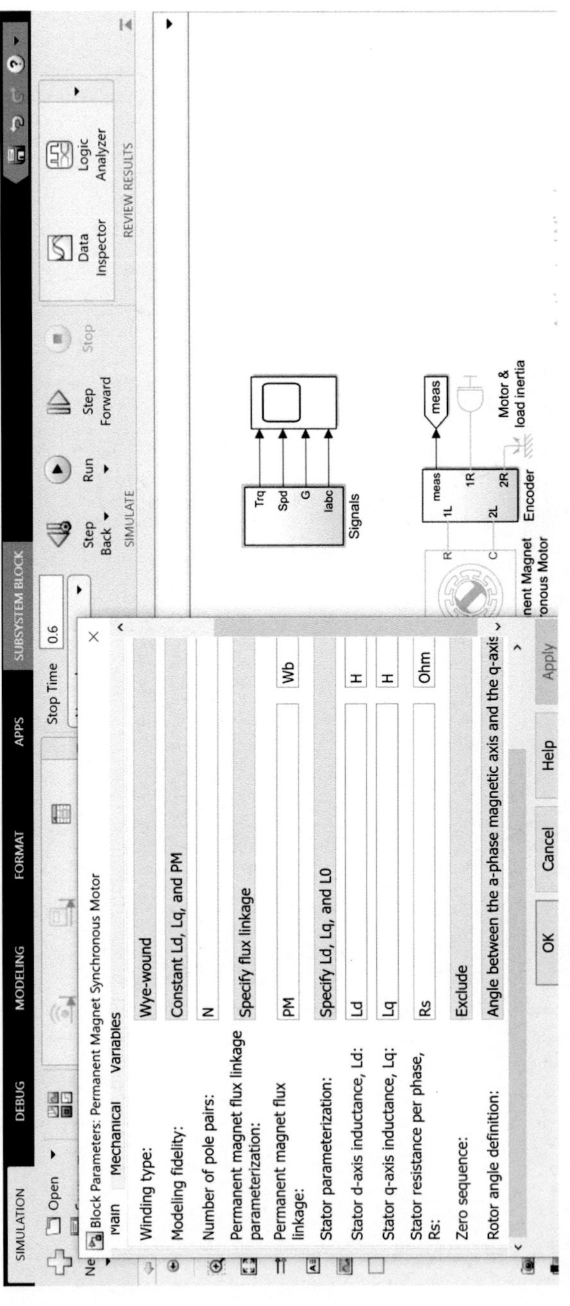

Figure 7.40 Block parameters—PMSM. *PMSM*, permanent magnet synchronous motor.

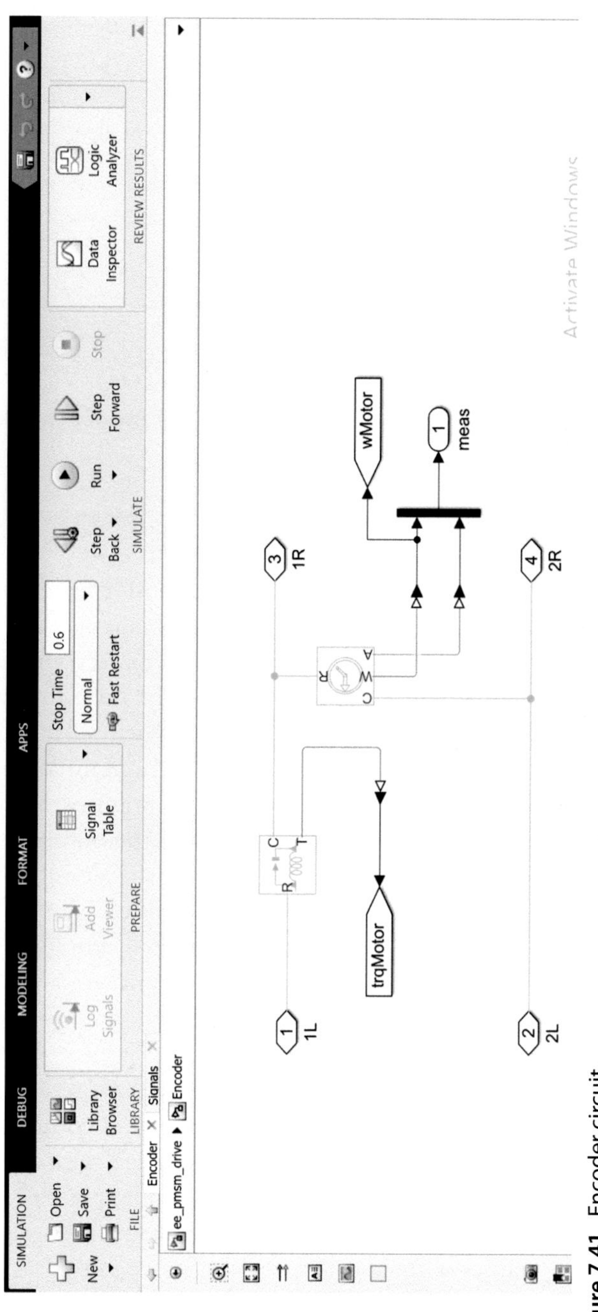

Figure 7.41 Encoder circuit.

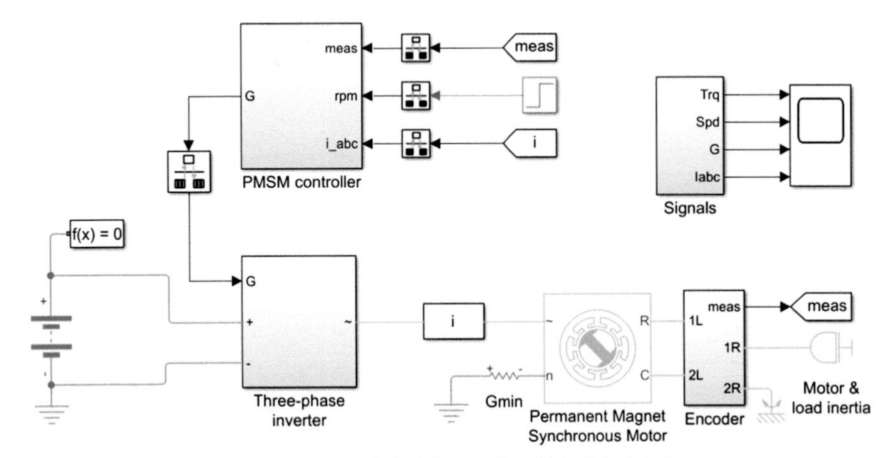

Figure 7.42 MATLAB/Simulink model of three-phase PMSM. *PMSM*, permanent magnet synchronous motor.

90 degrees depending on the rotor shaft angle. Two voltages can be used to calculate the position.

The position revolver, also known as rotating transformers, works on the transformer principle, as depicted in Fig. 7.51. The main winding is situated on the rotor, and the induced voltage at the two secondary windings of the transformer displaced by 90° varies depending on the rotor shaft angle. The two voltages can be used to calculate the position. The resolver is essentially a rotary transformer with two stator windings and one rotating reference winding (Vref). Because the reference winding is mounted to the rotor, it rotates in tandem with the shaft that passes through the output windings, as shown in Fig. 7.51. The sine and cosine voltages (Vsin, Vcos) are generated by two stator windings that are in quadrature (shifted by 90°) with one another. Both windings shall be referred to as output windings from now on. The relevant voltages are created by resolver output windings Vsin, Vcos as a result of the excitement given to the reference winding Vref and the angular movement of the motor shaft θ.

The generated voltages have the same frequency as the reference voltage, but their amplitudes change depending on the sine and cosine of the shaft angle θ. When one of the output windings is aligned with the reference winding, full voltage is generated on that output winding while zero voltage is generated on the other, and vice versa. These voltages in Fig. 7.52 can be used to calculate the rotor angle θ.

An inverse tangent function of the quotient of the sampled resolver output voltages Vsin, Vcos can be used to calculate the shaft angle (Fig. 7.53).

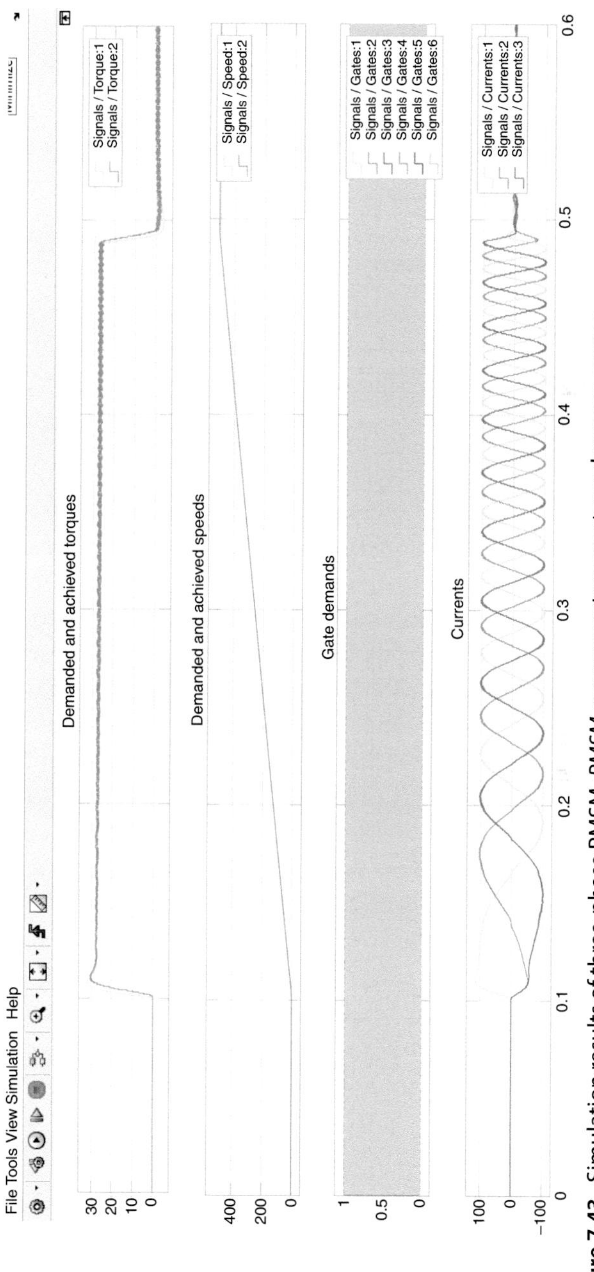

Figure 7.43 Simulation results of three-phase PMSM. *PMSM*, permanent magnet synchronous motor.

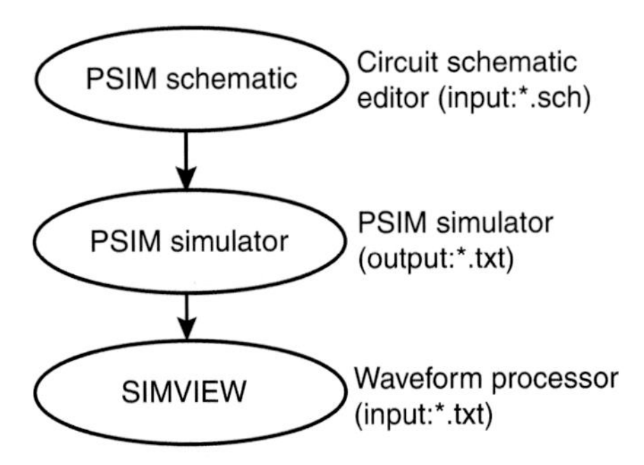

Figure 7.44 PSIM simulation process.

In terms of resolver output voltages, this determination can be represented as follows: $\theta = a\tan\left(\frac{U\ \sin}{U\ \cos}\right)$

7.7.1 Case studies

Simulation of vector control system of PMSM (Fig. 7.54).

A PMSM speed/current closed-loop simulation model was created. The simulation indicating that the simulation experiment platform and mathematical model are both effective. It also proves that the id=0 control algorithm and vector control are superior. The theory is presented for the research and design of the PMSM vector control system base.

Managing the d-axis component of the motor stator current is how the vector control system achieves its goal of controlling the motor. And the system functions smoothly, as predicted by the theoretical study, and has good dynamic performance and speed control characteristics comparable to those of a DC motor. The most significant advantage of the id = 0 control technique is that, like a DC motor, the motor torque is proportional to the amplitude of the stator current component. The research presented confirms the efficacy and excellence of MATLAB simulation, as well as providing a theoretical foundation for the real control of PMSMs. It's worth noting that, though many models are idealized, vector control of PMSMs in practice must nevertheless take into account a variety of parameters.

7.8 Summary

In this chapter, PMSM simulations using MATLAB/Simulink and PSIM are provided in step by step manner. The Simulink diagrams and results are clearly given here. It is easy for everyone to understand.

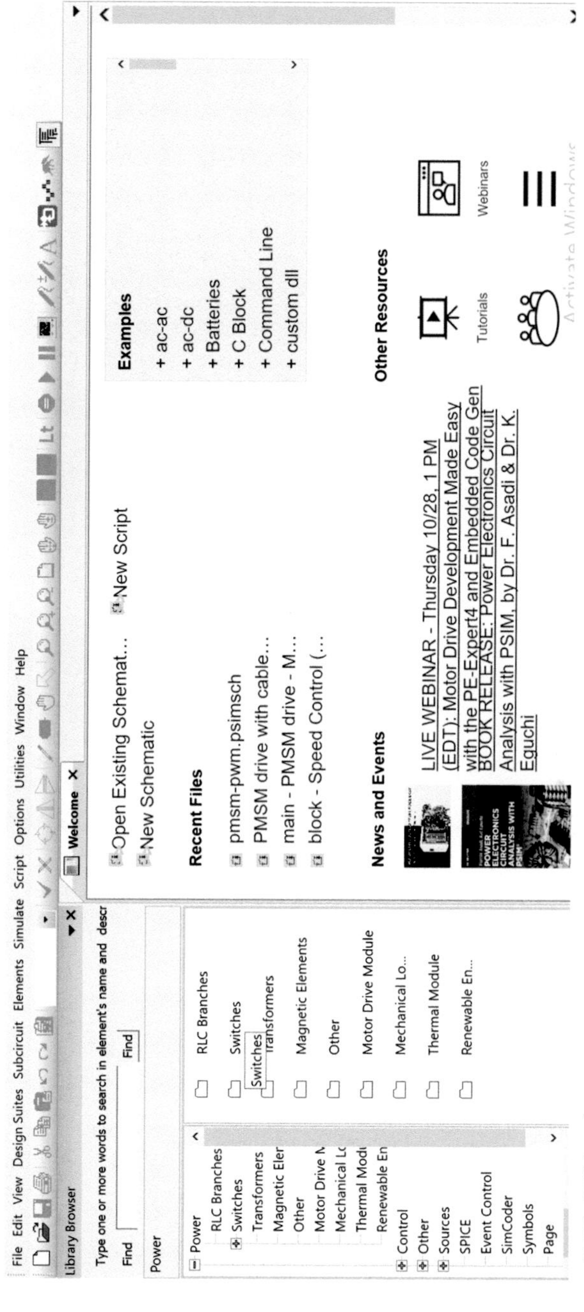

Figure 7.45 PSIM welcome screen.

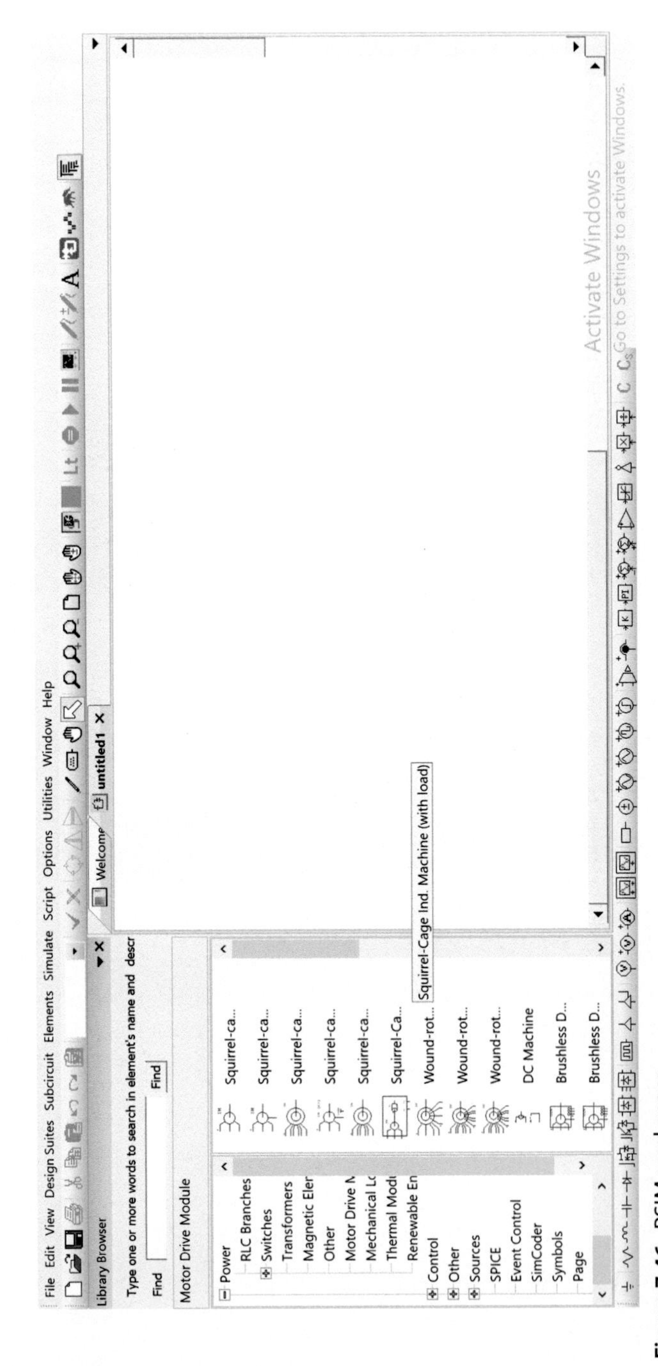

Figure 7.46 PSIM workspace.

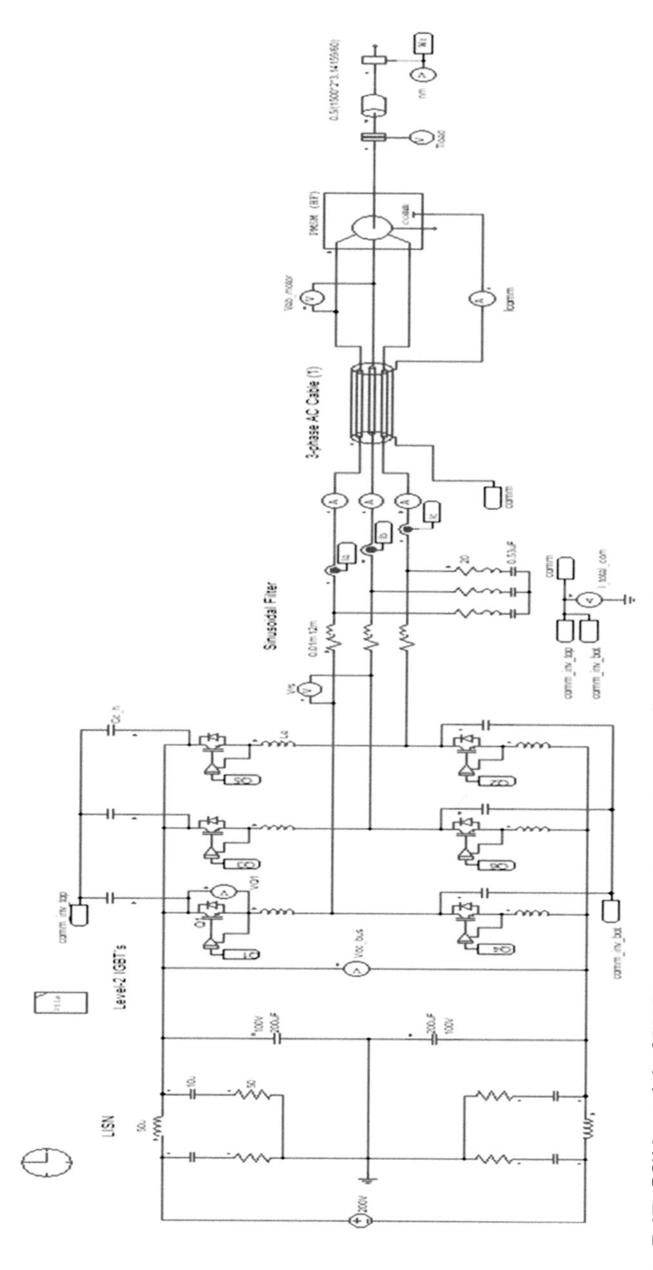

Figure 7.47 PSIM model of PMSM motor control with speed feedback. *PMSM*, permanent magnet synchronous motor.

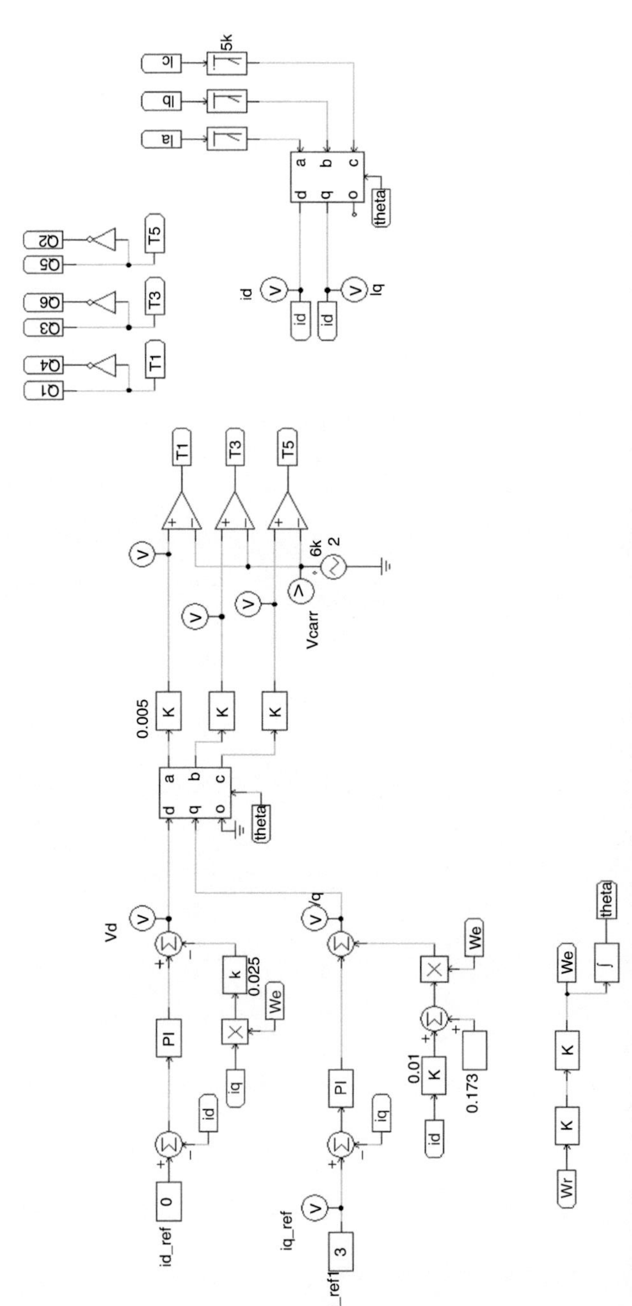

Figure 7.48 PSIM model of PMSM motor control with speed feedback subsystem. *PMSM*, permanent magnet synchronous motor.

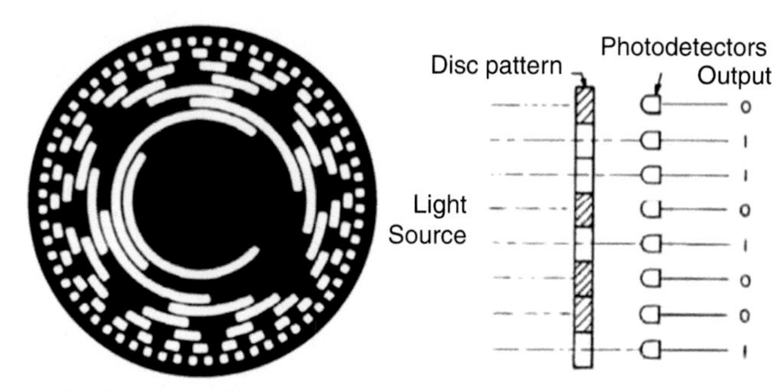

Figure 7.49 Absolute encoder.

Multiple Choice Questions

1. In which of the following permanent magnet synchronous motors, the permanent magnets are buried within the rotor?
   a) Projecting pole type permanent magnet
   b) Inset permanent magnet
   c) Interior permanent magnet
   d) Surface-mounted permanent magnet
   Answer: c) Interior permanent magnet
2. Which of the following motor would suit applications where constant speed is absolutely essential to ensure a consistent product?
   a) brushes DC motor
   b) disk motor
   c) permanent-magnet synchronous motor
   d) stepper motor
   Answer: c) permanent-magnet synchronous motor
3. The electrical displacement between the two stator windings of a resolver is
   a) 120°
   b) 90°
   c) 60°
   d) 45°
   Answer: b) 90°
4. PMSM stator construction is simillar to
   a) synchronous motor
   b) Induction motor
   c) Compound motor
   d) All of the above

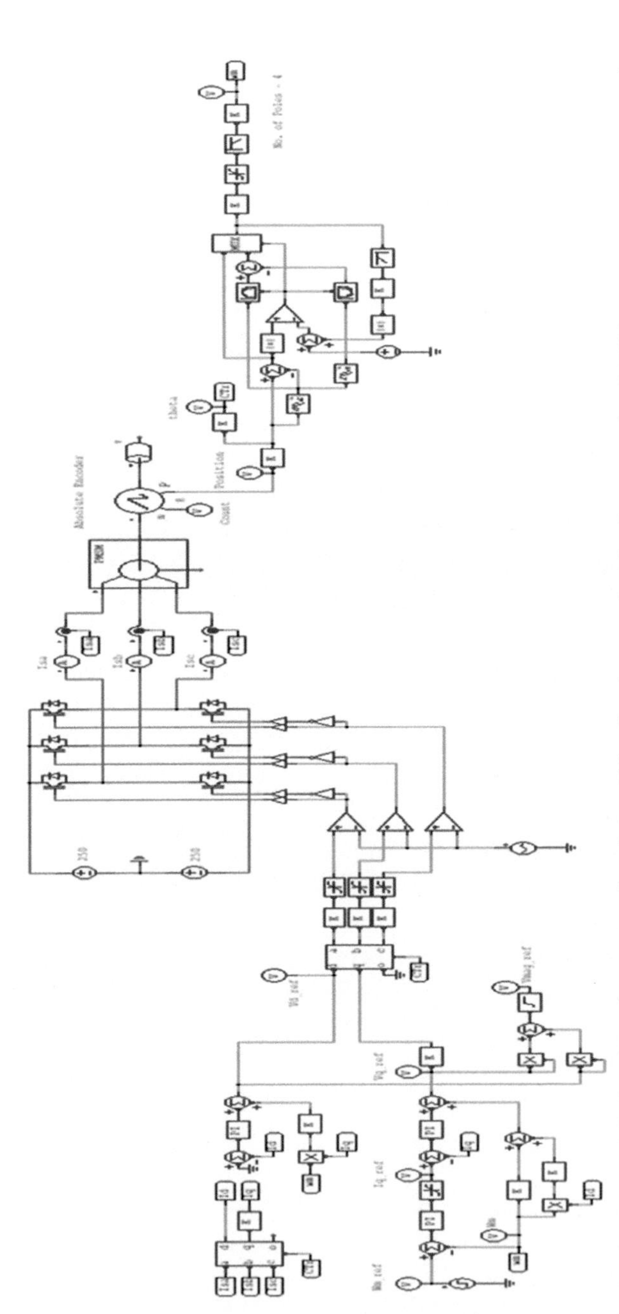

Figure 7.50 PSIM model of PMSM motor control with speed feedback using the absolute encoder. *PMSM*, permanent magnet synchronous motor.

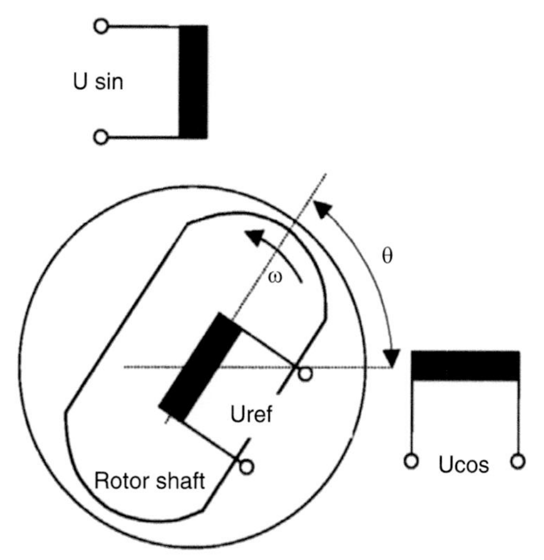

Figure 7.51 Resolver.

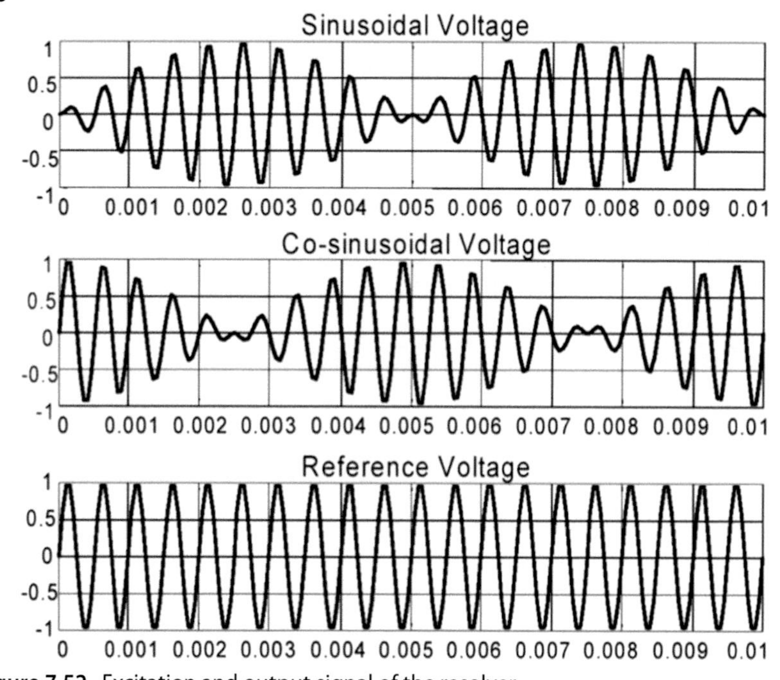

Figure 7.52 Excitation and output signal of the resolver.

Answer: a) synchronous motor

5. The main advantages of using permanent magnet in rotor
  a) Field winding Copper loss is reduced
  b) High efficient

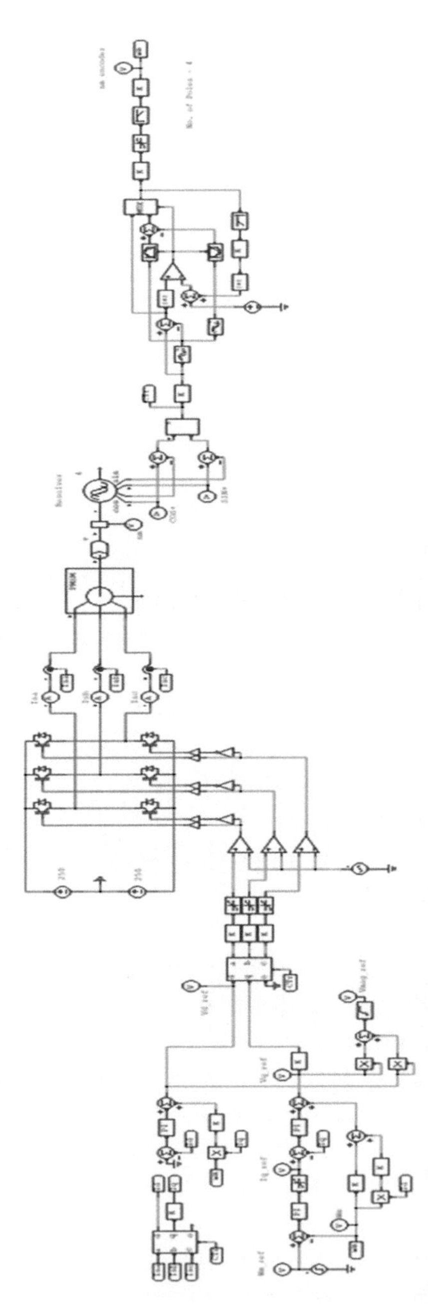

Figure 7.53 PSIM model of motor control with speed feedback using a resolver.

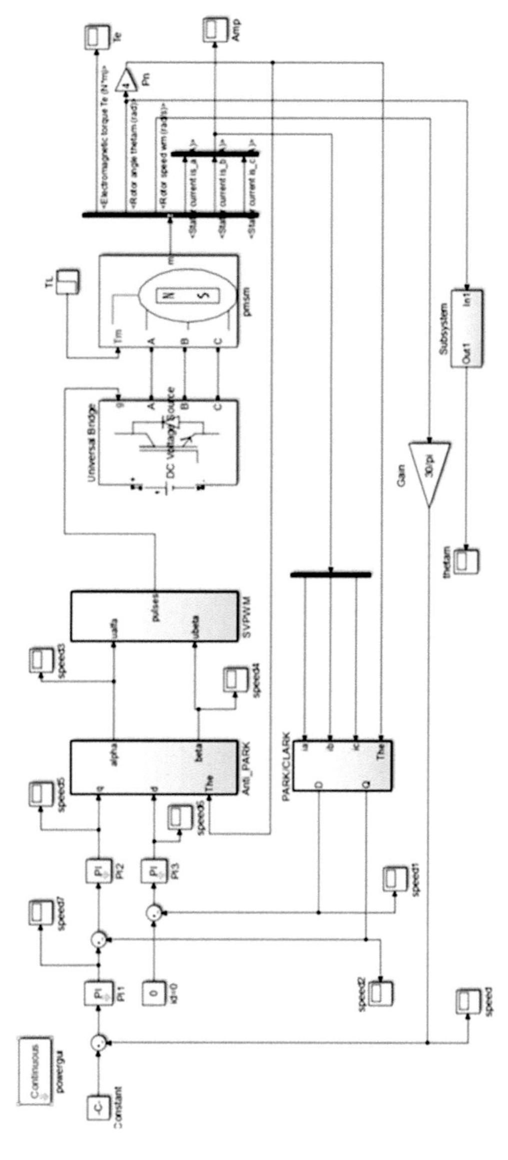

Figure 7.54 Simulation of vector control system of permanent magnet synchronous motor [3].

  - **c)** more losses
  - **d)** Option A & B

  Answer: d) Option A & B

6. PMSM working principle is
   - **a)** ohm's law
   - **b)** Ampire circuital law
   - **c)** Faraday's law
   - **d)** magnetic locking

   Answer: d) magnetic locking

7. In PMSM torque produced by two magnetic field one from PMs mounted in the rotor another one
   - **a)** from coils of the stator
   - **b)** from same in rotor
   - **c)** from field coils
   - **d)** none of the above

   Answer: a) from coils of the stator

8. Advantages of load commutation in permanent magnet synchronous motor
   - **a)** does not require commutation circuits
   - **b)** low frequency
   - **c)** required resonant circuit
   - **d)** none of the above

   Answer: a) does not require commutation circuits

9. Field oriented control the stator currents are transformed into a frame of reference moving with the rotor flux.
   - **a)** True
   - **b)** False

   Answer: a) True

10. PMSM are used in
    - **a)** Fiber spinning mills,
    - **b)** Rolling mills
    - **c)** Cement mills
    - **d)** all the above

    Answer: d) all the above

References

https://en.engineering-solutions.ru/motorcontrol/pmsm/.
https://in.mathworks.com.

Sharma, R.K., Sanadhya, V., Behera, L., Bhattacharya, S., 2008. Vector control of a permanent magnet synchronous motor. In: 2008 *Annual IEEE India Conference*, pp. 81–86.
Nagulapati II, K., Anitha Nair IIID, A.S., 2016. Sri Lakshmi permanent magnet synchronous motor control with speed feedback using a resolver. International Journal of Advanced Research in Education & Technology 3 (4), 74–78.

CHAPTER 8

Electric drives used in electric vehicle applications

Contents

8.1 Introduction

A road vehicle that uses electric propulsion is known as an electric vehicle (EV). The propulsion provided by an electric motor was used in the EV, which was powered by a portable and electrochemical energy source. The vehicle's powertrain is the electrochemical energy conversion linking system between the vehicle's energy source and the wheels. An EV's powertrain has both electrical and mechanical coupling [2]. Passenger vehicles are an important part of our everyday lives, yet due to tailpipe emissions from conventional internal combustion vehicles (ICEVs), they contribute to urban air pollution and greenhouse gas emissions, which contribute to global warming. Air quality has been shown to be decreasing around the world, with car emissions being one of the main contributors. The rise in automotive emissions is due to rising population, urbanization, and socioeconomic development, as well as the increased use of vehicles as a result.

Electric Motor Drives and Their Applications with Simulation Practices. DOI: https://doi.org/10.1016/B978-0-323-91162-7.00006-0

Fuel engines release greenhouse gases such as nitrous oxides (N_2O), methane (CH_4), carbon dioxide (CO_2), and a variety of pollutants such as nitrogen oxides (NO_x), sulfur dioxide (SO_2), hydrocarbon (HC), and particulate matter (PM). In 2004 and 2007, the transportation industry accounted for 23%–26% of global CO2 emissions and 74% of on-road CO2 emissions, respectively. Aging vehicles, a lack of appropriate road vehicle maintenance, high traffic congestion, fuel adulteration, and poor road infrastructure all contribute to rising pollution levels. Although heavy-duty diesel vehicles make up a smaller percentage of the fleet, their emissions have a substantial impact on air pollution. Acid deposition, stratospheric ozone depletion, and climate change are all exacerbated by road vehicle emissions. In order to reduce automotive emissions and improve air quality, industrialized countries enacted stringent legislation. Concerns about climate change have reached an all-time high, prompting EU members to pledge to cut emissions by 80% by 2050 in order to stabilize atmospheric CO2 at 450 parts per million, allowing them to work out a plan to keep global warming below 2 degrees Celsius. The endeavor to reduce emissions and global warming has been divided among various sectors, with the road transportation sector anticipated to reduce emissions by 95%. This tendency is also seen in other nations, such as Brazil, which established new CO (carbon monoxide), HC (hydrocarbon), and NOx (nitric/nitrogen oxide) emission limitations in Regulations 418/2011 and 315/2002. According to Steinberg, the cost of lowering each gramme of CO2/km has already increased from $17.03 (€13) to $65.50 (€50) before the 2020 objective of 159 g CO2/km has even been met. In terms of energy saving, zero emissions, and assuring oil supply security, pure EVs have superior advantages over traditional ICE vehicles, attracting a wide spectrum of automobile manufacturers and governments. Electric mobility has a number of advantages over ICE vehicles, including energy conservation, zero tailpipe emissions, and independence from oil supply. The major components of the various subsystems and their contributions to the overall system are depicted in Fig. 8.1. All of these systems work together to make EVs possible. A battery-powered car or a pure EVs use the electrical energy stored in batteries as a source of energy, and their motor drive system converts the battery's output power into rotational energy of the wheel, allowing the vehicle to operate. The working principle of a pure EV is to replace the internal combustion engine (ICE) and accompanying gasoline tank with an electric machine (electric motor) that uses an energy source (battery), and the vehicle's energy source is recharged as it is used to

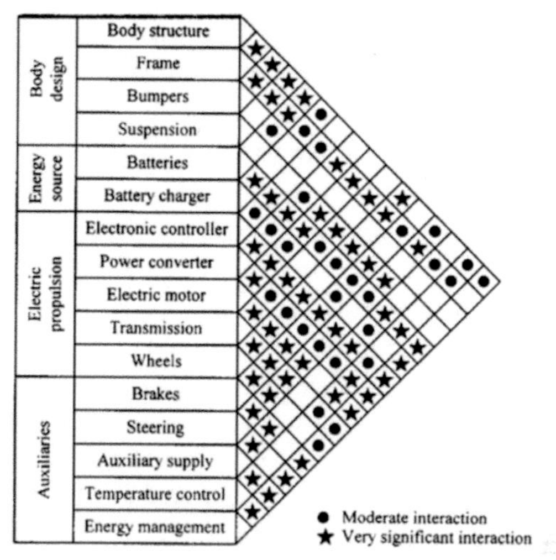

Figure 8.1 Different subsystems and their interaction with electric vehicles.

reclaim its energy source. In EVs, several subsystems work together in the same way they do in ICE vehicles, but without the fossil fuel engine and tail pipe. The EV works because of the interaction and integration of various subsystems, and multiple technologies can be used to operate the subsystems. When it comes to constructing EVs, there are two basic approaches: most EVs are converted from existing designs based on standard ICE vehicle styling, while some EVs are designed entirely from scratch. When EVs are designed utilizing a ground-up design technique, engineers have the ability to coordinate and integrate diverse EV subsystems so that they can function together efficiently [2–3]. Because an electric automobile has an empty space that can be used for baggage storage, its packaging requirements are different. To improve the outcome of the electric car design process, this total body system might be examined. To convey to the general public that an EV is fundamentally different from an ICE vehicle, the vehicle body elements should be created in accordance with the "EV technology" that has been employed, and as the technology evolves, further improvements should be reflected in the formal design.

8.2 Role of electric motor drives in EV's

The driving system of an electric car is identical to that of a vehicle powered by an ICE. The drive system is the component of an EV that distributes mechanical energy to the traction wheels, allowing the vehicle to move. An EV's components are vastly different from those found in a conventional car. A transmission is not required in an EV. A transmission in a typical car is used to change the gear input/output ratio inside the transmission to provide the vehicle a specified torque or power at certain speeds. The speed (RPM) at which the vehicle's power plant, or engine, rotates determines the change in gear ratio. Passengers generally feel a shock as speed is increased or decreased and the transmission moves to larger or smaller gears due to a mechanical transition from one set of gears to another.

The wheels of an EV are turned by an electric motor. Today's drive systems come in a variety of shapes and sizes. Vehicles having a single big electric motor linked to the rear wheels via a differential housing fall into this category. Other systems use two smaller motors to drive each wheel independently through separate drive shafts.

The most efficient system to date makes use of motors that are directly linked to the wheel. "Wheel motors" are what they are called. Mechanical losses between the motor and the wheels are kept to a minimum by eliminating drive-shafts and differentials. An EV's power system encompasses both the driving and control systems. The controller uses the batteries to power the motor. A gearbox transmits power from the motor to the drive wheels, allowing the vehicle to move [4].

Electrical energy is converted into mechanical energy by electric motors. In EVs, two types of electric motors are utilized to power the wheels. There are two types of motors: direct current (DC) and alternating current (AC). DC motors are made up of three major parts:

The magnetic forces that give torque are created by a collection of field coils around the periphery of the motor. A rotor or armature that spins inside the magnetic field formed by the field coils and is positioned on bearings. A commutator that reverses magnetic forces and causes the armature to turn, generating the mechanical force required to turn the drive wheels.

An AC motor is similar to a DC motor in that it contains a set of field coils and a rotor or armature, but it does not require a commutating mechanism because the current is reversed continuously (AC). Neither motor can be deemed superior to the other at this stage of development. They both have benefits and drawbacks, which are listed in Table 8.1.

Although an AC motor is less expensive than a DC motor, the expense of the complicated electronics connected with the AC converter and motor controller makes an AC system more expensive. In home appliances and machine tools, AC motors are the most often utilized motor. These motors

Table 8.1 Electric motor comparison.

AC motor	DC motor
Single-speed transmission	Multispeed transmission
Light weight	Heavier at equivalent power
Less expensive	More expensive
95% efficiency at full load	85%–95% efficiency at full load
More expensive controller	Simple controller
Motor/controller/inverter more expensive	Motor/controller less expensive

are extremely dependable, and because they only have one moving part, they should last the whole life of the vehicle with no maintenance.

8.2.1 Output characteristics of motor drives in EVs

Acceleration performance is measured by the time it takes the vehicle to accelerate from zero to a given speed (starting acceleration), or from a low speed to a given high speed (passing ability), gradeability is measured by the maximum road grade the vehicle can overcome at a given speed, and top speed is measured by the vehicle's top speed. Only the traction motor produces torque to the driven wheels in EVs. As a result, the torque-speed or power-speed characteristic of the traction motor determines the vehicle's performance entirely. A vehicle must operate totally in constant power in order to achieve its operational requirements, such as beginning acceleration and gradeability with minimum power. Any practical car, however, will not be able to operate completely on continual power. Fig. 8.2 depicts the desired output characteristics of electric motor drives for EVs. The EV motor drive is expected to provide high torque at low speeds for starting and acceleration,

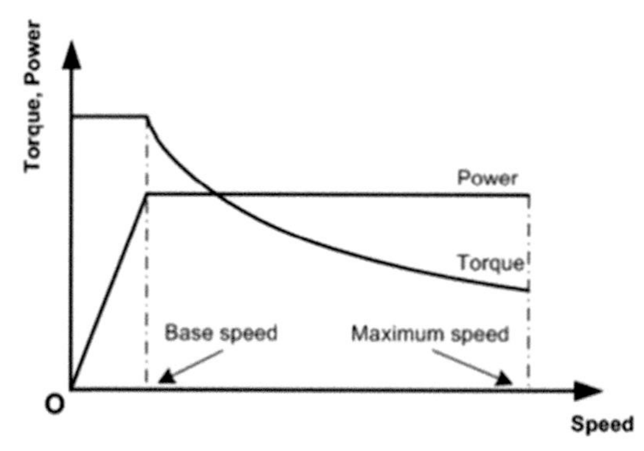

Figure 8.2 Desired output characteristics of electric motor drives in EVs. *EVs*, electric vehicles.

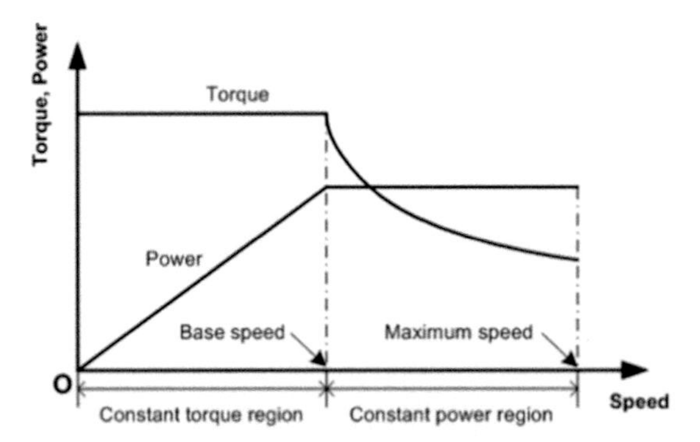

Figure 8.3 Typical performances of electric motor drives in industrial applications.

as well as high power at high speeds for cruising. Simultaneously, as wide a speed range as possible under constant power is sought. In an ideal world, eliminating the constant torque zone would result in the motor's lowest power rating, however, this is not physically possible.

The output performances of general electric motor drives in industrial applications are presented in Fig. 8.3. The electric motor drive may offer consistent rated torque up to its base or rated speed in typical mode of operation. The motor achieves its rated power limit at this speed. This continuous power zone is confined to operations above the base speed and up to the maximum speed. The constant power operation range is mostly determined by the motor type and control approach. A few electric motor drives, however, deviate from constant power operation beyond a specific speed and enter natural mode before reaching the maximum speed. In the natural mode of operation, the maximum attainable torque falls inversely with the square of the speed. Although the machine torque in the natural mode falls inversely with the square of the speed, the natural mode of operation is a significant element of the total power-speed profile for some exceptionally high-speed motors. Incorporating this natural mode into such motor drives could result in a reduction in total power consumption. The maximum motor speed requirement for indirect-drives is determined by the maximum vehicle speed, the wheel radius, and the gear ratio. The maximum motor speed required for direct-drives, on the other hand, is solely determined by the first two factors.

As a result of the output characteristics of electric motor drives for EVs, the following useful conclusions can be drawn. (1) As the constant power region ratio grows, the power demand (rated power) for acceleration performance (acceleration time and acceleration distance) lowers. (2) On

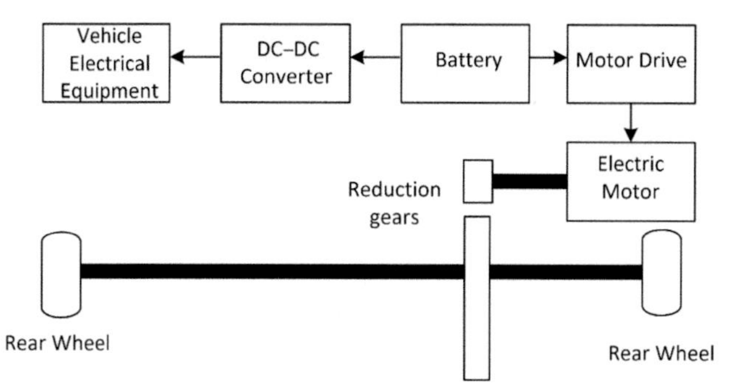

Figure 8.4 Typical schematic system of EVs. *EVs*, electric vehicles.

the other hand, as the constant power region ratio rises, the torque demand (rated torque) for acceleration rises. As a result, the motor's size and volume increase. (3) As the constant power region ratio grows, passing performance (passing time and distance) declines significantly. (4) A motor's maximum speed has a significant impact on the needed torque. The rated shaft torque of low-speed motors with a prolonged constant power speed range is much higher. As a result, they will require more iron and copper to sustain the increased flux and torque. (5) The needed torque increases when motor power lowers (due to extending the range of constant power operation). As a result, while the converter power demand (and thus the converter cost) decreases as the constant power range expands, the motor size, volume, and cost increase. (6) Increasing the motor's maximum speed reduces its size by allowing shaft torque to be increased through gearing. The maximum speed of the motor, on the other hand, cannot be increased endlessly without incurring additional costs and transmission needs. As a result, when the constant power range is extended, a slew of system-level conflicts arise.

8.3 Block diagram of EV

ICEs power conventional cars, which are also known as internal combustion engine vehicles (ICEVs). If an electric motor or several electric motors are used to drive the wheels of a vehicle, it is referred to as an EV. In addition, if a vehicle's wheels are propelled by both an electric motor and an ICE, the vehicle is called a hybrid electric vehicle. In this study, only electric cars are mentioned.

Fig. 8.4 shows a schematic of an EV system. The battery is the primary energy source in EVs, providing electric power to electric motor drives and other equipment such as lighting.

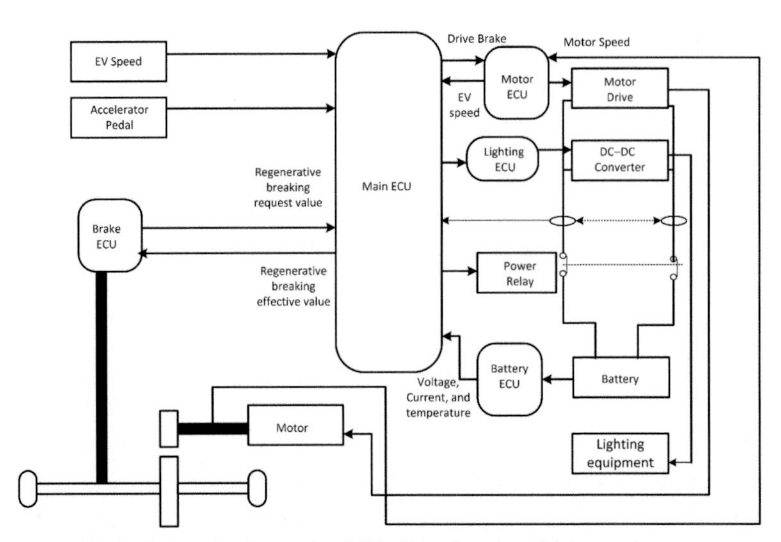

Figure 8.5 Typical control schematic of EVs. *EVs*, electric vehicles.

Fig. 8.5 shows a typical EV control schematic. The main ECU, motor ECU, battery ECU, brake ECU, and electric equipment ECU are the five electric control units (ECUs) that make up an EV's standard control system. The main ECU calculates the motor torque based on information such as the accelerator opening and car speed command to regulate the drive torque of an EV. The motor ECU receives the torque request value. The motor ECU controls the motor drive to create the desired torque based on the drive output value requested by the main ECU. Torque direct control can be achieved via the motor drive (DTC). The brake ECU manages the full brake torque produced by both the regenerative brake system and the traditional hydraulic brake system by coordinating the braking effort with the regenerative braking that is performed by the engine. The battery ECU keeps track of the battery's charging and discharging status. In general, the battery monitor covers leak detection, abnormal voltage detection, abnormal temperature detection, and abnormal current detection. The DC–DC converter is controlled by the electric equipment ECU, which generates a variety of DC voltage levels for lights and other equipment.

Electric motors come in a variety of shapes and sizes in industrial applications, as we all know. They are utilized to power a variety of industrial machines. When it comes to driving, EVs can be propelled by any electric motor. When electric motors are used in EVs, however, several performance indicators must be considered, such as efficiency, weight, cost, and dynamic characteristics of EVs.

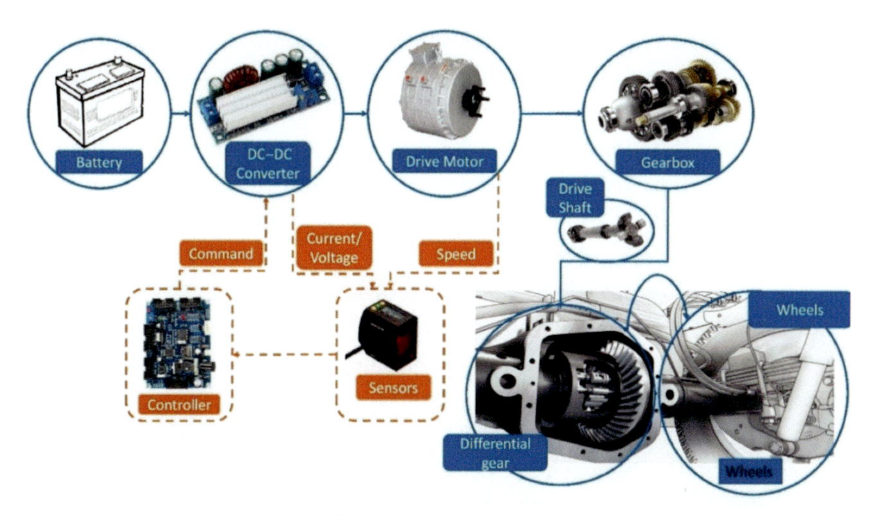

Figure 8.6 Basic configuration of DC motor drive.

8.4 DC motor for EVs

DC motor drives were commonly employed in EV propulsion systems. Because of their technological maturity and control simplicity, they were a good first choice for driving EVs. Permanent magnets (PMs) are used in DC motors, while brushes are used in rotors. For EV propulsion, the DC machine uses a high-power density that spins up to 5000 rpm before stepping down to 1000rpm using a fixed gear (FG) mechanism. Offering reverse rotation eliminates the need for a hefty, inefficient verse gear.

The basic motor drive train system is depicted in Fig. 8.6 with many subsystems ranging from the motor controller to the single-speed reducer differential and driving wheels. The stator houses the field winding or PMs, which aid in producing the magnetic field excitation, while the rotor houses the armature winding, which is switched by the commutator via carbon brushes. Fig. 8.6 depicts the basic setup of a DC motor drives system for controlling the DC machine's armature current and output torque. In typically, the motor speed is the only feedback control variable, whereas the armature current feedback is mostly used for protection.

8.4.1 Brushed DC drive control

Brushed DC motors are well-known for their capacity to generate high torque at low speeds and for having torque–speed characteristics that are excellent for traction. The motor speed is controlled by changing the voltage.

They have been employed on EVs since they are good at propelling a car and easy to manage. Depending on the power output and voltage, brushed DC motors can have two, four, or six poles, as well as series or shunt field windings. Shunt motors, on the one hand, are more controllable than series motors. Due to their decoupled torque and flux control properties, separately excited DC motors are intrinsically suited for field weakening operation.

Brushed DC motor drives, on the other hand, have a large design, low efficiency, low dependability, and a greater maintenance requirement, owing to the existence of a mechanical commutator and brushes. Brushed DC motors are tough to reduce. Brushed DC motors get heavier and more expensive as a result of this. Furthermore, the maximum motor speed is limited by friction between the brushes and the commutator.

DC–DC converters must be utilized to better manage the speed of the DC drive system for usage in EVs. For DC motor speed control, two methods are used: drive armature voltage control and flux-weakening control (FWC). For EV propulsion, pulse width modulation (PWM) is used to manage the armature voltage of the DC drive. When the armature voltage of a DC motor is reduced, the armature current and motor torque drop, lowering the motor's speed while raising the armature voltage and torque. When weakening the field voltage of the DC motor, the motor back EMF decreases. The motor back EMF drops as the field voltage of the DC motor is reduced. Due to low armature resistance, there is a considerable increase in armature current compared to its reduction in the field. As a result, the motor's torque raises the motor's speed. Characteristic features of individually excited DC motors and series DC motors are presented in Fig. 8.7A and B. The motors' inherent characteristics can function for any torque-speed characteristics with a constant slope against the change in speed as shown below. The maximum permitted armature current remains constant during armature voltage management, which takes use of the advantage of retaining maximum torque while keeping the maximum allowable current constant at all speeds.

Fig. 8.7A and B shows the functioning of a separately excited DC motor and a series DC motor during the weakening of the DC motor's field voltage. During this control phenomenon, the control slope of both dc motor characteristics varies, but it is also affected by the flux. The independent armature voltage control and FWC, which are applied to only separately excited DC motor drive systems to achieve a wide range of speed control.

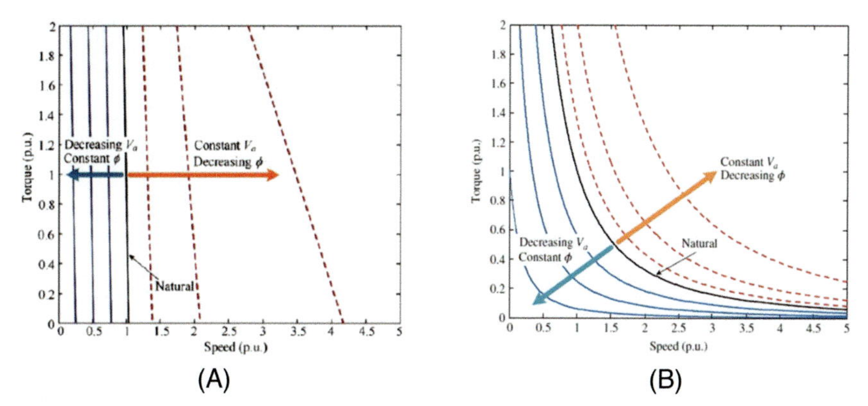

Figure 8.7 (A) Characteristics of separately excited DC motor control, (B) characteristics of series DC motor control.

8.4.2 Application of DC motor in electric vehicles

For EV drive systems, separately excited DC and series DC motor drives have been widely employed. Due to their poorer efficiency, lower power density, and constant wear and tear of the carbon brushes and commutator, DC motor drives are no longer used to power EVs.

8.4.2.1 Simulink library browser for electric vehicle

To simulate EV, it is necessary to know the key components availability in Simulink library browser. Go-to Simulink library in that select Simscape. In Simspace, driveline is to be chosen.

From Fig. 8.8, it is clear that in driveline, more blocks are available. Apart from this, we have to use some more components also for modeling EV which is been given in Fig. 8.9.

In vehicle dynamics lockset, more blocks like powertrain, sensors, vehicle body are available. In each block, more components are available.

8.4.3 Simulation of ideal electric vehicle using MATLAB Simulink

To draw the Ideal EV, first need to select voltage source from the Simulink library using the command mentioned earlier. The parameters of the voltage source are set as per Fig. 8.10.

Library:fl_lib/Mechanical/Mechanical Sensors/Ideal rotational motion sensors.

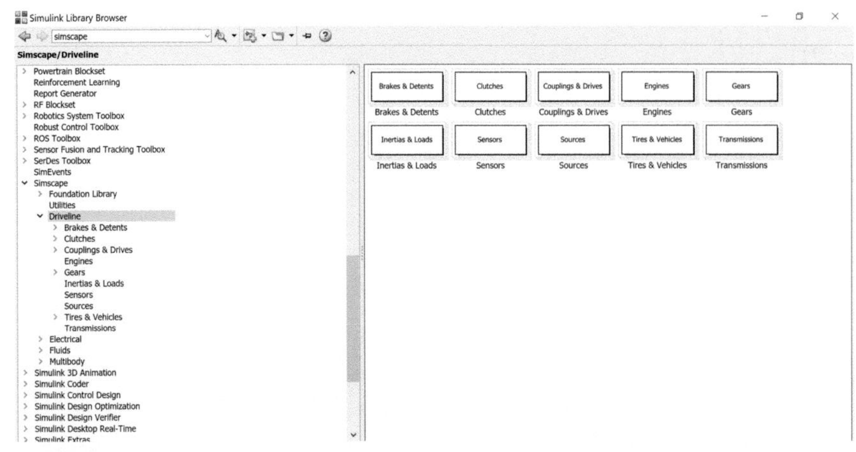

Figure 8.8 Driveline blocks in library.

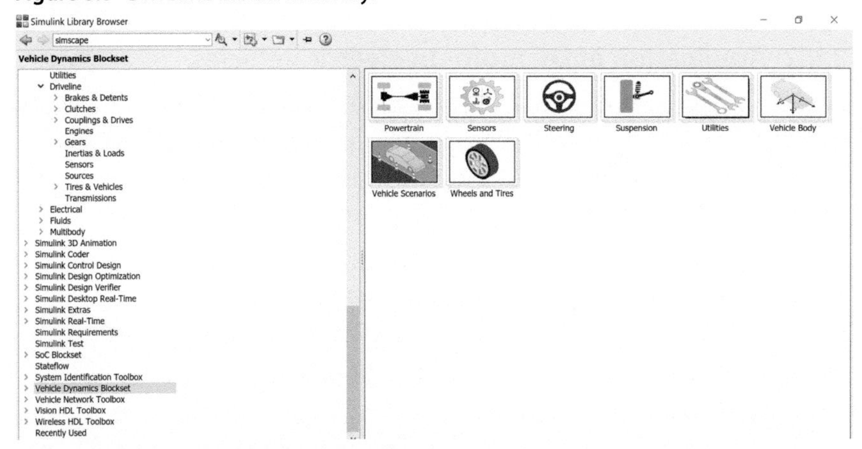

Figure 8.9 Vehicle dynamics blockset in library.

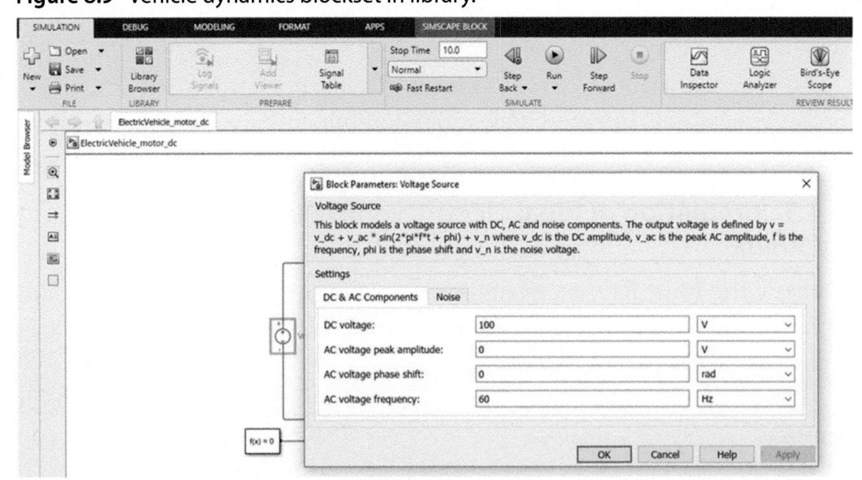

Figure 8.10 Voltage source parameter values.

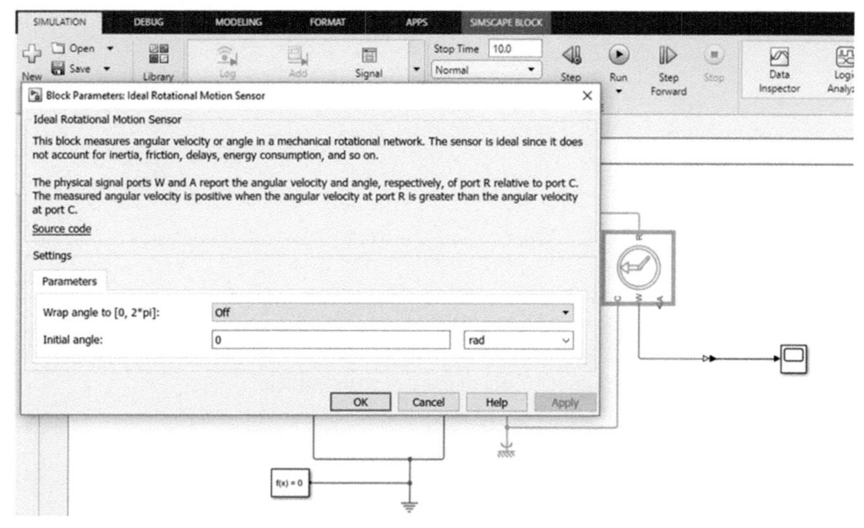

Figure 8.11 Block parameters—ideal rotational motion sensor.

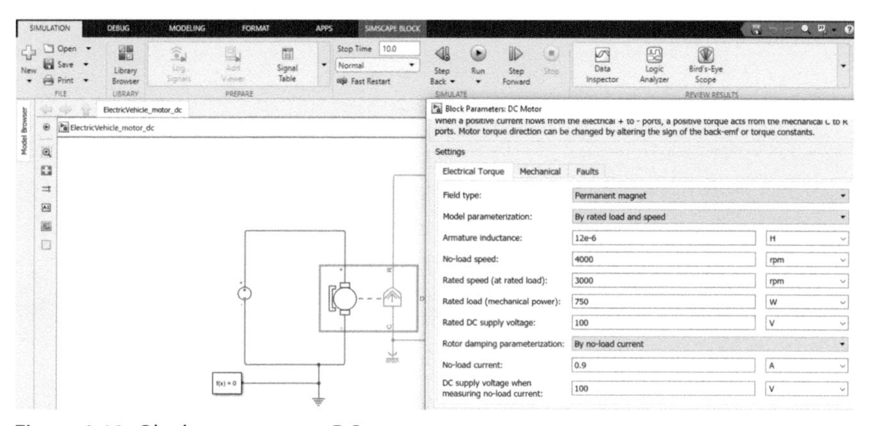

Figure 8.12 Block parameters—DC motor.

Using the above link, select ideal rotational sensors and give the parameters value as per Fig. 8.11.

DC motor is a very important component in EV, to select that use the link as Library: Electro Mechanical/Brushed Motors/DC motors. In that DC motors need to assign values for parameters like electrical torque, mechanical, and faults. As per Fig. 8.12, give values for each parameters.

After connecting the circuit, we will get the simulation circuit of ideal EV as per Fig. 8.13

After simulation, you can see the graph in scope by clicking it (Fig. 8.14).

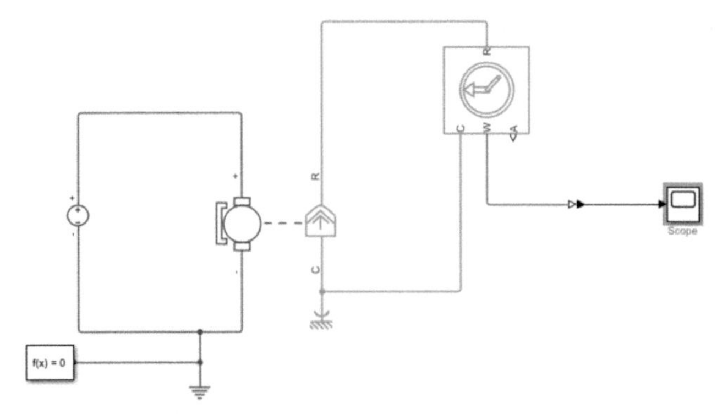

Figure 8.13 MATLAB/Simulink model of ideal electric vehicle.

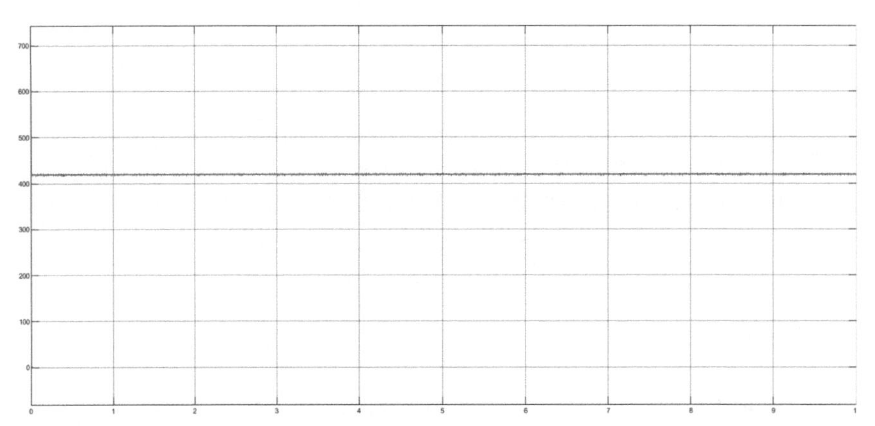

Figure 8.14 Simulation results of ideal electric vehicle.

8.5 Induction motor for EV's

Because of their dependability, low maintenance, low cost, and ability to function in harsh settings, induction motors are simple to build. Induction machines (IMs) are divided into two types: wound-rotor and squirrel-cage. Due to its high cost, requirement for maintenance, and lack of sturdiness, the wound-rotor induction motor is less appealing than its squirrel-cage counterpart—especially for electric propulsion in EVs. As a result, the squirrel-cage induction motor is also known as the induction motor for EV propulsion. Due to the lack of brush friction, the motors have the ability to enhance the maximum speed limit, improve the speed rating, and provide a high output. The speed of an induction motor can be changed by adjusting the voltage frequency. Induction motor torque control can be terminated using field-orientation control (FOC).

Simple design, reliability, ruggedness, low maintenance, low cost, and ability to function in harsh settings are all advantages of induction motors. The lack of brush friction allows the motors to increase their maximum speed limit, and the greater speed rating allows these motors to provide high output. The frequency of voltage is changed to change the speed of induction motors. Induction motors with FOC can isolate torque control from field control. As a result, the motor behaves similarly to an independently activated DC motor. This motor, on the other hand, does not have the same speed constraints as a DC motor with field-focused control, a properly built induction motor, such as a spindle motor, may reach a field weakening range of 3–5 times the base speed. Induction motor controllers, on the other hand, are more expensive than DC motor controllers. Furthermore, the presence of a breakdown torque restricts its ability to operate at constant power for long periods of time. The breakdown torque is attained at the critical speed. The essential speed for a traditional IM is typically two times the synchronous speed. Any attempt to run the motor at maximum current above, this speed will cause it to stall. Although FOC can extend constant power operation, it also causes an increase in breakdown torque, which leads to motor oversizing. Furthermore, due to the absence of rotor winding and rotor copper losses, efficiency at high speeds may decrease, in addition to the fact that IMs efficiency is fundamentally lower than that of permanent magnetic (PM) motors and switching reluctance motors (SRMs).

In induction motor control, the three major control system for induction motors are:

(a) Variable-voltage variable-frequency (VVVF) control

For frequencies below the rated frequency, it uses constant voltage control, and for frequencies beyond the rated frequency, it uses variable-frequency control with constant rated voltage. Voltage is raised at very low frequencies to recover the difference between the applied voltage and the induced EMF. There are three functioning zones visible in Fig. 8.15:

1. The motor delivers rated torque in the constant torque area below the rated speed.
2. The slip is gradually increased to its maximum value in the constant-power area with constant stator current, and the motor operates at rated power.
3. In the lower power area, when the stator current decreases and the torque capability decreases with the square of the speed, the slip remains constant.

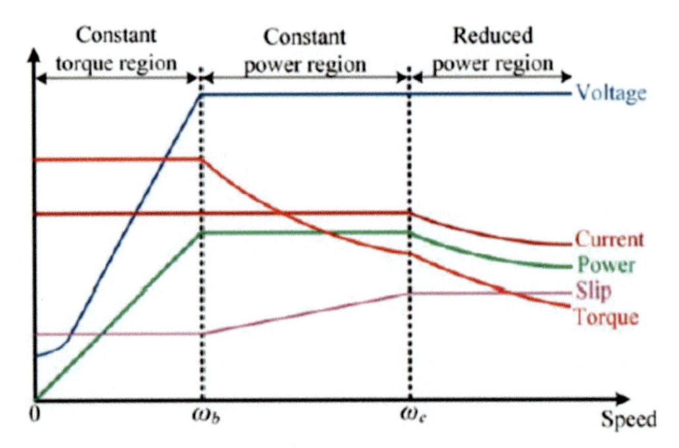

Figure 8.15 Capabilities of operating VVVF controlled induction motor.

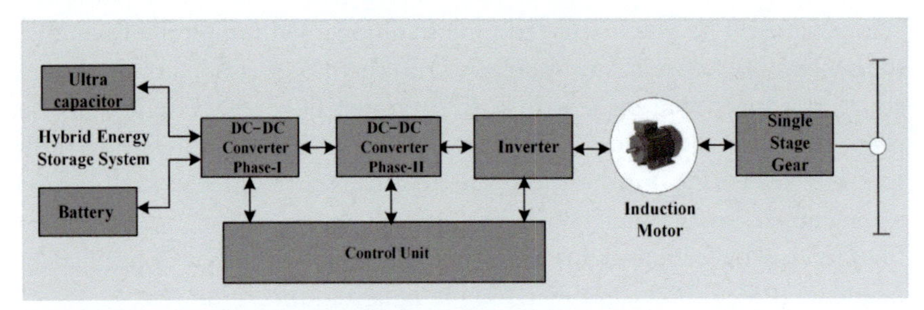

Figure 8.16 Induction motor in EV system. *EV*, electric vehicle.

(b) Field-oriented control (FOC)

FOC for the induction motor drive can be implemented by:

1. The direct FOC, also known as direct vector control, measures the air-gap flux from stator voltage or current to instantly identify the rotor flux linkage.
2. Induction motor drives for EVs have long used indirect FOC, often known as indirect vector control. The rotor flux linkage does not need to be identified for this procedure.

(c) Direct torque control (DTC)

For induction motor drives, the DTC gives equal performance. This technology directly controls the stator flux linkage and torque application of the IM in EVs by selecting the switching modes of the voltage-fed PWM inverter. Because the design is simple and stable, with more control and lower cost, IM motors are the most prevalent propulsion unit in EVs (Fig. 8.16).

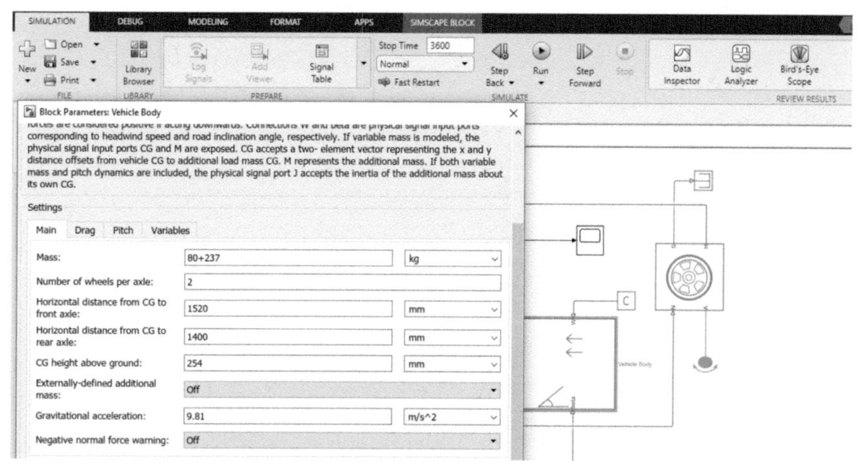

Figure 8.17 Block parameters—vehicle body.

Three-phase induction motors (IM) are used in EVs because of their high efficiency, precise speed control, and lack of a commutator to generate friction and loss. Low acceleration is a problem for the system. The stator winding receives a three-phase AC supply, which creates a revolving magnetic field. 3-ph IM is also known for its great dependability and low maintenance requirements. Multiple strategies for operating the motor with high-end controlling actions are accessible owing to power electronics. When compared to the base speed, vector control, which is one of the strategies for improving the dynamic performance of the electric drive framework, can be used to provide a wide range of speeds.

8.5.1 Simulation of two-wheeler electric vehicle body using MATLAB Simulink

A step-by-step approach of two-wheeler EV body simulation is given below. Along with circuit, the parameters value also mentioned in the figures.

Vehicle body is the heart of this circuit, select the vehicle body using the link.

Library:sdl_lib>> Tires & Vehicles >> Vehicle body. It is important to set the parameter values in the vehicle body. There are four settings are there as, Main, drag, Pitch, and variables. Set the parameters value as per Fig. 8.17.

Select tires using the link

Library:sdl_lib>> Tires & Vehicles >> Tire (Magic formula). Give the values as per Figs. 8.18 and 8.19 select 2 times tires or else we can do copy-paste to get two tires.

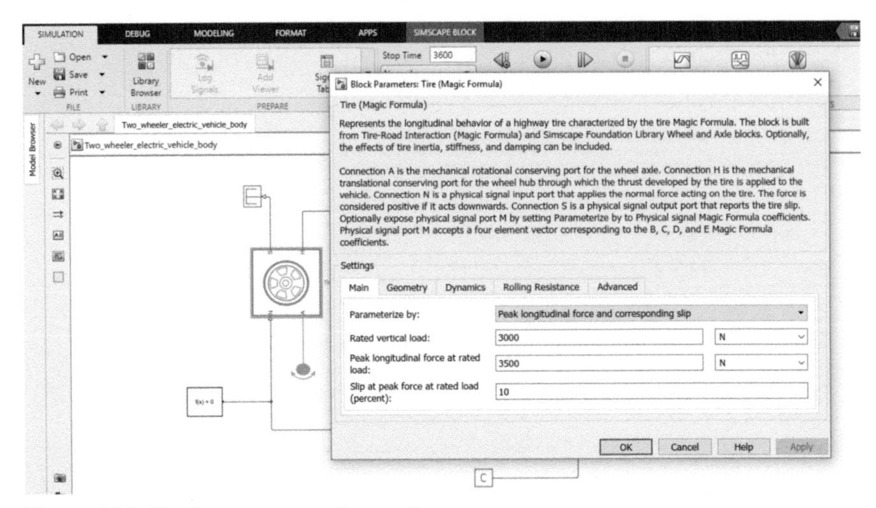

Figure 8.18 Block parameter—tire magic.

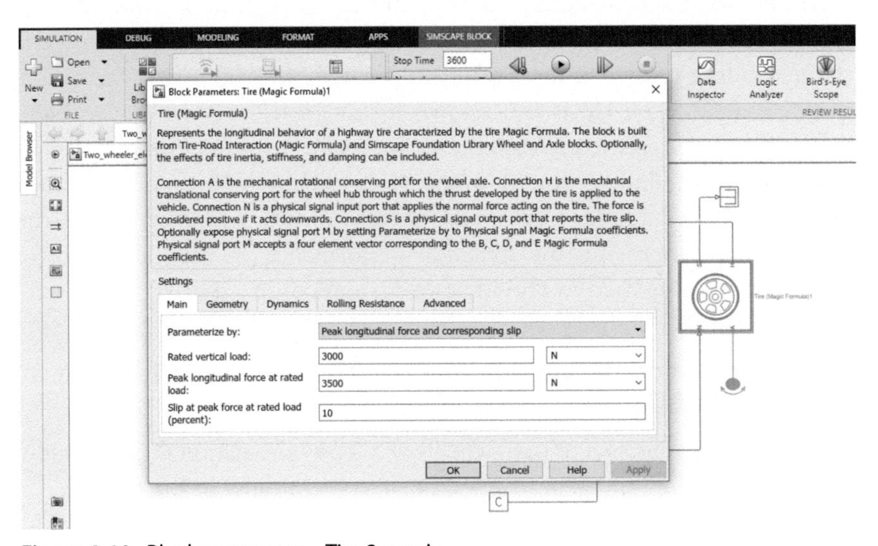

Figure 8.19 Block parameter—Tire 2 magic.

Select the physical signal constant and set the value as ((-20*pi)/180) (Fig. 8.20).

After selecting all the components, connect all as per Fig. 8.21.

After clicking the run button, the output waveform will be seen in the scope Fig. 8.22.

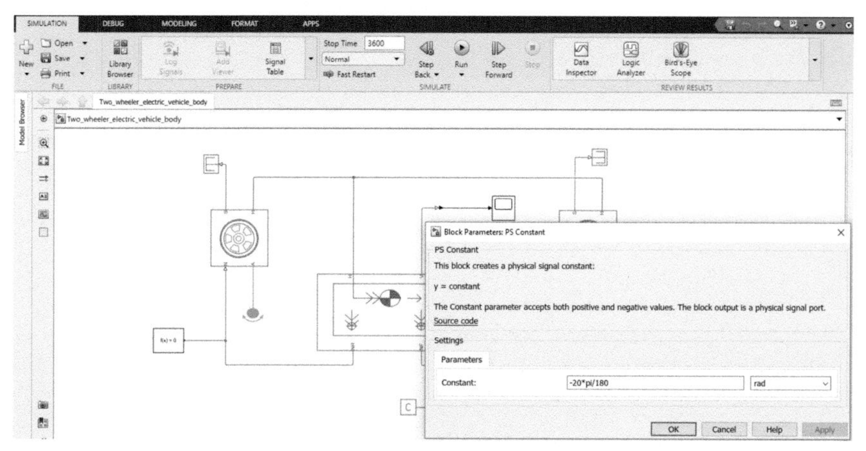

Figure 8.20 Block parameter—PS constant.

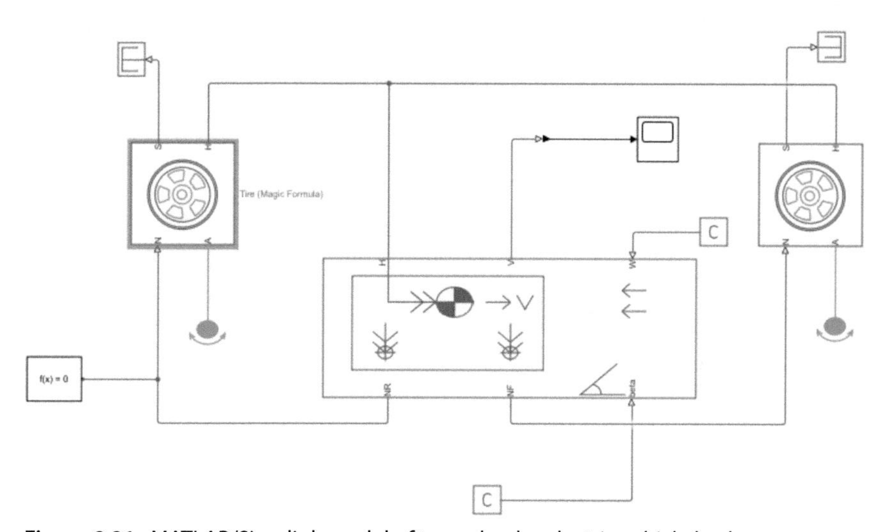

Figure 8.21 MATLAB/Simulink model of two wheeler electric vehicle body.

8.6 PMSM for EV's

8.6.1 Permanent magnet synchronous motor (PMSM)

Sinusoidal magnetomotive force (mmf), voltage, and current waveforms are found in permanent magnet synchronous motors. The machine runs as a synchronous machine when the air-gap flux and stator windings are distributed in a sinusoidal pattern. This motor drive's rare earth magnet material helps to boost flux density in the air gap, motor power density, and torque-to-inertia, allowing it to run at a wide constant power speed range.

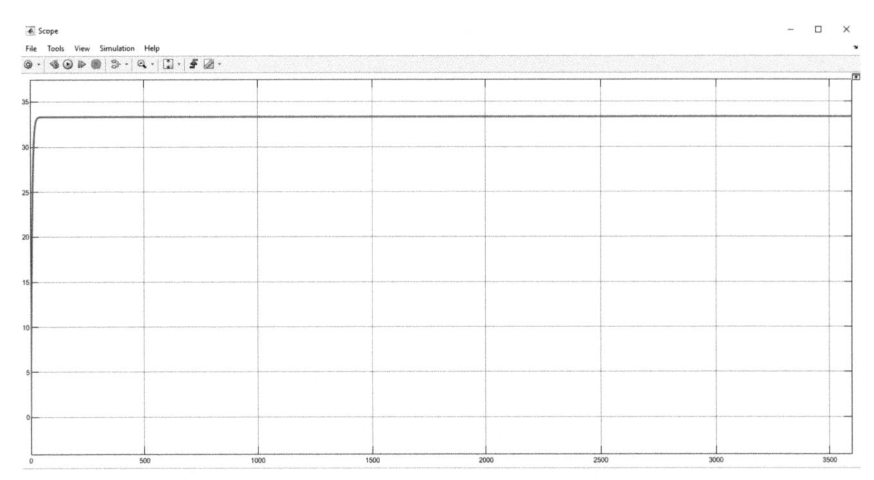

Figure 8.22 Simulation result of two-wheeler electric vehicle body.

Ferrites, samarium cobalt (SmCo), and neodymium-iron-boron (NEB) are the most popular magnet materials used in PM machines (NdFeB). The mechanism is the same as in BLDC moneodymium-iron-boron (NdFeB). Except for the sinusoidal wave shape of the back EMF, the working mechanism is identical to that of a BLDC motor. The following are the main benefits of PMSM: h higher efficiency than brushless DC motors, h no torque ripple when the motor is commutated, h better performance with higher torque, h reliable and less noisy, h high performance in both higher and lower speeds of operation, easy to control due to lower inertia of the rotor, h efficient heat dissipation, h smaller in size.

8.6.2 PMSM motor control

There are three different ways of controlling the PMSM motor, they are:

(a) Field-oriented control (FOC).

(b) Induction motor control systems such as FOC and direct torque control can be used with a PM synchronous motor. The FOC has been used to the PM synchronous motor for the purpose of operating EVs; nevertheless, PM field excitation in the PM synchronous motor differs from that of an induction motor.

(c) Flux-weakening control (FWC).

At PMSM's base speed, the terminal voltage equals the rated voltage. Because back EMF increases with speed, speed range expansion is only achievable when the air-gap flux is lowered, a process known as

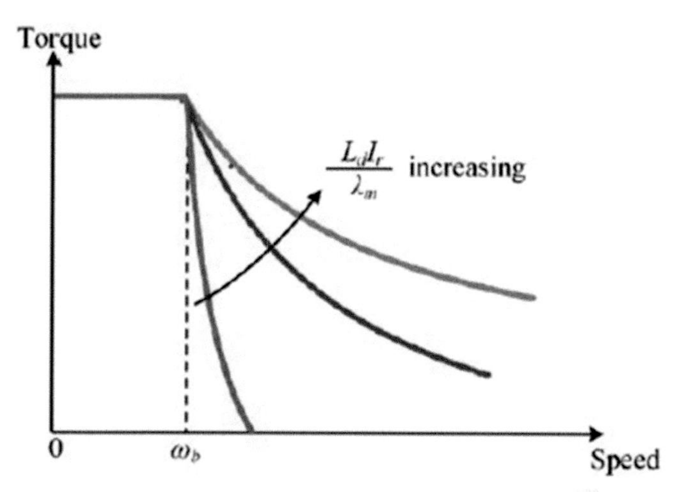

Figure 8.23 Torque-speed capabilities of flux-weakening control of MMSM.

flux-weakening. As a result, the torque reduces as the speed increases, resulting in constant power operation.

The torque-speed capabilities of the PM synchronous motor are shown in Fig. 8.23. It has been discovered that the higher the Ld Ir/m ratio used, the better the flux-weakening capability.

(d) Position sensorless control.

The PMSM's control system uses this position sensor, which is commonly an optical encoder. A PM synchronous motor with a position sensor is rarely used to power EVs.

8.6.3 Application of PMSM drive in electric vehicles

Due to its increased power density and efficient control system, PMSM motors have been the preferred choice in EV drive trains.

8.7 BLDC motor for EVs

PM BLDC motor drives are distinguished by their great efficiency and power density. The motors can eliminate the need for energy to manufacture magnetic poles by using permanent magnets. As a result, they are more efficient than DC motors, induction motors, and SRMs. In addition, heat is efficiently vented to the environment. If a conduction-angle control is utilized on a PM BLDC motor, the speed range can be expanded three to four times beyond the basic speed.

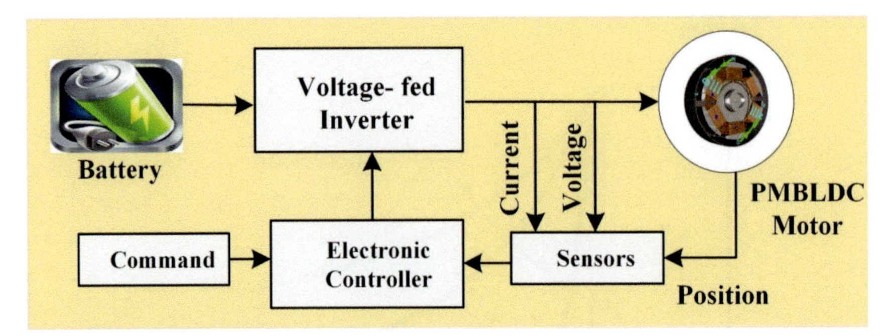

Figure 8.24 Basic set up of PMBLDC motor system.

Other disadvantages of PM BLDC motor drives include the high cost of the magnet and the difficulty of building a big torque into the motor due to the magnet's mechanical strength. Although PM BLDC motors do not have a brush to limit speed, there are still concerns about the magnet's fixing intensity, which limits the maximum speed if the motors are of the inner-rotor type. Furthermore, the field weakening capability of this motor is relatively limited. The presence of the PM field, which can only be diminished by producing a stator field component that opposes the rotor magnetic field, is the reason for this. Nonetheless, by extending the commutation angle, extended constant power operation is achievable.

PM brushless motor drives are primarily made of permanent magnets (PM). PM BLDC motors are PM AC machines with trapezoidal back-emf waveforms caused by the motor's focused windings. There is no rotor copper loss because there are no windings in the rotor, making it more efficient than induction motors. The motor drive is tiny, light, and reliable, and it produces more torque and specific power while dissipating heat more efficiently. In comparison to the DC brushed motor system, the PM BLDC motor system requires less maintenance and is more efficient.

The following are the main benefits of employing a PM BLDC motor: h high-energy PMs, lightweight, and low volume provide improved power density and efficiency due to the absence of copper loss; h better heat dissipation and cooling; h increased reliability due to decreased heating and manufacturing faults.

Fig. 8.24 shows the basic structure of the PM brushless DC machine

A voltage-fed inverter, an electronic controller, and sensors make up the single-PMBLDC motor architecture system. The current is synchronized with the flux with the help of a position sensor. Controlling the stator

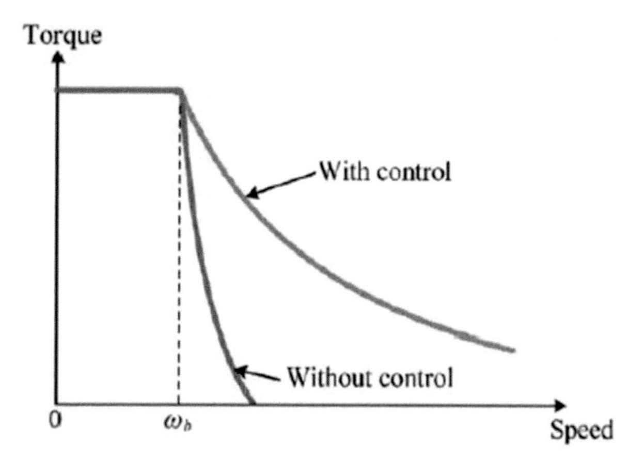

Figure 8.25 Torque-speed capabilities of phase advance angle control of PM BLDC motor.

currents to align the rectangular current with trapezoidal flux is a reasonably simple way to adjust the speed.

8.7.1 PM brushless DC motor control

The PM BLDC drive is powered by rectangular AC with substantial torque pulsation. To operate the PM BLDC motor, the stator and rotor fluxes are kept close to 90°, resulting in the highest torque per ampere in the constant-torque area. The phase-advance angle control (ACC) allows EVs to run at a steady power level. Due to the tiny difference between the applied voltage and back EM, the PM BLDC motor runs out of time to engage the phase current while operating at speeds greater than the base speed. The operating region with constant power can be extended by progressively increasing the phase-advance angle, as shown in Fig. 8.25.

8.7.2 Application of PM BLDC motor in electric vehicle

PM the BLDC motor has long been the preferred motor for EV applications. The majority of in-wheel hub BLDC motors may be found in two-wheelers, three-wheelers with FG drives, and EV conversion kits. BLDC motors are primarily employed in two-wheeler and three-wheeler vehicles these days. BLDC motors are divided into two types:

Out-runner type BLDC motor

In out-runner BLDC motors, the rotor is on the outside and the stator is on the inside. Because their wheel is directly coupled to the external

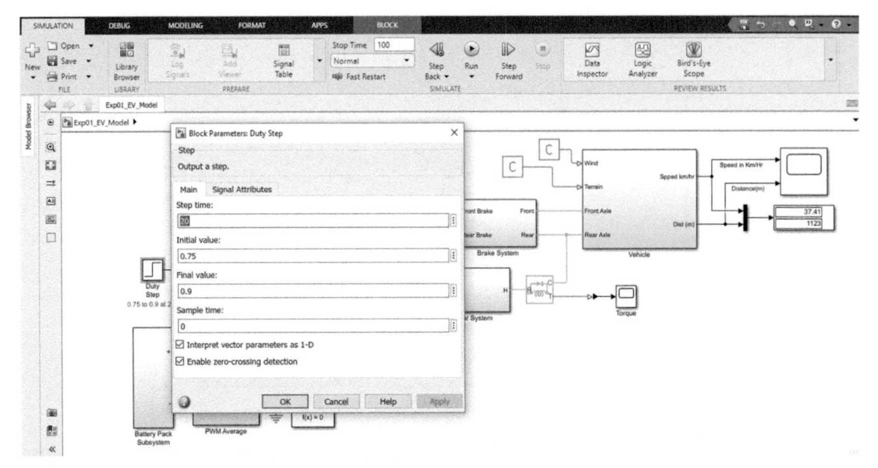

Figure 8.26 Block parameters—duty step.

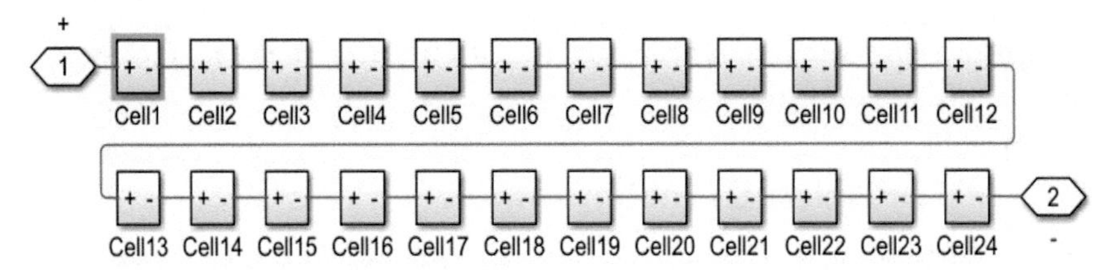

Figure 8.27 Battery pack subsystem.

rotor, they are also known as "hub motors." External gear systems are absent, although planetary gear systems are available in some cases. There is no requirement for space for mounting the motors because they are directly linked to the rotor. These are most commonly found in electric bicycles, scooters, and EVs with in-wheel drive.

In-runner type BLDC motor

Out-runner BLDC motors have the rotor outside and the stator inside, whereas in-runner BLDC motors have the rotor inside and the stator outside. These systems require external transmission systems, such as FG or chain drives, to send power to the wheels.

Simulation of EV:

Select step response by using link

Simulink/Sources/Step. Then set the parameters of step response is given in Fig. 8.26.

The battery pack subsystem and each cell subsystem are given in Figs. 8.27 and 8.28.

Here is the subsystem of PWM average (Fig. 8.29).

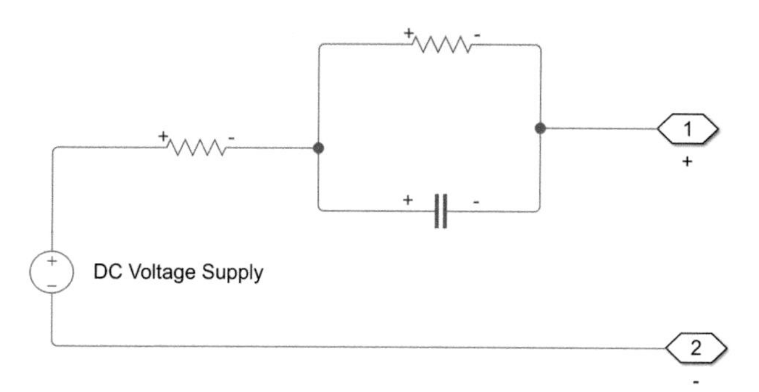

Figure 8.28 Cell subsystem.

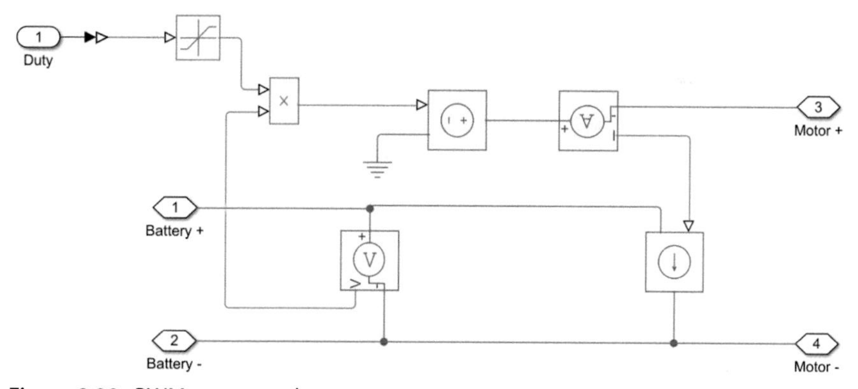

Figure 8.29 PWM average subsystem.

The DC motor parameters are given in Figs. 8.30 and 8.31.

The signal builder is given in Fig. 8.32

The subsystem of gear system and brake system are given in Figs. 8.33 and 8.34.

The vehicle body setup circuit is given in Fig. 8.35. At last, the MATLAB/Simulink Model of EV will get like Fig. 8.36.

8.8 Switched reluctance motor drives for EV's

Motor with reluctance switch SRMs, also known as doubly salient motors, are synchronous motors that use unipolar inverter-generated current to operate them. SRM motors operate on the variable reluctance principle. SRMs are typically used for high-speed operation with minimal mechanical failure. They are also suitable for driving EVs as in-wheel drive systems due to their great mechanical integrity [1]. However, they have lower torque

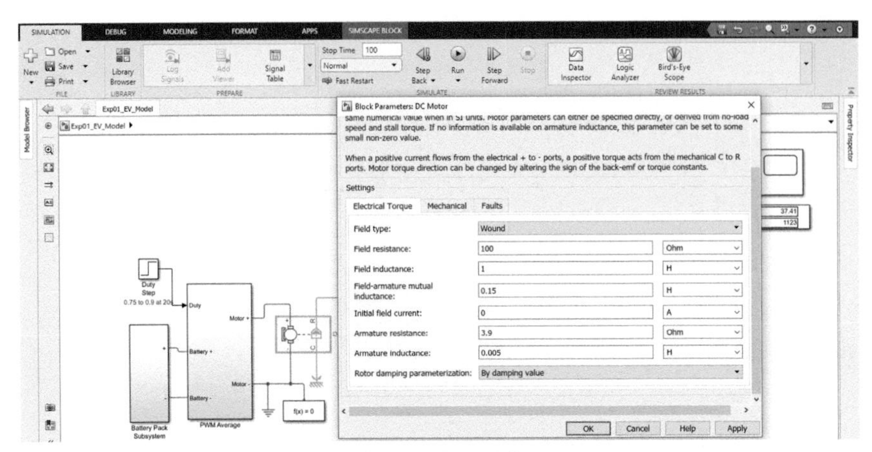

Figure 8.30 Setting of DC motor values—electrical torque.

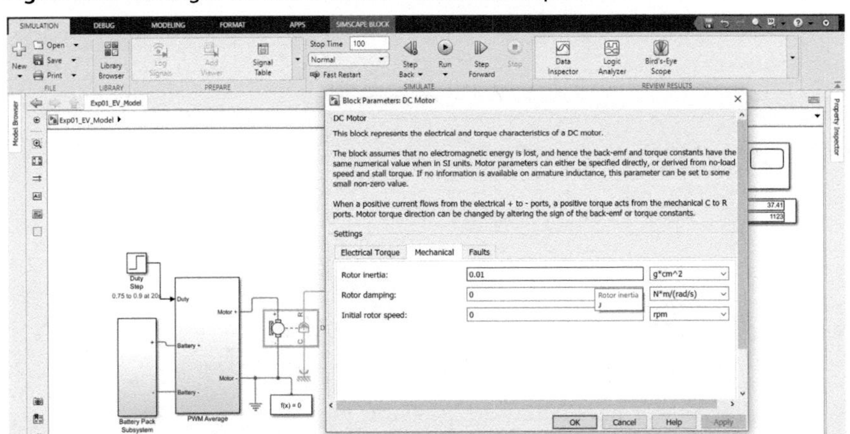

Figure 8.31 Setting of DC motor values—mechanical.

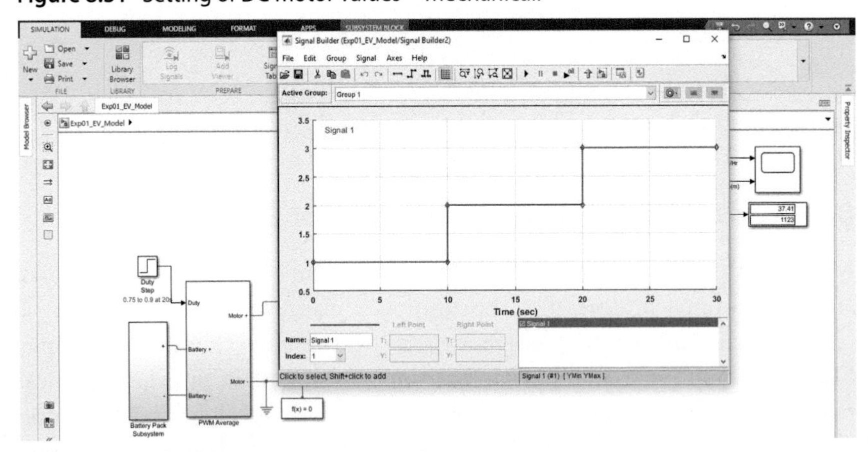

Figure 8.32 Signal builder.

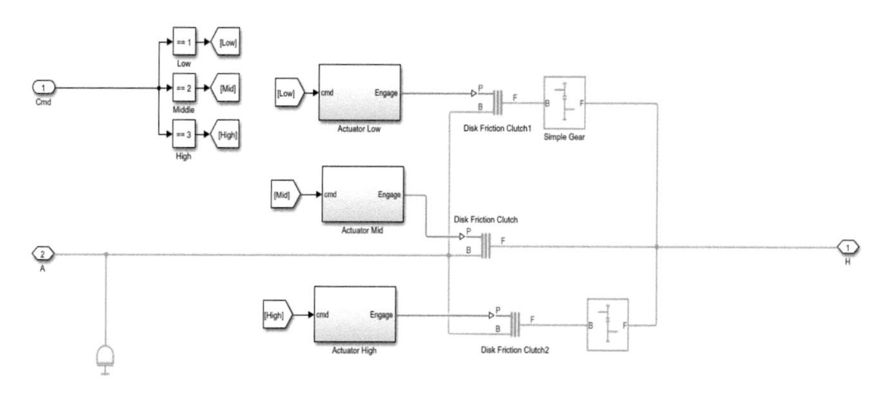

Figure 8.33 Gear system.

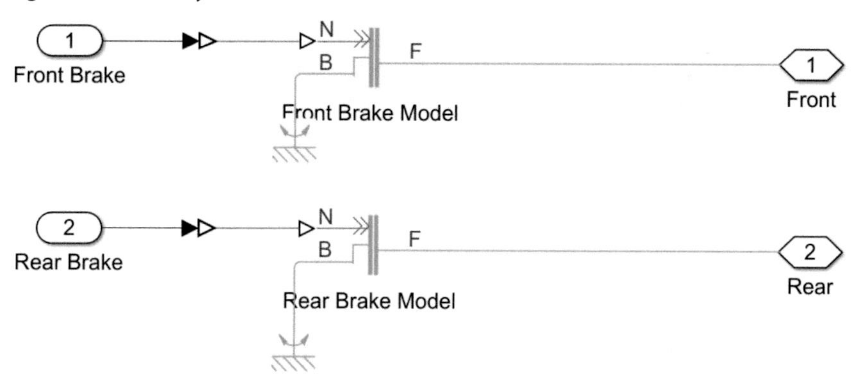

Figure 8.34 Brake system.

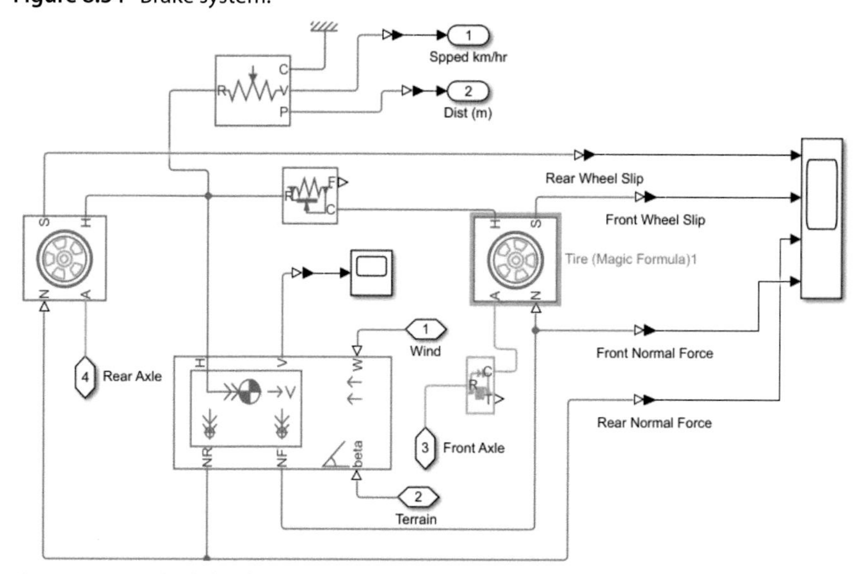

Figure 8.35 Vehicle body set up.

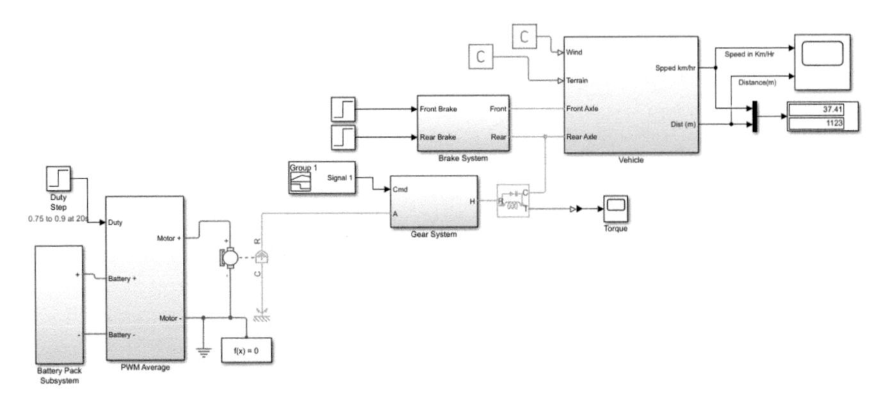

Figure 8.36 MATLAB/Simulink model of electric vehicle.

density, higher torque ripple, and more acoustic noise as drawbacks. The speed and torque responses are presented in Fig. 8.37. The scheme of SRM motor drive is given in Fig. 8.38.

Because the rotor construction is simple and does not require windings, magnets, commutators, or brushes, it can accelerate quickly and operate at extremely high speeds. Making it appropriate for EV propulsion without the use of gears. The two major control systems for SR motor control systems are current copping control (CCC) and AAC. The base speed, ωb, at which the back EMF equals the DC source voltage, is the speed boundary between these two control systems.

8.8.1 SRM control system

Because the back EMF is lower than the DC voltage below the base speed control system, the phase current can be regulated at the specified value, allowing for constant torque operation. The torque reduces when the back EMF is greater than the DC source voltage above the base speed. When the back EMF increases as the rotor speed decreases, the phase current torque decreases in inverse proportion to the rotor speed. In an AAC system operating in the constant power area, phase advancing is not achievable beyond the motor's critical speed, therefore it functions in natural mode.

Fig. 8.39 depicts the torque–speed profile in all three operating regions of SRMs in EVs: constant torque operation, constant power region, and natural operating region application. The first SR motor was created by Davidson in Scotland in 1838 for the purpose of driving a locomotive. In addition,

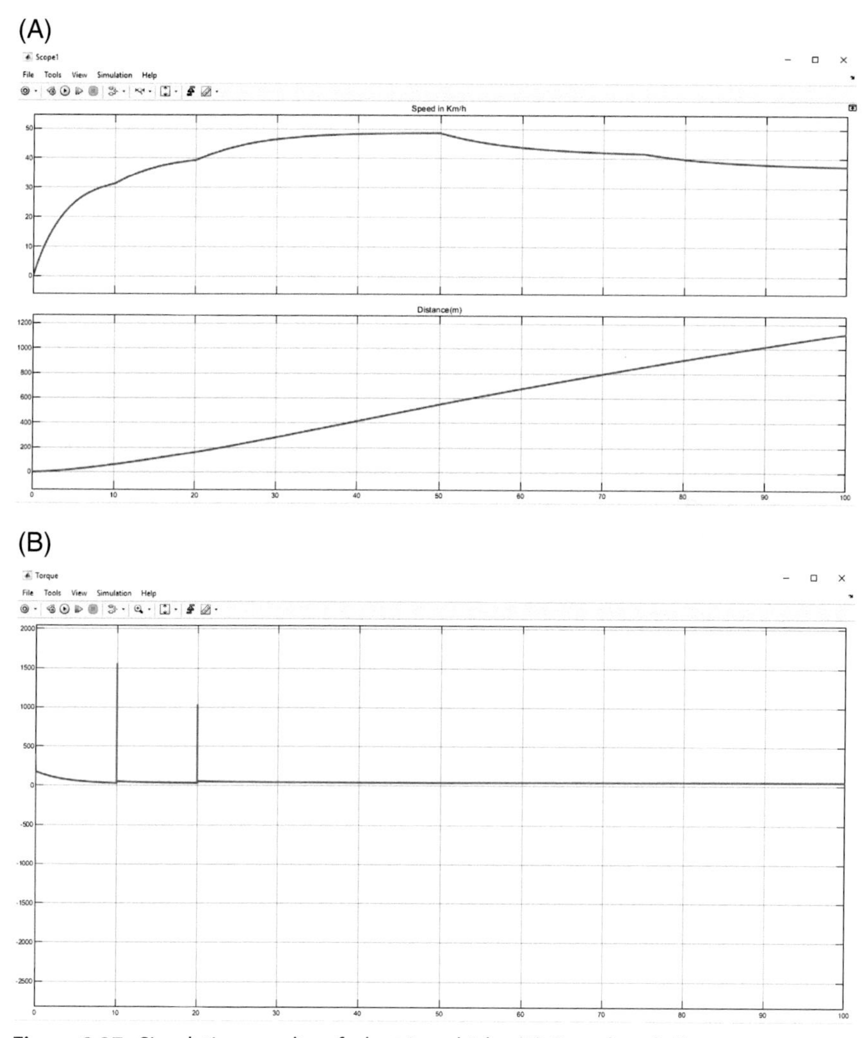

Figure 8.37 Simulation results of electric vehicle. (A) Speed and distance response, (B) torque response.

many research and development initiatives involving the use of SRM drive systems in electric propulsion systems are underway.

For driving EVs, two major types of SR motor drives are shown. The first is an SR with an internal high-speed capable motor and a planetary gear for speed reduction. The second is a low-speed SR motor for direct-drive in-wheel applications.

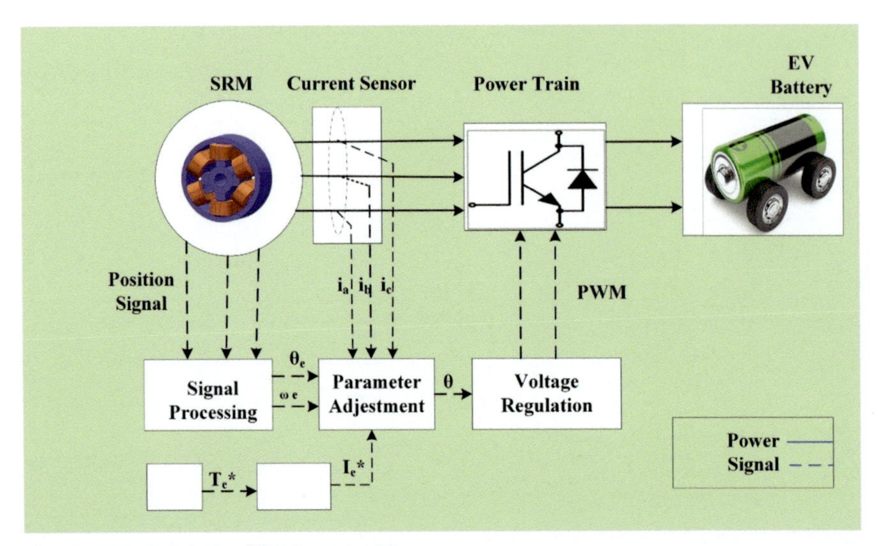

Figure 8.38 Scheme of SRM motor drive.

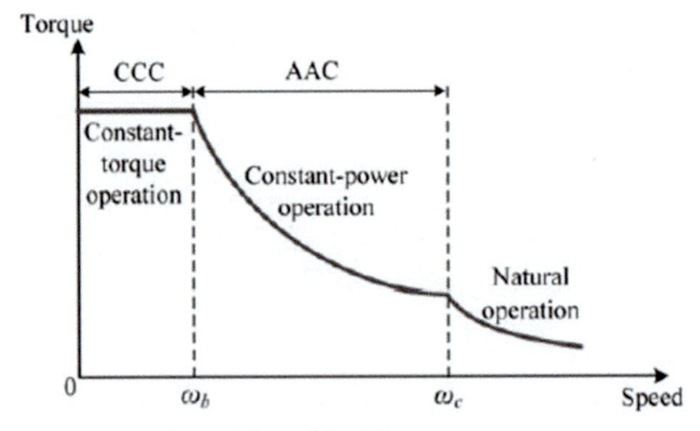

Figure 8.39 Torque-speed capability of the SR motor.

(a) SR motor drive with planetary gearing, the SR motor drive is designed for high-speed operation using a planetary gear to synchronize the motor speed to the wheel speed in a single-motor architecture system.

(b) To bypass the transmission gear and differential gear, the SR with an outer-rotor topology for low-speed operation is developed to directly drive each tyre while driving EVs.

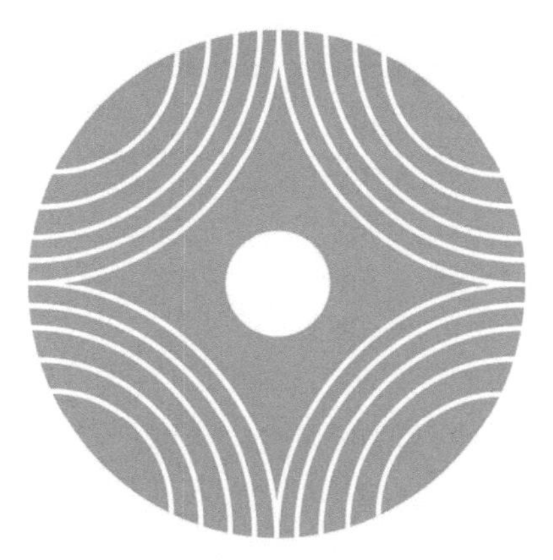

Figure 8.40 SynRM with axially laminated rotor.

8.9 Synchronous reluctance motor drives for EV's

A synchronous reluctance motor operates at synchronous speeds, combining the benefits of both PM and induction motors. They are as reliable and fault-tolerant as an IM, as efficient and tiny as a PM motor, and without the disadvantages of PM systems. They use a similar control method as PM motors. Controllability, manufacturing, and poor power factor are all issues with SynRM that prevent it from being used in EVs. However, research is ongoing, and some progress is being made, with the rotor design being the primary source of concern. One way to improve this motor is by increasing the saliency which provides a higher power factor. It can be achieved by axially or transversally laminated rotor structures, such an arrangement is shown in Fig. 8.40. Improved design techniques, control systems, and advanced manufacturing can help it make its way into EV applications. Increasing the saliency, which delivers a larger power factor, is one strategy to boost this motor. Axially or transversally laminated rotor architectures can be used to achieve this, as seen in Fig. 8.37. Improved design methodologies, control systems, and improved manufacturing could pave the road for it to be used in EVs.

8.9.1 PM assisted synchronous reluctance motor

By incorporating some PMs into the rotor of SynRMs, a PM-aided synchronous reluctance motor may be created with higher power factors. Despite the fact that it is identical to an IPM, the number of PMs employed and the flow linkages created by them are both lower. With minimal

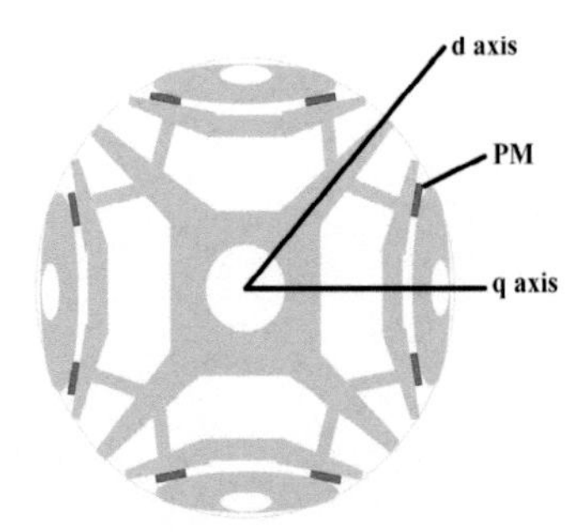

Figure 8.41 Permanent magnet (*PM*) assisted SynRM. Permanent magnets are embedded in the rotor.

back EMF and little modification to the stator, PMs placed in the proper amount to the core of the rotor improve efficiency. The issues related to demagnetization caused by overloading and high temperatures seen in IPMs are not present in this idea. This motor can perform similarly to IPM motors if an efficient optimization strategy is used. BRUSA Elektronik AG presented a PM-assisted SynRM suited for EV application (Sennwald, Switzerland). PM-assisted SynRMs, like SynRMs, can improve with better design methodologies, control systems, and advanced production technologies. Fig. 8.41 shows a demonstration of the rotor of PM-assisted SynRM.

8.10 Future trends of motor drives in EV applications

The adoption of EVs has opened up new options and ways to better both the vehicles and the systems that go with them, such as the power supply. EVs are being hailed as the future of transportation, while the smart grid appears to be the grid of the future. V2G is the link that connects these two technologies, and it benefits both of them. Other critical systems required for a sustainable EV scenario—charge scheduling, VPP, smart metering, and so on—come with V2G. Existing charging solutions must vastly improve in order for EVs to become universally adopted. To make EVs more flexible, the charging time must be drastically reduced. To advance EV technology, better batteries are required. There is a demand for batteries that are made of nontoxic materials, have a greater power density, are less expensive and weigh less, have a bigger capacity, and recharge in less time. Though technologies that are superior to Li-ion have been developed, they are not being explored commercially due to the high price of developing a working version.

Furthermore, Li-ion technology has a lot of room for improvement. Li-air batteries could be a viable way to extend the range of EVs. Permanent magnet motors made of rare-earth elements are unlikely to be used in EVs. Induction motors, synchronous reluctance motors, and SRMs are all options. Tesla now uses an induction motor in its models. Internal permanent magnet motors may be used indefinitely. The current cabled charging system is likely to be replaced by wireless power transfer technology. This characteristic was adopted by major automakers' concepts to emphasize their utility and convenience. The Vision Mercedes-Maybach 6 and the Rolls-Royce 103EX are two examples of this. Electric roadways for vehicle wireless charging may also appear. Though this is no longer a feasible option, it is possible that things will change in the future. Electrode, an Israeli startup, promises to be able to perform this feat in a cost-effective manner in recent work in this field. Trucks and other vehicles that follow a predetermined route along the roadway can acquire their power from overhead lines such as trains or trams. It will allow them to collect energy as long as their route is in close proximity to power lines, after which they will be able to continue using energy from on-board sources. Siemens tried a system like this on a highway in Sweden, using Scania diesel-hybrid vehicles. It is possible that new methods of recovering energy from the vehicle will emerge. Goodyear has developed a tyre that uses thermo-piezoelectric material to gather energy from the heat generated there.

Solar-powered automobiles are also a possibility. Until date, these have not proven to be useful because installed solar cells can only convert about 20% of the incoming electricity. Much research is being done to make the electronics and sensors in EVs more compact, durable, and affordable, which is leading to enhanced solid-state devices that can fulfill these goals with the promise of cheaper products if mass-produced. Gas sensors, smart LED drivers, smart drivers for automotive alternators, sophisticated gearboxes, and small and smart power switches for tough environments are other examples. The findings of some research could be useful in studies on fail-proof on-board power supplies for EVs. Future research will, of course, focus on making EVs more efficient, inexpensive, and convenient. A lot of study has already been done on how to make EVs more inexpensive and capable of reaching longer distances: energy management, construction materials, new energy sources, and so on. More research into better battery technologies, ultracapacitors, fuel cells, flywheels, turbines, and various individual and hybrid designs is anticipated to continue.

FCVs may receive substantial attention in military and utility-based studies, but academics interested in improving urban transportation systems may find the in-wheel drive arrangement for BEVs interesting. In the near

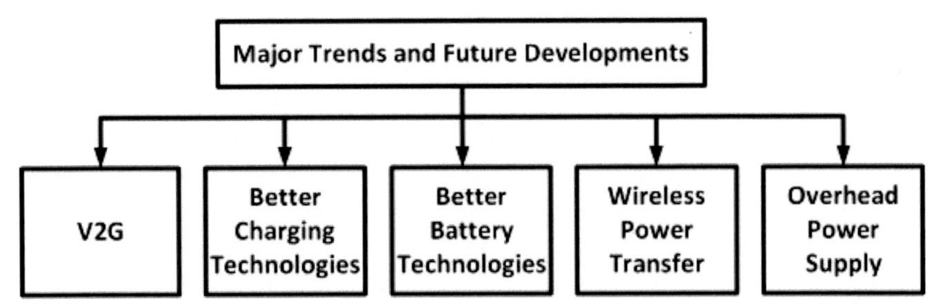

Figure 8.42 Major trends and sectors for future developments for EV. *EV*, electric vehicle.

future, better-charging technologies will remain a critical research topic. Wireless charging technologies are likely to gain more academics' attention because this is one of the areas where EV technology falls short. A great deal of research has already been done on EVs and the grid, including the problems and opportunities that EVs offer to the existing grid as well as the grid of the future. Research in these domains is projected to rise as more smart grids, distributed generation, and renewable energy sources are implemented. And, as research in the aforementioned fields expands, so will the search for improved algorithms to run the systems. Fig. 8.42 depicts the key trends and sectors for EV development in the future.

8.10.1 Electric motor with battery

The electric motor with battery is implemented using MATLAB Simulink is been given in Fig. 8.43 and its simulation result in Fig. 8.44.

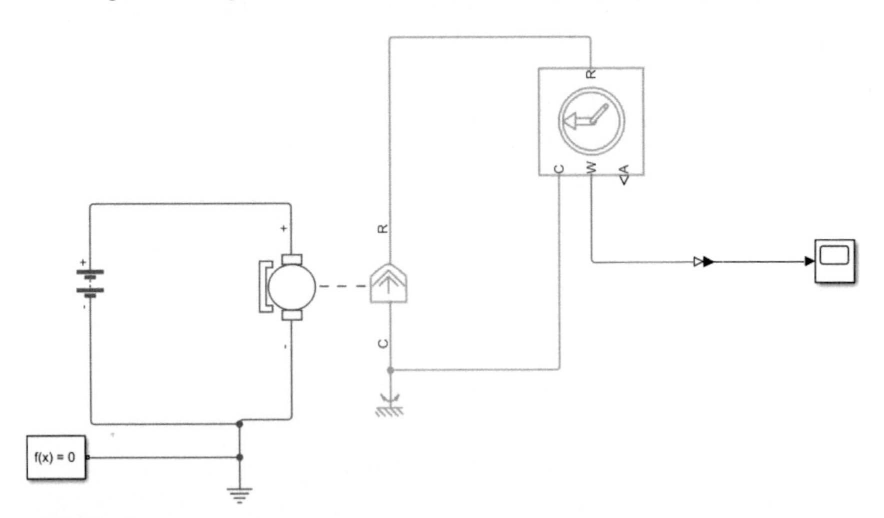

Figure 8.43 Electric motor with battery.

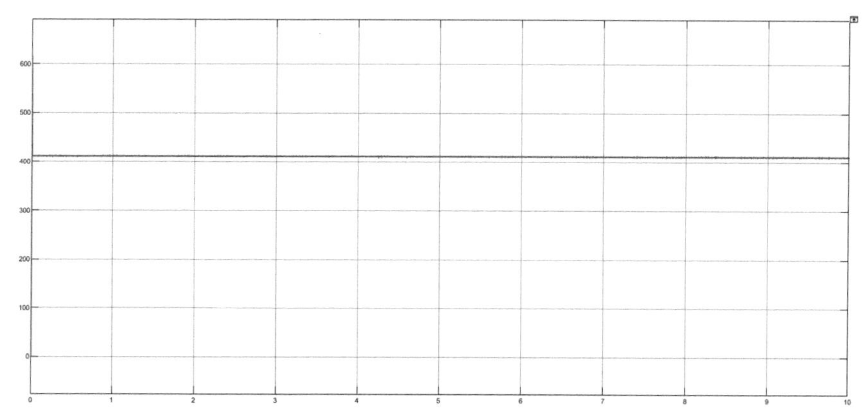

Figure 8.44 Simulation results of electric motor with battery.

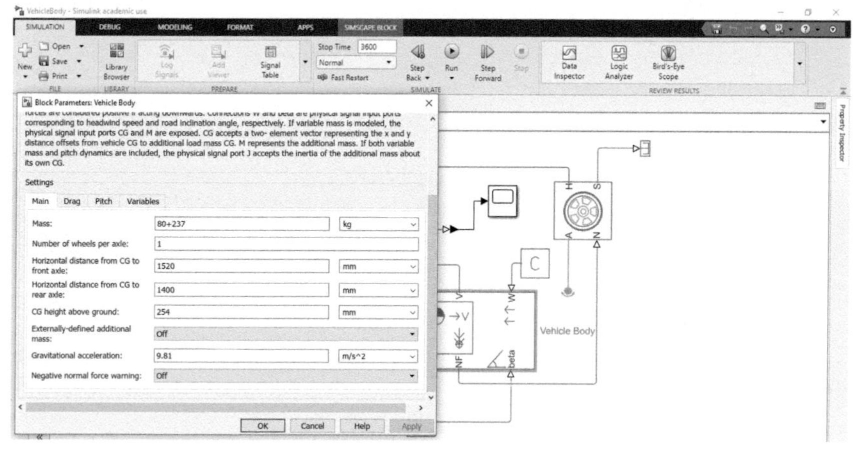

Figure 8.45 Block parameters—vehicle body—main.

8.10.2 Vehicle body

The MATLAB Simulink of vehicle body is been implemented and given in the below figures.

Once the vehicle body is been selected from the library, the parameters values are given as per Figs. 8.45 and 8.46.

The Tire (Magic formula) is been selected from the library, the parameters values are given as per Figs. 8.47–8.51.

The MATLAB/Simulink model of a vehicle body and its simulation results are given in Fig. 8.52.

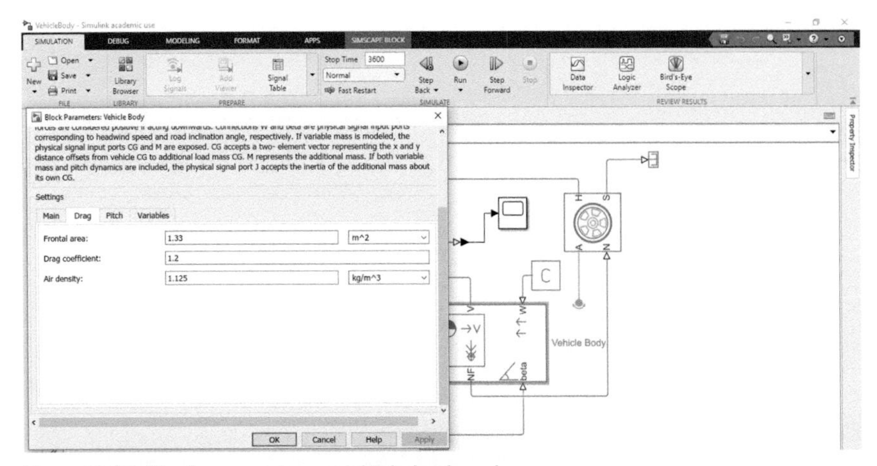

Figure 8.46 Block parameters—vehicle body—drag.

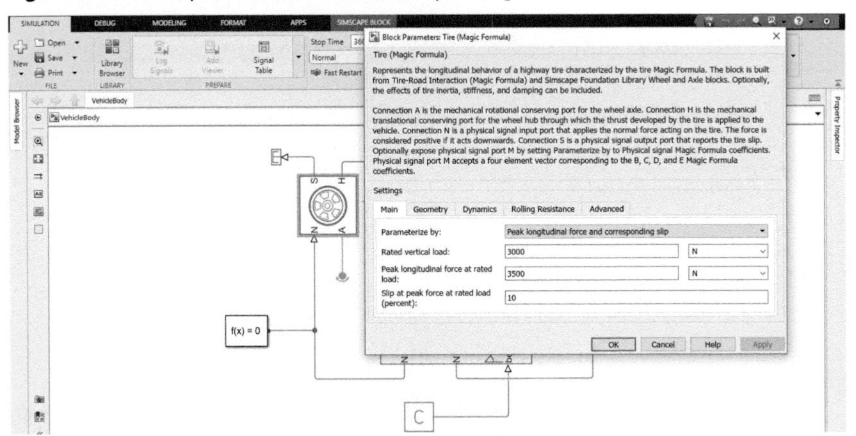

Figure 8.47 Block parameters—tire (magic formula)—main.

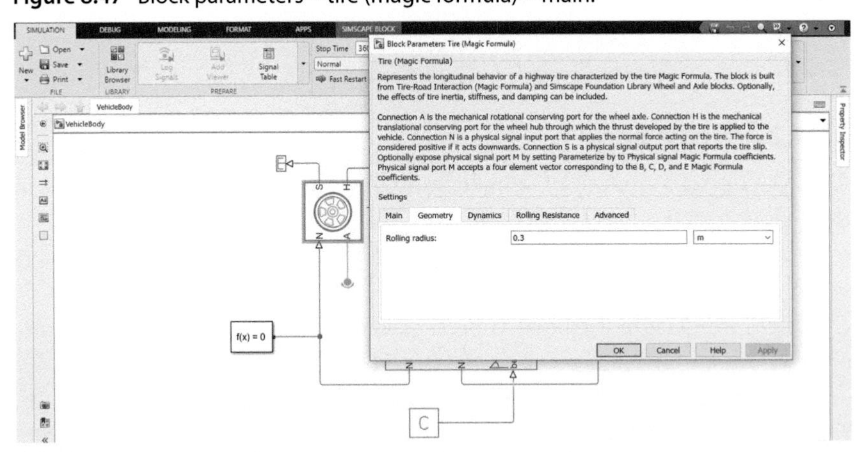

Figure 8.48 Block parameters—tire (magic formula)—geometry.

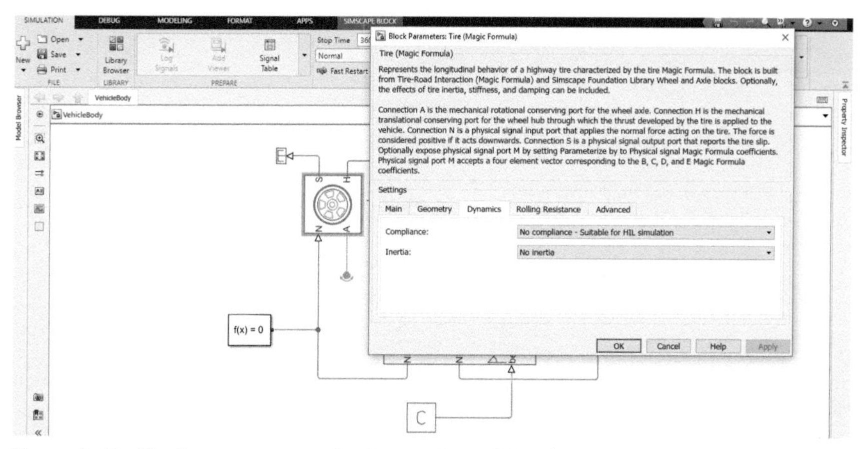

Figure 8.49 Block parameters—tire (magic formula)—dynamics.

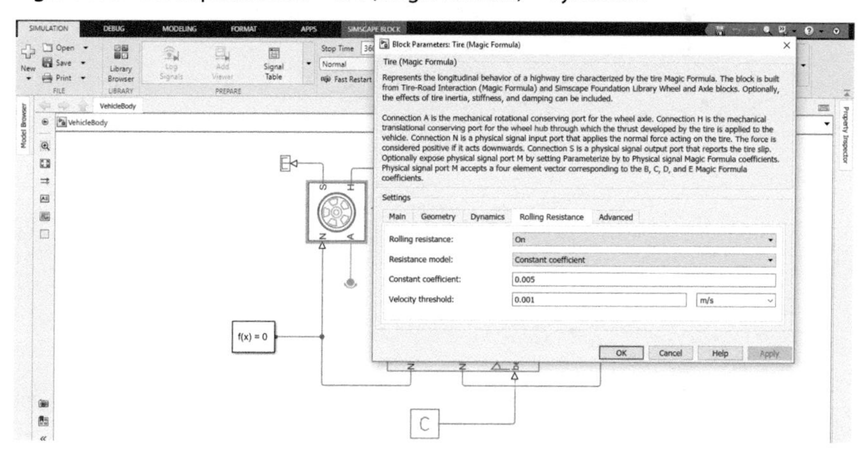

Figure 8.50 Block parameters—tire (magic formula)—rolling resistance.

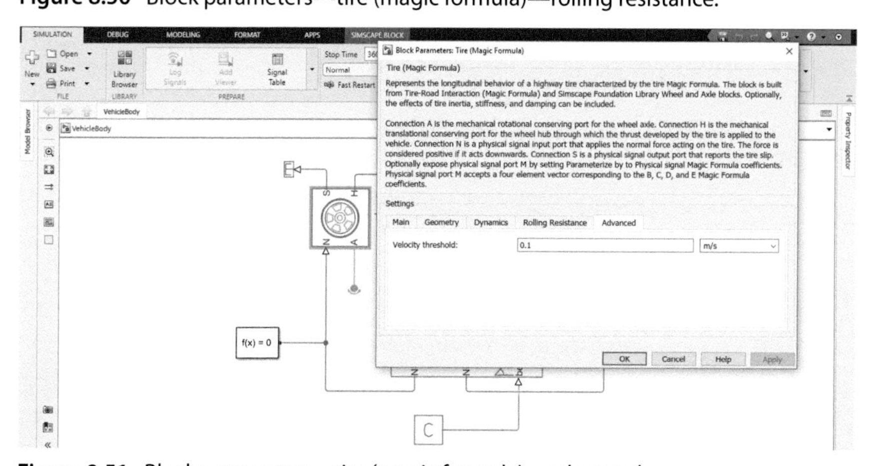

Figure 8.51 Block parameters—tire (magic formula)—advanced.

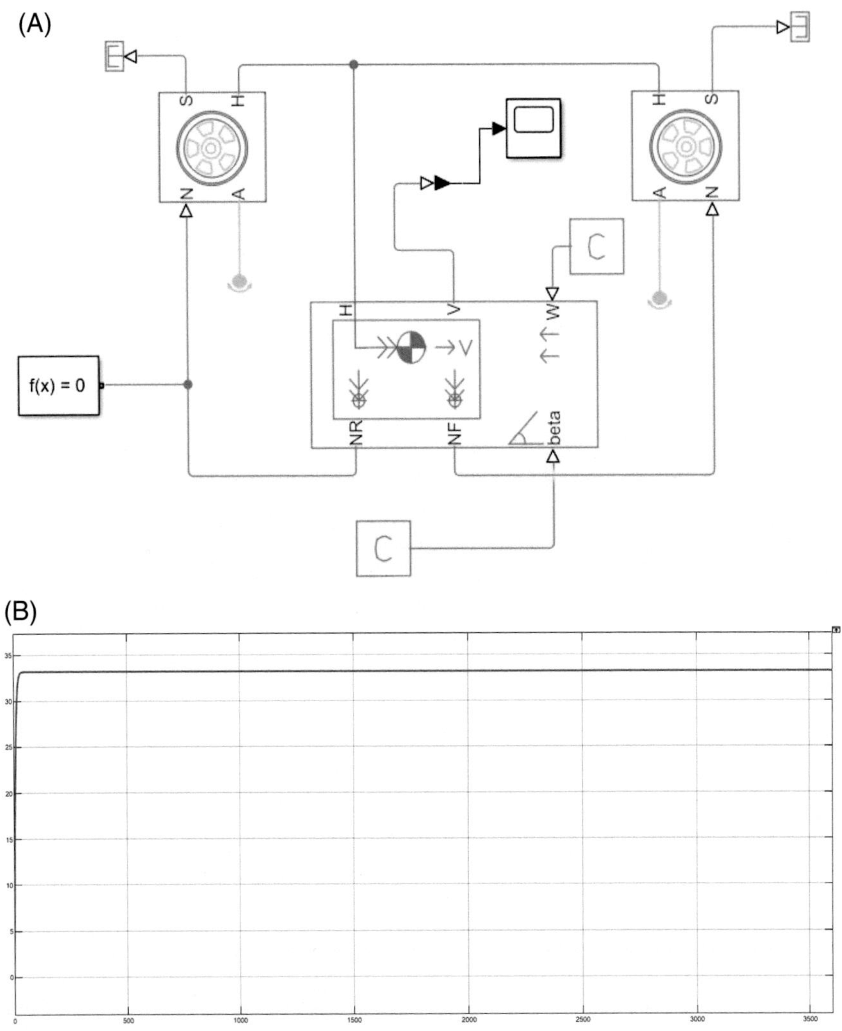

Figure 8.52 (A) MATLAB/Simulink model of vehicle body, (B) simulation results of vehicle body.

8.10.3 Simulation of electric vehicle with SOC

The battery pack subsystem is for the EV simulation is given in Fig. 8.53. The SOC estimation circuit is presented in Fig. 8.54.

The simulation of EV with SOC using MATLAB and its simulation results are given in Figs. 8.55 and 8.56.

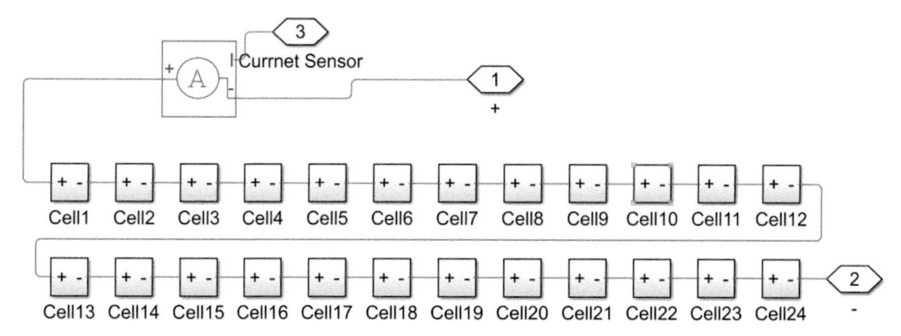

Figure 8.53 Battery pack subsystem.

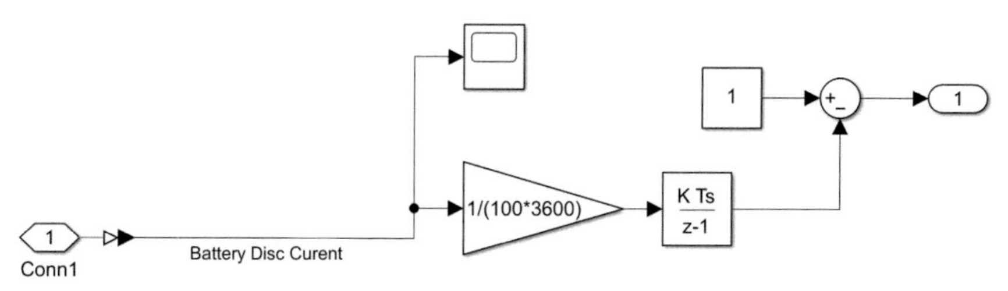

Figure 8.54 SOC estimation.

8.10.4 Advance magnetless motor

The absolute magnitude and unpredictability of the neodymium price have reignited research into advanced magnetless machines, posing a risk to PM machine development. The IM and the SR machine are magnetless machines because they do not have any PMs, but they do form their own families, and the term "advanced" is used to distinguish them from those magnetless machines that are still being developed or are relatively immature. The configuration of power electronics components in a battery EV is depicted in Fig. 8.57. The auxiliary supply provides the vehicle's equipment with the power it requires.

There are five key sophisticated magnetless motor drives that can be used for EV propulsion: synchronous reluctance (SynR) h DSDC (Doubly-salient DC) h flux-switching DC (FSDC) h vernier reluctance (VR) h vernier reluctance with two feeds (DFVR) magnetless motors with advanced technology have the potential to be used in a variety of applications. In terms of maturity, SynR and VR motor drives are the most mature technologies, having been developed for decades. DSDC and FSDC motor drives, on the other hand, are quite mature, having been developed for more than a decade, and are considered to be the most prominent magnetless motor

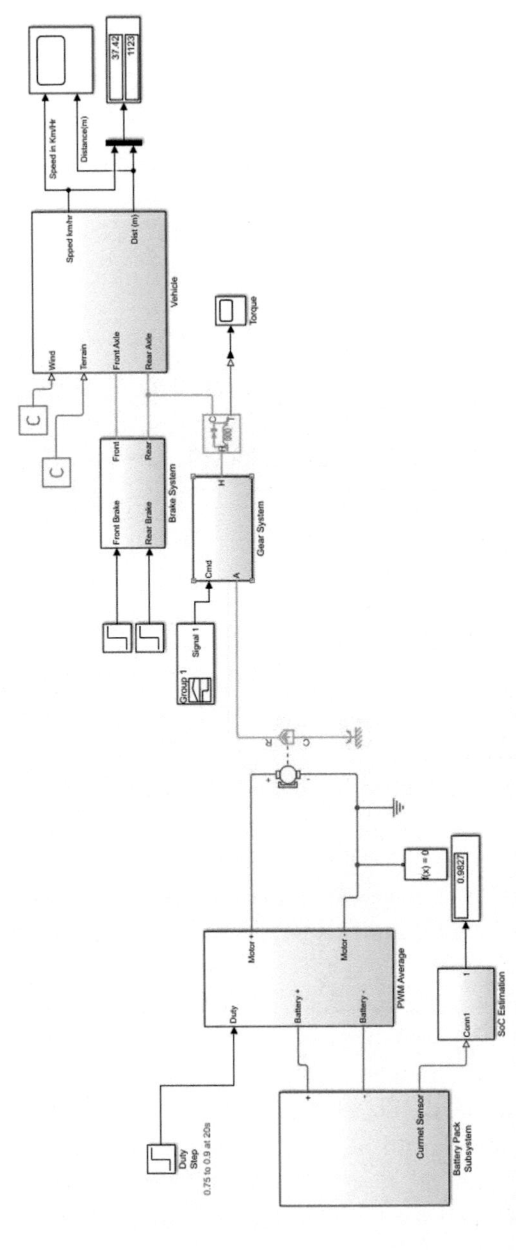

Figure 8.55 MATLAB/Simulink model of electric vehicle with SOC.

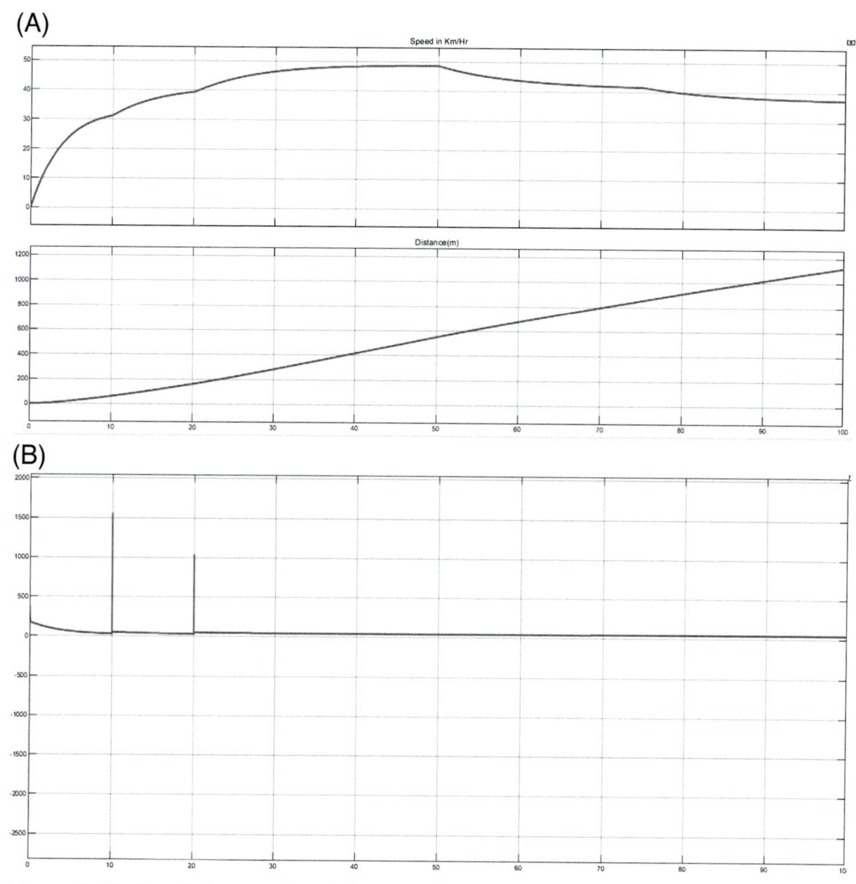

Figure 8.56 Simulation results of electric vehicle with SOC.

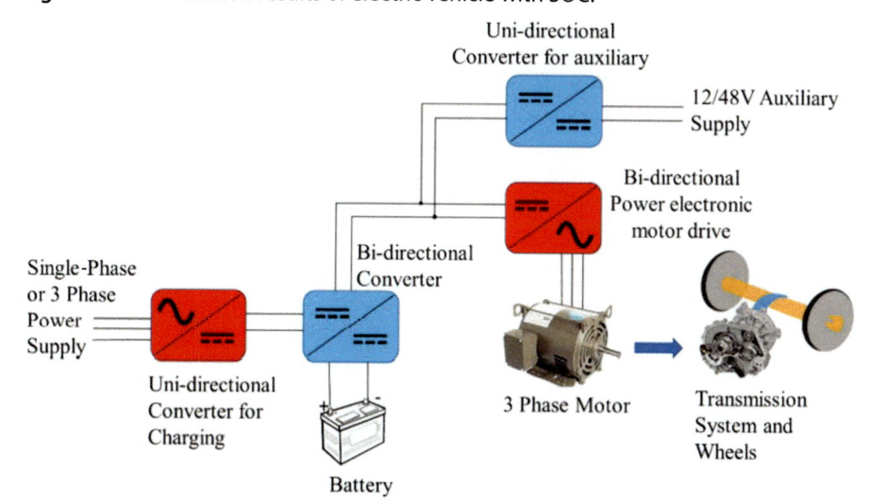

Figure 8.57 Power electronics in electric vehicles.

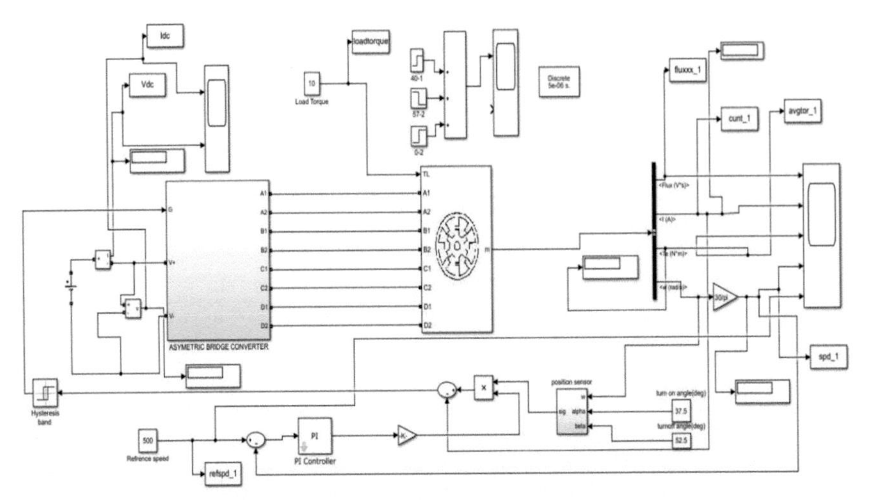

Figure 8.58 Speed control of 8/6 switched reluctance motor for light electric vehicle.

drives, namely the DSPM and FSPM. The AFM machines are immature, as is the DFVR motor drive, which was recently derived from the VR motor drive.

8.11 Case studies

1. Construct the Speed Control of 8/6 Switched Reluctance Motor for Light EV in Fig. 8.58.

8.12 Summary

In this chapter, EV simulations using MATLAB/Simulink are provided in step-by-step manner. The Simulink diagrams and results are clearly given here. It is easy for everyone to understand the EV circuits using MATLAB software.

Multiple choice questions

1. High voltage battery is used in which vehicle?
   - **a)** Electric Vehicle
   - **b)** Hybrid vehicle
   - **c)** Both Electric and Hybrid Vehicle

   Answer: c) Both Electric and Hybrid Vehicle

2. Which of the following vehicles produces zero emissions?
   a) Electric
   b) Hybrid
   c) Traditional
   Answer: c) Both Electric and Hybrid Vehicle
3. Electric Vehicles are generally powered by ___________
   a) Aluminum batteries
   b) Lead-acid batteries
   c) Sodium batteries
   d) Magnesium batteries
   Answer: b) Lead-acid batteries
4. What are the two main types of hybrid vehicle?
   a) The series hybrid vehicle and the mild hybrid vehicle.
   b) The parallel hybrid vehicle and the full hybrid vehicle.
   c) The series hybrid vehicle and the parallel hybrid vehicle.
   d) The full hybrid vehicle and the empty hybrid vehicle.
   Answer: c) The series hybrid vehicle and the parallel hybrid vehicle.
5. Which of these is a purpose of the power-split device?
   a) To split electrical energy into mechanical energy.
   b) To allow both the engine and electric motor to propel the vehicle.
   c) To recharge the battery while braking.
   d) To recharge the brakes while driving.
   Answer: b) To allow both the engine and electric motor to propel the vehicle.
6. Except _________, components are same in Electric and hybrid vehicles.
   a) Battery
   b) ECU
   c) Generator
   d) Internal combustion engine
   Answer: d) Internal combustion engine
7. What affect do petrol and diesel engines have on the environment?
   a) They help deplete natural resources and contribute to increasing CO2 emissions
   b) They help deplete natural resources and reduce CO2 emissions
   c) They contribute to increasing CO2 emissions but help reduce global warming
   Answer: They help deplete natural resources and contribute to increasing CO2 emissions

8. The term regeneration means
   - **a)** It's when electricity is generated during deceleration and braking
   - **b)** It's when the battery is charged during engine idling
   - **c)** It's when the battery is recharged from the mains supply

   Answer: a) It's when electricity is generated during deceleration and braking
9. What voltage is likely to be available from the battery of an electric vehicle or hybrid?
   - **a)** 12V
   - **b)** 24V
   - **c)** 300V

   Answer: c) 300V
10. The acceleration of vehicle does not depend on
    - **a)** Power delivered by the propulsion unit
    - **b)** Velocity
    - **c)** Aerodynamics of the vehicle
    - **d)** Road conditions

    Answer: b) velocity

References

Lan, Y., Benomar, Y., Deepak, K., Aksoz, A., Baghdadi, M.E., Bostanci, E., Hegazy, O., 2021. Switched reluctance motors and drive systems for electric vehicle powertrains: state of the art analysis and future trends. Energies 14, 2079.

Karki, A., Phuyal, S., Tuladhar, D., Basnet, S., Shrestha, B.P., 2020. Status of pure electric vehicle power train technology and future prospects. Appl. Syst. Innov. *3*, 35. https://doi.org/10.3390/asi3030035.

Un-Noor, F., Padmanaban, S., Mihet-Popa, L., Mollah, M.N., Hossain, E., 2017. A comprehensive study of key electric vehicle (EV) components, technologies, challenges, impacts, and future direction of development. Energies 10, 1217.

Xue, X.D., Cheng, K.W.E., Cheung, N.C., 2008. Selection of ELECTRIC mOTOR dRIVES for electric vehicles. In: 2008 Australasian Universities Power Engineering Conference, pp. 1–6.

CHAPTER 9

Electric drives for water pumping applications

Contents

9.1 Introduction

Water and wastewater treatment requires moving water through the different stages of treatment. To do that, pump stations are used. Each part or station of the water treatment process can require a different type of pump. While the pumps are different, they all share the common architecture of an electric motor and a method of controlling those motors.

According to the US EPA, pump station capacities range from 76 lpm (20 gpm) to more than 378,500 lpm (100,000 gpm). Prefabricated pump stations generally have a capacity of up to 38,000 lpm (10,000 gpm). Usually, pump stations include at least two constant-speed pumps ranging in size from 38 to 75,660 lpm (10 to 20,000 gpm) each and have a basic wet-well level control system to sequence the pumps during normal operation.

The process of moving water is extremely energy-intensive. In the United States, electric motor-driven devices, including pumps, use almost 65%–70% of all electricity produced in the country. Water and wastewater systems are known to utilize almost 50% of the energy in any municipality, of which 90% of the energy is used by pumps.

Electric Motor Drives and Their Applications with Simulation Practices.
DOI: https://doi.org/10.1016/B978-0-323-91162-7.00009-6

Liquid level switches and sensors trigger when a desirable water level is attained. Trapped air column, or bubbler system that senses pressure and level, is commonly used for pump station controls. Other control alternatives are electrodes placed at cut-off levels and float switches. These sensors and switches signal the pump motor control systems to keep the water treatment process flowing and achieving optimum process efficiencies.

Municipal water systems use pumps to draw raw water from resources, such as lakes or rivers, for treatment to meet regulatory standards for potable water for human consumption or use in cooling towers, boilers, and other industrial applications.

9.1.1 Types of pumps used in the water and wastewater industry

Water and wastewater management has become a priority in industries such as chemical manufacturing, energy production, and food and pharmaceutical processing. The quality of water treatment entirely depends on the type of process employed. These treatment plants employ primary, secondary, and tertiary processes that vary depending on the level of contaminants in the water. The following are some popular pumps largely used in the water and wastewater industry for water treatment.

- Positive displacement pumps.
- Centrifugal pumps.
- Submersible pumps
- Rotary lobe pumps.
- Peristaltic pumps.
- Progressive cavity pumps.
- Airlift pumps.
- Trash pumps.
- Water pressure booster pumps.
- Agitator pumps.
- Circulation pumps.

Proper pump, motor, and controls selection optimizes the performance of water treatment systems and can provide energy savings of 20%–50%. Selecting a pump with the correct characteristics is achieved by studying pump performance curves. Below is an example of an electric submersible pump (ESP) pump performance curve. Horsepower motor load is the determining factor when selecting the correct motor controls.

9.1.2 Types of motor controls used with water and wastewater pumps

Contactor: Contactors are components designed to switch on and off heavy loads in pump motors. These components feature main contacts (poles), auxiliary contacts, and an operating coil. They energize the contactor to switch on and off the main contacts. Auxiliary contacts are designed for controlling and signaling various circuit applications, whereas main contacts are the current-carrying parts of these contactors.

Typically these contactors feature 3-pole electrically operated switches, which take less space when installed inside electrical enclosures. The motors used in water treatment and wastewater treatment pumps are known to draw more energy at any voltage. The possibility of electric shock increases at high voltage and may cause heavy damage. However, AC and DC contactors are safe to use while starting the motor, as there is no current flow between the circuit powering a contactor and the circuit being switched.

The contactors are mounted so they do not touch the circuit that is being switched. Because these contactors use less power than the main switching circuit, they help reduce power consumption. Advanced motor contactors feature compact designs, which further help reduce the footprint of the device and its power consumption.

Overload relay: When a motor draws excess current, it is referred to as an overload. This may cause overheating of the motor and damage the windings of the motor. Because of this, it is important to protect the motor, motor branch circuit, and motor branch circuit components from overload conditions. Overload or overheating is one of the major reasons for pump failure. Overload Relays protect the pump's motor from these conditions.

Designed as electromechanical devices, overload relays are distinguished as bimetallic, melting alloy, or solid-state electronic relays on the basis of their construction.

Bimetallic overload relays are one of the most common types of overload protection devices, and they feature adjustable trip points. Bimetallic overloads are engineered for automatic reclosing and compensate to prevent ambient temperature changes. In addition, these overload relays protect motors in extreme temperature environments.

Advanced bimetallic overload relays feature manual or automatic reset and test modes and a stop button that enables better device management. Many of these relays possess single-phase sensitivity, which helps protect motors against phase loss conditions. These relays are provided in three trip class ratings:

1. *Class 10* is a quick trip rating, suitable for submersible pumps used in water and wastewater industries. This rating indicates that the bimetallic overload relay will trip automatically within 10 s of the overload condition.
2. *Class 20* is the standard and ideal for general motor applications.
3. *Class 30* is a slow trip rating, which is suited for motors that drive high inertia loads and require long starting periods.

These features help minimize energy consumption and increase motor efficiency.

Motor protection circuit breaker: Designed for protecting motors from short circuits, phase-loss, and overloads, motor protection circuit breakers (MPCBs) are alternatives to thermal overload relays, and are equipped with several advanced features.

Commonly used in many water pumping systems, these circuit protection components are used as manual motor controllers or paired with contactors in several multimotor applications. MPCBs are mainly distinguished as open or enclosed. The difference between these types is where the circuit breaker is secured, either inside an enclosure or open in the panel. Most advanced MPCBs offer space savings, as they are designed without individual motor branch circuit fuses, overload relays, or circuit breakers.

Direct-on-line (DOL) motor starter: As the name suggests, these devices are used to start electric motors of pumps and other electronic devices such as compressors, conveyor belts, and fans. A motor starter features various electronic and electromechanical devices such as a contactor paired with a MPCB or an overload relay. DOL starters are used to start small water pumps because they provide several advantages such as 100% torque during starting, simplified control circuitry, easy installation and maintenance, and minimal wiring. Enclosed DOL Motor Starters are also an option, where the entire starter assembly is placed inside an enclosure.

Programmable Logic Controller (PLC): The programmable logic controller or PLC is really an industrialized computer that operates without a keyboard or monitor. Originally, the PLC was a replacement for large panels of relays that switched on and off, controlling a machine operation. The programming language of the PLC mimicked the Relay Logic, making the transition from relays to PLC's an easy to understand process. Today's PLC's offer much more complex operational capabilities and communications via

Ethernet or proprietary networks. The ability to control multiple pumps in a coordinated fashion make PLC's a common component of water management systems.

Variable Frequency Drive (VFD): They are used for running an AC motor at variable speeds or to ramp up speed for smoother start-up. VFD's control the frequency of the motor to adjust the pump motor RPM's. VFD's are widely used to regulate water flow at a water treatment plant, allowing more control over the flow of the pump.

Soft Starter: Between the simplicity of a DOL motor starter and the complexity of a VFD sits the soft starter. Electric motors often require large amounts of electricity during their acceleration. A soft starter can be used to limit the surge of current torque of the electric motors, resulting in a smoother startup. Soft starters can protect an electric motor from possible damage and at the same time extend the lifespan of your electric motor by reducing the heat caused by frequent starting and stopping. Soft starters limit the large inrush current demands on the electrical supply system. Soft starters are used with pumps in a process that requires to bring them up slowly to reduce pressure surges in the water system.

9.1.3 Selection of motor controls for water and wastewater pumps

9.1.3.1 Contactors

These devices are often confused with relays, however, the main difference is contactors can easily switch higher currents and voltages, whereas the relay is used for lower current applications. Keep the following in mind when selecting contactors for your motors:

- Decide what amount of current, full load amperage will be required to power your pump motor.
- Select the coil voltage for AC or DC operation based on your motor horsepower, input voltage, and single or 3-phase. Coils are mainly offered in control voltages such as 24VAC, 230VAC, 400VAC, 24VDC, and so on.
- Selecting a contactor with an IEC utilization category of AC-3 is typical for pump applications requiring starting and switching off motors during run time.
- Determine if your pump operation will require reversing of the direction, in which case a reversing contactor will be required.
- Choose the auxiliary contact based on normally open or normally closed configurations.

- In addition to the above considerations, it is important to concentrate on ambient and environment temperatures, necessity of latching, interlocking, enclosures, overload, timers, or coil surge suppressors.

9.1.3.2 Overload relays

When used in water or wastewater industries, overload relays are governed by strict requirements. With so many designs available, choosing the right overload relay may become difficult. These factors will simplify the selection process:

- Choose the overload relay that provides the most thermal overload protection. Overload relays help protect the motor from dangerous overheating, which causes motor failure. This protection takes the electricity consumption of the motor into consideration and applies it to an overload model to simulate the thermal energy inside the motor. Most overload relays are designed on either of the two overload models: Two-body and I2T models. Most bimetallic relays use the I2T overload model, whereas many medium or large-voltage electric motors use the two-body model.
- Phase loss protection is another important factor to consider, as phase loss is one of the major causes of motor failure. Phase loss occurs when the value of one phase equals to zero amps due to a blown phase. When the motor remains in this phase for a long time, it may be damaged permanently. Overload relays are designed to detect the phase loss condition, therefore it is important to understand the type of phase loss protection offered by overload relays.
- Other important protection factors to consider include underload, ground-fault current, stall, jam, and power and voltage protection.

9.1.3.3 Motor protection circuit breakers

Most pump systems in the market today include basic motor protection built into the motor or control box. This protection however is designed to safeguard against only current problems, therefore additional motor protection should be considered. There are many options available to choose from, each presents slightly different performance characteristics under overload conditions. Factors to consider are:

- High fault short circuit current rating because it assures safety and reliability in extremely high fault applications.
- Trip indication helps determine the type of maintenance or service that may be required by identifying the cause of tripping—short circuit or overload.

- Self-protection of these devices assures excellent motor protection and helps eliminate the need for additional circuit breakers and upstream fuses.

9.1.4 Applications of motor controls in the water and wastewater industry

Overload relays, contactors, and MPCBs are largely used in the following applications.

- *Influent and effluent pumps:* The influent pumps are usually placed at the start of the wastewater treatment plant. The wastewater first enters into the influent pump, which pumps it to other parts of the treatment plant. Effluent pumps are used for treating water that may contain solids up to 3/4".
- *Booster pumps:* These pumps are used for boosting the water pressure for use in light industrial and commercial applications. The booster pumps are mainly used for potable water applications.
- *Digested sludge pumps:* Contaminated sludge is treated in wastewater treatment plants before being released into water bodies. The sludge contains solids that may block the pipeline, so special pumps are used for the purpose. Sometimes, sludge choppers are integrated into the pumps, or submersible centrifugal pumps are used for pumping fluids having high sludge content.
- *Submersible pumps:* As their name suggests, these pumps are completely submerged in the liquid. These pumps are used to drain the slurry or sewage and are mostly placed under the sewage or wastewater treatment plants. Their operation is controlled using advanced motors.

9.1.5 Applicable standards and compliances

All motor controls referenced in this document are designed in accordance with standards published by National Electrical Manufacturers Association (NEMA) or International Electrotechnical Commission (IEC). NEMA is primarily a North American Standard, whereas IEC is a global standard.

9.1.5.1 NEMA rating

The NEMA ratings of a starter depend largely on the maximum horsepower ratings given in the NEMA ISCS2 standard. The selection of NEMA starters is done on the basis of their NEMA size, which varies from Size 00 to Size 9.

The NEMA starter, at its stated rating, can be used for a wide range of applications, ranging from simple on and off applications to plugging and

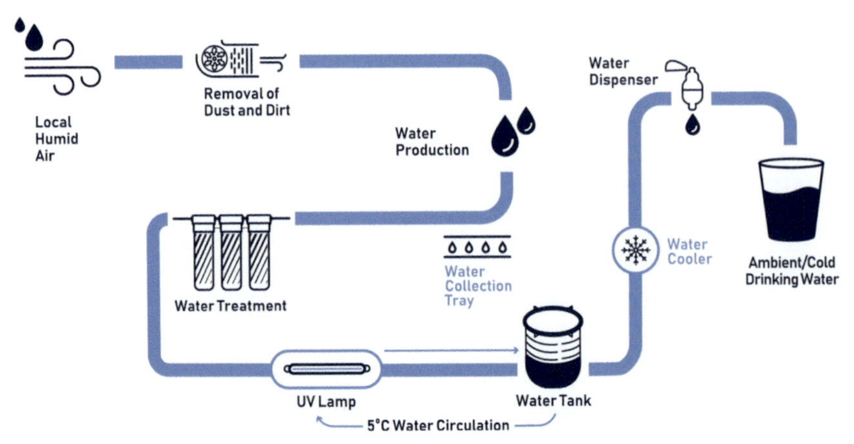

Figure 9.1 Drinking water production.

jogging applications, which are more demanding. It is necessary to know the voltage and horsepower of the motor when selecting the proper NEMA motor starter. In the case where there is a considerable amount of plugging and jogging involved, then derating a NEMA-rated device will be required.

9.1.5.2 IEC rating

The IEC has specified the operational and performance characteristics for IEC devices in the publication IEC 60947. Standard sizes are not specified by the IEC. The typical duty cycle of IEC devices is defined by utilization categories. As far as general motor starting applications are concerned, AC3 and AC4 are the most common utilization categories.

Unlike NEMA sizes, they are typically rated by their maximum operating current, thermal current, HP and/or kW rating.

9.2 Requirement of drives in drinking water production

Whether surface water or groundwater is used for water production, AC drives can help optimize the process and reduce energy and maintenance costs. A typical application is the control of deep well pumps, where the integrated advanced minimum speed monitor secures sufficient lubrication to protect the pump. Energy savings are achieved by selecting wells based on water level height. The drive is also widely used in the desalination industry for controlling inlet and high-pressure pumps as well as booster pumps in relation to energy recovery. The new method of water production is presented in Fig. 9.1.

9.3 Requirement of drives in drinking water distribution

Water distribution systems consist of an interconnected series of components.

- pipes,
- storage facilities, and
- components that convey drinking water.

Water distribution systems meet fire protection needs for:

- Cities.
- Homes.
- Schools.
- Hospitals.
- Businesses.
- Industries.
- Other facilities.

Public water systems depend on distribution systems to provide an uninterrupted supply of pressurized safe drinking water to all consumers. Distribution system mains carry water from either:

- the treatment plant to the consumer; or
- the source to the consumer when treatment is absent.

Distribution systems span almost one million miles in the United States. They represent the vast majority of physical infrastructure for water supplies. Distribution system wear and tear can pose intermittent or persistent health risks.

Within the water supply, drinking water distribution is typically the largest energy consumer. At the same time, 25%–50% leakage is not unusual. By dividing the water distribution into pressure zones, the average pressure can often be reduced by 30%–40%. The drive is very widely used in boosting pumping stations to regulate pressure in each pressure zone. Integrated application software functions, such as sleep mode, dry run detection, automatic energy optimization, Cascade control, automatic flow compensation, and ramp functions help simplify the installation and make it more reliable, reduce the risk of water hammer, control pressure, and reduce energy consumption.

9.4 Benefits of VFD drives in irrigation pumping

When using VFD, there are many benefits and cost-saving possibilities. Not only does this modern electrical device allow the possibility of unlocking electrical savings, but VFD also for the first time enable the farmer to take

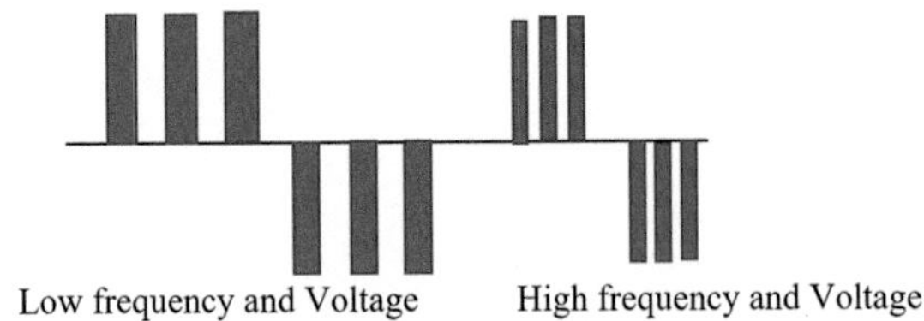

Figure 9.2 PWM-based variable frequency controller. *PWM*, pulse width modulation.

control of the pump. It now becomes possible to manage the operations of the pumping unit, such as ramping periods, controlling the flow or pressure according to various variables. The pump/s can also be started and controlled remotely, via GSM (cell phone modem technology) and RF (radio frequency) devices with ease.

Briefly, VFDs deliver the following benefits to our industry:

- Electricity savings.
- Control of startup and shutdown procedures.
- Reduces motor burnouts.
- Prevents shaft breaks.
- Reduces wear on pumps and motors.
- Keeps pumps working at best efficiency points (BEPs) on curve.
- Minimizes water hammer in pipelines due to controlled acceleration/deceleration.
- Reduces voltage drawdown on power lines at startup.
- Eliminates the need for pump control valves in many instances.
- Controls pump to deliver accurate flows or constant pressures.

Frequency inverter is perhaps a better term to use when looking for a definition for this device which is used for controlling electric motors. This is a fair mouthful, however, what is important to note is that it is a solid-state device for controlling electric motors and therefore has no contactors which can burn out. Power factor correction is built into most VFDs of the conventional range, in some situations single phase incoming power can be used with certain limitations. Essentially, in the VFD AC power is internally converted to DC and using a system called pulse width modulation (PWM) the frequency of the electrical wave) is changed which gives us different speeds on the motor (Fig. 9.2).

The VFD, therefore, has as its main function the ability to vary the speed of the motor it is coupled to. Furthermore, it has inbuilt electrical protection devices such as phase failure, phase recognition, over/under voltage protection, amperage protection. It is capable of recognizing the direction of rotation of the motor which is important in automatic restarts.

When using a VFD coupled to an electric motor which is in turn coupled to a centrifugal pump the Laws of Affinity come into play. These laws define that, if the impeller size of the pump is kept constant and the speed is varied:

- *Flow rate* is proportional to the shaft speed;
- *Pressure head* is proportional to the square of the shaft speed;
- *Power required* is proportional to the cube of the shaft speed.

The benefits with regards to pumping water are that we can now use the same pump and impeller for varying flows and heads, and still keep it as close as possible to the BEP on the pump's curve. Best of all we will only use the electricity required for the work done with no unnecessary energy losses due to throttling valves etc.

Centrifugal Pumps

Always start and stop against a closed valve

- Run a pump according to its BEP on the curve.
- Choose the driver motor size (kW) according to nonoverload for the impeller size.
- Always fit a nonreturn valve on the delivery side of the pump after the main valve in such a manner that the load of the thrust which is applied upon it from the water returning following an uncontrolled shutdown is dissipated as soon as possible into thrust blocks. Never allow the thrust be taken up by the pipework or volute casing.

With a VFD fitted to the motor of the pump, the VFD can be programmed to start from zero RPM to the maximum required speed over a specified period of time which will allow the column of water to start moving slowly. The reverse is also true when stopping the pump. The pump can start automatically after power failures and can easily be started using remote control via GSM or Rf.

9.4.1 Applications using a VFD in the pumping of irrigation water

9.4.1.1 Single pump with multihead duty points

This is possibly the most common application for a VFD in pumping applications. In a situation where a pump has a fairly constant flow, but a varying head due to the topography of the land we are able to change the pressure of the pump by changing the rotational speed of the pump. In doing this it is possible to unlock considerable energy savings.

Figure 9.3 Single pump with multihead duty points.

This system can be used in most of the different irrigation systems from micro/drip to sprinklers and center pivots. In fixed block systems such as microirrigation, the irrigated area can be divided into different height zones and each zone can have a different pressure requirement at the pump station. For instance, if there are five zones requiring from 3, 4, 5, 6, and 7 bar, respectively, to fulfill the highest duty point, the VFD will then ramp the pump speed up or down at the requirement of the valves as they operate in their respective zones (Fig. 9.3). This can be either manual or automatic. With a center pivot for instance, if the land is sloping, it too can be cut into sectors that require different pump pressures. These points on the circle of the pivot can be activated by means of limit switches which are activated and deactivated as the pivot traverses its arc. The signals can be relayed either by wires or by using wireless signals sent to the VFD to activate the different set points.

The point of the exercise is to strive to create only the energy required at the different duties by the system to work optimally and not to have to waste energy due to using throttling devices. These devices can be pressure reducing valves; flow control valves; orifice plates and similar. It is seldom possible to be exact and practical, so most of the time we cannot discard these devices entirely but rather use them for the "fine tuning."

The VFD operates on PID (a proportional–integral–derivative controller is a generic control loop feedback mechanism) using a 4–20 milliamp pressure transducer fitted in the discharge line of the pump. This pressure transducer will give feedback on the actual line pressure and the VFD will compare this to the pressure setting on the VFD as set by the potentiometer (setpoint).

9.4.1.2 Pumping with multiflow

In a scenario such as where we have a pump with a fairly large capacity and operating a number of irrigation blocks or number of sprinkler lines, we are able to allow the flow to vary and maintain a constant operating pressure. In other words, when cutting back on flow, the pump will reduce in speed and the flow will reduce, but still keep as close as possible to the BEP on the curve. As the system calls for a larger volume of water the pump will supply more but keep a constant pressure.

If there are two pumps in parallel, it is possible to use two VFDs and allow them to cascade. This means when pump 1 exceeds its flow and the pressure condition can no longer be met as it is pumping too far to the right on the pump curve, the second VFD can be called in and then the two pumps will balance out in flow and supply what is required at the determined pressure. As the flow increases or decreases, the pumps will ramp accordingly until the flow decreases and the second pump is no longer needed and it will fall out and only pump number 1 will operate.

A similar way of dealing with varying flows at constant heads can be solved using one VFD and conventional Star Delta starters on the other motor/s. There are positives and negatives in each of these scenarios which require investigation on a case-by-case scenario.

9.4.1.3 Serial pumping

Two or more pumps in tandem can also be controlled with VFDs. Using either a VFD at each pump (ultimate flexibility) or one VFD coupled with Star Delta starters on the subsequent motor/s.

As the first pump ramps to its maximum hertz and maintains it for a determined time the second VFD can be pulled in and then the two operate together. These can be programmed so that either both float together or one operates at maximum and the other fills in the difference required. Here too a Star Delta combination can be used. The Star Delta can be fitted to the second pump and when the VFD is at its maximum it can bring in the Star Delta and then ramp down to meet the exact requirement of the system. Again there are positives and negatives in each of these scenarios which require investigation on a case-by-case scenario (Fig. 9.4).

9.4.1.4 Pumping from a borehole

With many boreholes, the dynamic water level and correctly tested flow rate are not known. This results in many pumps working well to the right of the curve (or too far left). A VFD is able to control the pumping from

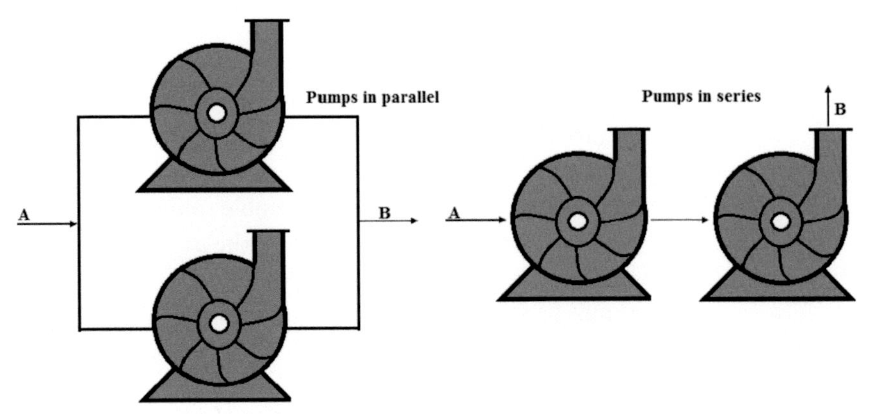

Figure 9.4 Serial and parallel pumping.

a borehole by pressure or flow control. This will extend the life of many pump units as they will operate closer to their BEPs on the curve. If the borehole is pumped at the correct volume according to the 48 h constant flow tests it will last longer than when poorly managed by over-pumping (Fig. 9.5).

When using a VFD, there are many benefits and cost-saving possibilities for pumping applications in the irrigation sector. It is important to look at a VFD application for a motor in conjunction with the curve of the pump. The hydraulics of the system need to be understood and best is to plot the curve onto a simulation program that can show how the varying speed influences the characteristics of the pump. This coupled with the change in kilowatts power consumed at the different duty points will allow the customer to make an informed decision.

Fresh, clean and well-treated water is a basic element of civilization; vital for agriculture and important for industries. As the world's most water-focused AC drive company, we have a thorough understanding of all water-based applications and processes. We provide you with AC-drive solutions that improve process control, water quality, and asset protection, reduce energy and maintenance costs, ensure higher reliability and performance from your plant, and increase the sustainability of water usage.

Reduced leakage and energy consumption: The energy used in the water distribution system typically represents 60%–80% of the total energy consumption for the whole water supply system. By adapting the pressure to the real need using pressure zones and boosting pumping stations, energy savings of 25%–40% can be achieved. At the same time, water leakage can be reduced by 30%–40%.

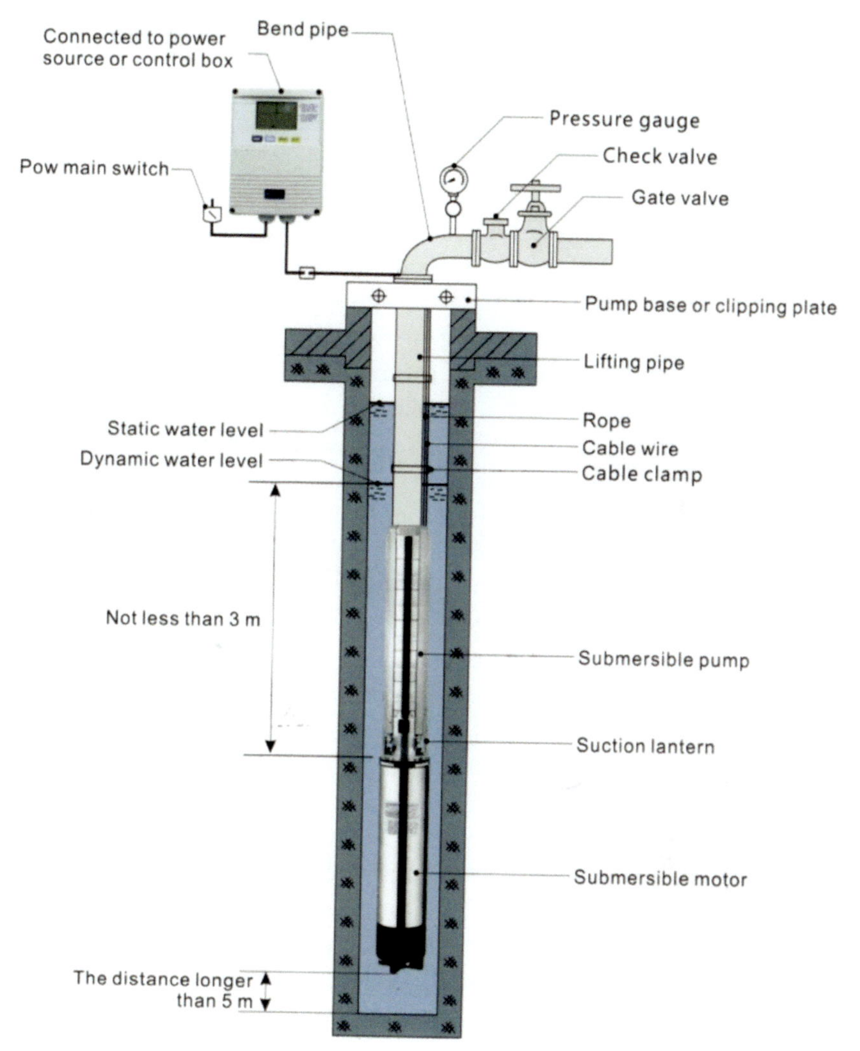

Figure 9.5 Pumping from a borehole.

Regulating the pressure in the network can also provide the following benefits:

1) a 40%–55 % reduction in the amount of new pipe breaks,

2) a reduction in maintenance costs and expensive pipe and road repair,

3) limited risk of bacteria and contamination of tap water (infiltration),

4) extended service life of the network,

5) postponed investment in plant upgrades, and

6) reduced risk of water hammer.

9.5 Requirement of drives in wastewater canalization system

The use of AC drives in the water industry is constantly increasing. Today, drives are widely used all over the industry from water production and distribution to wastewater pumping in sewage canalization systems and wastewater treatment plants both within municipalities and the industrial side of the business. AC drives control the speed of motors in a wide range of pump, blower, mixer, and dewatering applications to help enhance performance and value while saving energy and costs.

Typically, water and wastewater treatment processes account for 25%–40% of a municipality's electricity bill and are the equivalent of 8% of global electricity consumption. Water and wastewater facilities are therefore normally the single-largest electricity consumer for a municipality.

With extensive use of AC drives, energy-efficient components, and real-life online process control combined with energy production based on the methane from a wastewater plant's digester, the first full-scale facilities are now in operation on a completely energy-neutral basis. This is obtained without adding external carbon. The energy neutrality covers the whole water cycle, from water production and distribution to wastewater pumping and treatment.

Wastewater pumps and variable speed drives: Install a variable speed pump controller. This may accomplish the desired goals of matching pump performance to changing job requirements and saving energy. In most wastewater applications with little static head, this may save more energy than trimming the impellers and also provide increased pump and motor life due to the soft-start and soft-stop features of a variable speed system. Reducing pump speed reduces starting torque and thrust as well as eliminating the high starting current created using a standard across the line starter.

9.5.1 Wastewater treatment plants

The latest buzzwords in wastewater treatment plants are "bio refinery" and "water resource recovery facilities." These cover the general acceptance that wastewater has to be considered as a resource from which both energy and valuable resources can be gained. In some of the most advanced cases, energy production has now (even without adding external carbon resources) got to a level where the energy recovered from wastewater treatment process not only covers the plant's own needs, but also the energy needs for drinking water production and distribution as well as wastewater pumping. In other words, the whole water cycle can be considered energy neutral.

A precondition for energy neutrality is a control handle in the form of an AC drive available for all rotating equipment, so that the fully computer-controlled facility can adapt to the changing load. Typically, 30%–60% of energy is used in the biological process; AC-drive control can usually cut energy consumption by 20%–40%. Sludge age control by control of the RAS pumps is similarly important to both limit energy consumption and also avoid "burning" carbon, which is needed in the digester for generating gas for energy production. The VLT AQUA drive has been chosen for some of these most advanced facilities, where high installed efficiency of the drive and greater reliability and ease of operation have been key selection criteria.

9.5.2 Efficient wastewater treatment and surplus energy production

In wastewater treatment plants, the high energy consumption is related to the energy-intensive processes and the continuous operation cycle: 24/7, 365 days a year. Focus in the industry has been on developing new processes and control strategies to reduce energy consumed per liter of water processed. However, the demands for improved wastewater treatment quality, for example, based on demand for enhanced nutrient removal, increase net energy consumption. This creates an even greater need to cut energy consumption based on advanced process control.

Blowers or surface aerators typically consume 40%–60 % of the total energy used in wastewater treatment plants. Controlling the aeration equipment with Danfoss AC drives can deliver energy savings of 30%–50%.

Additionally, efficient control of the sludge balance and sludge age with AC drives reduces energy consumption. It also increases the amount of carbon to the digester, which can then produce more gas that can be used for energy production.

The most advanced facilities are therefore able to both clean the wastewater to a very high level and at the same time produce surplus amounts of energy. A precondition for this is normally a fully computer-controlled facility where it's possible, via the installation of AC drives on more or less all rotating equipment, to regulate all parameters.

9.5.3 Industrial water and wastewater

The importance of a clean and stable water supply as well as taking advantage of opportunities to make major cost reductions on the wastewater-handling side has been realized by many industries. Some industries are even moving toward a zero, or very close to zero, water use. On the clean water side, this requires utilization of advanced water treatment technologies such as RO and ultrafiltration. The drive helps improve process reliability and reduce operation costs. On the wastewater handling side, the trend is similar to the municipalities' vision; wastewater should be seen as a resource from which energy and other resources can be obtained.

9.5.4 Generating surplus power from wastewater treatment (case study)

Since 2010, Marselisborg wastewater treatment plant has transformed its focus beyond minimizing energy consumed, to maximizing net energy surplus. Nowadays the facility has net production of both electricity and heat, supplying the district heating system in Denmark's second-largest municipality, Aarhus. The carbon footprint has been reduced by 35% accordingly.

Water and wastewater treatment facilities are normally the single largest electricity consumer for a municipality. Typically, water and wastewater treatment processes account for 25%–40% of the municipal electricity consumption. The high consumption is related to the energy intensive processes but also its continuous operation cycle, 24/7 and 365 days annually.

Over the years focus has been on developing new processes and control strategies to reduce energy consumed per liter of water processed. However, at the same time the increasing demands upon wastewater treatment quality, for example in nutrient removal, in turn increase net energy consumption.

Energy balance optimization: water and wastewater treatment processes are characterized by high load variation during the 24-hour cycle and seasonally throughout the year. The use of frequency converters has therefore steady increased in order to control blowers, pumps, and other motorized equipment, to adapt to the changing demand.

Since 2010 Aarhus Water has worked intensively together with water environment consultants to improve the energy balance for Marselisborg wastewater treatment plant.

Key steps in the strategy:

- Optimization of the nitrogen removal process using online sensor control. The frequency converter adapts the level of aeration precisely to the need. This control system reduces energy consumption and increases the amount of carbon left in the system.

- Blower technology upgrade to a high-speed turbo blower. The upgrade achieves further reduction of energy consumption in the aeration process.
- Aerobic sludge age control as a function of temperature and load on the plant. Here frequency converter control of the return sludge pumps is the key to achieving energy reduction and increased retained carbon in the system.
- Upgrade of combined heat and power process for energy production, with 90 % energy efficiency.

These changes together with improvements including the effective coproduction of electricity and heat based on methane gas extracted from the aerobic sludge digestion process have created the impressive results of:

- 130 % electricity production (30% excess electricity).
- Excess heat production of about 2.5 GWh/year.

9.6 Induction motor drive for PV array fed water pumping

In PV pumping systems, an IM shows good performance as compared to other commercial motors because of its greater robustness, lower cost, higher efficiency, lower maintenance cost, and availability in local markets. In addition to the power exchange from PV to the IM, a DC–DC boost converter is introduced. For good utilization of PV panels, it is necessary to extract maximum power from the PV panels. For this reason, variable step size incremental conductance-based maximum power point tracking control is implemented due to its performance [1]. The concept of fuzzy direct torque control (FDTC) is similar to classical DTC. It consists in regulating the flux and torque of the IM by the selection of one of the eight voltage vectors generated by each voltage inverter; this choice is made by switching the table. However, this table is elaborated by the reasoning of fuzzy logic from the flux error, torque error, and position of flux vector. This technique is characterized by good precision, stability, a good torque response, robustness, and simplicity of implementation. The principle of FDTC for the IM is shown in Fig. 9.6.

9.7 Solar PV-based water pumping using BLDC motor drive

The schematic of proposed grid interfaced PV fed brushless DC motor-driven water pumping system is shown in Fig. 9.7. A PV array of 6.4 kW, possessing a sufficient power to run the water pump at its full capacity under the standard climatic conditions, feeds a BLDC motor via a boost converter

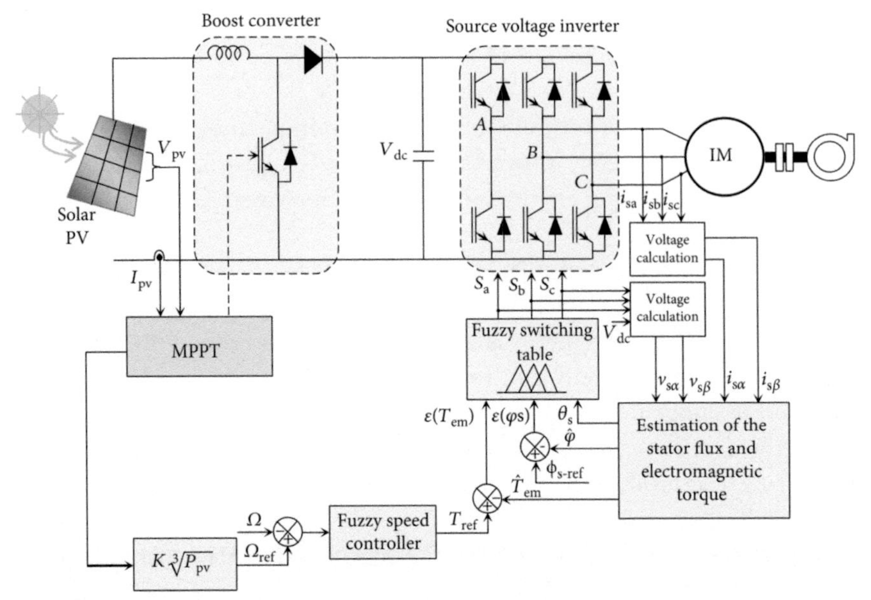

Figure 9.6 Induction motor drive for PV array fed water pumping.

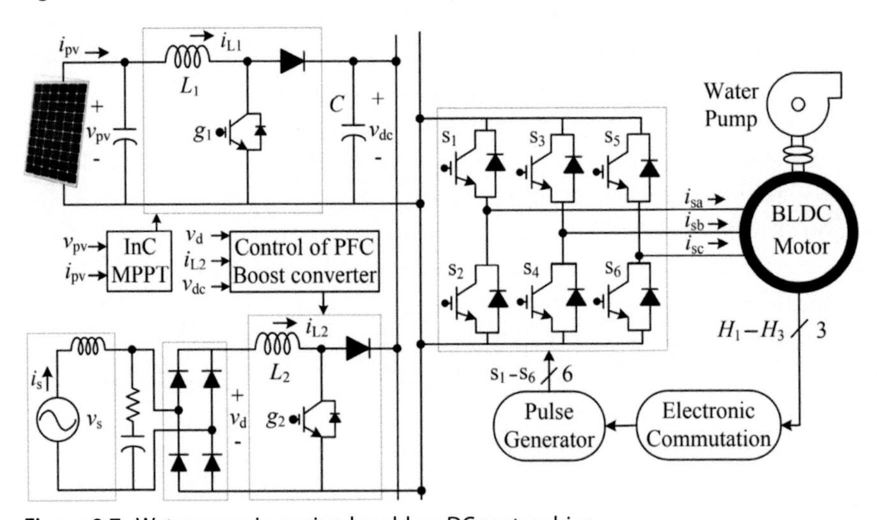

Figure 9.7 Water pumping using brushless DC motor drive.

and a VSI. The DC–DC boost converter and the VSI, respectively, carry out the MPPT of PV array and an electronic commutation of the motor. Three Hall effect sensors are used to generate the commutation signals. A 6 pole, 5.18kW BLDC motor which has a rated speed of 3000 rpm at 310 V is used to run the water pump. A single-phase utility grid support is provided, via a bridge rectifier and a PFC boost converter, at the common

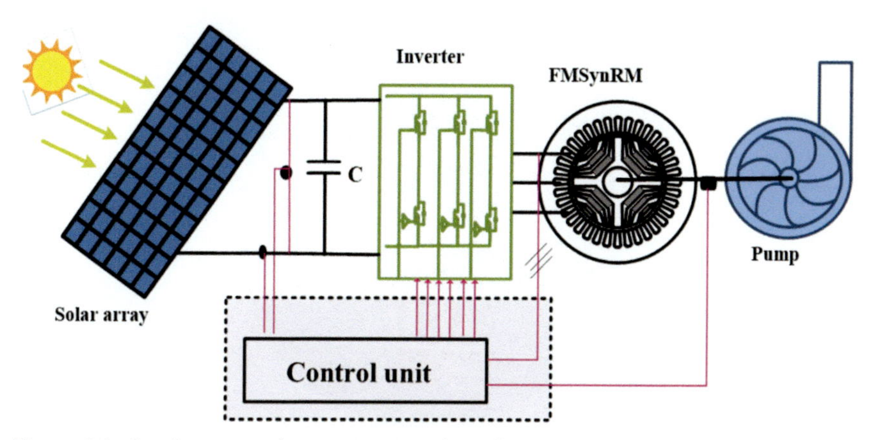

Figure 9.8 Synchronous reluctance motors-based water pumping system.

DC bus of VSI. The power transfer is controlled by operating the PFC converter through a unidirectional power flow control. The developed control enables a power transfer from utility grid to the DC bus if a PV-generated power is insufficient to meet the power demand, otherwise no power is transferred from the utility [6].

9.8 Solar array fed synchronous reluctance motor-driven water pump

Unlike the conventional PVWPS, in this system, a solar array supplies the motor via a three-phase voltage source inverter (VSI). Besides, a ferrite magnet synchronous reluctance motor drives centrifugal pump. Moreover, a proposed control system is implemented to increase the PV output power and also to drive the motor properly. It is clear that the FMSynRM power varies with the current angle; there is a current angle at which the power of the motor is maximum, that is, called optimal current angle (Fig. 9.8). Further, the optimal current angle value is not a fixed value; it depends on the winding current level due to core saturation [2]. This means that using a fixed current angle is not correct and it results in a reduction in the motor power. Consequently, using the optimal current angle during the control of the motor is important to maximize the motor output power. Moreover, the power of the motor at the rated current and optimal current angle reaches the target motor power for the application (i.e., about 6.5 kW). Fig. 9.9 shows the motor power factor versus the current angle at various winding currents. One can see that the power factor of the designed FMSynRM reaches a unity power factor in contrast to the pure reluctance machines;

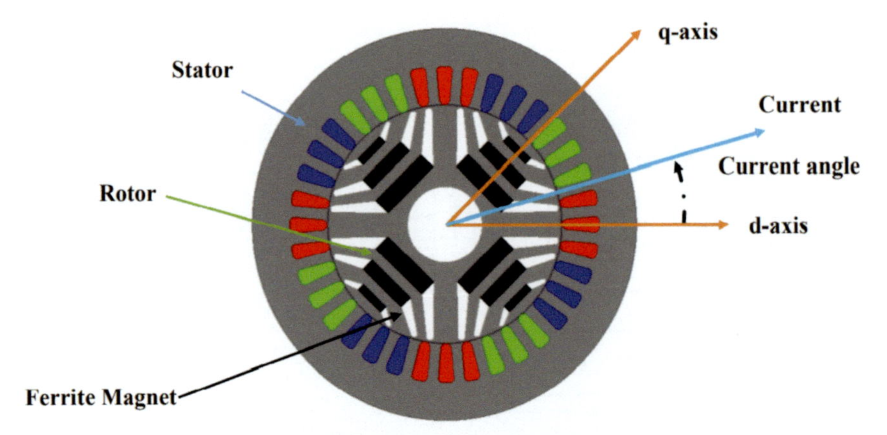

Figure 9.9 Cross-sectional view of synchronous reluctance motors.

this is thanks to adding the ferrite magnets. In addition, controlling the FMSynRM to work at the maximum power point reduces the power factor to about 0.8 and 0.78 at 25% and 100% of the rated stator winding current respectively. However, this power factor is still comparable to induction machines.

9.9 Permanent-magnet synchronous motor-driven solar water-pumping system

Fig. 9.10 shows the schematic diagram for the stand-alone solar PV-based PMSM drive for water pumping system. The proposed system consists of solar PV panel, a boost converter, a three-phase VSI and a PMSM coupled with a centrifugal water pump. A PV or solar cell is the basic building block of a PV system. An individual PV cell is usually quite small, typically producing about 1 or 2W of power. To increase the power output of PV cells, these cells are connected in series and parallel to assemble larger unit called PV module. The PV array is connected to the DC-to-DC boost converter to increase the output voltage level. An insulated gate bipolar transistor based VSI is used for DC to AC conversion and connected to the PMSM drive. The constant DC voltage is converted to the AC output using a VSI. Reference speed of PMSM is a function of solar irradiation.

The basic block diagram of the presented system is shown in Fig. 9.11. It exhibits the PV array with MPPT achieved by the P&O control algorithm with a boost converter. A three-phase VSI is used to drive the pump coupled with PMSM [4]. The utility grid is connected to the same dc link through a VSC. The single-phase VSC is responsible for bidirectional power flow

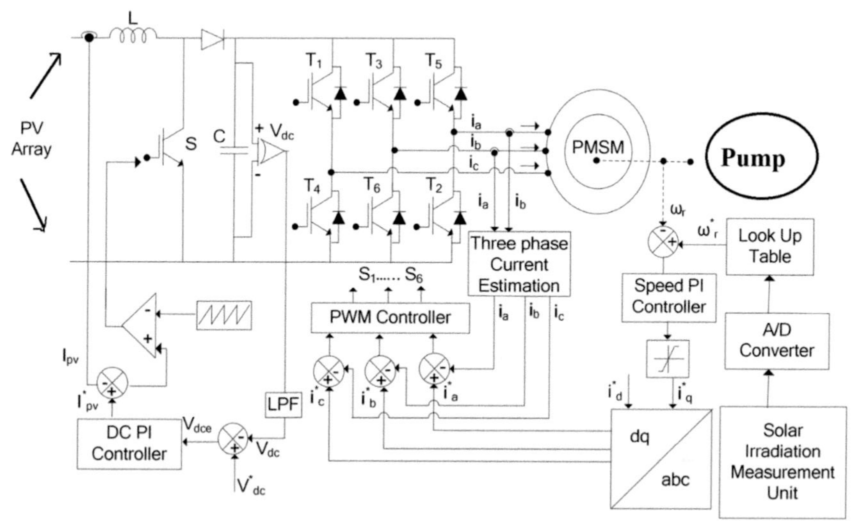

Figure 9.10 PV-based permanent-magnet synchronous motor water pumping system.

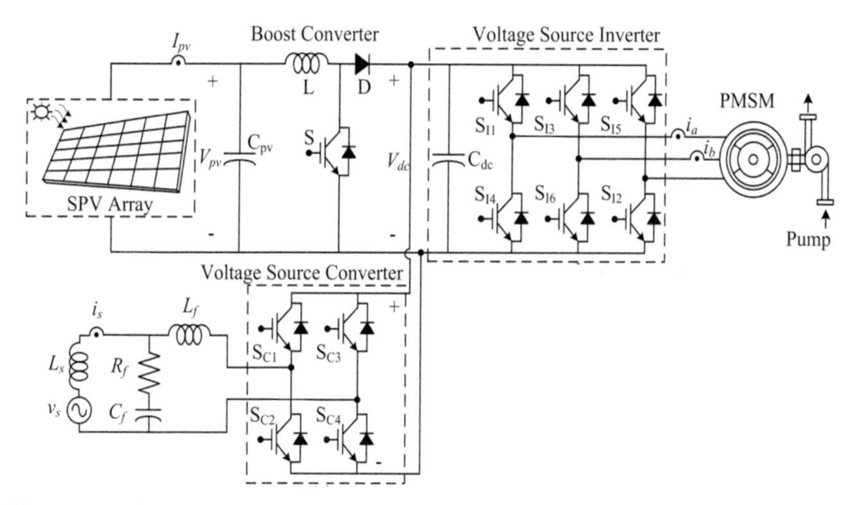

Figure 9.11 Permanent-magnet synchronous motor water-pumping system.

through a DC-link capacitor. An interfacing inductor is introduced in series of the single-phase grid to limit the current ripples in the supply. Moreover, an RC filter is connected in parallel to serve the purpose of higher-order harmonics mitigation in the supply voltage. The system is operated in two modes. One is the grid-connected mode, whereas the other one is an islanded mode of operation. Three conditions can be realized in the grid-connected mode of operation. The first one is when the only grid

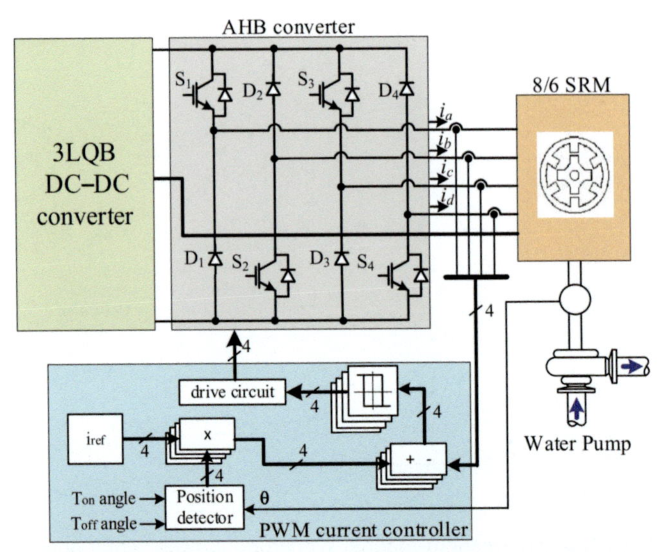

Figure 9.12 PWM current control solution of SRM drive. *PWM*, pulse width modulation; *SRM*, switched reluctance motor.

is feeding the pump. During the night when solar insolation is not available and pumping operation is required, the grid feeds the pump. The second one is when both the grid and solar PV array are feeding the pump. When solar insolation is not enough to operate the pump at rated speed, the required partial power is fed from the grid. The third one is when the solar insolation is available and no pumping operation is required. During this condition, the solar PV energy is fed into the grid [5]. During the first two conditions, the utility grid acts as a source, whereas during the third condition, the grid acts as a sink. Islanded condition refers to the grid outage. During this condition, the pumping operation is solely dependent on the availability of solar insolation. The speed of the pump is proportional to the intensity of solar insolation.

9.10 Switched reluctance motor drives for water pumping applications

The proposed solution combines several systems, namely a solar PV array, a DC–DC converter with MPPT control algorithm, an asymmetrical half-bridge (AHB) converter and a centrifugal pump powered by a four-phase 8/6 switched reluctance motor (SRM), as presented in Fig. 9.12. Due to the proposed DC–DC converter, the four phases of the 8/6 SRM can have a

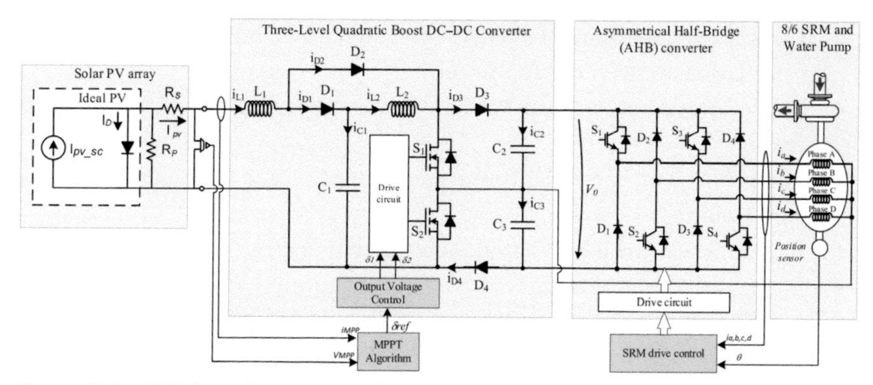

Figure 9.13 SRM-based water pumping system. *SRM*, switched reluctance motor.

common star winding connection, and consequently the AHB converter is composed by four branches only. To describe the SRM operating principle, it is necessary to recall that the reluctance of the machine magnetic circuit and its variations are dependent on the rotor position. The variation of the magnetic reluctance in the SRM is essential to have torque (reluctance torque), and it can be properly controlled by accurately controlling the time of energizing and de-energizing the stator phases.

A water pumping system, powered by solar photovoltaic panels, employing a SRM drive and a three-level quadratic Boost (3LQB) DC–DC converter was presented in this paper. The DC–DC converter provides quadratic static gain and reduced number of active and passive components [3]. This converter can create a dual-output balanced voltage with opposite polarity considering a common connection point that is used to connect to the middle point of the SRM drive. The use of this DC–DC converter allows to use an AHB converter also with reduced number of active power switches. An MPPT algorithm that is based in the variation of the power and voltage with respect to time was also proposed, resulting in a robust and very easy practical implementation. The combination of all the proposed devices and systems allows to fully exploit the solar energy to provide as much as possible a continuous water flow in isolated pumping solutions (Fig. 9.13).

Practice questions

1. Compare the performance of different types of drives used in water pumping applications.
2. Draw and explain the working of SRM drives used in water pumping applications.

3. Describe the working of permanent-magnet synchronous motor-driven solar-based water pumping system.
4. Write short notes on solar PV-based water pumping using BLDC motor drive.
5. How solar array fed synchronous reluctance motor drive operates in the water pump.
6. Explain the working of variable frequency drives.
7. What is the role of electric drives in Irrigation?
8. List the types of pumps used in the water and wastewater industry.
9. Name the applications of motor controls in the water and wastewater industry.
10. Give the details of standards and compliances followed in electric drives.

References

Cordeiro, A., Pires, V.F., Foito, D., Pires, A.J., Martins, J.F., 2020. Three-level quadratic boost DC-DC converter associated to a SRM drive for water pumping photovoltaic powered systems. Sol. Energy 209, 42–56.

Dubey, M., Sharma, S., Saxena, R., "Solar PV stand-alone water pumping system employing PMSM drive" 2014 IEEE Students' Conference on Electrical, Electronics and Computer Science.

Errouha, M., Derouich, A., El Ouanjli, N., Motahhir, S., "High-performance standalone photovoltaic water pumping system using induction motor" Hindawi International Journal of Photoenergy Volume 2020, Article ID 3872529 doi:http://doi.org/10.1155/2020/3872529.

Ibrahim, M.N., Rezk, H., Al-Dhaifallah, M., Sergeant, P., 2020. Modelling and design methodology of an improved performance photovoltaic pumping system employing ferrite magnet synchronous reluctance motors. Mathematics 8, 1429.

Kumar, R., Singh, B., "Grid interfaced solar PV based water pumping using brushless DC motor drive" 978-1-4673-8888-7/16/$31.00 ©2016 IEEE.

Singh, B., Murshid, S., 2018. A grid-interactive permanent-magnet synchronous motor-driven solar water-pumping system. IEEE Trans. Ind. Appl. 54 (5).

Index

Page numbers followed by "*f*" and "*t*" indicate, figures and tables respectively.

F

G

H

I

L

M

O

P

Printed in the United States
by Baker & Taylor Publisher Services